Springer Laboratory

Goerg H. Michler

Electron Microscopy of Polymers

With contributions by
Dr. R. Godehardt · Dr. R. Adhikari · Dr. G.-M. Kim
Dr. S. Henning · DP V. Seydewitz · DP W. Lebek

 Springer

Prof. Dr. Goerg H. Michler
Institut für Physik
Institut für Polymerwerkstoffe e.V.
Martin-Luther-Universität Halle-Wittenberg
06099 Halle
Germany
goerg.michler@physik.uni-halle.de

ISBN 978-3-642-07166-9   e-ISBN 978-3-540-36352-1

DOI 10.1007/978-3-540-36352-1

Cover design: WMXDesign, Heidelberg, Germany

Printed on acid-free paper

9 8 7 6 5 4 3 2 1

springer.com

Preface

Electron microscopy and atomic force microscopy have developed into powerful tools in the field of polymer science. By using different techniques and methods, morphological details at length scales from the visible (0.1 mm) up to a few 0.1 nm can be detected. Consequently, the microscopic techniques used in polymer research support the tendency, over the last two decades, to shift the level of interest from the µm-scale to the nm-scale region. Systems with at least one structural dimension below ~100 nm are now considered to comprise a new class of materials, the so-called *nanostructured polymers* or *nanocomposites*. In addition, the influence of several parameters can be studied by changing the morphology of the material. In particular, the influence of the actual, local morphology on mechanical loading effects can be determined. The micromechanical properties or mechanisms that occur at nano- and microscopic levels form the bridge between structure, morphology and mechanical properties. Therefore, electron microscopy and atomic force microscopy directly contribute to a better understanding of structure–property correlations in polymers.

Part I offers an overview of electron microscopy and atomic force microscopy techniques and summarises distinctive applications of polymeric materials. The wide variety of preparation methods used to study polymers with the different microscopic techniques are presented and illustrated with typical micrographs in the chapters of Part II. Each technique is discussed in detail, highlighting its application for solving specific problems arising in the characterisation of materials. The applicability of the microscopic techniques and preparation methods described in Parts I and II to the main classes of polymers is documented in Part III. All relevant groups of solid polymers used domestically, industrially, in research and in medicine are mentioned. The characteristic features and also the variety of structures and morphologies of the different polymer classes are illustrated with typical micrographs. In particular, the application of different microscopic techniques is shown to reveal similar polymeric structures, enabling laboratories that possess only some of the techniques to use them beneficially. As well as descriptions of characteristic morphologies and micromechanical properties the most commonly occurring defects and failures are also illustrated.

The volume is directed at polymer scientists from research institutes and industry, and aims to demonstrate the widespread possibilities enabled by the application of microscopic techniques in polymer research and development. Each of these techniques allows one to solve a number of problems, as even for the specialist it is not always evident which technique is best suited to solving a given problem. The monograph is also directed at research and applied technicians, since it provides a basic understanding of the principles of the different microscopic techniques and exhausts all of the possibilities of using these techniques to solve specific research problems. All of the preparation methods applied for the study of a variety of polymeric materials using different techniques are described in depth, which will also aid laboratory assistants or students that are new to microscopy, as well as those that wish to improve their skills. Finally, the book will be also helpful for students of polymer physics, chemistry and engineering, as well as those researchers interested in the micro- and nanoscopic world of polymers.

This volume draws upon the experiences and studies of the working groups of the editor in research institutes, industry, and academia in the period from 1970 onwards (i.e. over three decades). The authors or coauthors of the various chapters are:

Dr. R. Godehardt (Chaps. 2, 3, 4, 5)

Dr. R. Adhikari (Chaps. 5, 19)

Dr. G.-M. Kim (Chaps. 18, 21, 22, 24)

Dr. S. Henning (Chaps. 9, 11, 16, 23)

DP V. Seydewitz (Chap. 4)

DP W. Lebek (Chaps. 7, 10, 12, 13)

For additional contributions and remarks, I thank Prof. Dr. F.J. Baltá-Calleja, Instituto de Estructura de la Materia, CSIC, Madrid, Dr. W. Erfurth, Max Planck Institut für Mikrostrukturphysik Halle (in Chap. 4), Dr. J. Lacayo-Pineda, Continental AG, Hannover (in Chap. 20), DI St. Scholtyssek (in Chaps. 16 and 24) and DI M. Buschnakowski (in Chap. 11). My former or current coworkers DI (FH) I. Naumann, DI (FH) H. Steinbach, Mrs I. Schülke, Dr. J. Starke, DP J. Laatsch, DI (FH) S. Goerlitz and Mrs C. Becker are gratefully acknowledged for providing many of the examples of microscopic investigations of different polymers and micrographs referred to in this book. I also thank DI W. Schurz for image processing many of the electron micrographs, Mrs B. Erfurt for typing many of the chapters, and DP W. Lebek for his valuable technical help during the completion of the manuscript.

Finally, I also wish to gratefully acknowledge the coworkers at Springer-Verlag for their understanding and help during the preparation of the manuscript.

Halle/Merseburg, March 2008   *Goerg H. Michler*

Abbreviations

General techniques

AFM	Atomic force microscope
AM	Amplitude modulation
BSE	Backscattered electrons
CCD	Charge-coupled device
CRT	Cathode ray tube
EDX(A)	Energy dispersive X-ray analysis
EELS	Electron energy-loss spectroscopy, electron energy-loss spectrometer
EFTEM	Energy-filtered transmission electron microscopy
ELNES	Energy-loss near-edge structure
EM	Electron microscope
ESI	Electron spectroscopic imaging
ESD	Electron spectroscopic diffraction
ESEM	Environmental scanning electron microscope
FEG	Field-emission gun
FIB	Focussed ion beam
FM	Frequency modulation
GIF	Gatan imaging filter
GSED	Gaseous secondary electron detector
HRTEM	High-resolution transmission electron microscope
HVTEM	High-voltage transmission electron microscope
ID	Interparticle distance
LFM	Lateral force microscope
LVTEM	Low-voltage transmission electron microscope
M_W	Molecular weight
MCA	Multichannel analyser
MDS	Minimum-dose systems
MOS	Metal-oxide semiconductor
PE	Primary electrons
PEELS	Parallel electron energy-loss spectroscopy
PFM	Pulsed force mode
ROI	Region of interest
SE	Secondary electrons

SEM	Scanning electron microscope
SFM	Scanning force microscope
SNOM	Scanning near-field optical microscope
SPM	Scanning probe microscope
STEM	Scanning transmission electron microscope
STM	Scanning tunnelling microscope
TEM	Transmission electron microscope
T_g	Glass transition temperature
TM	Tapping mode
TMAFM	Tapping-mode atomic force microscope
WD	Working distance
WDX(A)	Wavelength dispersive X-ray analysis

Materials/polymers

ABS	Acrylonitrile-butadiene-styrene
ASA	Acrylonitrile-styrene-acrylate
BR	Butadiene rubber
AN	Acrylonitrile
COC	Cyclic olefin copolymer
EB	Ethylene-butadiene copolymer
EOC	Ethylene/1-octene copolymer
EPDM	Ethylene propylene diene rubber
EPR	Ethylene propylene rubber
HDPE	High-density polyethylene
HIPS	High-impact polystyrene
iPP	Isotactic polypropylene
LDPE	Low-density polyethylene
LLDPE	Linear low-density polyethylene
MMT	Montmorillonite
MWCNT	Multiwalled carbon nanotube
NBR	Acrylonitrile-butadiene rubber
NR	Natural rubber
OsO_4	Osmium tetroxide
PBMA	Poly(n-butylmethacrylate)
PB	Polybutadiene
PC	Polycarbonate
PCH	Polyvinylcyclohexane
PE	Polyethylene
PET	Polyethylene terephthalate
PEB	Ethylene-butylene copolymer
PFS	Poly(ferrocenyl-dimethylsilane)

PI	Polyisoprene
PMMA	Polymethylmethacrylate
PnBA	Poly(n-butylacrylate)
PNC	Polymer nanocomposite
POSS	Polyhedral oligosilsesquioxane
PP	Polypropylene
PS	Polystyrene
PTFE	Polytetrafluoroethylene
PVC	Polyvinylchloride
PVDF	Polyvinylidenefluoride
RuO_4	Ruthenium tetroxide
PVP	Poly(2-vinylpyridene)
SAN	Polystyrene-acrylonitrile
SIS	Polystyrene-*block*-polyisoprene-*block*-polystyrene
SBS	Polystyrene-polybutadiene-polystyrene triblock copolymer, polystyrene-*block*-polybutadiene-*block*-polystyrene *block* copolymer
TPE	Thermoplastic elastomer
UHMWPE	Ultrahigh molecular weight polyethylene
VLDPE	Very low density polyethylene

Table of Contents

Part III Main Groups of Polymers

Introduction

Polymers form one of the most important groups of materials used in modern industry. The materials in this group encapsulate great diversity in terms of their characteristics, resulting in various polymeric forms such as plastics, rubbers, fibres or dyes. Polymers are applied in almost all sectors of daily life, from households to medicine, agriculture, the automotive industry, up to microelectronics and space research. Polymers are typically composed of very large molecules, known as *macromolecules*, which usually consist of thousands of repeating units, termed *monomers*. There is huge variety in terms of the types, numbers, arrangements and combinations of monomers found in polymers, which therefore leads to an extremely wide variety of different polymeric materials. As well as synthetic polymeric materials, there are also natural and biological polymers, including proteins, silk, and cellulose.

The wide range of polymeric materials that have good processability, environmental stability, lightweight characteristics and easy machinability make polymers very useful materials. This usefulness is reflected in the worldwide production of all types of polymers, thermoplastics, resins, and rubbers, which has increased enormously since 1950, with annual average growth rates of about 15% [1]. This growth continues even now, with growth rates currently about 5–10% per annum, and polymer production reached around 230 million t in 2005 globally. The production of steel (by volume) was surpassed by polymer production in the year 1989 by about 100 million m^3 [2], in such a way that today polymers are produced in greater amounts than any other group of materials. Steady growth in polymer production is also expected in the near future, based on the abovementioned characteristics and several other developments: better functionality of polymeric parts or a wider range of uses, the same or better performance achieved with lower amounts of raw material (oil) and energy, and better recycling of these materials, thus enhancing sustainable development. It is remarkable that more than 80% of all polymers are currently based on so-called mass polymers or commodities, e.g. polyethylenes, polypropylenes, polystyrene, polyvinylchloride and rubber. This renaissance of mass polymers is due to improved polymerisation, controlled molecular weight and macromolecular design, better macromolecular regularity, many modifications of the arrangements of known monomers and polymers, and modification with fillers. Examples include the broad fields of polymer blends, high-impact polymers, block copolymers or composites. A shift of interest to smaller and smaller details, from the former µm level to the now increasingly interesting nm level, is currently occurring. This increasing tendency to make structural modifications has also pushed polymer research to improve morphological control through the use of electron microscopy. There is increasing interest in achieving accurate correlations between synthesis, molecular structure, morphology and properties; see Fig. I.1. The most important properties of many applications of polymers include their mechanical properties. Here, the micromechanical processes of deformation and fracture provide the bridge between structure/morphology and mechanical properties. To gain a better knowledge of structure–property correlations, much effort must be expended in studying structures, morphology and micromechanical properties. When used for polymer characterisation, electron microscopic techniques have a tremendous advantage over other methods, as they can provide a direct "view" with a high local resolution of the material of study. Other techniques do not provide direct pictures comparable to those yielded by electron microscopy, but average information about larger volumes instead. With the shift in interest to smaller and smaller structural details, electron microscopy and atomic force microscopy are becoming more and more important. The recent increased availability of high-resolution electron microscopy and atomic force microscopy is now making it possible to view molecular arrangements, and this should lead to further advancements in our understanding of the forms and structures of all types of polymer systems.

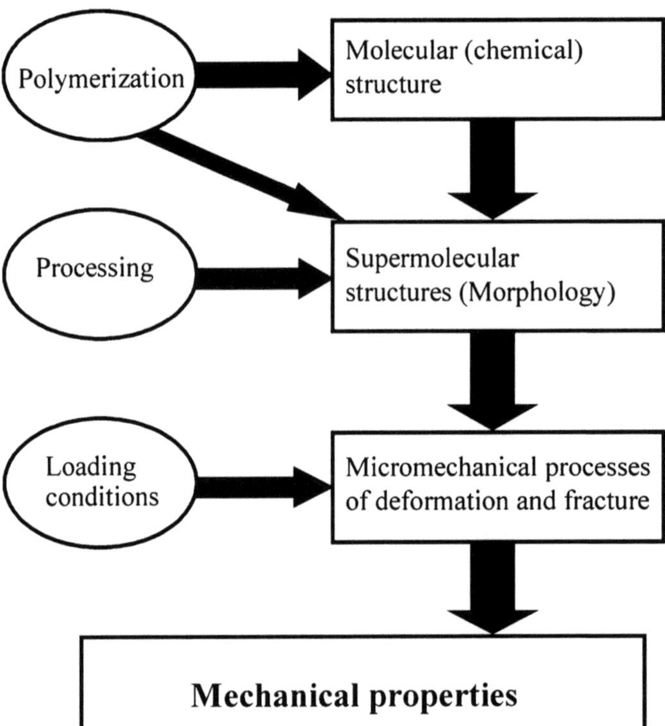

Fig. I.1. Correlations between structure, morphology, influencing parameters and mechanical properties of polymers

An additional trend in microscopy is to reveal not only structural details with improved resolution but also changes in morphology under the action of influencing parameters, such as physical and thermal ageing, outdoor weathering and mechanical loading. In particular, the influence of mechanical loading on changes in the structure and morphology of polymers – in other words, their micro- or nanomechanical properties and mechanisms – can be revealed by electron and atomic force microscopy with an otherwise unattainable accuracy. A better understanding of structure–property correlations enables the defined modification of polymeric structure or morphology and thus the improvement of polymers. While a huge variety of macromolecular and supramolecular structures exist, not all of them are of equal relevance for property improvements. Usually, only a few of the structures dictate the mechanical behaviour of the polymer; these are called "property-determining structures" [3]. A detailed knowledge of these structures and the underlying micromechanical mechanisms associated with them enable criteria to be defined for the modification and production of polymers with specifically improved or new properties [4]. This is known as the "microstructural construction of polymers" [5]. In summary, the technique of electron microscopy plays a decisive role in aiding our understanding of the structure–property correlations encountered in polymer research as well as in the polymer industry.

References

1. Menges G (1990) Werkstoffkunde Kunststoffe. Carl Hanser Verlag, München
2. Editor (2005) Kunststoffe 95:35
3. Michler GH (1992) Kunststoff-Mikromechanik: Morphologie, Deformations- und Bruchmechanismen. Carl Hanser Verlag, München
4. Michler GH, Baltá-Calleja FJ (eds) (2005) Mechanical properties of polymers based on nanostructure and morphology. CRC Press, Boca Raton, FL
5. Michler GH (1998) Polym Adv Technol 9:812

Techniques of Electron Microscopy

1 Overview of Techniques

This introductory chapter provides an overview of the various techniques of microscopy that are available. Starting with the improvement in the resolution of microscopes during their historical development from optical microscopes up to high-resolution transmission electron microscopes and scanning tunnelling microscopes, the field of microscopy is classified based on the principles of imaging. The main techniques used to study the structures of surfaces, the internal structures of polymers and their chemical compositions are listed and discussed in detail in the subsequent chapters in Part I. An overview of additional techniques used to study the morphologies and micromechanical properties of polymers closes the chapter.

The structures and morphologies of polymers have been under investigation for more than 60 years. Early applications of transmission electron microscopy were concerned with the study of the spherulitic crystallisation of natural rubber and low-density polyethylene. After 1957 lamellar crystals of polyethylene crystallised from dilute solution were studied by transmission electron microscopy by Keller and Bassett and the folded chain hypothesis was advanced [1,2]. This was followed by a large number of contributions on the morphology of crystalline polymers. Techniques that allowed the use of transmission electron microscopy to investigate relatively complex structures, such as spherulites, polymer blends and block copolymers, were developed in the 1960s and 1970s. Scanning electron microscopy was introduced in the 1960s, and since then it has been used to investigate fracture surfaces, phase separation in polymer blends and crystallisation of spherulites. The recent availability of high-resolution electron microscopy, coupled with image processing techniques, or atomic force microscopy is now making it possible to view structures down to the molecular level.

The primary reason for developing electron microscopes was to improve the resolution of microscopes, and so one of the most important aspects of each microscope is its resolution power, which is the minimum distance between two adjacent object points that can still be imaged separately. It is well known that the resolution of optical microscopes is limited by the wavelength of visible light (it is half of the wavelength, about 0.2 μm). After the first development of optical microscopes in the seventeenth century, the improvements made mainly by Abbe, Zeiss and Schott at the end of the nineteenth century yielded microscopes with this resolution and physical evidence of the resolution limit. The development of electron microscopes in the 1930s resulted in 1934 in the creation of the first transmission electron microscope with a better resolution than an optical microscope. Since then there have been enormous improvements in resolution; see Fig. 1.1. Recent advances in techniques used as well as in interpreting and processing the images have allowed resolutions of the order of 0.1 nm (= 1 Å) or better to be achieved for inorganic crystal structures.

However, the best resolution achieved in polymers is, in practice, poorer than this because of polymer-specific problems with high electron irradiation sensitivity and low contrast. The next jump in resolution came with the development of scanning tunnelling microscopy or, in general, scanning probe microscopy, which enabled the first ever three-dimensional imaging of solid surfaces with atomic resolution. Scanning probe microscopes do not belong in the field of electron microscopy. However, since they are usually used in close connection with electron microscopy and artefact-free evaluations of structures are easier to achieve when their results are compared with those from electron microscopes, they are discussed in this volume too. Field-ion microscopy, a special surface technique with atomic resolution, is not applied to polymers.

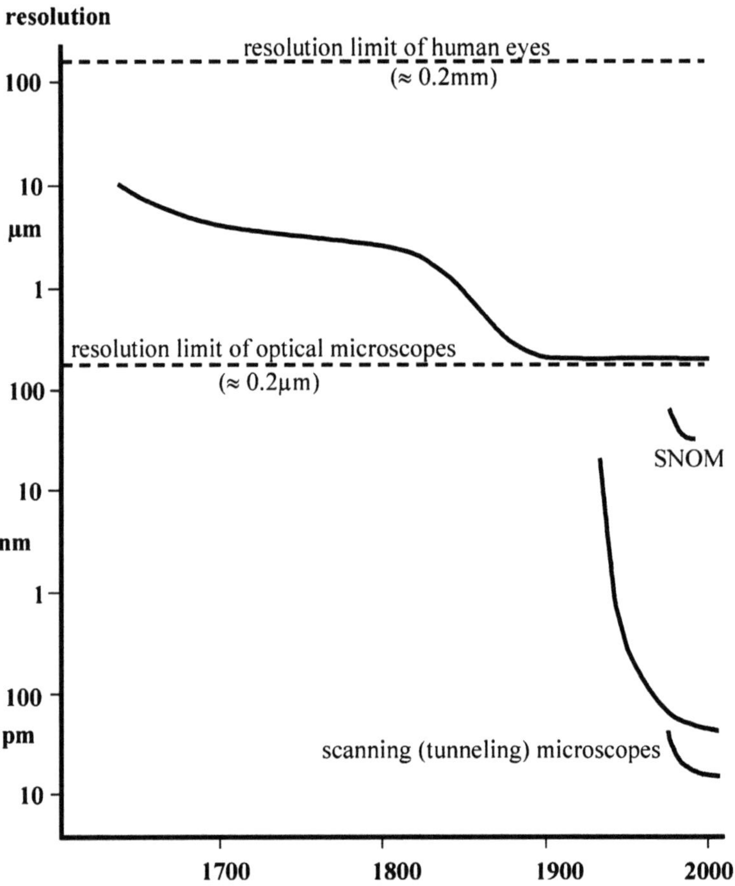

Fig. 1.1. Improvements in the resolution of microscopy

Electron microscopy (EM) can be divided into the techniques of transmission electron microscopy (TEM) and scanning electron microscopy (SEM). A comparison in terms of resolving power shows that scanning electron microscopes are somewhat intermediate between optical microscopes (OM) and transmission electron microscopes. A significant advantage of using SEM compared to TEM is that the former can image the surfaces of bulk samples with a large depth of focus. This large depth of focus also allows SEM to be used at low magnifications instead of optical microscopes.

It is necessary to use all of the microscopic techniques if we wish to study the large variety of morphologies and structures of polymeric materials, i.e. from the sizes and shapes of grains or powders up to crystalline structures; see Fig. 1.2. In general, all of the different types of microscopes can be classified according to whether imaging is

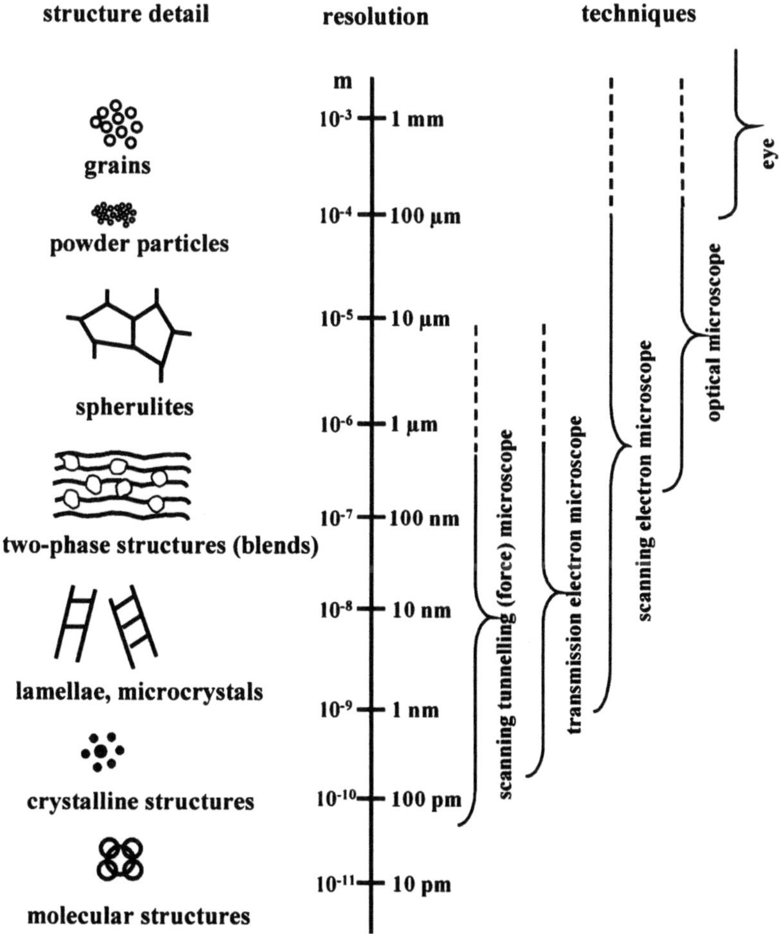

Fig. 1.2. Scheme of possible structures of different sizes present in polymers and resolutions attainable with the different microscopic techniques

achieved by irradiating the object with a "lamp" or to feeling the surface with a "finger" or "needle" (see Fig. 1.3):

1. A fixed beam of light or electrons is transmitted through the (thin) specimen (as a transmitted beam) in the transmission mode of the optical microscope and in transmission electron microscopes.
2. A stationary beam is reflected off the (bulk) specimen surface (as a reflected beam) in the reflection mode of the optical microscopes or in electron mirror microscopes (the latter are not used for polymers).
3. A focussed beam is scanned across the specimen, passing through the (thin) specimen (scanning transmission EM) or resulting in a reflected beam (as in confocal laser scanning microscopy) or secondary or backscattered electrons in scanning electron microscopes.

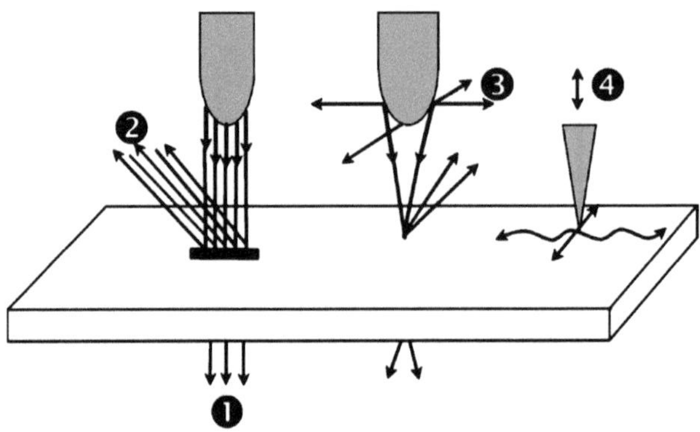

❶ **_Transmitted beam_**
 ⇨ light microscope (transmission mode)
 ⇨ transmission electron microscope

❷ **_Reflecting beam_**
 ⇨ light microscope (reflection mode)
 ⇨ electron mirror microscope

❸ **_Scanning beam_**
 ⇨ scanning electron microscope
 ⇨ confocal laser scanning microscope

❹ **_Scanning tip_**
 ⇨ scanning tunnelling microscope
 ⇨ atomic force microscope

Fig. 1.3. Schematic representation of the principles of different types of microscopes (see text)

4. A mechanical tip is scanned across the specimen in order to make use of different physical properties in tunnelling microscopes and atomic force microscopes.

When studying the bulk material, the material's surface or its interior is the target of the microscopic investigations; see Fig. 1.4. The surface can be studied directly with scanning electron microscopy (SEM), atomic force microscopy (AFM) and, indirectly after replication, with transmission electron microscopy (TEM). Ultra- and semithin sections from the interior can be used for TEM and thicker sections for analytical TEM and SEM or for AFM.

The traditional electron microscopy technique is stationary-beam TEM, which has been applied to a wide range of materials, including polymers (Chaps. 2 and 3). The main limitation of this approach is that a transparent thin foil that is resistant to damage by the electron beam must be prepared. In addition to this conventional TEM method, special equipment has been developed to achieve high resolution (high-resolution TEM, HRTEM), to be able to use high (high-voltage TEM, HVTEM) or low accelerating voltages (low-voltage TEM, LVTEM), for scanning transmission (STEM), for holography and for spectroscopy or emission of X-rays in analytical microscopes (ATEM).

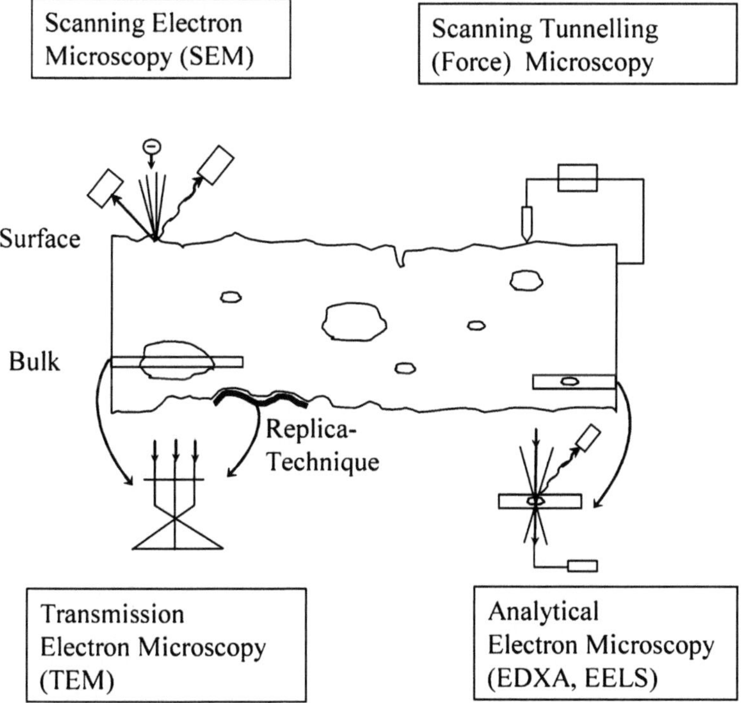

Fig. 1.4. Application of different microscopic techniques to study the surface and interior of a bulk polymeric material

Scanning electron microscopy (SEM) is currently the most popular of the microscopic techniques (see Chap. 4). This is due to the user-friendliness of the apparatus, the ease of specimen preparation, and the general simplicity of image interpretation. The obvious limitation is that only surface features are easily accessible. With SEM, the chemical analysis of different elements is usually possible (energy dispersive or wavelength dispersive analysis of X-rays, EDXA, WDXA).

As mentioned above, scanning probe microscopy techniques cannot be classified as types of electron microscopy. However, because of the wide application of atomic force microscopy (AFM) and the fact that it is often used in combination with electron microscopic techniques, it is discussed in Chap. 5. There are some other techniques of electron microscopy, such as emission EM, mirror-EM, field-electron or field-ion microscopy, which cannot be applied to polymers.

In the past, the central reason for using EM was structure and morphology determination, but it is also currently of importance for investigating different processes, i.e. changes in the material caused by interactions with several factors, such as heat, electric or magnetic fields and environmental liquids or gases. Of particular interest is the study of micromechanical processes of deformation and fracture, as discussed in Chap. 6 (on in situ microscopy).

Valuable methods for improving images and quantitatively estimating structures are included in the final chapter of Part I, Chap. 7 (on image processing).

There are a number of reviews and monographs that discuss the different techniques in more detail, e.g. [3–11], and their application to polymers, e.g. [12–14]. A good overview of microscopically determined structures present in polymers is provided by A.E. Woodward [15].

Besides the techniques of EM and AFM discussed in this volume, there are other techniques that are used to study the morphologies and mechanical properties of polymers [16]. The most important of these techniques are:

- Optical microscopy with magnifications of up to about 1000×, which is very useful for gaining an overview and as a first step in any morphology analysis
- Laser scanning and optical near-field microscopies (scanning beam techniques; see Fig. 1.3), which have improved resolution compared to optical microscopes
- Macromolecular orientations can be visualised using optical birefringence;
- Acoustic microscopy (ultrasound microscopy) is based on the reflection of ultrasound in the sample, which yields information on density differences, microvoids, cracks, etc., with sizes of less than 1 µm
- Small-angle light scattering (SALS) can be used if the polymers are capable of scattering light due to density or birefringence fluctuations of the order of the wavelength of the light, and is a useful method for studying textures that are larger than about 1 µm, e.g. spherulites in semicrystalline polymers
- Classical methods of light-optical interference are used to detect small details on the order of the wavelength of the light in transparent materials; in particular, the micromechanics of crazes at crack tips in transparent glassy polymers can be investigated

- Small-angle X-ray scattering (SAXS), the traditional technique used to study periodicities in semicrystalline polymers (e.g. long periods of fibrils or lamellae) and also to detect microcavities (e.g. microvoids between craze fibrils and interfibrillar spacing)
- Wide-angle X-ray scattering (WAXS) yields information on the crystallinity (type and size of crystals and lattice defects) in polymers
- Rheo-SAXS experiments using X-ray radiation from a synchrotron source allow us to measure in situ changes in structure and crystallinity and to perform real-time measurements at low speeds and frequencies
- Small-angle neutron scattering (SANS) characterises the fluctuations in the density, concentration, and magnetic properties of the material and yields information on the conformation, size and mobility of the macromolecular coils, but is far from a routine technique
- Infrared (IR) and Raman spectroscopy characterise the type and constitution of the macromolecules present
- Rheo-optical methods include a mechanical test performed under static conditions which is carried out simultaneously with optical measurements; among other optical methods, Fourier transform infrared spectroscopy (FTIR) has become one of the most frequently applied tools in rheo-optics, since it enables changes in molecular orientations in polymers and different types of macromolecules in polymer blends to be identified
- Dynamic mechanical analysis (DMA) is a very helpful tool for determining relaxation processes, glass transition temperatures and mixing or phase separations of different polymer constituents in blends and copolymers
- Differential scanning calorimetry (DSC) measures melting or crystallisation temperatures and degrees of crystallinity
- Microindentation hardness gives information on crystallisation behaviour, local mechanical properties, and micromechanical processes
- Positron annihilation spectroscopy (PAS) allows us to estimate the local concentration of free volume or the size of nanovoids
- Electron spin resonance (ESR) and nuclear magnetic resonance (NMR) measure, for instance, radical formation and macromolecular mobility.

The advantage of all of these techniques is that they can be used to investigate larger material volumes and to give integral parameters for the measured structure. On the other hand, their major disadvantage is that they cannot clarify variations in structural or morphological parameters, such as the size distributions of lamellae or particles, orientation differences, local concentrations of additives, deviations in phase separation, and many others. In order to gain an understanding of mechanical properties, in particular strength, elongation at break or toughness, it is not sufficient to measure the "average morphology", since these properties depend strongly on the variation in structural elements and, for instance, on extreme values of them. Therefore, microscopic techniques with high local resolution are exceptionally important for polymer research and the development of materials with improved mechanical properties. To maximise the information obtained about a particular material, a combination of electron microscopy (with its high local resolution of structures and variations in details) with some of other integral techniques that provide average values of structures should be utilised if possible.

References

1. Keller A (1958) In: Growth and perfection of crystals (Proc Int Conf Crystal Growth, Cooperstown, NY). Wiley, New York, pp 499
2. Bassett DC, Frank FC, Keller A (1963) Philos Mag 8:1753
3. Glauert AM (1973–1991) Practical methods in electron microscopy, vols 1–13. Elsevier Science, Amsterdam
4. Bethge H, Heydenreich J (eds) (1987) Electron microscopy in solid state physics. Elsevier, Amsterdam
5. Williams DB, Carter CB (1996) Transmission electron microscopy: a textbook for materials science. Plenum, New York
6. Amelinckx S, van Dyck D, van Landuyt J, van Tendeloo G (1997) Electron microscopy: principles and fundamentals. Wiley-VCH, Weinheim
7. Goodhew PJ, Humphreys FJ, Beanland R (2000) Electron microscopy and analysis, 3rd edn. Taylor & Francis, London
8. Zhang X-F, Zhang Z (eds) (2001) Progress in transmission electron microscopy 1: concepts and techniques. Springer, Berlin
9. Shindo D, Oikawa T (2002) Analytical electron microscopy for materials science. Springer, Tokyo
10. Fultz B, Howe J (2003) Transmission electron microscopy and diffractometry of material, 2nd edn. Springer, Berlin
11. Li ZhR (ed) (2003) Industrial applications of electron microscopy. Marcel Dekker Inc., New York
12. Sawyer LC, Grubb DT (1987) Polymer microscopy. Chapman and Hall, London
13. Roulin-Moloney AC (ed) (1989) Fractography and failure mechanisms of polymers and composites. Elsevier, London
14. Bassett DC (1984) Electron microscopy and spherulitic organisation in polymers, vol 12. CRC Press, Boca Raton, FL, p 97
15. Woodward AE (1988) Atlas of polymer morphology. Hanser, Munich
16. Michler GH (2001) J Macromol Sci Phys B 40:277

2 Transmission Electron Microscopy: Fundamentals of Methods and Instrumentation

In this chapter, after a brief history of transmission electron microscopy, the fundamentals of electron optics and instrumentation are described. Following the path of the electron beam, the microscope can be split into the following parts: electron gun, illumination system, objective lens and specimen stage, image-forming system and viewing chamber/image recording. These units are described in detail before the fundamentals of image formation are discussed. The differences between the scattering mechanisms that occur in amorphous and crystalline materials lead to an explanation of image contrast. However, the intensity distribution of the image depends not only on the interaction of the electron beam with the object, but also on the illumination conditions and in particular the action of the objective lens and arranged apertures. This electron-optical imaging process, including microscope aberrations, is described in detail based on the wave-mechanical theory of contrast formation.

2.1 A Brief History

Recently, key events in the history of electron microscopy have been documented extensively [1], and a few reviews [2–8] have been published that are excellent sources of individual reminiscences and further information concerning the origins and the historical development of electron optics and electron microscopy. The year that saw the birth of the transmission electron microscope (TEM) appears to be a little vague. In 1932 Knoll and Ruska published their results obtained with magnetic lenses and by applying two-step imaging in *Zeitschrift für Physik* in a rather detailed form, and they designated one chapter of their paper "Das Elektronenmikroskop" (see, e.g., [5]), and so 1932 is usually said to be the year that the electron microscope was invented. A few years before this, the investigations of Busch had shown that rotationally symmetrical magnetic and electric fields possess some characteristics of lenses, and that the simple lens equation known from optics holds when imaging with such lenses. In 1934 Brüche and Scherzer published the first monograph on geometrical electron optics. The first images from an electron microscope were demonstrated in 1931, and in 1935 the first reliable images in which the resolution of the optical microscope had been surpassed were made available. Soon after, in 1936, Scherzer demonstrated that the spherical and chromatic aberration coefficients of electron lenses are intrinsically nonvanishing, and hence cannot be eliminated by skilful design. Within the next three years, TEMs were developed by commercial companies and electron microscopy was established in countries other than Germany. Furthermore, the first scanning transmission electron microscope (STEM) was constructed by von Ardenne in 1938. This pioneering period that laid the foundations for electron microscopy ended with the beginning of World War II.

In the second period, which lasted to the end of the 1950s, the number of electron microscopists increased enormously and TEMs became widely available from several sources. Both the preparation of the specimens to be investigated and the electron microscopes were improved significantly. The principle of the stigmator was developed by Bertein (1947), Hiller and Ramberg (1947) and Rang (1949). The wobbler, a focussing aid, was invented by Le Poole in 1949. Almost all the methods of correcting the spherical aberration of electron lenses were listed by Scherzer in 1947, and the first studies on image formation and resolving power in wave-optical terms were carried out by Scherzer in 1949 and Glaser in 1950. The use of an electron biprism to obtain interference fringes was demonstrated by Düker and Möllenstedt in 1954.

The third period, which covers the time from the late 1950s right up to the present, brought further continuous improvement in all of the operational systems of the TEM as a result of technological progress in general. In particular, progress in electronics combined with the replacement of vacuum tube technology by solid-state circuitry and later on by integrated circuit technology was a driving force for significant improvements in the critical parts of the TEM, such as the power supplies for the lens current and the high voltage. Furthermore, important improvements were made in the objective lens, such as the use of lenses exploiting superconductivity (first proposal by Laberrigue and Levinson in 1964), the introduction of the condenser objective lens (Riecke and Ruska in 1966), and the introduction of the high-quality multipurpose objective lens (Mast et al. in 1980). This initially led to the development of high-resolution TEMs (HRTEMs) with a voltage of about 100 kV in the 1970s. Then, in the 1980s, instruments capable of a resolution of less than 0.2 nm became available for the intermediate voltage range of 300–400 kV. On the other hand, interest in high-voltage TEMs (HVTEMs) increased. In 1960 Dupouy published the first pictures he had obtained with the 1.2 MV TEM, constructed in Toulouse. Ten years later, the same author presented the first results obtained with the Toulouse 3 MV TEM. Further high-voltage projects in Cambridge and Japan were completed in the middle of the 1960s. In the 1990s, it became possible to construct HVTEMs with sufficient accelerating voltage stability to allow investigations to be performed at a resolution of 0.1 nm, as shown by a 1250 kV TEM from JEOL working at the Max Planck Institute in Stuttgart [9].

Furthermore, the range of applications of TEMs could be significantly extended by adding supplementary equipment for microanalytical investigations. The first description of an analytical microscope combining a TEM and a microprobe analyzer was given by Duncumb in 1968. Although, this "EMMA" was not a commercial success, it was the first step towards the next generation of analytical TEMs based on energy dispersive X-ray (EDX) analysis. One year later in 1969, a post-column electron energy-loss spectrometer (EELS) for use with a TEM was presented by Wittry, and the development of parallel EELS (PEELS) attachments was started in 1977 following results reported by Jones et al. Besides these external attachments for analytical investigations, the implementation of an in-column imaging electron energy filter was an important step forward in analytical electron microscopy. In this context the Ω-filter introduced by Rose and Plies in 1974 was a milestone in the development of the energy-filter TEM (EFTEM). Furthermore, the development of STEM was an especially important milestone in the evolution of analytical electron microscopy. Crewe et al. presented a high-resolution STEM based on the first major use of a field-emission gun in 1968. The first basic study of the mechanism of image formation in the STEM was carried out by Thomson and Zeitler in 1970, and Crewe presented images of single atoms obtained by Z-contrast in the same year. High-quality STEMs have been available commercially since 1974.

Revolutionary developments in computer techniques have also considerably improved the design, adjustment and working routines of TEMs since the start of the 1990s, and computer-assisted and computer-controlled microscopy has become the state of the art. Furthermore, due to the development of high-resolution digital recording media, such as image plates and slow-scan CCD cameras, digital image recording and processing have become increasingly popular. One offshoot of this development is improved data acquisition for a tilt series, which provides the basis for the three-dimensional reconstruction of an object by electron tomography. Originally pioneered in the life sciences, this technique has recently become a powerful imaging and analytical tool in materials science [10–16].

Furthermore, since the end of the 1990s the doors have opened to the new aberration-corrected world of both conventional and scanning transmission electron microscopy [17–25]. The breakthrough came with the use of non-round lenses. The principle for this had already been outlined by Scherzer in 1947, as mentioned above. These multipole lenses are capable of generating a negative value of the spherical aberration coefficient C_s which can then cancel the positive C_s of the round lens. The resolution of a spherical aberration-corrected TEM or STEM is limited by the energy spread of the incident electron beam and by the uncorrected chromatic aberration. Schemes for the correction of the latter have been proposed by Rose [26]. On the other hand, the energy spread of the primary beam can be significantly reduced by using a cold field-emission gun, and additional reduction is possible with the aid of a monochromator. Moreover, the application of a monochromator allows high-resolution EELS spectroscopy. Very recently, a new corrected (S)TEM platform has been developed that is capable of the highest TEM and STEM performance; it permits a lateral resolution of far better than 0.1 nm and an energy resolution of down to 0.1 eV to be attained [24, 25].

2.2 Fundamentals of Electron Optics and Instrumentation

2.2.1 Some Fundamental Properties of Electrons

Although Stoney had already introduced the term electron to designate an elementary charge more than ten years before, it was not until the experiments of Thomson in 1895 that the term electron was used with its present-day meaning. Experiments like those performed by Thomson to measure the ratio of charge to mass (em^{-1}) for the electron, as well as the series of famous "oil drop" experiments of Millikan in 1917 to determine the electronic charge e, showed incontrovertibly that electrons act like particles. On the other hand, de Broglie stated the principle of the wave-like nature of matter in 1924, as reflected in the famous relationship

$$\lambda = \frac{h}{p}, \tag{2.1}$$

where λ is the associated wavelength, h is Planck's constant and p is the magnitude of the particle momentum $\mathbf{p}$. This relation between the particle nature and wave-like nature of matter can also be expressed in the following form

$$\mathbf{p} = h\mathbf{k}. \tag{2.2}$$

Here, $\mathbf{k}$ is the wave vector, the magnitude of which can be written as

$$k = |\mathbf{k}| = \frac{1}{\lambda}. \tag{2.3}$$

In the case of the electron, the physical existence of the matter wave was demonstrated by Davisson and Germer in 1927 in the first electron diffraction experiment.

An electron passing across a large potential difference V is accelerated to a velocity v, which may well approach the velocity c of the light in vacuum. Therefore, relativistic effects must be taken into account. With m_0 denoting the rest mass of the electron, the relativistic change of the electron mass m in relation to the velocity v is given by the well-known equation

$$m = \frac{m_0}{1 - \frac{v^2}{c^2}}. \tag{2.4}$$

The change in energy of the electron caused by its transit across the potential difference V can be expressed by

$$mc^2 = m_0c^2 + eV. \tag{2.5}$$

The momentum p is estimated by combining Eqs. 2.4 and 2.5:

$$p = mv = \left[2eVm_0 + \left(\frac{eV}{c}\right)^2 \right]^{\frac{1}{2}}. \tag{2.6}$$

Thus, using Eq. 2.1, the wavelength of the electrons depends on the potential difference, or in other words on the accelerating voltage in the following way:

$$\lambda = \frac{h}{p} = h\left[2eVm_0 + \left(\frac{eV}{c}\right)^2 \right]^{-\frac{1}{2}}. \tag{2.7}$$

Using the fundamental constants given in Table 2.1, the properties of electrons as a function of the accelerating voltage are represented in Table 2.2.

Table 2.1. Fundamental constants

Charge of the electron $(-e)$	-1.602×10^{-19} C
Rest mass of the electron (m_0)	9.109×10^{-31} kg
Velocity of light in vacuum (c)	2.998×10^{8} m s^{-1}
Planck's constant (h)	6.626×10^{-34} N m s

Table 2.2. Electron properties as a function of accelerating voltage

Accelerating voltage V [kV]	Velocity of electrons v [ms^{-1}]	Ratio of electron velocity to light velocity vc^{-1}	Wavelength of electrons λ [pm]
10	5.83×10^{7}	0.194	12.20
100	1.64×10^{8}	0.548	3.70
200	2.08×10^{8}	0.695	2.51
500	2.59×10^{8}	0.863	1.42
1000	2.83×10^{8}	0.941	0.87
3000	2.97×10^{8}	0.989	0.36

2.2.2 Electron Lenses

In electron optics (see, e.g. [27–31]) we use the same principles as used in traditional light optics to describe the formation of images by lenses, and we also use corresponding terms to characterise the action and aberrations of a lens. The ray diagrams in Figs. 2.1 and 2.2 show two fundamental features of image formation by an ideal lens, where the lens is assumed to be a so-called "thin" lens (which means it is thin enough that its action on the paths of electron rays through the lens can be illustrated by refractions of the electron rays in the principal plane of the lens). In Fig. 2.1 a so-called "self-luminous object" is assumed, and the radiation is emanating from an off-axis point in the object plane. Based on the action of apertures in an electron microscope, a limiting diaphragm restricts the angular spread of the electrons entering the lens.

All electron rays passing through the lens are refracted in its centric plane in such a way that they form a point image at their crossover. This is the first fundamental action of an ideal lens. Only two of all of the electron rays that meet at the image point are needed to find the location of this point when drawing a ray diagram. Fortunately, we use two special ray paths for this purpose, as shown by the two bold rays in the figure. The first one is a ray passing through the lens at its centre, the direction of which is not changed by the action of the lens. The second is a ray with a path that is parallel to the electron-optical axis before entering the lens. This one is refracted by the action of the lens in such a way that it crosses the electron-optical axis at the focus in the back focal plane of the lens.

In Fig. 2.2, the image formation of a finite object (represented by an arrow) is illustrated. Starting at three different points in the object plane (at the point, at the middle and at the end of the arrow), in each case a set of three electron rays are drawn in the ray diagram. The set of rays at each starting point consists of a ray parallel to the electron-optical axis and two rays with the same inclination to the latter. Following Fig. 2.1 we find, for each of the selected starting points in the object plane, a corre-

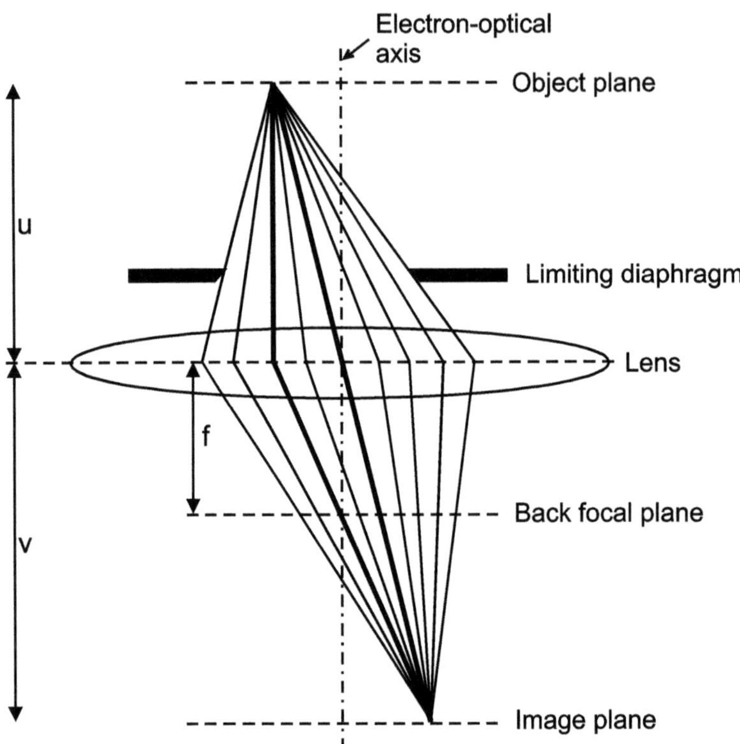

Fig. 2.1. Formation of the image of a point source. A diaphragm restricts the angular spread of electrons entering the lens and contributing to image formation. The crossover of two special ray paths is used to find the image of the point object

sponding image point by finding the crossover point for the set of three rays in the image plane. The image plane is conjugate to the object plane, which means that the planes are electron-optically equivalent, and thus rays leaving a point in one plane are brought to a point in the conjugate plane and vice versa. Using the distances u and v (as labelled in Fig. 2.2), the magnification M of the imaged object can be expressed by

$$M = \frac{v}{u},$$ (2.8)

and, introducing the focal length f of the lens, Newton's well-known lens equation

$$\frac{1}{u} + \frac{1}{v} = \frac{1}{f}$$ (2.9)

is valid, as described in standard textbooks on light optics. Figure 2.2 also illustrates the second important feature of image formation by a lens: parallel rays starting at different object points are focussed in the back focal plane of the lens, and the distance of this focus from the electron-optical axis increases with increasing oblique incidence of the parallel beam. The special case of rays entering the lens parallel to

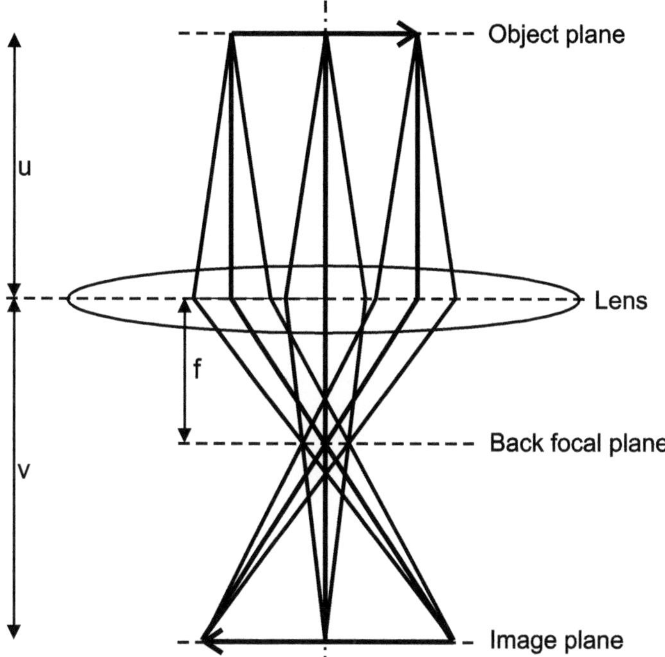

Fig. 2.2. Illustration of the two fundamental features of image formation using a ray diagram for a finite object (represented as an *arrow*). The three important distances in the diagram are labelled *u, v* and *f*. All rays emerging from a point in the object plane are gathered by the lens and converge to the conjugated point in the image plane. On the other hand, all parallel rays starting at different points in the object plane are focussed in the back-focal plane of the lens

the electron-optical axis results in a focus that is positioned on this axis; this is known as the focus of the lens.

The Lorentz force which a moving electron experiences in electric and magnetic fields and the resulting deflection of the electron provide the physical basis for electron lenses. With an electric field strength $\mathbf{E}$ and a magnetic flux $\mathbf{B}$, the Lorentz force $\mathbf{F}$ is

$$\mathbf{F} = -e(\mathbf{E} + \mathbf{v} \times \mathbf{B}) \tag{2.10}$$

where $-e$ and $\mathbf{v}$ are the charge and velocity of the electron, respectively. It is worth noting that the magnetic field action expressed by the vector cross-product of $\mathbf{v}$ and $\mathbf{B}$ results in a force vector that is normal to $\mathbf{v}$ and $\mathbf{B}$. Inserting Eq. 2.10 into Newton's law of motion

$$m\ddot{\mathbf{r}} = \mathbf{F} \tag{2.11}$$

yields the law of particle optics.

Although one can in principle use either electrostatic or magnetic lenses to focus a beam of electrons, magnetic rather than electrostatic lenses are preferred because they are more convenient to use and have lower aberrations, so only magnetic lenses will be considered here. A homogeneous magnetic field already acts as a weak electron lens for rays with a small inclination with respect to the field direction. For practical purposes, however, magnetic lenses with short focal lengths are obtained by concentrating the magnetic field by means of pole pieces. As illustrated in Fig. 2.3, a conventional magnetic lens consists of a coil of copper wire wound symmetrically around the electron-optical axis. The coil is enclosed by a soft iron shield apart from a narrow gap between a pair of pole pieces across which the focussing field appears. The lens is energised by passing current through the windings, and the magnitude of this current determines the strength of the lens expressed by the focal length.

The focussing action of the lens arises as follows. Electrons travelling initially parallel to the electron-optical axis experience upon entering in the field a tangential force due to the interaction of the axial velocity with the radial component of the magnetic field. Therefore, the subsequent direction of travel of the electrons is inclined to the electron-optical axis, and the tangential velocity component interacts with the axial component of the field to produce a radial force toward the axis. The resulting electron paths are helical and, as the field action becomes stronger with increasing distance from the electron-optical axis, a crossover of the electron paths in the lens focus results. An image rotation (marked by the angle ϕ in Fig. 2.3) caused by the helical paths of the electrons within the focussing magnetic field is a typical feature of the action of a magnetic lens.

2.2.3 Electron-Optical Aberrations and Resolution

In addition to the simple image formation discussed before, a more realistic one must take into account electron-optical aberrations. Unfortunately it is not possible to can-

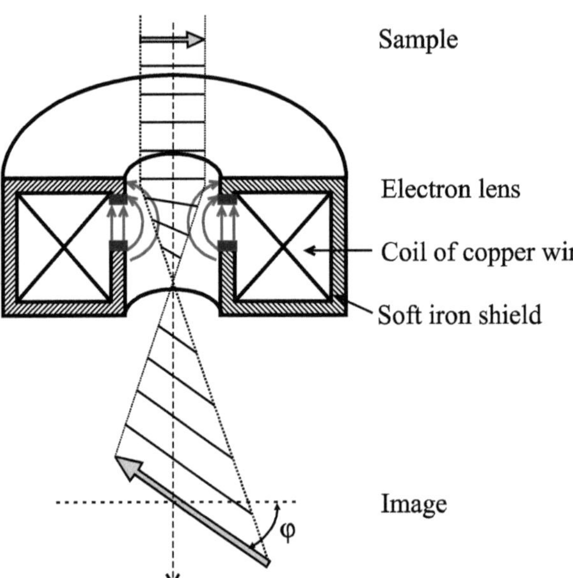

Fig. 2.3. Concentration of a rotationally symmetric magnetic field in the gap between a pair of lens pole pieces and its principle action on an electron beam

cel out or correct aberrations in electron optics simply by combining positive and negative elements of different refractive indices, as is done in light optics. Instead, it is necessary to choose operating conditions such that the influence of aberrations is minimised. The aberrations arise, on the one hand, from aberrations of the lens itself and, on the other hand, from the so-called diffraction error, which is a consequence of the presence of diaphragms. Aberrations limit the resolution or resolving power of a microscope, which is the ability to make out points which are close together in the object seen in the image. In the following, only a simplified method is used to describe the resolving power of a TEM. It is related to the image of two adjacent self-luminous object points, the radiations from which are completely independent of each other (incoherent illumination). A single point source will not be imaged as a point, even when no lens aberrations are present. The finite size of the lens results in the diffraction of the rays at the outermost collection angle of the lens, usually defined by a limiting aperture. This diffraction results in a point being imaged as a disc, usually called an *Airy disc* since the Airy function describes its intensity profile. The Rayleigh criterion is used to define the closest distance between two object points for which the points are still distinguishable in the image. The Rayleigh condition is fulfilled when the maximum from one source lies in the first minimum of the other source. Under these circumstances the distance between the two incoherent point sources is defined as the theoretical resolution r_{th} of the lens and is given by the radius r_d of the Airy disc

$$r_{th} = 0.61 \frac{\lambda}{\alpha_0} \tag{2.12}$$

where λ is the wavelength of the imaging electrons and α_0 is the semi-angle of collection of the lens (compare Fig. 2.4). As aberrations can always be expected in electron microscopes, it is necessary to modify the resolution derived solely for diffraction effects shown in Eq. 2.12. Reimer classified lens defects by ten kinds of lens aberrations [32]. In practice, however, only spherical aberration, chromatic aberration and astigmatism limit the microscope performance substantially. Because of the technical difficulties involved in correcting electron lenses, one must use electron beams with very small aperture angles in order to avoid image disturbances. Since the object region imaged in the TEM is extremely small, the imaging electron rays are likewise only small distances from the electron-optical axis, so that out of all of the geometrical aberrations only the spherical aberration, which is independent of the distance of the ray from the electron-optical axis, has to be taken into account. Compared to other geometrical aberrations, which depend on the distance from the electron-optical axis in a linear (coma), quadratic (off-axis astigmatism) or cubic (distortion) manner, only the spherical aberration causes a blurring of the image point on the electron-optical axis, thus limiting the achievable resolution.

Spherical Aberration

Spherical aberration is the inability of a lens to focus all incident rays from a point source to a point. This defect is caused by the lens field acting inhomogeneously on the off-axis rays. The effects of spherical aberration are shown in Fig. 2.4.

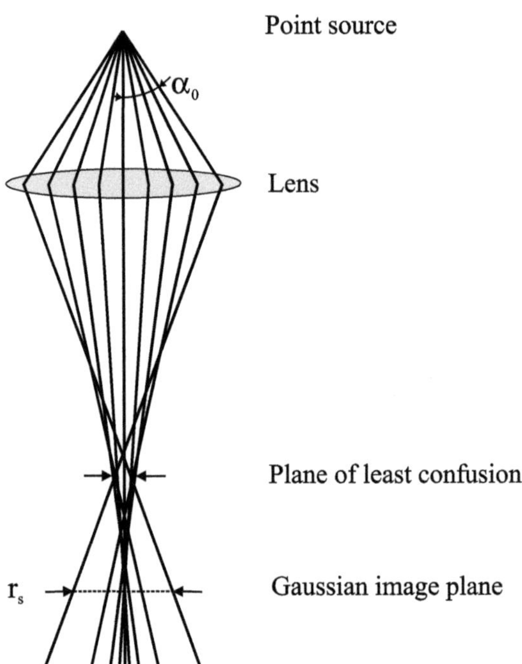

Fig. 2.4. Formation of a disc of confusion due to spherical aberration

For reference we define the true or so-called Gaussian image plane as the image plane for paraxial imaging conditions. In paraxial imaging conditions, the rays are near the electron-optical axis and make only small angles with respect to the electron-optical axis. On the other hand, Fig. 2.4 shows that the further off-axis the electron is, the more strongly it is bent back toward the electron-optical axis and its focal length decreases. Therefore, an image disc is obtained of an object point in the Gaussian image plane. The radius r_s of the aberration disc in the Gaussian image plane depends on the aperture angle α_0, and is expressed by

$$r_s = C_s \alpha_0^3 .\tag{2.13}$$

The quantity C_s is the coefficient of spherical aberration, which is a characteristic constant with the dimensions of a length that determines the quality of an electron lens. A value of about 3 mm is typical of C_s for an objective lens in a TEM, while in a HRTEM the value is lowered to about 1 mm. For an electron lens arranged behind the objective lens, the aperture of the incident ray is smaller than the aperture of rays incident into the objective by a factor given by the magnification of the objective lens. For this reason only the spherical aberration of the objective lens needs to be taken into account when evaluating the resolution limit of the microscope.

Chromatic Aberration

The other aberration that affects the performance of an electron microscope is chromatic aberration. By analogy with the corresponding effect in light optics whereby the focal length of a lens varies with the wavelength of the light, the focal length of an electron lens varies with the energy of electrons. The lens bends electrons of lower energy more strongly, and thus electrons from a point in the object once again form a disc image in the Gaussian image plane. The radius r_c of this disc is given by

$$r_c = C_c \frac{\Delta E}{E} \alpha_0 , \qquad (2.14)$$

where C_c is the chromatic aberration coefficient of the lens (like C_s it is a length), ΔE is the deviation of the electron energy from its mean value E, and α_0 is again the aperture angle of the lens. The chromatic aberration coefficient C_c of a magnetic objective lens is usually slightly smaller numerically than the focal length. Chromatic aberration is a lens defect that degrades the image whenever electrons in the beam cease to be monoenergetic. This may be the result of electrons starting from the gun with a spread of energies, or of the accelerating voltage fluctuating with time, or of the electron beam losing energy through collisions when passing through the specimen. In modern instruments the stability of the accelerating voltage and also that of the lens current, which has a similar influence on chromatic aberration, is so good that we don't have to worry about the chromatic aberration caused by the illumination system, as it is insignificant when compared with the energy losses associated with the electrons that are transmitted through a sample. Inelastic scattering of the high-energy electrons by plasmon excitations is a common way for electrons to lose 10–20 eV, and for thick samples the energy-loss spectrum is additionally broadened by multiple energy losses. A rule of thumb provided by Sawyer and Grubb is that, for biological and polymeric specimens, the resolution limit is about one-tenth of the specimen thickness [33]. Therefore, thin specimens have to be used to minimise the blurring of TEM images caused by chromatic aberration.

Astigmatism

Astigmatism is another lens defect that can degrade the resolution of an electron lens. This defect occurs when a lens does not have perfect cylindrical symmetry. It arises from lack of perfection in the machining of the lens pole pieces, in particular a lack of circularity in the bores and the flatness of the pole faces, and also from asymmetry in the magnetic material of the lens itself. The characteristic feature of astigmatism is that beams leaving the object in two perpendicular planes (containing the electron-optical axis) intersect in different image planes. The difference between the focal lengths of these planes that lie perpendicular to each other is used to measure the astigmatism. Fortunately, astigmatism is one of the few defects in electron lenses that can be corrected. Two lenses of the TEM require routine corrections for astigmatism using a "stigmator". On the one hand, the first condenser lens must be stigmated to produce a circular incident beam on the specimen. On the other hand, an objective stigmator is necessary to cancel the objective astigmatism in order to get better resolution. A stigmator introduces a balancing cylindrical lens field perpendicular to the astigmatic effect of the lens and hence compensates for this effect. The stigmator can be either electromagnetic or electrostatic in nature, the essential requirements being that the magnitude and direction of the compensating field should be independently variable.

Given a thin specimen and ideal conditions during the operation of the microscope, such as stable object position, adequate alignment of the electron-optical system, accurate focussing, etc., the ultimate resolution limit attainable in electron microscopy depends, on the one hand, on the diffraction aberration and, on the other hand, on the spherical aberration of the objective lens. Since according to Eq. 2.12 the diffraction aberration decreases with increasing aperture angle, and according to Eq. 2.13 the spherical aberration increases with the third power of the aperture angle, we would like to find the optimum aperture angle at which the combined aberration caused by diffraction and spherical aberration has a minimum. If we take the combination of the diffraction and spherical aberration discs in quadrature

$$r(\alpha_0) = \sqrt{r_d^2 + r_s^2} = \sqrt{\left(\frac{0.61\lambda}{\alpha_0}\right)^2 + (C_s\alpha_0^3)^2} \,, \tag{2.15}$$

the compromise value exists when

$$\frac{dr}{d\alpha_0} = \frac{-0.74\lambda^2\alpha_0^{-3} + 6C_s^2\alpha_0^5}{2r} = 0 \,. \tag{2.16}$$

From this equation the optimum (compromise) value of the aperture is obtained as

$$\alpha_{0,\mathrm{opt}} = 0.77\left(\frac{\lambda}{C_s}\right)^{\frac{1}{4}} = A_1\left(\frac{\lambda}{C_s}\right)^{\frac{1}{4}} \,. \tag{2.17}$$

If this expression for $\alpha_{0,\mathrm{opt}}$ is substituted into Eq. 2.12, we can calculate a minimum value for the theoretical resolution limit $r_{\mathrm{th,min}}$:

$$r_{\mathrm{th,min}} = 0.79\left(C_s\lambda^3\right)^{\frac{1}{4}} = A_2\left(C_s\lambda^3\right)^{\frac{1}{4}} \,. \tag{2.18}$$

Because of the only semi-quantitative significance of Eqs. 2.17 and 2.18, the numerical values should be replaced by the constants A_1 and A_2, as then the equations also correspond to the results of wave-mechanical calculations, which lead to somewhat different numerical values (compare Sect. 2.4.3, Eq. 2.47). The value for $r_{\mathrm{th,min}}$ is typically about 0.25–0.3 nm, but for high-resolution instruments it decreases to about 0.15 nm.

2.2.4 Vacuum System

The vacuum system of an electron microscope is necessary for two reasons. On the one hand, it is essential to remove most of the air molecules from the column of the microscope in order to minimise the scattering of the electron beam by gas molecules so that the electrons can travel from the gun to the specimen and from the specimen to the viewing screen or camera. Considering how the mean free path of the electrons depends on the reduced pressure, this requirement is easily met by using a relatively modest vacuum system that provides pressures of about 10^{-3} Pa or better in the microscope column. On the other hand, a vacuum of sufficiently low pressure is needed to prevent the specimen, the apertures and the parts of the electron gun becoming contaminated by contaminants such as hydrocarbons and water vapour created by the interaction of the electron beam with molecules of the residual gas. In general, a better vacuum results in lower contamination. Since a very good vacuum is not needed in all parts of the microscope, the present trend is towards a localised higher and cleaner vacuum of about 10^{-6} Pa, as provided by sputter ion pumps, in the specimen chamber to reduce the contamination of the sample and in the electron source chamber for operating LaB_6 sources or field-emission sources. In addition to the action of the vacuum pumps, the vacuum around the specimen is usually improved by applying an internally supplied cryo pump in the form of metal parts close to the specimen that are cooled by liquid nitrogen. By applying differential pumping the ultrahigh vacuum parts of the microscope can be separated in a dynamic way from the high vacuum parts by apertures of appropriate diameters. Generally, the operating vacuum in electron microscopes is provided by a cascade system of different pumps, as each different types of pump only works over a limited vacuum range. The principles and features of the vacuum pumps employed in electron microscopes are summarised in Table 2.3. A rough vacuum is achieved with a rotary mechanical pump, and an oil diffusion pump backed by the rotary pump is usually used to achieve a working pressure of about 10^{-3} Pa. Other types of pumps sometimes used include turbo molecular pumps, sputter ion pumps and cryo pumps, which have the advantage of giving a cleaner vacuum, in particular less contamination from hydrocarbons. Each operational step of the complex vacuum system of an electron microscope is realised by performing special pumping sequences of the different pumps combined in the system. In practice, however, the user rarely needs to be aware of these pumping arrangements since the pumping sequence is automatic and safety devices ensure that the microscope cannot operate until an appropriate vacuum is reached.

2.3 The Instrument

Although the basic principles of operation of a TEM have not changed since the beginning of electron microscopy, 70 years of electron microscopic development has resulted in a lot of improvements and refinements, such that modern instruments really are wonders of technical innovation. Figure 2.5 shows the appearance of a modern conventional TEM. It is additionally employed with special attachments for analytical investigations, which are discussed in Sect. 3.9.

The arrangement of basic components in a TEM is illustrated in Fig. 2.6. Electrons emitted from a thermionic or a field-emission source are accelerated in the gun by a high voltage produced via a high-voltage generator. The electron beam is formed with the aid of condenser lenses and sometimes a condenser mini-lens (if the objective lens is a condenser/objective lens), a condenser lens aperture, a condenser lens stigmator and beam tilt and translate coils for alignment, and then it enters the objective lens and strikes the specimen in appropriate ways. The specimen is usually

Table 2.3. Principles and features of vacuum pumps employed in electron microscopes (from [34]; reproduced with permission from Springer)

Vacuum pump	Principles	Working vacuum range	Features and notes
Rotary pump (RP)	Pump sucks, compresses, and evacuates gas by rotating a rotor in a chamber kept sealed and lubricated with oil.	Atmosphere to 10^{-2} Pa	As the pump works at atmospheric pressure, it is employed for rough pumping of the TEM. It is also used for back-line evacuation of oil diffusion pumps (DP) and turbo molecular pumps (TMP). Whenever the pump stops, the pump chamber should be set to atmospheric pressure to prevent the oil from flowing back.
Oil diffusion pump (DP)	An oil vapour jet is effused from a nozzle by heating oil. Gas molecules are swept away with the oil jet.	10^{-1} to 10^{-8} Pa	As the pump works at a lower vacuum level than the RP and has a high evacuation speed, it is used to evacuate large-capacity camera chambers in which much gas is generated. Increased the back-line pressure (decreasing the vacuum) streams oil vapour back into the fore-line. Therefore, the back-line should be evacuated continuously with an RP.
Sputter ion pump (SIP)	Ions generated by magnetron discharges are sputtered onto the surface of a titanium wall. The active molecules generated trap gas molecules, which are adsorbed on the thin wall.	10^{-2} to 10^{-9} Pa	As the pump is oil-free, it is called a dry pumping system and is used to evacuate electron gun chambers and columns. Because it adsorbs residual gas, it is not suited to areas with lots of residual gas. It is better used to maintain high vacuum in the system. It is impossible to adsorb inert gas molecules, such as helium and argon, using an SIP. The pumping power is recovered by maintenance, during which the pump is baked and evacuated by a DP.
Turbo molecular pump (TMP)	Gas molecules are evacuated by rotating a metal rotor fin at high speed.	10^{-2} to 10^{-8} Pa	As the pump works in the range from low to high vacuums and is oil-free, it is used to evacuate columns. To avoid vibrations, a magnetic buoyant-type rotor is employed. The back-line is evacuated by an RP.
Cryo pump (CP)	Gas molecules are adsorbed on the surface of a metal fin cooled with a coolant such as liquid nitrogen.	10^{-2} to 10^{-13} Pa	The pump adsorbs all gas molecules, including inert gases. It is possible to attain an ideal vacuum. An anti-contamination fin installed in the specimen chamber is considered to be a kind of cryo pump.

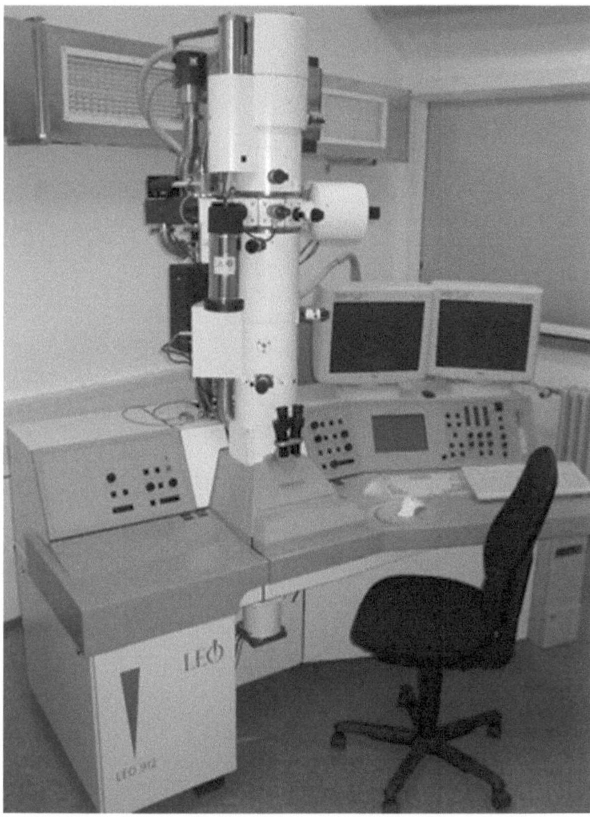

Fig. 2.5. The LEO 912 OMEGA, which has an imaging energy filter, is an example of a modern TEM used for polymer investigations

placed via the specimen holder and the goniometer stage within the objective lens between the upper and the lower pole pieces. After passing through the specimen, the electrons form an image through the action of the objective lens and an objective aperture in the back focal plane of the lens. The image is corrected by an objective stigmator and enlarged by an image-forming system consisting of a series of intermediate and projector lenses and alignment units, and thus finally a highly magnified image becomes visible on the viewing screen or can be recorded by a camera or an electron-sensitive film.

Following the path of the electron beam, the microscope can be split into the following parts: electron gun, illumination system, objective lens and specimen stage, image-forming system and viewing chamber/image recording. These units are described in the following sections.

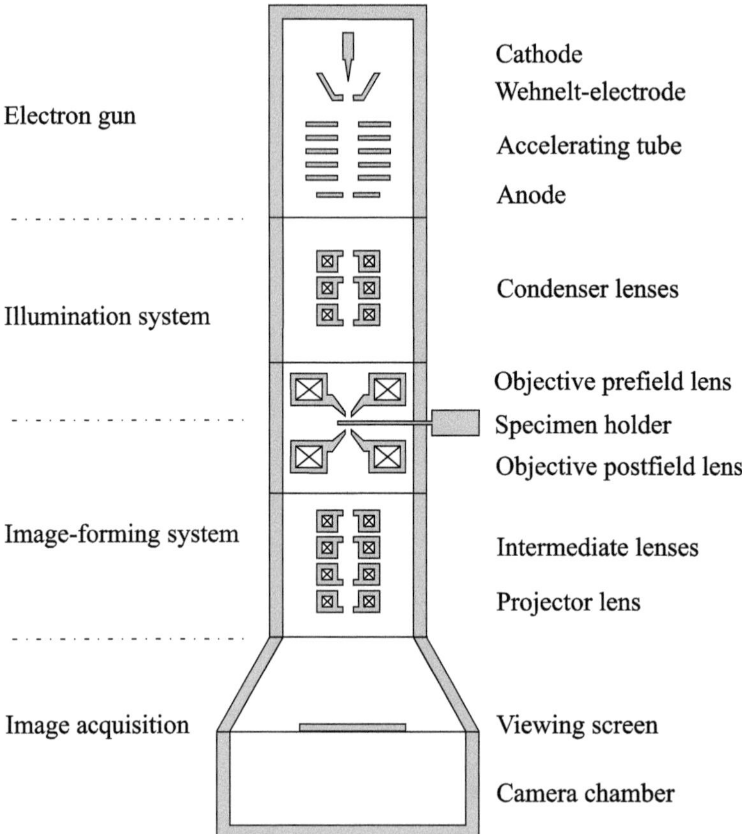

Fig. 2.6. Basic components of a TEM

2.3.1 Electron Gun

The electron gun is located at the top of the microscope column. It generates electrons at the negatively biased cathode and accelerates them between the cathode and anode in such a way that the paths of the accelerated electrons form a crossover which acts as a virtual electron source for the first condenser lens of the illumination system discussed in the next section. For many years a sharply bent tungsten wire, as shown in Fig. 2.7a, was widely used as the electron source in conventional TEMs. When heated electrically, this tungsten hairpin filament gets hottest at its sharp tip, where the highest resistance to current flow is found. If the temperature is high enough, some electrons receive sufficient thermal energy to surmount the work function of the tungsten/vacuum interface and leave the tungsten cathode. Increasing the temperature also increases the thermionic emission, but unfortunately also leads to evaporation of the filament material and a decrease in filament lifetime. The conventional thermionic emission from a tungsten hairpin cathode is limited in terms of temporal coherence by an energy spread of the emitted electrons of the order of a few eV

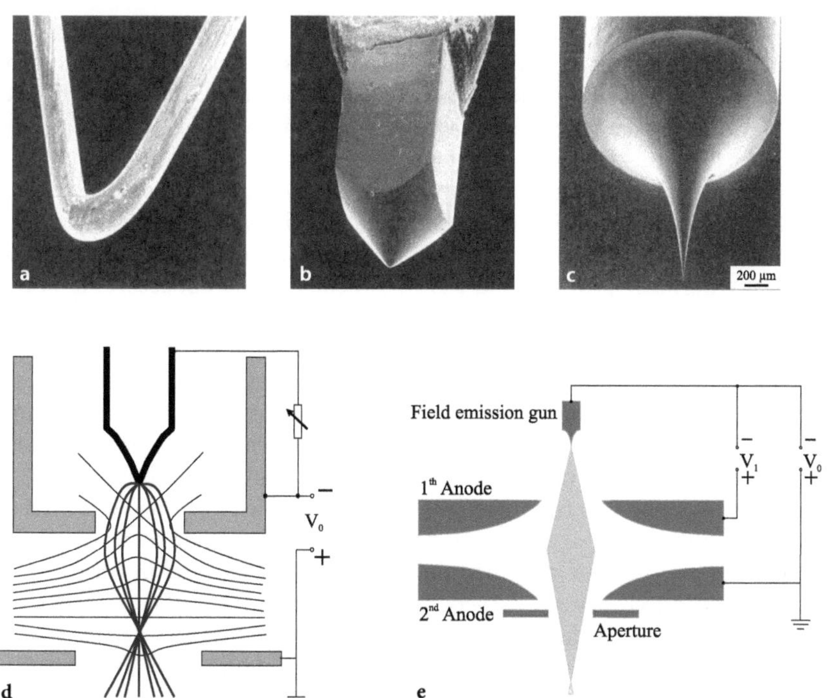

Fig. 2.7. Various electron sources [tungsten filament (**a**), LaB$_6$ emitter (**b**) and tungsten field-emitting tip (**c**)] and schematic diagrams of a thermionic gun (**d**) and a field-emission gun (**e**)

(electronvolts) and in terms of spatial coherence by the gun brightness. Brightness, defined as current density per unit solid angle of the source, is an important property of the electron source. A smaller source size gives higher brightness and better spatial coherency, but often less stability. Therefore, cathodes of LaB$_6$ (or other borides) and field-emission cathodes provide alternatives that yield reduced energy spread and increased gun brightness. The lower work function of LaB$_6$ has made it the preferred material for thermionic electron sources. This lower work function more than overcomes its lower operating temperature compared to tungsten. LaB$_6$ cathodes consist of small pointed crystals, as shown in Fig. 2.7b. They require indirect heating because their electrical resistance is too high for direct current heating.

LaB$_6$ crystals are more susceptible to thermal shock than tungsten, so it is important to take care when heating and cooling an LaB$_6$ source, and as LaB$_6$ is a highly reactive material, the gun vacuum has to be 10–100 times better than that for tungsten cathodes.

As shown in Fig. 2.7d, a thermionic electron gun is a triode system consisting of the cathode connected to a highly negative potential, the (grounded) anode aperture, and the Wehnelt electrode in-between, which has a potential that is hundreds of volts more negative than the cathode potential. Therefore, electrons emitted from the point of the electrically heated cathode are not only accelerated towards the anode but are also strongly influenced by the action of the Wehnelt electrode. The negatively biased Wehnelt electrode serves to restrict the electron emission, and also serves to focus the electrons to a crossover as they are accelerated to the anode. Usually the Wehnelt electrode is "self-biased" electrically, as the electron current in the microscope flows through the bias/emission potentiometer.

This has the added effect of stabilising the emission current against fluctuations in filament temperature, since any tendency for the emission to increase results in a more negative bias voltage which acts against the increase. Several values of bias resistors are normally provided and can be switched in and out according to how much current is required from the gun.

In a field-emission gun (FEG), the thermionic emission is replaced by electron extraction through quantum tunnelling. Under ultrahigh vacuum conditions, a material subjected to a sufficiently strong electric field will emit electrons in the region of maximum field strength. A cold FEG employs tungsten with a (310) plane surface as an emitter, which works at room temperature without heating. As shown in Fig. 2.7c, such a tungsten cathode is fabricated with a sharply pointed shape and a 100 nm radius of curvature to localise the electric field. As well as a cathode with a sharpened tip, an arrangement of two anodes is used in a FEG, as illustrated in Fig. 2.7e. The first anode is positively biased by several kV with respect to the cathode. This extraction voltage V_1 generates the field strength at the cathode that pull electrons out of the tip. The second anode accelerates the electrons to the final energy determined by the voltage V_0 between the cathode tip and the grounded second anode. A cold FEG is an excellent point source of illumination, and may not even require demagnification action from the first condenser lens. Another important advantage of cold FEG is the absence of thermal energy spread, so that the beam can be highly monochromatic. On the other hand, the emission current is destabilised by a contamination of the tip, and so regular maintenance (the so-called "flashing procedure") is necessary. The electric field requirements and the demanding vacuum requirements of FEG can be reduced significantly by heating the tip. In a thermal FEG the emitter is heated to a temperature that is essentially lower than that of thermal emission. Owing to the reduction of the work function due to the strong electric field, the Schottky effect is the basis for electron emission. Usually, the work function of the emitter is also reduced by covering the tungsten tip with a thin layer of ZrO. The performance characteristics of a cold FEG are better than that of a thermal FEG, but the latter has the advantage of less emission noise and provides a stable emission current which is not influenced by contamination layers.

A comparison of the characteristics of various electron guns is given in Table 2.4. The performance characteristics of cold and thermal FEGs make them a good choice for modern analytical TEMs used in materials science, while a thermionic gun with a LaB_6 emitter is a less expensive alternative for use in a modern TEM applied to investigate polymeric materials.

Table 2.4. Comparison of the characteristics of various electron guns

Emission	Thermionic	Thermionic	Schottky	Field emission
Material	W	LaB_6	ZrO/W	W
Work function (eV)	4.5	2.7	2.8	4.5
Working temperature (K)	2800	1400–2000	1400–1800	300
Emission current (Acm^{-2})	1–3	25–100	1000	10^5–10^6
Gun brightness ($Acm^{-2}\,sr^{-1}$)	10^5	5×10^5–5×10^6	10^8	10^8–10^9
Probe current range	1 pA–1 μA	1 pA–1 μA	1 pA–5 nA	1 pA–300 nA
Crossover diameter (μm)	20–50	10–20	0.1–1	0.01–0.1
Energy spread (eV)	1–3	0.5–2	0.3–1	0.2–0.5
Vacuum requirements (Pa)	10^{-2}–10^{-3}	10^{-3}–10^{-5}	10^{-7}–10^{-8}	10^{-8}–10^{-9}
Cathode lifetime (h)	100	200–1000	>2000	>1000

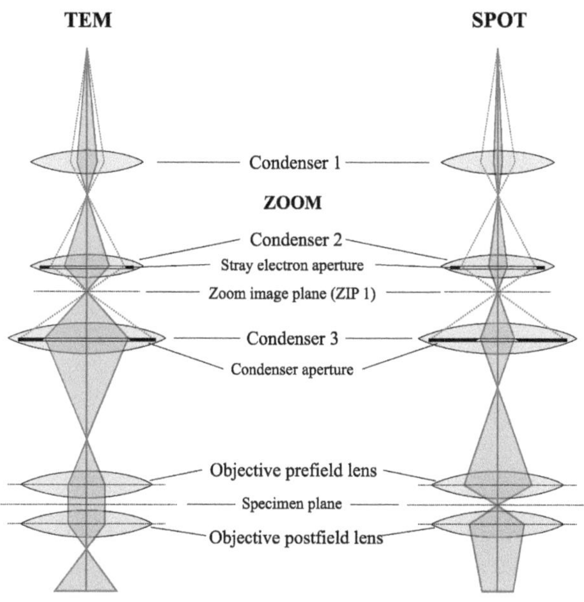

Fig. 2.8. Arrangement of lenses and apertures in the illumination system, and illustration of the corresponding TEM (*left*) and spot (*right*) illumination for the microscope shown in Fig. 2.5 (according to the ray diagram in the manual of the LEO 912 OMEGA; reproduced with the permission of Carl Zeiss SMT AG)

2.3.2 Illumination System

The illumination system transfers electrons from the gun crossover to the specimen under various illuminating conditions, ranging from a broad and essentially parallel beam to a focussed probe. In principle, a single, rather weak, condenser lens with a magnification of unity can project the electron crossover in the gun to a focussed spot in the specimen plane or to an unfocussed broader disc. However, the resulting focussed spot with a diameter of some tens of micrometres is much larger than is needed to illuminate the specimen for high-magnification operation; for example, at a magnification of 100 000× the extension of the imaged specimen area is only about 1 μm. However, the irradiation of the specimen area should correspond as closely as possible to the viewing screen, whatever the magnification. This reduces specimen drift due to heating and limits the radiation damage and contamination in unirradiated areas. Therefore, a TEM has at least a double condenser system with a strong first lens to reduce the gun crossover to an image about 1 μm in diameter and a weaker second lens to project this image onto the specimen plane. Usually the number of lenses in the illumination system in modern microscopes is increased beyond this, however. This is because, on the one hand, a defined range of illumination apertures is required in the normal TEM mode, from around 1 mrad for medium magnifications to $\approx$ 0.1 mrad for high-resolution and phase contrast microscopy to $\approx$0.01 mrad for small-angle electron diffraction and holographic experiments. On the other hand, modern TEMs are often also used for analytical investigations, and so with the aid of the illumination system a small electron probe appropriate for X-ray microanalysis, convergent beam electron diffraction techniques or scanning transmission mode must be produced. Therefore, a typical illumination system in a modern TEM consists of three condenser lenses and also includes the upper part of the objective as an objective prefield lens, and sometimes an additional condenser minilens is arranged between the third condenser lens and the objective prefield lens. As an example, Fig. 2.8 schematically illustrates the arrangement of lenses and the beam formation for TEM (left) and spot (right) illumination in the illumination system of the microscope shown in Fig. 2.5. Condensers 1 and 2 together form a zoom lens system which images the crossover of the gun, which is variably reduced in the constant zoom image plane. Stray electrons are blocked from passing through the zoom system by a fixed aperture provided in the principal plane of condenser 2. For spot illumination, the third condenser images the crossover in the entrance image plane of the objective prefield lens, which produces crossover image on the specimen that is reduced by approximately 20×, and the variable aperture placed in the principal plane of condenser 3 acts as condenser aperture for the focussed spot. In the TEM illumination mode, the third condenser transmits the crossover image of the zoom system to the front focal point of the objective prefield lens. The illumination of each object point is thus parallel with the electron-optical axis, and the diameter of the illuminated specimen area is determined by the variable aperture in the principal plane of the third condenser. In Fig. 2.9 a more detailed ray diagram of the section between condenser 3 and the specimen plane illustrates that the illuminating cones for each object point have the same illuminating aperture α_i, given by the radius of the crossover image r_{co} divided by the focal length f_{pf} of the objective prefield lens. Therefore, the illumination aperture can be varied by changing the excitation of the zoom system consisting of condensers 1 and 2.

It is necessary to have a convenient system for aligning the electron beam such that the latter passes the elements of the illumination system in a proper manner. The alignment steps needed are, on the one hand, the translation of the beam without altering its inclination and, on the other hand, the tilting of the beam such that it does

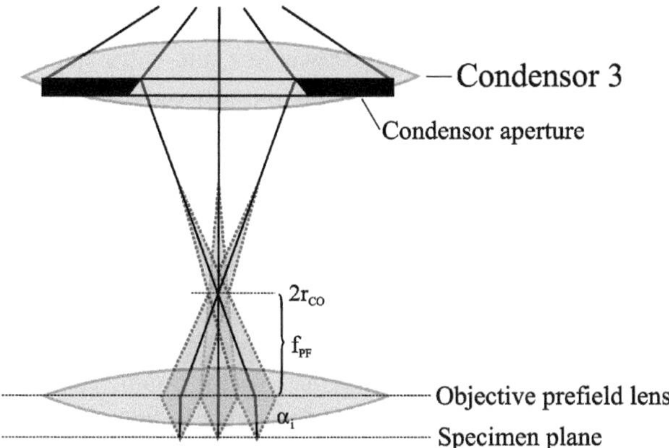

Fig. 2.9. Illustration of the illuminating cones for several object spots using corresponding ray diagrams (according to the ray diagram in the manual of LEO 912 OMEGA; reproduced with the permission of Carl Zeiss SMT AG)

not change the beam's position in the reference plane. Both of these motions can be obtained by a pair of scan coils which cause appropriately balanced beam deflections in opposite directions. Such coils are also used to scan or rock the electron probe in the scanning mode or for special diffraction techniques, respectively, and to generate a dark-field TEM mode.

2.3.3 Objective Lens and Specimen Stage

The combination of the objective lens and the specimen stage is of particular importance for the microscope. As already noted above, the aberrations of the objective lens determine the resolution in a fundamental manner, and so its astigmatism must be corrected by a stigmator placed behind the objective lens, and the region of lens field action must be as small as possible to reduce spherical aberration. The stage is used to clamp the specimen holder into the correct position such that the specimen is positioned in the pole piece gap with great adequacy and stability and the objective lens can form images and diffraction patterns in a reproducible manner. Additionally, the stage must provide orthogonal x and y movements in the specimen plane (typically ≈ 1 mm from its centred position on the microscope axis) to select specimen sites of interest for imaging. Furthermore, defined orientations of crystalline materials and tomographical investigations of specimens require a eucentric goniometric stage with tilt rotation or double tilt motion. There are goniometer stages that permit the specimen plane to be tilted by up to $\approx 60°$, combined with rotation by $360°$ about the lens axis, or alternatively, simultaneous and independent tilts about two axes at right angles. The space between the pole pieces of the objective lens is however very limited, and particularly in TEMs equipped with special high-resolution pole pieces the narrowness of the gap in the objective lens limits the specimen tilt to significantly smaller values. There are two quite distinct types of specimen stages in common use: top-entry stages and side-entry stages. The specimen holder in a top-entry stage enters the bore of the objective lens from above. In a side-entry stage the specimen holder takes the form of a flattened rod and is introduced into the pole piece gap from the side. Top-entry stages are less sensitive to vibrations and thermal drift and are therefore used in a few HRTEMs, despite some drawbacks (limited tilting, unsuitability for analytical attachments, difficult to combine with airlocks and anticontamination). Modern side-entry stages have performances similar to those of the best top-entry stages and are the most common type used. They are easy to operate with an airlock and with a cryoshield around the specimen for reducing specimen contamination, and can also be well designed for analytical investigations and in situ manipulations (heating, cooling, straining) of specimens. Usually, polymeric ultrathin films for TEM investigations are fixed on a support consisting of a fine metal grid 3 mm in diameter (see Chap. 10). The grid with the specimen can easily be transferred with the help of tweezers into a conventional side-entry specimen holder, where it is fixed in place by a special holding bow. The application of special holders for in situ heating, cooling and specimen straining are described in Chap. 6.

2.3.4 Image-Forming System

The image-forming system of a TEM consists of at least three lenses: the objective lens, the intermediate lens and the projector lens. The function of the objective lens, to form an image of the illuminated specimen area, results from the postfield of the lens at the back side of the specimen, while the objective prefield lens is included in the illumination system, as explained above and illustrated in Fig. 2.8. The intermediate lens can magnify the first intermediate image, which is formed just in front of this lens (Fig. 2.10a), or the first diffraction pattern, which is formed in the focal plane of the objective postfield lens (Fig. 2.10b) by reducing the excitation. One of these images produced by the intermediate lens is enlarged by the subsequent projector lens and displayed on the viewing screen.

The bright-field mode (Fig. 2.10a) with a centred objective aperture in the back focal plane of the lens is the typical TEM mode, and the scattering contrast caused by this aperture (see Sects. 2.4.1 and 3.1) is widely used for TEM investigations of polymers. In the so-called selected area diffraction mode (Fig. 2.10b), a diffraction pattern of a reduced specimen area is usually projected onto the viewing screen. This is carried out by selecting the area of interest via a corresponding aperture (SAD aperture) in the plane of the first intermediate image formed by the objective lens.

Real imaging lens systems in TEM instruments are far more complicated than the simple three-lens model of Fig. 2.10. An increased number of lenses and additional alignment elements tuned to each other with the aid of computer control are necessary to change the magnification across a wide range from about 50× up to 1 500 000× in an appropriate manner, to compensate for image rotations and to provide different imaging modes. To understand the optics of any particular microscope, it is necessary to consult the ray diagrams in the manufacturer's operating manual.

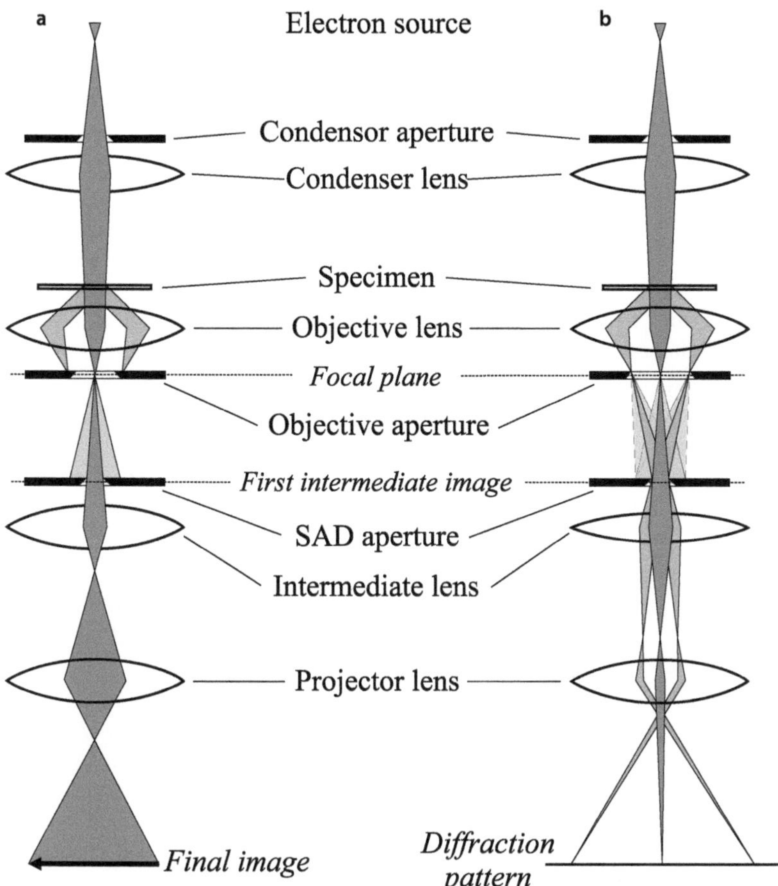

Fig. 2.10a,b. Two basic operation modes of the TEM image-forming system: **a** bright-field image mode, **b** selected area electron diffraction mode

2.3.5 Viewing Chamber and Image Acquisition

The final electron microscopic image is projected onto a luminescent screen in the viewing chamber, which can be viewed via a window made of lead glass.

An attached pair of binoculars is usually used to observe details in the image. For a long time the recording of images in a TEM was based on the exposure of electron-sensitive negative films. This has changed recently with development of slow-scan CCD cameras and image plates. Characteristic features of the different recording techniques are briefly described below.

Negative Film
Traditionally, electron-sensitive photographic emulsion layers on polymer (formerly glass) support bases have been used for image recording. The sensitivity of the film to

electrons is similar to their sensitivity to light. The camera system is placed within the microscope immediately below the viewing chamber. A light-tight dispensing magazine which contains about 50 cassettes with film sheets of size, e.g., 60 mm × 90 mm is loaded into the camera chamber. In order to record an image, one of the cassettes is transferred from the light-tight box into the position below the screen where the electron image falls on it when the screen is tilted out of the electron beam, and an electrically operated shutter is opened for a preset exposure time. After exposure, the film sheet is transferred into the same or another storage box, from where it can be retrieved later for processing to a permanent negative image. Although film is the oldest recording medium, it still retains sufficient advantages to find use in current systems. Compared to digital recording media, film has significantly better resolution, which depends on the type of the film and is around 10 μm in most cases. With its high resolution and large area, film gives the largest field of view. Film is also an excellent archiving media, and we know from experience that the original image information can be stored in film for a long time. Unlike film, digital images are stored with today's technologies, including data formats, storage media and reading software, and hardware devices. Ensuring the survival of these images is by no means a simple task, and this is still an unsolved challenge.

A quantitative analysis of an image stored in a film can be done with the aid of a film scanner, by measuring the optical density (which, however, falls within a small range from 0.01 to 0.2 $e\mu m^{-2}$), which is proportional to exposure. The total dynamic range is also rather limited (about 10^3). These properties permit normal imaging applications, but limit the application of film to low-dose imaging and for recording images with very strong contrast, as typically observed for diffraction patterns.

Slow-Scan Charge-Coupled Device (SSCCD) Cameras

SSCCD cameras have high sensitivity, a linear response over a wide range of intensities (four orders of magnitude), and offer a direct path to digital data acquisition. However, with a pixel size of between 15 μm and 25 μm and typically 1024 × 1024 or 2048 × 2048 pixels, a much smaller image area is covered than in image recording with negative films. In principal, incident high-energetic electrons could be directly detected by a CCD, but the direct exposure of the CCD to the electron beam causes radiation damage, and due to the high number of thermalised electrons caused by each of the high-energetic electrons, the CCD will be saturated by a relatively small number of incident electrons. Therefore, current SSCCD cameras used in a TEM record an electron image in three stages: (1) incident electrons are converted into light by an yttrium-aluminium-garnet (YAG) single-crystal scintillator or a powder phosphor; (2) released photons are transported to the CCD array via a fibre optic plate or lens couplings; (3) the light is converted to an electron charge that is temporarily stored in each channel of the CCD, where the CCD is an array of metal-oxide-semiconductor (MOS) capacitors. By controlling the gates of the MOS capacitors, charge can be accumulated and transferred. For the SSCCD, the charge is transferred row by row to the serial register placed next to the CCD array, and then transferred pixel by pixel and measured by an analogue/digital converter. Usually a SSCCD camera is cooled by liquid nitrogen or by an electronic Peltier cooler in order to reduce the dark current (which produces noise). Images can be acquired, viewed and processed immediately with a SSCCD camera. Thus the quality of images can be assessed during the operation of the microscope, which is very useful when image recording depends critically on the microscope operation conditions.

Image Plate (IP)

The image plate used in standard film cassettes is a reusable flexible sheet. It consists of a layer of storage phosphor, where tiny crystals of (typically) $BaFBr:Eu^{2+}$ are embedded in a resin. This active layer is coated onto a polyester support sheet and covered by a polymeric protective layer. The electron irradiation results in $Eu^{2+} \rightarrow Eu^{3+}$ transitions and excites the crystals at their luminescence centres to a semi-stable state. The image information created by this excitation is stable for many hours and decays within days. By irradiating them with red laser light with a wavelength of 635 nm, the crystals are stimulated to release the stored information as a blue luminescence signal with a wavelength of 390 nm. The storage process itself is highly localised, but due to the lateral scattering of the incident electrons within the active layer of the IP, it is limited to a few μm. Finally, however, the spatial resolution of an IP is determined by the read-out process. The IP must be transferred to a special external read-out device, where it is scanned and the stored information is read out by a moving unit containing the laser and a detection system for the luminescence signal released. Depending on the instrument, the read-out pixel sizes of an IP with an effective area of 80 mm × 90 mm are between 15 μm and 50 μm. The luminescence signal emitted from the plate while reading is directly proportional to the exciting electron dose, and the dynamic range is more than six orders of magnitude.

2.3.6 Alignment and Operation of Transmission Electron Microscopes

No TEM will give its best performance unless its electron-optical elements are aligned with each other, the astigmatism of the illumination system and the objective lens is corrected by corresponding stigmators, and the beam-limiting diaphragms are accurately centred on the electron-optical axis. The lenses and other components of the column of a modern electron microscope are accurately aligned mechanically by the manufacturer. However, there is always the need for minor adjustments, which are performed by the operator using sets of double deflection coils placed at strategic points in the column. Alignment procedures differ from microscope to microscope, but in principle they are based on the following steps [35]. First, without the presence of a specimen, the emission current of the electron gun must be adjusted. The focussed beam is projected onto the screen, and under unsaturated conditions (corresponding to certain settings of the filament temperature and bias voltage) a symmetrical filament pattern is formed with the aid of gun tilt coils. Then the illumination system must be aligned. This includes correctly positioning the condenser aperture and correcting for astigmatism using the condenser stigmator. Typically, a condenser alignment procedure centres the focussed beam at different lens currents of the first condenser lens (variation of spot size), and this procedure is repeated until the beam is found to occur at the centre of the screen (no shift) irrespective of the spot size.

After the alignment of the condenser lens system the specimen is placed in the reference plane, where the image on the screen is in focus at the optimum value of the objective lens current. This reference plane is also the eucentric plane of a side-entry holder; this means that a point on the electron-optical axis does not move laterally when tilted around the holder axis. An image wobbler is usually used to find the exact position of the reference plane by varying the z-control of the sample holder. This produces a superposition of two images on the screen as long as the specimen is placed below or above the reference plane, and the distance between the corresponding image structures in the double image increases with increasing distance between specimen plane and reference plane. With the specimen in the reference plane, the rotational centre of the objective lens can be checked. The illumination is tilted if wobbling the strength of the objective lens results in an image shift. The adjustment of the illumination tilt is however often undertaken by varying the beam voltage with a wobbler (adjustment of the voltage centre), as recent instruments show little instability in the objective lens current, so image broadening is mainly caused by the change in wavelength due to the inelastic scattering of electrons in the specimen. The astigmatism of the objective lens is adjusted with the objective lens aperture inserted. As the astigmatism depends on the aperture size and its position, the chosen aperture is inserted and centred in the selected area diffraction mode, and then, after switching back to the normal image mode, the lens and aperture can be stigmated as a unit. To correct the astigmatism at magnification factors of up to several hundreds of thousands, the features of the Fresnel fringes at thin edges or small holes of an amorphous specimen (e.g. carbon) are usually observed at different focus conditions. For high-resolution imaging at magnifications of >200 000×, the streaking in the image is used for astigmatism adjustment. Other alignments in subsequent parts of the image forming system are less crucial and are less likely to change from day to day. If, for example, a feature of interest located at the centre of the viewing screen does not remain there or even moves out of the viewing area as the magnification is varied, this is a typical indication of the misalignment of the intermediate or projector lenses. Correction is achieved using a set of one or more deflection coils for image shifts below the objective lens. This misalignment predominantly appears at magnification steps where continuous changes in magnification can only be achieved by changing the lens currents of some lens via a computer. These magnification changes are construction-dependent features of the particular microscope used, and so such adjustments must be carried out according to special routines described in the microscope's manual.

2.4 Fundamentals of Image Formation

Image contrast is caused by the different scattering mechanisms that occur in amorphous and crystalline materials. However, the intensity distribution of the image depends not only on the interaction of the electron beam with the object, but also on the illumination conditions and in particular on the action of the objective lens and the arranged apertures.

2.4.1 Scattering Mechanism and Contrast Formation

In general, electron scattering (see, e.g. [32,36–38]) can be divided into two types: inelastic and elastic scattering. Inelastic scattering is an electron–electron interaction where electrons scattered by the atomic shell electrons suffer a considerable loss of energy. Elastic scattering can be considered to be an electron–nucleus interaction in which the atomic electron cloud merely screens the Coulomb potential. However, for small scattering angles θ, a negligible amount of energy is transferred to the nucleus and so the electrons are scattered with no appreciable energy loss. This small-angle elastic scattering is the most important of all interactions in polymeric materials that cause contrast in the electron image. For larger scattering angles, and also for high electron energies, an appreciable energy transfer from the electron to the nucleus results from the elastic electron–nucleus interaction process. These electron–nucleus collisions can lead to specimen damage if the energy transfer exceeds the threshold energy (on the order of 10–30 eV), such that an atom can be displaced from its original position in the crystal to an interstitial site. A knowledge of the angular distribution of the elastically scattered electrons is the basis for contrast interpretation. Elastic electron scattering is illustrated in Fig. 2.11 in the particle model (a) and in the wave model (b). Electrons that pass through an element of area $d\sigma$ of the parallel incident beam will be scattered via Coulomb forces into a cone of solid angle $d\Omega$ in the particle model. The inclination of the scattered electrons is expressed by the scattering angle θ. In the wave model, the incident electron beam is described by a (time-invariant) plane wave

$$\psi = \psi_0\, e^{2\pi i k_0 z} \tag{2.19}$$

with an amplitude ψ_0 and a magnitude k_0 of the wave vector whose direction coincides with the coordinate z. Using a spherical coordinate system with its origin at the scattering centre, the scattering event is described by a spherical scattered wave

$$\psi_{sc} = \psi_0 f(\theta)\frac{1}{r}e^{2\pi i k r} \tag{2.20}$$

with a magnitude k of the wave vector and an amplitude that depends on the scattering angle θ. Far from the nucleus, the total wave field is expressed as the superposition of the undisturbed plane incident wave and the spherical scattered wave

$$\psi_s = \psi_0\, e^{2\pi i k_0 z} + i\psi_0 f(\theta)\frac{1}{r}e^{2\pi i k r} \tag{2.21}$$

where the phase shift of 90° between the scattered and the incident wave is taken into account by the factor i of the second item. In general, the scattering amplitude $f(\theta)$ is a complex quantity and can be expressed by

$$f(\theta) = |f(\theta)|e^{i\eta(\theta)}\,. \tag{2.22}$$

This means that there is not only the mentioned phase shift of 90° but also an additional phase shift $\eta(\theta)$, which depends on the scattering angle.

a

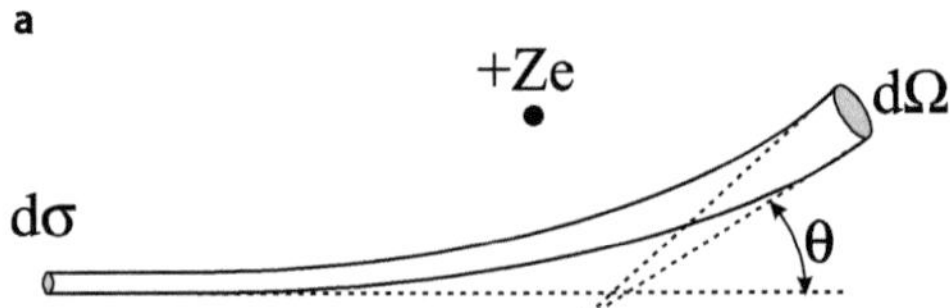

b

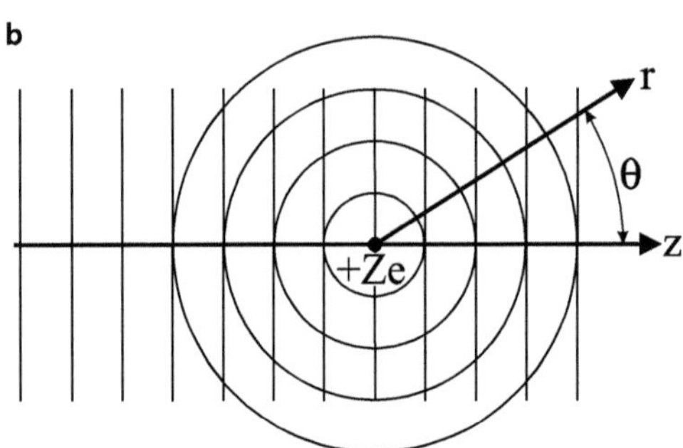

Fig. 2.11. Elastic electron scattering in the particle model (**a**) and scattering in the wave model with the superposition of a plane incident wave and a spherical scattered wave (**b**)

The most convenient quantity for characterising the angular distribution of scattered particles is the differential cross-section: the ratio $d\sigma/d\Omega$. This cannot be calculated exactly from the classical particle model, but using the wave model the relation

$$\frac{d\sigma_{el}}{d\Omega} = |f(\theta)|^2 \tag{2.23}$$

between the elastic differential cross-section $d\sigma_{el}/d\Omega$ and the scattering amplitude $f(\theta)$ of a single atom can be derived. The scattering amplitude $f(\theta)$ can be calculated from the wave-mechanical Schrödinger equation. However, a detailed treatment of this problem is beyond the scope of this book, and the reader is referred to textbooks on transmission electron microscopy (see e.g. [32]), where theoretical and experimental results are discussed.

There is a simple geometric relationship between the scattering angle θ and the solid angle Ω

$$\Omega = 2\pi(1 - \cos\theta) \tag{2.24}$$

and therefore

$$\frac{d\Omega}{d\theta} = 2\pi \sin\theta \tag{2.25}$$

is obtained when Eq. 2.24 is differentiated. Using Eq. 2.25, the following differential relation for $d\sigma_{el}$ can be written:

$$d\sigma_{el} = \frac{d\sigma_{el}}{d\Omega}\frac{d\Omega}{d\theta}d\theta = \frac{d\sigma_{el}}{d\Omega}2\pi\sin\theta\,d\theta. \tag{2.26}$$

By integrating Eq. 2.26 over θ from α to π, the number of electrons elastically scattered through angles $\theta \geq \alpha$ can be calculated. This gives the partial elastic cross-section $\sigma_{\mathrm{el}}(\alpha)$:

$$\sigma_{\mathrm{el}}(\alpha) = \int_{\alpha}^{\pi} \frac{\mathrm{d}\sigma_{\mathrm{el}}}{\mathrm{d}\Omega} 2\pi \sin\theta \, \mathrm{d}\theta. \tag{2.27}$$

The total elastic cross-section σ_{el} is obtained if the lower limit of integration is zero, i.e.,

$$\sigma_{\mathrm{el}} = \int_{0}^{\pi} \frac{\mathrm{d}\sigma_{\mathrm{el}}}{\mathrm{d}\Omega} 2\pi \sin\theta \, \mathrm{d}\theta. \tag{2.28}$$

Taking into account the corresponding expression for inelastic scattering, the total scattering cross-section σ_t of an atom is given by

$$\sigma_t = \sigma_{\mathrm{el}} + \sigma_{\mathrm{inel}}, \tag{2.29}$$

where σ_{inel} denotes the total inelastic cross-section. The total scattering cross-section σ_t determines whether or not scattering occurs and can be regarded as the effective target area presented by the scatterer. The description of scattering by extended specimens must take into account possible mutual interference of the electron waves scattered at the atoms. Therefore, the contributions of the individual scatterers to the intensity in the image plane depend on the atomic arrangement, and so amorphous objects and crystalline objects – the two limiting cases of disorder and order, respectively – contribute in different ways to the interaction events. For an (ideal) amorphous object, the electrons are scattered incoherently, i.e. systematic correlations of the phases do not have to be taken into account, and in order to determine the intensity in the final image plane the intensities and not the wave amplitudes of the individual scattering events are summed. Assuming that the specimen consists of several layers of thickness $\mathrm{d}x$ where the scatterers do not overlap, there will be

$$N = \frac{N_A}{A} \rho \, \mathrm{d}x \tag{2.30}$$

atoms per unit area in each layer. Here N_A is Avogadro's number, A is the atomic weight and ρ is the density of the specimen. As the N scatterers block an area equal to $N\sigma_t$, the intensity I of the incident electron beam is lowered by an amount $\mathrm{d}I$ in each layer, and this can be expressed as

$$\frac{\mathrm{d}I}{I} = -\frac{N_A}{A} \sigma_t \rho \, \mathrm{d}x \tag{2.31}$$

The integration of Eq. 2.31 over a finite thickness x yields the following relation for the intensity I as decreased by scattering

$$I = I_0 \, e^{-\frac{N_A}{A} \sigma_t \rho x}, \tag{2.32}$$

where I_0 denotes the intensity of the incident beam. In the bright-field mode, as described in Sect. 2.3.4, the diaphragm in the focal plane of the objective lens acts as a stop that absorbs all electrons scattered through angles $\theta \geq \alpha$, where α is determined by the aperture hole. To obtain the decrease in transmission for this case, the total cross-section σ_t in Eq. 2.32 must be replaced by the sum of the partial cross-sections $\sigma_{el}(\alpha)$ and $\sigma_{inel}(\alpha)$ for elastic and inelastic scattering, respectively (compare Eq. 2.27). Scattering contrast is caused by the local density ρ and the local specimen thickness x in the same way, and so the product of the density and the thickness is often referred to as the mass thickness and the corresponding contrast is called the mass-thickness contrast. Although in practice an ideal amorphous object with complete random disorder does not exist, this contrast describes the image intensity for low and medium magnifications well, unless a highly coherent electron beam and large defocusing are employed. However, at high magnifications and electron-optical imaging conditions, which permit the resolution of details that are smaller than 1 nm, a contrast interpretation must take into account that according to the wave-optical theory of imaging it is necessary to sum over the (complex) wave amplitudes and obtain the image intensity by squaring the wave amplitude in the final image plane. This results in a prevailing phase contrast of a thin amorphous weak scattering object, as described in more detail in Sect. 2.4.3.

2.4.2 Electron Diffraction and Diffraction Contrast

The electron beam interacts with the specimen in a very specific way when the specimen is crystalline. In crystalline structures, the constituent atoms are geometrically arranged in a regular pattern, based on a unit cell which is repeated in all three dimensions throughout the crystal. Normal to well-defined directions within the lattice there are sets of equidistant, parallel lattice planes. Each set of corresponding lattice planes is characterised by a triple of Miller indices h, k, l and an interplanar distance d_{hkl} between neighbouring lattice planes. Unlike the incoherent scattering in amorphous specimens, in a crystalline object the incident electrons are scattered coherently in well-defined directions at the atomic arrays. This scattering phenomenon is called *diffraction*, and the interference maxima occur according to the Bragg law:

$$n\lambda = 2d_{hkl} \sin \Theta . \qquad (2.33)$$

Here λ is the wavelength of the incident electrons, n is an integer and determines the order of diffraction, and Θ is the Bragg angle, which is half of the scattering angle θ, i.e. $\theta = 2\Theta$. Relative to the lattice planes, the angle of incidence and the Bragg angle Θ are equal. This means that diffraction in crystalline materials, which is strictly an interference phenomenon, can be interpreted as a reflection of the incident electrons at the lattice planes. As the wavelength λ of the electrons used in a TEM is about two orders of magnitude lower than typical values of d_{hkl}, the corresponding values of Θ are very small. This means that, in practice, an electron beam will only be strongly diffracted from lattice planes which are almost parallel to the electron beam. On the one hand, this factor makes the geometry of electron diffraction patterns much simpler than that of X-ray diffraction patterns, for which Θ can be very large. On the other hand, the accuracy of a structural analysis performed via electron diffraction is rather limited by the relatively small values of the Bragg angles. As described in

Sect. 2.3.4, the diffraction pattern is formed in the focal plane of the objective post-field lens, where each diffracted beam is focussed into a corresponding spot according to its appearance as an oblique parallel beam. With the aid of the subsequent intermediate lens and the projector lens, this diffraction pattern is enlarged and displayed on the viewing screen. Usually, the selected area diffraction mode (Fig. 2.10b) is applied to display a diffraction pattern of the reduced specimen area of interest.

When the TEM is used in the image mode, some of the diffracted electrons can be removed from the image plane by the objective aperture. In the bright-field image this aperture is used to stop all diffracted electrons and only undiffracted electrons are allowed to contribute to the image. A bright-field image of a perfect, unbent crystal of uniform thickness does not produce any contrast. Contrast is formed if the diffraction conditions for the electrons vary locally, e.g., due to variations in the sample thickness, as a result of bending the crystal, or due to the presence of crystal defects. It is often more informative to use diffracted electrons rather than the transmitted beam, since these electrons have actually interacted with the specimen and image contrast is usually also enhanced. There are two ways to produce these dark-field images. The first and simplest way is to displace the objective aperture sideways so that the image is formed with a corresponding off-axis diffracted beam. This easy mode of operation has the disadvantage that off-axis rays are used for imaging and image quality is adversely affected by aberrations in the lens. Therefore, in the alternative second mode the objective aperture is kept centred on the electron-optical axis, and the incident beam is moved off-axis by tilting the illumination.

The main interest in diffraction contrast arises from the ability to make crystal defects (such as dislocations, stacking faults, grain boundaries and precipitates) visible. The kinematical or the dynamical theory of electron diffraction must be applied in order to understand the contrast that arises from defects. Often a contrast interpretation based on the two-beam approximation can be used, where only the primary beam and one diffracted beam are strongly excited. The kinematical theory assumes that the amplitude of a wave diffracted according to the Bragg law is small compared to that of the primary wave. This approach is only applicable to very thin specimens. The dynamical theory, based on the Schrödinger equation, also describes wave propagation in thick crystals. In the periodic potential of a crystal lattice, the electron waves propagate as a Bloch-wave field, which exhibits the same periodicity as the lattice. With the aid of the dynamical theory, the Bloch-waves can also be calculated for the so-called n-beam case, which considers the transmitted beam and $n-1$ diffracted beams and is used for high-resolution imaging of crystal lattice structures. Visualising defects by diffraction contrast and high-resolution imaging modes are very important applications of TEM investigations in the field of inorganic materials (see, e.g.,[32, 36–44]). However, due to the electron beam sensitivity of polymers, diffraction contrast does not play a major role when studying crystalline polymers by means of TEM. Therefore, a more detailed description of the contrast formation, the special investigation techniques and the theoretical models used to interpret experimental results are beyond the scope of this book.

2.4.3 Fundamentals of the Imaging Process

The image contrast obtained is mainly determined by two processes: the electron scattering in the sample (Sects. 2.4.2 and 2.4.3) and the electron-optical imaging process, including the microscope aberrations. The latter can be described in detail on the basis of the wave-mechanical theory of contrast formation (see, e.g., [32, 36, 40, 45–47]). As illustrated in Fig. 2.12, the overall process of image formation can be roughly explained as follows. The interaction of the incident electron beam with the specimen results in an object wavefunction $\psi_{ob}(x)$ at the exit surface of the specimen (object plane). Here and in the following, for sake of a simplified notation, only the functions for a one-dimensional object are taken into consideration. Far away from the object, the scattered electron wave forms a Fraunhofer diffraction pattern, which can be observed using the objective lens in its back-focal plane. This step is mathematically described by the Fourier transform of the object wavefunction $\psi_{ob}(x)$, which leads to the wavefunction $\Psi_d(u)$, where u is the spatial frequency (i.e. the coordinate in the reciprocal space), which corresponds to the scattering angle θ according to the relation

$$\theta = u\lambda. \tag{2.34}$$

The propagation of the wave from the back-focal plane to the image plane can be described by the inverse Fourier transform, if for simplicity the magnification factor is not taken into consideration. Thus, the action of an ideal lens without imperfections would result in an image wavefunction $\psi_i(x)$, which directly corresponds to $\psi_{ob}(x)$.

However, taking into account real imaging conditions, the effect of the objective lens on the propagating wave can be expressed by multiplying the wavefunction $\Psi_d(u)$ in the back-focal plane of the lens by the transfer function $T(u)$, which is composed of three multiplicative parts:

$$T(u) = A(u)C(u)D(u). \tag{2.35}$$

The value of the real aperture function $A(u)$ is directly related to the geometrical position and size of the diaphragm, i.e. $A(u) = 1$ for the open parts of the aperture, and $A(u) = 0$ for the opaque parts. The specimen-independent properties of the electron-optical imaging system are characterised by the contrast transfer function $C(u)$. In the case of perfectly coherent axial illumination, $C(u)$ describes the phase shift of the electron wave due to the influence of the spherical aberration and the defocussing of the objective lens by the following equation:

$$C(u) = e^{i\chi(u,\Delta_f)} = e^{\frac{\pi i}{2}(C_s\lambda^3 u^4 - 2\lambda\Delta_f u^2)}. \tag{2.36}$$

Here, C_s denotes the spherical aberration coefficient, Δ_f the defocus and λ the wavelength of the incident electrons.

In reality, electron sources are not coherent; their finite chromatic aberrations and beam divergences result in a damping of the high-resolution information. In general,

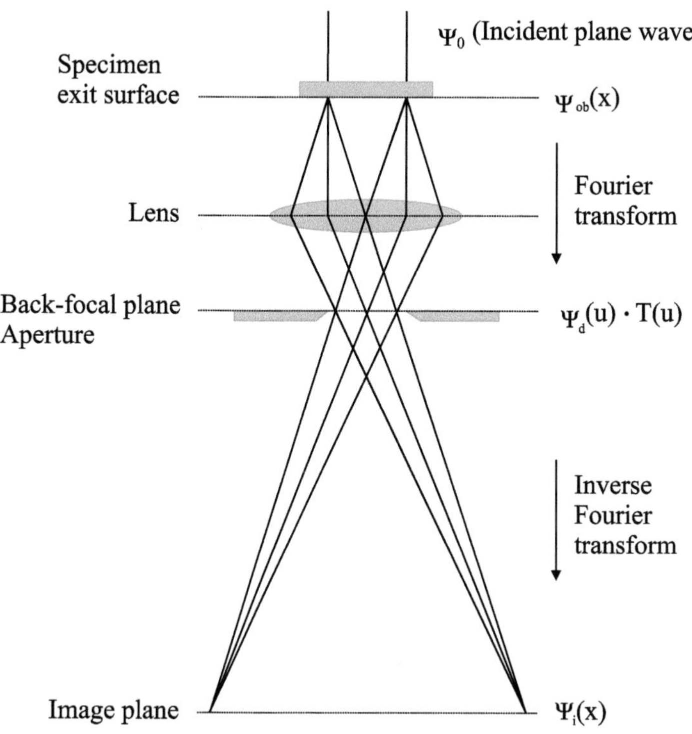

Fig. 2.12. Schematic representation of the Abbe formulation of the imaging process

envelope functions are a useful means for describing the effect of damping. Thus the damping function $D(u)$ can be written as

$$D(u) = D_1(u)D_2(\alpha_i, u). \tag{2.37}$$

Here, D_1 is the damping envelope due to focus spread and D_2 is the damping due to beam divergence, as denoted by the illumination aperture α_i. In practice it is usually sufficient to take into consideration a certain spread of defocus δ_f. With this standard deviation δ_f in the range of 2–15 nm, the envelope function D_1 is given by

$$D_1(u) = e^{-\frac{1}{2}\pi^2\lambda^2\delta_f^2 u^4}. \tag{2.38}$$

$D_1(u)$ damps high spatial frequencies significantly and limits the spatial resolution attainable.

In order to calculate the image wavefunction $\psi_i(x)$ and the image intensity $I(x)$ given by

$$I(x) = |\psi_i(x)|^2, \tag{2.39}$$

it is useful to take advantage of the spread function $t(x)$, which is the Fourier transform of the transfer function $T(u)$. Since the Fourier transform of the product of two functions is the convolution of their Fourier transforms, $\psi_i(x)$ in Eq. 2.39 can be replaced by the convolution of the functions $\psi_{ob}(x)$ and $t(x)$, so that the image intensity becomes

$$I(x) = |\psi_{ob}(x) * t(x)|^2 \tag{2.40}$$

where $*$ is the convolution operator.

A relatively simple relationship which is better at showing the influence of the transfer function can be derived if a very thin specimen is taken into consideration. This acts as a pure phase object, and the object wavefunction $\psi_{ob}(x)$ at the exit surface of the specimen is related to the incident wave ψ_0 as follows:

$$\psi_{ob}(x) = \psi_0 \, e^{-i\sigma\Phi(x)} \, . \tag{2.41}$$

Here, $\Phi(x)$ is the projection in the beam direction of the electrostatic potential distribution in the specimen and σ is the interaction constant given by

$$\sigma = \frac{\pi}{\lambda V_0} \tag{2.42}$$

where V_0 is the accelerating voltage. If the amount of phase shift is small, the exponential function in Eq. 2.41 may be expanded and higher order terms neglected. This results in a normalised function $\psi_{ob,n}(x)$ with

$$\psi_{ob,n}(x) = \frac{\psi_{ob}(x)}{\psi_0} = 1 - i\sigma\Phi(x) \tag{2.43}$$

which is known as the weak phase object approximation (WPOA). Taking into account that the complex spread function $t(x)$ is composed of a real part denoted by $c(x)$ and an imaginary part denoted by $s(x)$, Eq. 2.40 reduces to

$$I(x) = 1 + 2\sigma\Phi(x) * s(x) \tag{2.44}$$

if $\psi_{ob}(x)$ is replaced by the function $\psi_{ob,n}(x)$ and terms of second order in $\sigma\Phi(x)$ are omitted. Thus, the image intensity directly represents the projected potential of the sample, smeared out by convolution with the imaginary part $s(x)$ of the spread function $t(x)$, which therefore determines the resolution in the image. The optimum resolution is given when $s(x)$ is a sharp positive or negative peak [48]. As $s(x)$ is the Fourier transform of the imaginary part of the transfer function $T(u)$, this will occur when $\sin\chi(u)$ is equal to $+1$ or -1 over as large a region of the reciprocal space as possible. Owing to the dependence of $\chi(u)$ on the defocus Δ_f (see Eq. 2.36), the overall form of $\sin\chi(u)$ and therefore the passband of spatial frequencies which contribute to the image are controlled by focus-setting. Using the parameters $V_0 = 120\,\text{kV}$ (i.e. $\lambda = 3.348\,\text{pm}$) and $C_s = 2.7\,\text{mm}$, the defocus dependence of the contrast transfer function is demonstrated in Fig. 2.13 for defocus values of $-30\,\text{nm}$ (a), $-110\,\text{nm}$ (b) and $-1000\,\text{nm}$ (c).

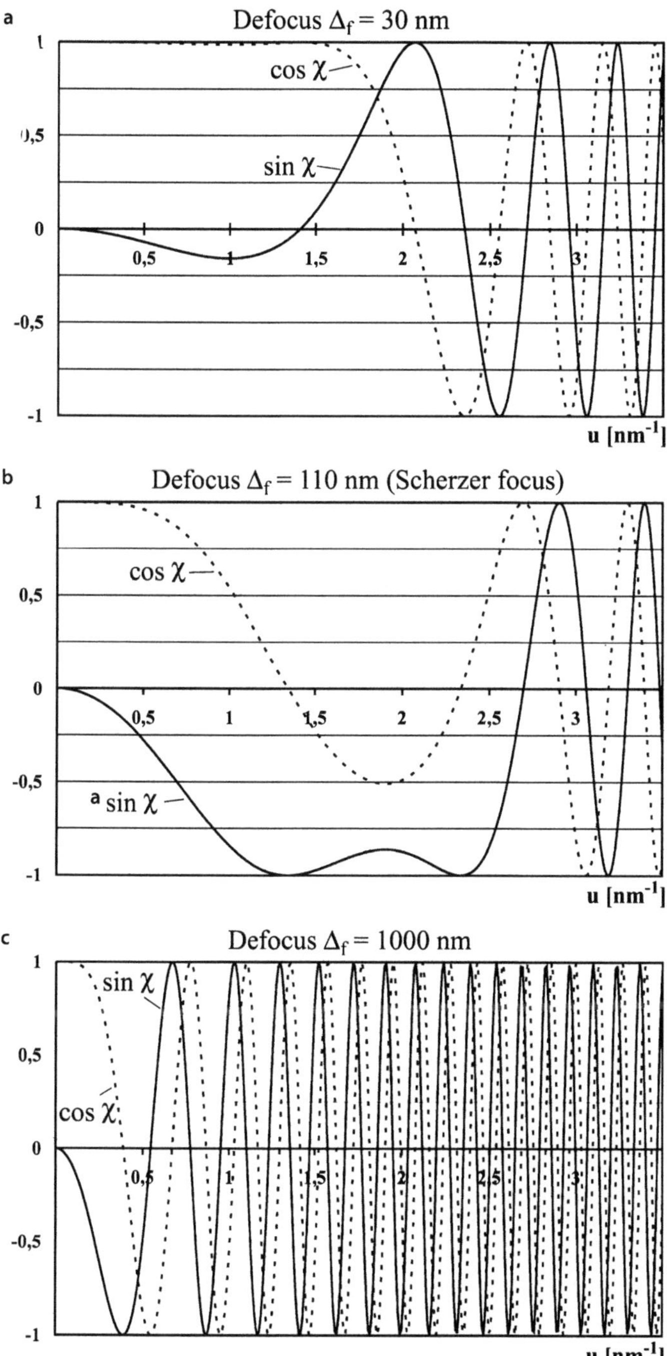

Fig. 2.13. The functions $\sin\chi(u)$ and $\cos\chi(u)$ for 120 keV electrons with $C_s = 2.7$ mm for defocus values of 30 nm (**a**), 110 nm, i.e. Scherzer focus (**b**), and 1000 nm (**c**)

Scherzer was the first to describe the focus setting for optimum phase contrast transfer [48]. Plot (b) in Fig. 2.13 represents the contrast transfer function at the so-called Scherzer focus which, in accordance with [36], is given by

$$\Delta_f = -\left(\frac{4}{3}C_s\lambda\right)^{\frac{1}{2}} .$$

(2.45)

At this slight underfocus of the objective lens, $\sin\chi(u)$ decreases to -1 and is nearly equal to -1 for a relatively broad range of spatial frequencies u before increasing to zero at u_z with

$$u_z = 1.51\left(C_s\lambda^3\right)^{-\frac{1}{4}} .$$

(2.46)

Since for higher values of u the signs of the contributions to the image are reversed and than oscillate rapidly, an objective aperture must be inserted to cut off all values of u that exceed u_z. In practice, this is often not necessary, as high spatial frequencies are significantly damped according to the damping envelopes of $D(u)$.

The reciprocal Δ_x of the first zero u_z defines the "interpretable resolution" under optimum conditions, which corresponds to "point resolution" in the case of nonperiodic objects. At the Scherzer focus condition, this resolution is given by

$$\Delta_x = 0.66\left(C_s\lambda^3\right)^{\frac{1}{4}} .$$

(2.47)

References

1. Haguenau F, Hawkes PW, Hutchison JL, Satiat-Jeunemaître B, Simon GT, Williams DB (2003) Microsc Microanal 9:96
2. Mulvey T (1962) Brit J Appl Phys 13:197
3. Cosslett VE (1979) J Microsc 117:1
4. Ruska E (1980) The early development of electron lenses and electron microscopy (Trans Mulvey T). S Hirzel Verlag, Stuttgart
5. Bethge H (1982) Ultramicroscopy 10:181
6. Hawkes PW (ed) (1985) The beginnings of electron microscopy. In: Advances in electronics and electron physics. Academic, New York
7. Hashimoto H (1986) J Electron Microsc Tech 3:1
8. Marton L (1994) Early history of the electron microscope, 2nd edn. San Francisco Press, San Francisco, CA
9. Phillipp F, Höschen R, Osaki M, Möbius G, Rühle M (1994) Ultramicroscopy 56:1
10. Midley PA, Weyland M (2003) Ultramicroscopy 96:413
11. Leapman RD, Kocsis E, Zhang G, Talbot TL, Laquerriere P (2004) Ultramicroscopy 100:115
12. Thomas JM, Midley PA, Yates TJV, Barnard JS, Raja R, Arslan I, Weyland M (2004) Angew Chem 116:6913
13. Weyland M, Midgley PA (2004) Materials Today 7:32
14. Sugimori H, Nishi T, Jinnai H (2005) Macromolecules 38:10226
15. Kübel C, Voigt A, Schoenmakers R, Otten M, Su D, Lee T-C, Carlsson A, Bradley J (2005) Microsc Microanal 11:378
16. Bals S, Van Tendeloo G, Kisielowski C (2006) Adv Mater 18:892
17. Hayder M, Uhlemann S, Schwan E, Rose H, Kabius B, Urban K (1998) Nature 392: 768
18. Lentzen M, Jahnen B, Jia CL, Thust A, Tillmann K, Urban K (2002) Ultramicroscopy 92:233
19. Batson PE, Dellby N, Krivanek OL (2002) Nature 418:617

20. Batson PE (2003) Ultramicroscopy 96:239
21. Krivanek OL, Nellist PD, Dellby N, Murfitt MF, Szilagyi Z (2003) Ultramicroscopy 96:229
22. Bleloch A, Lupini A (2004) Materials Today 7:42
23. Hetherington C (2004) Materials Today 7:50
24. Freitag B, Kujawa S, Mul PM, Ringnalda J, Tiemeijer PC (2005) Ultramicoscopy 102:209
25. van der Stam M, Stekelenburg M, Freitag B, Hubert D, Ringnalda J (2005) Microsc Anal 19:17(EU)
26. Rose H (2004) Nucl Instrum Methods Phys Res A 519:12
27. Glaser W (1952) Grundlagen der Elektronenoptik. Springer, Wien
28. El-Kareh AB, El-Kareh JCJ (1970) Electron beams, lenses, and optics. Academic, New York
29. Grivet P (1972) Electron optics, 2nd edn. Pergamon, Oxford
30. Hawkes PW (1972) Electron optics and electron microscopy. Taylor & Francis, London
31. Hawkes PW, Kasper E (1989) Principles of electron optics, vol 1: Basic geometrical optics; vol 2: Applied geometrical optics. Academic, London
32. Reimer L (1993) Transmission electron microscopy: physics of image formation and microanalysis, 3rd edn. Springer, Berlin
33. Sawyer LC, Grubb DT (1987) Polymer microscopy. Chapman and Hall, New York
34. Shindo D, Oikawa T (2002) Analytical electron microscopy for materials science. Springer, Tokyo
35. Chescoe D, Goodhew PJ (1990) The operation of transmission and scanning electron microscopes (Royal Microscopical Society Microscopy Handbooks 20). Oxford University Press, New York
36. Buseck PR, Cowley JM, Eyring L (eds)(1988) High-resolution transmission electron microscopy and associated techniques. Oxford University Press, Oxford
37. Fultz B, Howe J (2002) Transmission electron microscopy and diffractometry of material, 2nd edn. Springer, Berlin
38. Ernst F, Rühle M (eds)(2003) High-resolution imaging and spectrometry of materials. Springer, Berlin
39. Bethge H, Heydenreich J (eds)(1987) Electron microscopy in solid state physics (Materials Science Monographs 40). Elsevier, Amsterdam
40. Spence JCH (1988) Experimental high-resolution electron microscopy, 2nd edn. Oxford University Press, New York
41. Williams DB, Carter CB (1996) Transmission electron microscopy: a textbook for materials science. Plenum, New York
42. Shindo D, Hiraga K (1998) High-resolution electron microscopy for materials science. Springer, Tokyo
43. Goodhew PJ, Humphreys FJ, Beanland R (2001) Electron microscopy and analysis, 3rd edn. Taylor & Francis, London
44. Zhang X-F, Zhang Z (eds)(2001) Progress in transmission electron microscopy 1: Concepts and techniques; 2: Applications in materials science. Springer, Berlin
45. Lenz F (1971) Transfer of image formation in the electron microscope. In: Valdré U (ed) Electron microscopy in materials science. Academic, New York, p 540
46. Erickson HP (1973) The Fourier transform of an electron micrograph: First order and second order theory of image formation. In: Barer R, Cosslett V (eds) Advances in optical and electron microscopy. Academic, London, p 163
47. Coley JM (1981) Diffraction physics, 2nd rev edn. North-Holland, Amsterdam
48. Scherzer O (1949) J Appl Phys 20:20

3 Transmission Electron Microscopy: Conventional and Special Investigations of Polymers

Studies of polymeric materials using transmission electron microscopy are often directed towards observations of the morphologies of these materials at low-to-medium magnifications. Therefore, in practice most investigations use conventional transmission electron microscopy and take advantage of the mass-thickness contrast of selectively stained polymer samples. Nevertheless, a lot of investigations of polymers have also been carried out by applying special investigation techniques. This chapter reviews these special methods and techniques, including electron diffraction, high-resolution transmission electron microscopy, phase contrast transmission electron microscopy, electron holography, low-voltage transmission electron microscopy, high-voltage transmission electron microscopy, electron tomography, analytical transmission electron microscopy and scanning transmission electron microscopy, as well as their application to polymeric samples.

3.1 Conventional Investigations Utilising Mass-Thickness Contrast

The main purpose of performing studies using TEM is to observe the morphologies of polymeric materials at low-to-medium magnifications. Therefore, in practice most TEM investigations do not use special techniques and make use of mass-thickness contrast for contrast formation. Many examples of this approach are given in this book in Chaps. 15–24, which deal with investigations of different kinds of polymeric materials. Figure 3.1 illustrates the two typical features of mass-thickness contrast by showing a TEM micrograph of a test specimen with both small gold particles and much larger polystyrene (PS) spheres distributed on a thin carbon support film. Apart from the intensity modulations within the particles due to diffraction contrast, the small gold particles appear dark owing to the high density of gold. The influence of a locally increased specimen thickness on contrast formation is shown by the PS particle in the centre of the image. Even though the thickness of the PS particle is much larger than that of the gold particles, as schematically illustrated by the side view of the specimen, it does not appear darker than the gold particles, as the density of PS is about twenty times lower than that of gold.

Generally, polymeric materials are composed of only low atomic number elements with similar density. Therefore, significant contrast due to inherent variations

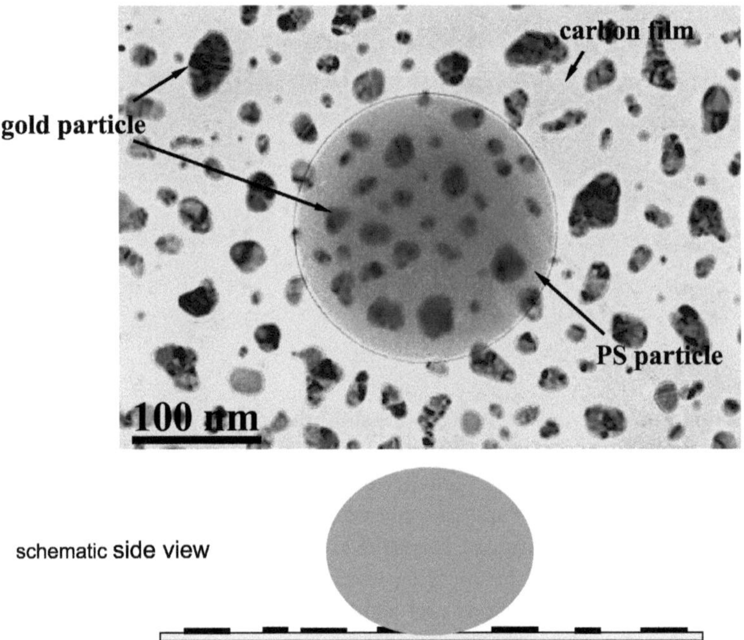

Fig. 3.1. Illustration of mass-thickness contrast: TEM micrograph of a test specimen with small gold particles and polystyrene latex spheres distributed on a thin carbon support (*top*) and schematic side view of the specimen (*bottom*)

of the local density cannot be expected in an image focussed in the Gaussian image plane. Mass-thickness contrast in chemically untreated polymers can be caused by a locally varying specimen thickness. On the one hand, it is known that contrast appears between different parts of heterogeneous polymers under the action of the electron beam due to beam damage, if the thickness is reduced in one part by the evaporation of volatile fractions of polymer chains [1]. On the other hand, so-called strain-induced contrast can result in a thin section of originally homogeneous thickness if the sample is stretched in a tensile stage [2]. Areas which are differently strained will be made visible by a correspondingly different change in thickness; typical structures are bright-appearing local shear bands, homogeneous or fibrillated crazes, or even voids in the vicinity of modifier particles.

Mass-thickness contrast due to locally varying specimen thickness is also the basis for imaging the surface relief of bulk materials using replicas. Often the replication technique is combined with metal shadowing to enhance the contrast. Shadowing is formed by coating the specimen surface with a thin metal layer (typically consisting of platinum–palladium) such that the metal vapour meets the surface at oblique incidence, so that the metal is preferentially condensed on the high points of the sample surface (see Sect. 9.5).

Selective staining is a very useful preparation technique that makes the morphology of polymer samples visible by enhancing mass-thickness contrast in TEM micrographs. It takes advantage of the different depositions of heavy-metal compounds, e.g. RuO_4 or OsO_4, in the amorphous and crystalline regions of a polymer or of the different affinities of the stain due to the chemical and physical natures of the components of the polymer. These preparation techniques are described in more detail in Chap. 13.

3.2 Electron Diffraction

3.2.1 Selected Area Diffraction

As described in Sect. 2.3.4, the diffraction pattern is formed in the focal plane of the objective postfield lens, where each diffracted beam is focussed into a corresponding spot according to its appearance as an oblique parallel beam. With the aid of the subsequent intermediate lens and the projector lens, this diffraction pattern is enlarged and displayed on the viewing screen. Usually, the selected area diffraction mode (Fig. 2.10b) is applied to display a diffraction pattern of a reduced specimen area of interest. Figure 3.2 shows typical sets of TEM micrographs and corresponding diffraction patterns for changes in the supramolecular structure in fractions of linear polyethylene (PE) with increasing molecular weight. While highly oriented lamellae produce a diffraction pattern composed of individual spots, irregularly arranged crystalline lamellae correspond to a polycrystalline material and cause a ring pattern in the diffraction plane. The ring pattern becomes arced with increasing orientation of the lamellae. Therefore, electron diffraction is sometimes used as a means to evaluate the orientation of the structural entities (lamellae, fibrils, microfibrils) of a polymer.

3.2.2 Structure Analysis

Direct structure analysis of small crystals is possible when a goniometer stage is also used, since it allows three-dimensional sampling of the crystal through the use of specimen tilt. Therefore, electron diffraction seems to be a very useful technique for studying crystal structures of small dimensions in polymers, e.g. crystalline lamellae of semicrystalline polymers. However, radiation damage is the main limitation on those applications and a diffraction pattern of a polymeric single crystal usually disappears in a very short time under conventional image conditions. Nevertheless, a few research groups have successfully applied electron diffraction analysis in order to study polymer crystal structures. Recent reviews of these special investigations of polymers via TEM have been given by Dorset [3–5] and Voigt-Martin [6]. However, it should be mentioned that during diffraction work on sensitive samples the experimentalist must do everything to reduce radiation damage, for example by cooling the specimen down to very low temperatures and working at very low electron illumination levels (see Sect. 8.2).

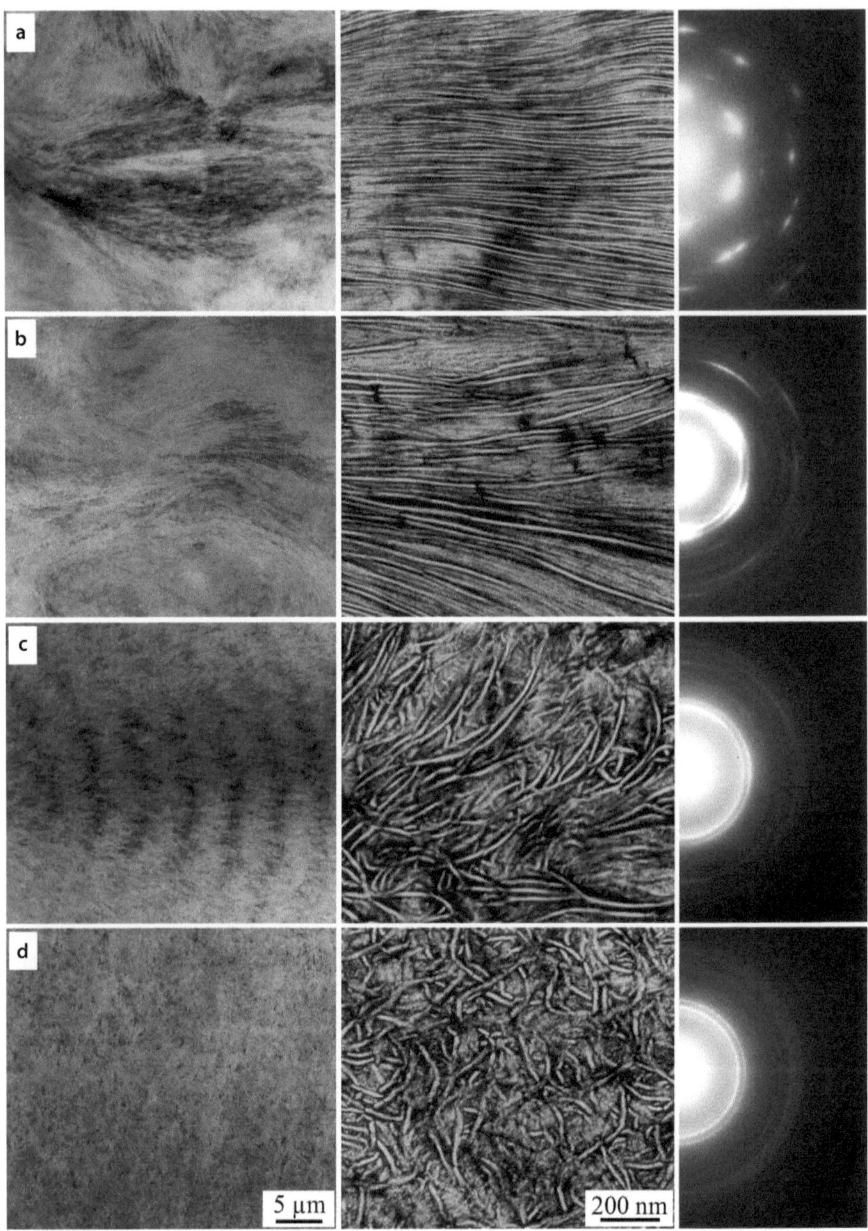

Fig. 3.2a–d. Changes in the supramolecular structure in fractions of high-density polyethylene (HDPE) with increasing molecular weight M_W as revealed by HVTEM bright-field images of thin sections cut by means of a cryoultramicrotome (*left column*), TEM bright-field images of stained ultrathin sections (*middle column*), and diffraction pattern (*right column*); **a** M_W = 8700: sheaflike structure, formed by long, parallel lamellae; **b** M_W = 34 000: bundle of long lamellae; small variations in local orientation; **c** M_W = 143 000: spherulites with concentric rings; rings are constructed from short lamellae which periodically change their orientations; minor local orientation; **d** M_W = 579 000: very short lamellae are irregularly arranged; no preferred orientation (from [60], reproduced with the permission of Carl Hanser Verlag)

3.3 High-Resolution Transmission Electron Microscopy

3.3.1 Introduction

While a resolution of below a couple of nanometres should not be expected for conventional TEM investigations of polymers, high-resolution transmission electron microscopy (HRTEM) corresponds to imaging at a resolution sufficient to resolve, for example, the local packing of molecules into a crystalline lattice. On the one hand, this requires a TEM with adequate equipment and perfect beam alignment and, on the other, a properly prepared specimen with a thickness that is small enough to be treated as a weak phase object. Additionally, optimum defocus conditions (Scherzer focus) must be applied. In recent years, the application of HRTEM to the study of crystal structures in various inorganic materials has become very common. However, the use of this technique for most polymeric materials has been limited by the rapid degradation of the material in the electron beam.

3.3.2 Evaluation and Reduction of Radiation Damage

The radiation damage caused to polymers (see also Sect. 8.2) during electron beam exposure depends on details of the interactions between the electrons and the molecules. Examples of these include chain scission, free radical formation, crosslinking and the destruction of crystal order. Usually, the sensitivity of polymers to electron-beam illumination is quantitatively expressed by the total end-point dose (TEPD) J_e, which is the incident electron dose needed for the complete disappearance of all crystalline reflections in the electron diffraction pattern. For HRTEM investigation, it is important to restrict exposure at the sample to less than $J_e/3$ to avoid imaging the damaged lattice [7]. A related quantity is the critical end-point dose J_c. Taking into account that the loss of intensity of a Bragg reflection as a function of dose can be described by an exponential decay function, J_c is defined as the characteristic dose at which the intensity of a Bragg reflection is reduced by a factor of e^{-1}. Either J_e or J_c is a useful measure of the sensitivity of polymeric materials for HRTEM investigations. Kumar and Adams have found an exponential dependence of J_c on the melting temperature for a broad range of melting temperatures of polymers from 300 K up to 1000 K [8]. In their very recent review of HRTEM of ordered polymers and organic molecular crystals, Martin et al. have shown an updated version of the graph showing the correlation between beam stability and thermal stability [9]. Furthermore, it is well known that J_e and J_c are dependent on the accelerating voltage and on the temperature. According to the variation in the inelastic cross-section with the velocity v of the incident electrons, the TEPD can be assumed to be proportional to v^2 [10], and this dependency has been roughly confirmed by experiments [11–16]. Therefore, a gain of about two in the TEPD can be obtained if the accelerating voltage is increased from 100 kV to 200 kV, but an increase to 1000 kV results in a gain of only about three in the TEPD. Tsuji and Kohjiya have reported results revealing the temperature dependence of the TEPD of various polymer crystals [17]. They measured the values of J_e for an accelerating voltage of 200 kV at room temperature and at 4.2 K, and found for example $J_e = 3.9 \times 10^2$ electrons per nm^2 at room temperature and $J_e = 6.0 \times 10^3$ electrons per nm^2 at 4.2 K for PE. The enormous increase in the TEPD—a factor of 15—shows the effectiveness of so-called cryoprotection. According to the dependence of the TEPD on the accelerating voltage and on the temperature, there are two ways to extend the lifespan of a polymer crystal under electron beam irradiation in a TEM; namely, increase the high voltage and cool the specimen to a low temperature. HRTEM investigations of polymers in a modern cryogenic transmission electron microscope (e.g. JEM 4000SFX from JEOL) can use both of these approaches by applying an accelerating voltage of 400 kV and cooling the specimen to the temperature of liquid helium [18–20]. Additionally, most modern TEMs serve so-called low-dose units (LDU) or minimum-dose systems (MDS) in order to reduce the illumination dose in the area of interest to a value needed for actual investigations. LDU and MDS enable the beam to be aligned and focussed at a location slightly away from the region of interest, where the beam is exclusively switched for image recording.

3.3.3 Application of HRTEM

One of the more insidious problems encountered during HRTEM imaging is the effect of sample drift, as a drift of more than 0.2 nm during exposure can easily ruin a micrograph [7]. It must be taken into account that a mechanical stage drift occurs after specimen movement, and also that changes in imaging conditions can cause a drift due to hysteresis effects of the lenses. Furthermore, a sample drift is caused by the heating-up and charging-up of the polymer sample. The coating of the specimen with a thin carbon film and the utilisation of carbon- or metal-coated microgrids (perforated films) are recommended methods for suppressing the specimen drift [21].

Even if the experimental conditions are properly accommodated to the different requirements of HRTEM investigations of polymers and image recording is optimised, information stored in the micrographs will be buried due to a poor signal-to-noise ratio. Therefore, computer processing and computer simulations of the image contrast play an important role when interpreting experimental results in the field of HRTEM. This is illustrated in Fig. 3.3 for the lattice imaging of a poly(tetrafluoroethylene) (PTFE) film, as reported in [22]. The HRTEM micrograph (a) and the corresponding microdiffraction pattern (b) show the obtained experimental results. The information, which is buried in the noisy micrograph (a), is revealed in the Fourier-filtered image (c) and has been interpreted by the computer-simulated image (d).

For further information, including the application of commercially available software packages, the reader is referred to the reviews [6, 7, 9, 17, 21]. All of the specific aspects of HRTEM of polymers are discussed in these reviews in more detail, the historical development in this field is reported, and a lot of examples and corresponding references are presented. Although HRTEM of polymers is not as common as that of

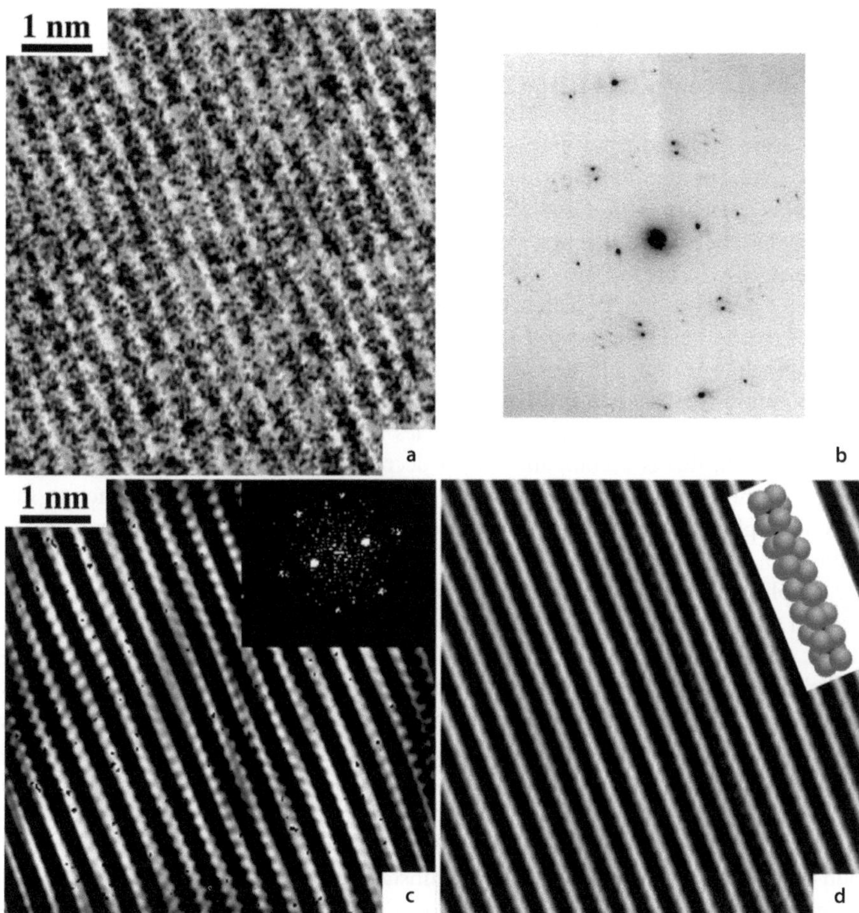

Fig. 3.3. **a** Selected region of a HRTEM micrograph (100 nm defocus) and **b** the corresponding microdiffraction pattern; **c** is a Fourier filtered image reconstructed from the fast Fourier transform of **(a)** (shown in the *inset*) and **(d)** is a simulated image for similar operating conditions and an assumed sample thickness of 20 nm (the *inset* shows the 15_7 helical conformation used to obtain this image, not to scale). (Reprinted from [22] with the permission of Springer-Verlag)

inorganic materials, a lot of studies of polymer microstructures have provided several new and important insights. It has been established that, just like low-molecular-weight solids, polymer crystals contain classical lattice defects such as dislocations and grain boundaries. Defect-mediated curvature and twisting in polymer crystals have been directly visualised by HRTEM [23]. The sizes, shapes, relative orientations and internal perfections of certain polymer crystals have now been determined [7], and an analysis of displacement fields near dislocation cores in ordered polymers has also been carried out [24].

3.4 Phase Contrast Transmission Electron Microscopy

3.4.1 Phase Contrast at Large Defocus Values

As described above, the optimum conditions for phase contrast transfer, which enable to a high resolution to be obtained, occur at the Scherzer focus. However, low spatial frequencies do not significantly contribute to the image at this focus setting if they belong to the first part of the contrast transfer function, where the values of $\sin \chi(u)$ are not sufficiently close to –1 (compare Fig. 2.13b). Lamellar structures of semicrystalline polymers or periodic patterns of block copolymers are not imaged at the Scherzer focus if their long periods correspond with the low spatial frequency region outside the passband of spatial frequencies contributing to image contrast. As shown in Figure 2.13c, the passband of useful values of u can be shifted towards lower spatial frequencies by a fairly large amount of underfocus. However, this is accompanied by a decrease in the bandpass width and also a shift of the first zero of $\sin \chi(u)$ at u_z, i.e. a loss in resolution. Nevertheless, this tuning of phase contrast at large defocus values can be used to image structures of interest.

Applications of this phase contrast technique to image the morphology of untreated polymeric materials were first introduced by Petermann and Gleiter [25]. Later, detailed studies were carried out to reveal the use and misuse of the defocus electron microscopy of multiphase polymers [26–28], and recently the technique was successfully applied to directly image spherulites with a sheaf-like appearance in thin films of isotactic PS [21, 29]. Furthermore, a TEM equipped with a so-called Lorentz lens, which is a minilens in the lower pole piece of the objective lens, has been applied to image at large defocus values the structures of modifier particles in high-impact PS (HIPS) and lamellar structures of styrenic block copolymers [30]. A result of the latter is shown in Fig. 3.4a. The unstained polystyrene-*block*-polyisoprene-*block*-polystyrene (SIS block copolymer) was recorded in Lorentz mode at a strong defocus of some tenths of a micron to visualise the lamellar structures by optimised phase contrast.

3.4.2 Phase Contrast by Means of Phase Plates

In the case of optical observations, phase contrast optical microscopy has been widely applied by using a Zernike phase plate to provide phase contrast. Though there were different attempts to introduce this principle of phase contrast into TEM in the 1970s, the Zernike phase contrast method has not yet been put into practice for transmission electron microscopy due to technical problems. Recently, however, such problems were overcome by a technique described in [31, 32]. The Zernike plate, which takes the form of a thin carbon film with a central hole about 1 μm in diameter, is positioned in the back-focal plane of the objective lens. The central unscattered electron beam passes through the hole, while the scattered electrons that pass through the film outside the hole are retarded in phase by $\Delta\varphi = -\pi/2$. This is realised by

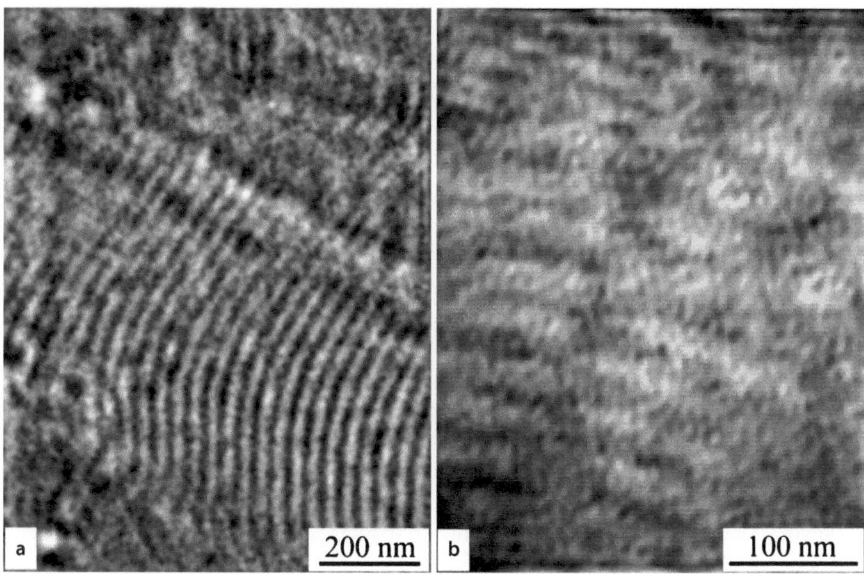

Fig. 3.4. Lorentz micrograph (**a**) and electron phase image reconstructed from an electron hologram (**b**) of polystyrene-*block*-polyisoprene-*block*-polystyrene copolymer (SIS-S50) with a lamellar
periodicity of 34–36 nm (**a**) and 29 nm (**b**). (Reprinted from [30] with the permission of Wiley)

using a Zernike plate of an appropriate thickness, which can be determined by the
following relativistic formula for the additional phase delay $\Delta\varphi$ [33]:

$$\Delta\varphi = -\pi \frac{h_{\mathrm{pl}}}{\lambda} \frac{V_{\mathrm{pl}}}{V_0} \frac{1 + 2aV_0}{1 + aV_0}. \tag{3.1}$$

Here, h_{pl} and V_{pl} are the thickness and the inner potential of the Zernike plate,
respectively, V_0 is the accelerating voltage, λ is the electron wavelength and $a =
0.9785 \times 10^{-6}\,V^{-1}$ is a constant. The phase delay $\Delta\varphi = -\pi/2$ must be added to $\chi(u)$,
and so, according to

$$\sin\left[\chi(u) - \frac{\pi}{2}\right] = -\cos\chi(u), \tag{3.2}$$

the sine function is replaced by the cosine function in the imaginary part of the contrast transfer function, which is responsible for the phase contrast in the image. The
dotted-line plots in Fig. 2.13a obviously show that there is a relatively broad passband
beginning at $u = 0$ where $\cos\chi(u)$ is close to 1. The resolution, again determined by
the first zero of $\cos\chi(u)$, is only somewhat worse than in the case of the Scherzer
focus for high-resolution imaging.

First applications of phase contrast imaging with the aid of a Zernike plate to
polymeric materials are recently described in [34]. The experiments were carried out
in a JEM-3100FFC TEM from JEOL equipped with a field-emission gun, a cryogenic
specimen stage cooled with liquid helium, an energy filter and a CCD camera. The

objective lens was specially designed to accommodate the phase plate accurately on the back-focal plane. The reported results also include the application of another kind of phase contrast imaging. By replacing the Zernike plate by a semicircular phase plate, the so-called Hilbert differential contrast is produced [35]. Although this contrast is not directly related to the three-dimensional shape of the specimen, a quasi-topographic image of the sample seems to appear.

3.5 Electron Holography

3.5.1 Introduction

Phase contrast imaging by electron holography (see, e.g., [36–38]) provides a further, alternative means of recovery phase modulation in a TEM for weakly scattering systems such as unstained polymeric materials. Historically, in the first article of Gabor [39], who later called the new method "holography" [40], the actual purpose was to invent an electron-optical technique for strongly magnified images, the aberrations of which could be eliminated afterward by light optical processing.

In electron holography, the phase shift of the electron wave due to the interaction with the specimen is determined by causing that wave to interfere with a reference wave of known phase and amplitude distribution. A lot of different schemes for producing electron holograms are known. According to the initial proposal of Gabor, electron shadow imaging techniques can be used to form the hologram. Here, the incident electron beam comes from a very small bright source which is close to the specimen, and the transmitted unscattered wave serves as the reference wave. The hologram is, in this case, a Fresnel diffraction image of the object.

On the one hand, this kind of electron holography has been realised by a modern, simple experimental set-up, using a single-atom nanotip as the electron point source, a low working voltage (50–300 V) and a piezomechanical nanodisplacement system for controlling both the position of the specimen and the tip–specimen distance. The results of applying this Fresnel projection microscopy to carbon and polymer fibres have been reported in [41]. On the other hand, the optics of the projection microscopy can also be realised in a modern STEM equipped with a field-emission gun, using a fixed and slightly defocused electron probe. However, this technique has not progressed beyond the demonstration of feasibility.

3.5.2 Image Plane Off-Axis Holography

The most advanced holographic technique used today in a TEM is image plane off-axis holography. Here, the illuminating wave is split into the object wave and the reference wave. The reference wave ψ_{ref} moves outside of the specimen through the vacuum, and so is not affected by the sample. The object wave propagates through the specimen, resulting in an object wavefunction $\psi_{\mathrm{ob}}(x)$ at the exit surface of the specimen. Its subsequent modulation by the action of the objective lens via the transfer function is described in Sect. 2.4.3. An electron biprism, i.e. a positively charged wire, is arranged behind the objective lens and superimposes the object and the reference waves in the image plane. The plane reference wave with an amplitude equal to 1 and a tilt corresponding to $u = u_c$ can be expressed in the following form:

$$\psi_{\text{ref}} = e^{2\pi i u_c x}. \tag{3.3}$$

Thus, the intensity distribution of the interference pattern (the so-called hologram) is given by the equation

$$\begin{aligned}
I_{\text{hol}}(x) &= |\psi_{\text{ob}}(x) * t(x) + \psi_{\text{ref}}|^2 \\
&= 1 + |\psi_{\text{ob}}(x) * t(x)|^2 + 2\,|\psi_{\text{ob}}(x) * t(x)|\,\cos[2\pi u_c x + \varphi_{\text{ob}}(x)]
\end{aligned} \tag{3.4}$$

where $t(x)$ is the spread function described in Sect. 2.4.3 and $\varphi_{\text{ob}}(x)$ is the phase of the object wavefunction. The first term "1" and the second term correspond to the intensities of the reference and the image waves, respectively. Additionally, Eq. 3.4 contains the encoded image wave in the third term. This term describes a set of cosinoidal interference fringes at the given spatial carrier frequency u_c with phase shifts $\varphi_{\text{ob}}(x)$ and amplitudes $2|\psi_{\text{ob}}(x) * t(x)|$ that represent, respectively, the phase and the amplitude of the image wave.

Highly monochromatic and sufficiently coherent illumination by a field-emission gun is necessary to achieve a high contrast of the hologram fringes. Furthermore, taking into account that hologram fringes must have a spacing of less than one third of the achievable resolution limit, the electrical and mechanical stability of the instrument has to be very high. Usually, after being magnified by subsequent lenses of the TEM, the hologram is recorded by a CCD camera. Then the digitised hologram is fed into a computer, where the electron wave is reconstructed numerically by means of wave optical image processing techniques. This involves the application of a Fourier transform to the intensity distribution of Eq. 3.4, resulting in

$$\begin{aligned}
F\{I_{\text{hol}}(x)\} = {}&\delta(u) + \Psi_i(u)T(u)\Psi_i^*(u)T^*(u) \\
&+ \delta(u + u_c)[\Psi_i(u)T(u)] \\
&+ \delta(u - u_c)[\Psi_i^*(u)T^*(u)]
\end{aligned} \tag{3.5}$$

where the symbols have the same meanings as in Sect. 2.4.3 and conjugate complex quantities are marked with a star. In this expression, the first two terms (first row) form the central band, which agrees with the normal diffraction pattern of the specimen. The most interesting advantage over a conventional diffractogram is the additional information found in each of the sidebands formed by the third (second row) and fourth terms (third row) of Eq. 3.5. They represent, properly isolated from the remainder, the complex Fourier spectra of the image wave and its conjugate complex, respectively. Thus, the spectrum is very convenient for further optical wave processing by means of a computer. If the third term is selected by using a suitable aperture and multiplied by $T^*(u)$, the resultant image is just $\psi_{\text{ob}}(x)$, corrected for aberrations. Thus, the real and imaginary parts of $\psi_{\text{ob}}(x)$, or its amplitude and phase components, can be derived separately. It should be noted that the phase component, and hence the phase contrast, is dominated by the real part of the spread function and accordingly by the $\cos\chi(u)$ term [42], as was described for Zernike phase contrast in the previous section.

3.5.3 Examples

There are only a limited number of publications on electron holography of polymers, which all deal with styrenic materials. Using off-axis transmission electron holography, the mean free path for inelastic electron scattering in PS nanospheres has been determined [43], and the particles could be imaged by quantitatively interpretable phase contrast [42]. Charged PS latex particles have been used to study electric charging at different image conditions with the aid of electron holography at low magnification [44]. Furthermore, results of electron holography investigations of HIPS and styrenic block copolymers have been reported recently [30, 45, 46]. An example from [30] is given in Fig. 3.4b. The micrograph shows the electron phase image reconstructed from an electron hologram of an unstained SIS block copolymer with lamellar morphology.

3.6 Low-Voltage Transmission Electron Microscopy

3.6.1 Introduction

There is a decrease in resolution that is proportional to $\lambda^{-3/4}$ (compare Eqs. 2.18 and 2.47) if the accelerating voltage of the TEM is lowered. On the other hand, it is well known that the contrast in the TEM increases with decreasing electron energy and that this increase in contrast is significant if the accelerating voltage is on the order of kV. According to the results reported in [47], an enhanced contrast at those voltages that is nearly 20 times higher than that for a voltage of 100 kV is observed, while the resolution decreases only threefold.

Therefore, in the 1960s and 1970s, a few authors considered the construction of a low-voltage TEM (LVTEM) [48–51], but several technological problems blocked attempts to build a practically usable instrument.

3.6.2 A Dedicated Low-Voltage TEM and its Application

However, taking advantage of the great progress made in the last few years in the construction of electron microscopes, a commercially available LVTEM (LVEM5) has recently been developed by the Czech company Delong Instruments s.r.o. [47, 52]. The special goal of the company was to create an instrument which could conveniently provide enhanced image contrast for low atomic number specimens without a substantial loss of resolution. Based on a completely unconventional approach, the instrument comprises two basic parts: the first one is a miniaturised 5 kV TEM equipped with an ultrahigh vacuum chamber, a Schottky field-emission electron source, a simple electron-optical system and a fluorescent YAG screen, where the image is formed with a relatively low magnification (25–250×). The second part is a standard light microscope (with several hundred times magnification) for the observation and recording of the image that appears on the fluorescent screen. The projection system, which consists of a YAG fluorescent screen and a light microscope, results in high image brightness, and so experiments can be carried out at low illumination dose. Furthermore, the LVEM5 is capable of operating in TEM mode and in STEM mode, and electron diffraction experiments can also be carried out.

Recently, results have been reported which show that a broad range of unstained polymer samples, including polymer single crystals [53], blends of both amorphous and semicrystalline polymers [54–56], block coploymers [57] and electrospun nanofibres [57], can be successfully imaged with the LVEM5. However, not all mechanisms of image contrast formation are fully understood. Thus, an interpretation of contrast which can explain the differences found during the formation of contrast in imaged blends in LVTEM and LVSTEM is still the aim of further studies [54]. Furthermore, the relatively limited penetration capability of about 50 nm obtained for an accelerating voltage of 5 kV presents new challenges to sample preparation, which have been discussed in [55–57] for example. Additionally, it must be taken into account that the high image contrast obtained at low accelerating voltages results from a very strong interaction of the electron beam with the specimen, and this strong interaction, on the other hand, is also the main cause of the beam damage at those voltages.

3.7 High-Voltage Transmission Electron Microscopy

3.7.1 Introduction

Reductions in the electron wavelength and in the scattering cross-sections are the most important effects of increasing the accelerating voltage, and so these effects have been the driving forces for the development of high-voltage transmission electron microscopes (HVTEMs), aimed at improving the resolution and the penetration power of the imaging electron beam. Additionally, the gain in lifetime (TEPD) obtained at increased accelerating voltages [11–16], as described in Sect. 3.3, is a positive effect of HVTEM investigations of polymers. Due to the technical problems that arise with the increase in the accelerating voltage of a TEM, the development of high-resolution TEMs for applications in materials science has mainly covered the intermediate voltage range of 200–400 kV so far.

Therefore, the main goal of most of HVTEMs working at a voltage of 1 MV or even higher than this is the possibility of using specimen thicknesses that are some 3–10 times larger than possible with a conventional TEM.

3.7.2 Advantages and Applications of HVTEM

Because of the low density of polymeric materials, specimens that are up to several micrometres in thickness can be investigated with sufficient resolution if an HVTEM with an accelerating voltage of about 1 MV is used [1, 58–60]. Relatively thick specimens are advantageous in several respects. On the one hand, it is much easier to prepare semi-thin sections from bulk material than ultrathin sections for investigations in a conventional TEM. Due to the significantly increased stability, the handling of thicker samples is also easier. On the other hand, the investigation of samples with increased thickness is of particular importance if structures in the micron range (e.g. particles in HIPS) and their size distributions within the polymer material are to be studied. Micrographs produced by HVTEM investigations of semi-thin sections directly reveal the desired information, whereas conventional TEM investigations of ultra-thin sections give smaller size distributions, since smaller sections of the larger particles are imaged (the so-called "tomato salad" problem, see Chap. 10) [1, 58, 60].

The large overall dimensions of an HVTEM (required for strong magnetic excitation of the lenses and for X-ray protection) provide an advantage during the installation of specimen-treatment devices for in situ experiments in which the specimen is (for example) heated, cooled or strained whilst under observation. In situ tensile tests which directly reveal deformation structures and mechanical microprocesses should preferably be carried out in an HVTEM, as the mechanical properties of polymer bulk materials are significantly better represented by those of semi-thin sections than by those of ultrathin sections. Furthermore, thicker tensile specimens can be prepared much more easily, as described in Chap. 12. The application of HVTEM to in situ tensile tests of polymers has been reviewed in [1, 58–62], and results from recent deformation experiments of different polymers with the aid of an HVTEM have been reported in [63–67]. Numerous examples are provided in this volume, particularly in Part III (see Chaps. 15, 18, 21). It should be mentioned that the high velocities of the imaging electrons in an HVTEM usually cause the exposure time of photographic films to increase. Therefore, it is advantageous to use photographic material with a higher sensitivity for HVTEM investigations [68].

3.8 Scanning Transmission Electron Microscopy

3.8.1 Introduction

A scanning transmission electron microscope (STEM) combines some advantages of a TEM with those of a scanning electron microscope (SEM), which is the topic of discussion in Chap. 4. The electron beam is focussed to a spot that is as small as possible, and it is scanned across the specimen area to be investigated while the transmitted electrons are collected to form the signal. This combination has been realised in a very simple form for investigations of very thin specimens in conventional SEMs by making use of an attached detector for transmitted electrons or by detecting the signal of transmitted electrons with the aid of the secondary electron detector of the SEM. Features of the image contrast in the transmission mode of a SEM have been described in [69] and the results of applying this special investigation technique to polymeric samples have been reported in, e.g., [54, 70].

However, common STEM investigations are carried out over a range of accelerating voltages that are typical of transmission electron microscopy. Most modern TEMs can easily be configured with scanning attachments, thus making them STEM as well as TEM instruments. STEM originally used electron microscopes specifically designed for scanning transmission imaging with the best resolution. Such a dedicated STEM with a field-emission gun was introduced by Crewe et al. [71], and commercial versions of dedicated STEM instruments were available from the company Vacuum Generators (which is now no longer trading).

3.8.2 Similarities and Differences between STEM and TEM

In many respects the STEM is similar to the TEM, while in others it is distinctly and even uniquely different. Thus, a dedicated STEM without a normal TEM capability does not need a set of post-specimen lenses to magnify the image or the diffraction pattern, as the transmitted electrons only need to be collected by an appropriate detector. On the other hand, STEM and TEM have an important aspect in common: applying the reciprocity principle to the ray paths of the microscopes, Cowley [72] showed that there is a close relationship between STEM and TEM in terms of contrast formation. The different illuminations of the specimen in TEM mode and STEM mode (spot mode) are described in Sect. 2.3.2 and illustrated in Fig. 2.8. The incident electron beam in the specimen plane is almost a parallel beam of normal incidence in the TEM mode. The illuminating cones for each object point (illustrated in the more detailed ray diagram in Fig. 2.9) have an illuminating aperture of the order of $\alpha_i = 0.1$–1 mrad. In STEM mode, the electron beam is focussed to a spot in the specimen plane, resulting in a relatively high probe aperture (semi-angle of the cone of the focussed beam) of $\alpha_p = 5$–20 mrad. This value is the same as that of the objective aperture α_{ol} in the TEM mode, which is the semi-angle of the cone of electrons emanating from the same specimen point and entering the objective lens to be focussed at the corresponding image point. Therefore, there is an analogy between both microscope modes with respect to image formation if the electron detector acceptance angle α_d of the STEM mode is chosen to be equal to the illumination angel α_i of the TEM mode. Thus, a ray diagram that is completely identical to that of the TEM will result if the direction of the path of rays is reversed in the STEM. According to the reciprocity theorem, the contrast of a microscope is maintained if the electron source and detector are interchanged, i.e. if the path directions of the rays are reversed. This means that STEM and TEM show analogous contrast phenomena for $\alpha_p = \alpha_{ol}$ and $\alpha_d = \alpha_i$. Theoretical studies [73] and practical demonstrations [74] show that this analogy involves scattering, diffraction and phase contrast. However, it is only valid in the limit of the weakly scattering object approximation and at medium resolution. If inelastic scattering becomes appreciable, i.e. for increasing specimen thickness, reciprocity does not apply.

Despite the analogy between the image processes in STEM and TEM due to the reciprocity principle, there are decisive differences between the image formation in both microscopes. In contrast to STEM, where the electrons pass through the lenses responsible for imaging before the electron–specimen interaction takes place, this process occurs after the interaction in TEM. Therefore, chromatic aberration due to the electron–specimen interaction, which can significantly limit the resolution in TEM mode imaging, has no influence on the image quality in STEM mode. Because of this advantage, STEM permits the investigation of thicker samples. Thus, the much lower costs and far fewer installing difficulties associated with STEM can make it an advantageous alternative to an HVTEM. However, a loss in resolution due the top-bottom effect caused by multiple scattering in thicker samples has to be taken into account.

Further important advantages of the STEM mode are the production and positioning of small electron probes for X-ray analysis and electron energy-loss spectroscopy of small specimen areas.

The results from an investigation performed by STEM and energy dispersive X-ray (EDX) analysis carried out at an accelerating voltage of 200 kV with a JEOL microscope of type JEM 2010 equipped with a scanning unit and a Voyager II X-ray quantitative microanalysis system from Noran Instruments are shown in Fig. 3.5. The STEM micrograph shows a segregated polystyrene/poly(styrene-*co*-4-bromo-styrene) blend. The bromostyrene phase appears bright in the micrograph, as shown by the EDX line profiles from the K- and L-radiation of bromine in comparison with the grey level of the micrograph along the line, which is marked in the image by its dark contamination track. The additionally recorded line profile of the K-radiation of carbon does not show a correspondence with the grey-level profile, as it arose in both blend phases in the same way. The microscopic investigations of this polymer blend also directly revealed differences in the beam damage caused in STEM and TEM mode. While in STEM mode the segregated phases of the blend could be imaged in recorded micrographs and EDX investigations could also be carried out, when the same microscope was used in TEM mode the corresponding visualisation failed, as the original segregation disappeared in a very short time, even at low electron beam current densities. This experience corresponds to the results of comparative studies of beam damage in paraffin single crystals in STEM and TEM reported in [75]. The authors of these studies also found a significant increase in observation time for STEM investigations.

3.8.3 Application of Bright-Field and Dark-Field Modes

A modern STEM instrument is usually equipped with an axial and an annular electron detector, resulting in two different contrast modes. The bright-field image is generated with the aid of the axial detector, which collects transmitted electrons and electrons scattered inside the small detector acceptance angle α_d. If a sufficiently thin specimen is used, the reciprocity principle is valid and the image contrast resembles that of a conventional TEM, as described above. For thicker specimens, the detected signal corresponds to a large fraction of inelastically scattered electrons that are concentrated into a limited angular range.

A dark-field image is preferably generated by means of a high-angle annular dark-field (HAADF) detector, which collects electrons scattered or diffracted into an angle greater than the axial detector acceptance angle α_d. This high-angle signal mainly results from electrons that have experienced nuclear interactions. Therefore, the signal is incoherent and the contrast does not result from either diffraction or phase contrast. When high-angle elastic scattering (which occurs near individual

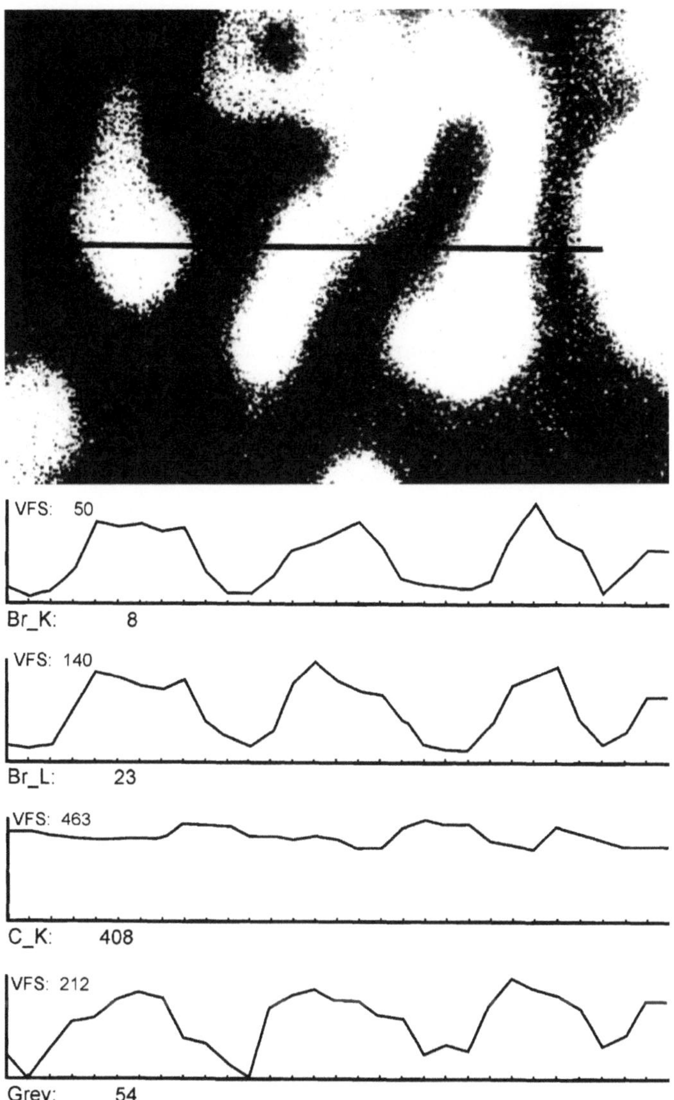

Fig. 3.5. STEM and EDX investigations of a segregated PS/poly(styrene-*co*-4-bromostyrene) blend. STEM micrograph is shown with a dark contamination track after the EDX line scan and line scans along this track of the bromine K- and L-radiation, the carbon K-radiation and the grey-level of the STEM bright-field image

atoms) is utilised, the HAADF imaging mode has emerged as the primary imaging technique.

STEM imaging in the HDAAF mode enjoys several advantages over conventional TEM imaging: (1) the spatial resolution is somewhat better; (2) it is sensitive to the atomic number of the imaged atoms; and (3) it provides a positive definite transfer of the specimen's spatial frequencies, allowing the direct interpretation of results with fewer ambiguities [76]. Using the HAADF image mode, a point-to-point resolution of better than 0.1 nm has been convincingly demonstrated with a 120 kV C_s-corrected STEM [76, 77].

In the field of polymers, this advantageous STEM imaging mode has been successfully applied for investigations of ionomers [78–85]. Typical ionomers are random copolymers with a minority of ionising monomeric units (typically acids) and a majority of nonionising monomeric units. The ionising groups can be partially or fully neutralized with ions. The resulting ionic groups microphase-separate from the nonionic monomeric units to create ionic aggregates with dimensions at the nanoscale. For example, a poly(ethylene-*ran*-methacrylic acid) melt neutralized with zinc (Zn-EMMA) contains randomly distributed spherical ionic aggregates with an average diameter of 2.1 nm, whereas a poly(styrene-*ran*-methacrylic acid) solution neutralized with caesium (Cs-SMMA) contains randomly distributed vesicular ionic aggregates with a diameter range from 5 to 45 nm and a shell thickness of about 3 nm [84]. Using a conventional TEM, nanoscale crystalline particles are usually imaged by positioning a particle in a Bragg condition and using diffraction contrast to generate an image. However, the nanoscale aggregates in ionomers are amorphous and their detection by mass-thickness contrast is an interpretive challenge because the phase contrast contributions of the surroundings have a similar appearance [78, 79]. However, the nanoscale aggregates in ionomers have been clearly imaged by STEM performed in both HDAAF mode [79, 80, 82, 83] and bright-field mode [79, 81, 82]. The additional use of tilt [80] or double tilt [81] enabled the series shapes, sizes and aspect ratios of the aggregates to be determined. Small aggregates (down to 1 nm) could be confidently detected and analysed by means of EDX [83]. Comparative studies of gold nanoparticles on PS support films revealed that the specimen thickness should not exceed 50 nm for optimum imaging of such small aggregates [85]. Furthermore, it has been reported that the application of image processing by deconvoluting raw STEM images of ionomers greatly decreases the noise level in the images, and can provide ways to improve investigations of the morphologies of ionomers [84].

3.9 Analytical Transmission Electron Microscopy

3.9.1 Introduction

Electron energy-loss spectroscopy (EELS) and X-ray microanalysis are the analytical techniques that can be used in TEM and STEM to investigate local elemental compositions. X-ray microanalysis in TEM and STEM mainly relies on EDX spectroscopy. X-ray quanta are emitted isotropically, but, owing to the restricted instrumental nature of a TEM, only a small solid angle (of the order of 10^{-2} sr) is collected by an EDX detector [10]. Furthermore, the fluorescence yield, i.e. the probability that a characteristic X-ray quantum is emitted rather than an Auger electron, is very low for light elements. For this reason, applications of this technique to polymer investigations like those presented in Fig. 3.5 have only rarely been reported. Therefore, X-ray micro-analysis of polymers is not considered in this section, but it is described in Chap. 4 in conjunction with SEM investigations that generally yield better signal-to-noise ratios.

EELS is the analysis of the energy distribution of electrons (with an initial energy E_0) that have passed through the specimen and experienced an energy loss ΔE due to inelastic interactions with it. To discriminate the electrons according to their energy $E = E_0 - \Delta E$, magnetic fields are applied in electron energy-loss spectrometers which deflect incident electrons as a function of energy. The spectrometer is usually the final component of an analytical microscope. Unlike the emission of characteristic X-rays, the electron energy-loss spectrum does not show corresponding limitations for light elements. Another advantage of EELS is that electrons that have been inelastically scattered by ionisation processes are concentrated within small scattering angles, resulting in a correspondingly high collection efficiency of the spectrometer.

There are two ways of detecting the electron energy-loss spectrum. A serial detector (serial EELS) uses a single detector and the spectrum is scanned across the slit in front of it by varying the strength of the magnetic prism, so that each energy is detected in turn. If a position-sensitive detector is used, the whole spectrum can be detected at once in a parallel spectrometer (parallel EELS, PEELS).

Originally, the spectroscopic capabilities of EELS provided the basis for a highly effective materials analysis technique in TEM and STEM. Although EELS techniques have advanced notably in recent years, the development of imaging energy-filters and advances in corresponding instrumentation have largely made energy-filtered TEM (EFTEM) a powerful tool for the chemical analysis of materials at the nanometre scale. EFTEM offers, on the one hand, the mode of zero-loss filtering, where only the unscattered and elastically scattered electrons contribute to the image. On the other hand, a wide range of modes of electron spectroscopic imaging (ESI) and electron spectroscopic diffraction (ESD) result when only electrons that have energy losses corresponding to the selected energy-loss window are permitted to pass through the energy filter. This acquisition of maps that elucidate the spatial distribution of any feature in the electron energy-loss spectrum can be considered an extension of the classic EELS technique in two dimensions.

There are currently two alternative EFTEM techniques in use. Dedicated EFTEMs with in-column filters (improved versions of the Ω-filter introduced by Rose and Plies [86]), for example the LEO 912 TEM shown in Fig. 2.5, were developed by Zeiss and LEO and later by JEOL. Alternatively, post-column energy filters called Gatan imaging filters (GIFs) [87] that can be attached as the final part of the electron-optical column of almost any TEM have been manufactured by Gatan. Both types of imaging energy filters can form images at a user-defined energy loss ΔE with electrons from a small energy range δE, which is determined by an adjustable slit aperture in the energy-dispersive plane of the spectrometer. If the slit aperture is omitted, all of the electrons contribute to the image (global mode), just as in a conventional TEM without an energy-loss filter. High spatial resolution electron energy-loss spectra (image

EELS) can be obtained from any position in the imaged specimen region by recording an extended series of images of the region (each image is successively shifted by ΔE) and subsequently reading the intensity as a function of ΔE from a particular position of interest.

A scheme for an electron energy-loss spectrum is given in Fig. 3.6. The plot of the inelastically scattered intensity as a function of the energy loss illustrates that a number of scattering processes contribute to the electron energy-loss spectrum, but it also shows a significant decrease in the mean intensity with increasing energy loss. Therefore, a corresponding intensity enhancement is necessary to visualise characteristic features of the spectrum. The zero loss is located at the origin of the spectrum and corresponds to unscattered and elastically scattered electrons ($E = E_0$, $\Delta E = 0$). Electrons which excite valence electrons, inter- or intraband transitions, or plasmons (collective oscillations of conduction-band and valence-band electrons) contribute to the low-loss region of the spectrum with $0\,\mathrm{eV} < \Delta E < 50\,\mathrm{eV}$. Beyond this low-loss region is a smoothly falling background with the ionisation edges of atoms whose absorption energies for inner-shell ionisation are reached in this region superimposed on it. The onset energies of these edges are element-specific, enabling one to perform elemental analysis. Each absorption edge has its own characteristic fine structure, and information on chemical bonding, molecular structure and dielectric constants may be obtained from a detailed study of this energy-loss near-edge structure (ELNES). The absorption edge maximum is usually followed by a tail of smoothly falling intensity due to the acceleration of bound electrons in the continuum. A quantitative interpretation of energy-loss spectra and element distribution images requires a knowledge of the local specimen thickness x. The spectrum itself yields the corresponding information according to following relation

$$x = \lambda_\mathrm{i} \ln \left(\frac{I_\mathrm{t}}{I_0} \right). \tag{3.6}$$

Here I_t is the total integrated spectral intensity and I_0 is the intensity of the zero-loss peak. The total inelastic mean free path λ_i is a material constant that depends on the incident-electron energy and the collection acceptance angle.

The following consideration is restricted to the features of electron energy-loss spectra and related investigation modes that are relevant for polymer investigations. Detailed descriptions of instrumentation and special EELS and EFTEM techniques are beyond the scope of this book, and readers interested in obtaining this information are referred to the broad assortment of recent books that cover this field, e.g. [88–93].

3.9.2 EELS Investigations

Using a TEM equipped with a post-column energy-loss spectrometer (Gatan PEELS 666 or GIF), electron energy-loss spectra were acquired with an energy resolution of about 1 eV by Varlot et al. [94–98] in order to investigate the low-loss regions and ELNES fine structure of poly(ethylene terephthalate) (PET) [94, 95], PS [96]

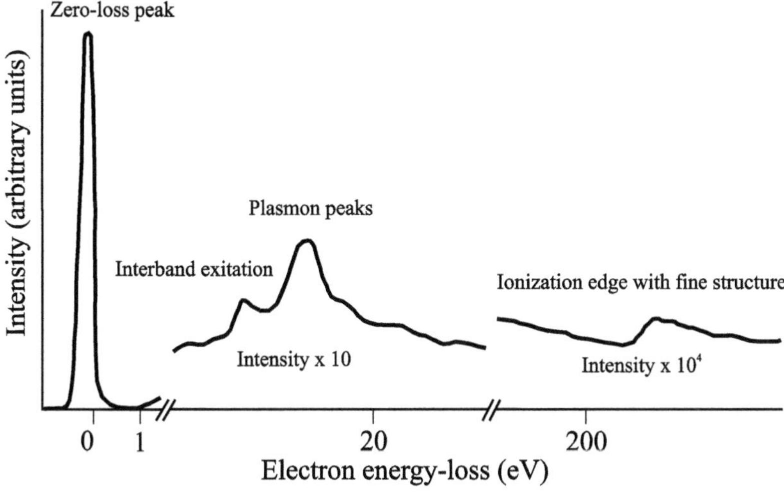

Fig. 3.6. Scheme of an electron energy-loss spectrum

and poly(methyl methacrylate) (PMMA) [97,98]. The authors performed these EELS analyses of different polymers in order to evaluate the possibility of obtaining chemical information on polymers at the nanometre scale. It was possible to assign the ELNES fine structure in the acquired spectra to different chemical bonds in agreement with molecular orbit calculations and the results of corresponding X-ray absorption spectroscopy (XANES) experiments. Moreover, the experiments carried out with the specimen cooled to liquid nitrogen temperature showed highly visible changes in the spectra as soon as the electron dose exceeded 10^3 Cm^{-2} (about 160 e^- nm^{-2}) for PET, 10^4 Cm^{-2} for PS, and 10^2 Cm^{-2} for PMMA when a large probe size was used. These results confirm that PMMA is very sensitive to electron beam irradiation, although it was also found that a high dose rate in a nanometre-diameter electron beam is less destructive, and spectra from far less degraded PMMA could even be obtained at 10^7 Cm^{-2} in this way [98]. The O/C ratio in PMMA and peaks in the low-loss regions of PS and PET are quite sensitive to irradiation, and so their variations have been used to determine the critical electron dose for radiation damage.

Sometimes the characteristic peaks of aromatic carbon bonds in the low-loss region are used as fingerprints for the corresponding polymer component in heterogeneous polymers in order to distinguish it from the other components. Using a VG HB501 STEM equipped with a Gatan model 666 PEELS detector, Hunt et al. [99] visualised the different phases of a blend consisting of PS and low-density polyethylene (LDPE) be means of spectrum imaging and integration of the spectrum intensity at 6.7 ± 1.6 eV. This spectral region between the large peaks of zero loss and plasmon loss covers the small peak which corresponds to the $\pi \rightarrow \pi^*$ excitation of the phenyl ring of PS. LDPE with only saturated carbon bonds does not show an increased energy loss in this region. Imaging the blend by means of the raw data resulted in micrographs with low contrast between the blend components. However, by using PS and LDPE reference spectra as basis functions for multiple least-squares fits to spectra with unknown compositions at each pixel in the spectral image, quantitative maps of the distributions of PS and LDPE in the blend have been obtained with good quality.

Chou et al. [100] studied the influence of RuO_4 staining of PS on the peak of the $\pi \rightarrow \pi^*$ excitation of the phenyl ring by means of a Philips CM 20 TEM/STEM with an attached Gatan model 666 PEELS detector. They found that the peak disappeared after staining, indicating that the stain had covalently reacted with the aromatic rings in the specimen. Furthermore, as in the case of OsO_4-stained dienes, RuO_4-stained aromatics appear more brittle than when unstained. Taken together, these observations suggest that RuO_4 both opens the aromatic ring and serves as a crosslinker between adjacent aromatic structures, similar to the manner in which OsO_4 is believed to crosslink unsaturated diene rubbers. The latter was confirmed by Ribbe et al. [101] by a quantitative analysis of the staining reaction of OsO_4 with a polyisoprene-*block*-polystyrene copolymer using EELS, as performed with a JEOL 2000FX TEM equipped with a Gatan filter. Thin 40-nm sections were exposed to OsO_4 vapour for various exposure times. Thereafter, electron energy-loss spectra of the films were acquired with an energy width of $\delta E = 50$ eV at the carbon K-edge at 284 eV and the oxygen K-edge at 532 eV. The ratio of the oxygen atoms N^{oxygen} (i.e. absorbed OsO_4) to the carbon atoms N^{carbon} (i.e. polystyrene and polyisoprene) present in the selected sample area was determined in relation to the staining time using the following equation:

$$\frac{N^{\text{oxygen}}}{N^{\text{carbon}}} = \frac{\sigma_{\text{K}}^{\text{carbon}}(\alpha, \delta E)\, I_{\text{K}}^{\text{oxygen}}(\alpha, \delta E)}{\sigma_{\text{K}}^{\text{oxygen}}(\alpha, \delta E)\, I_{\text{K}}^{\text{carbon}}(\alpha, \delta E)}. \tag{3.7}$$

$I_{\text{K}}^{\text{oxygen}}$ and $I_{\text{K}}^{\text{carbon}}$ are the background-corrected integrals within the energy window δE under the oxygen K-edge and the carbon K-edge, respectively, for a collection aperture angle α as determined by the angle of the limiting objective aperture. The partial inelastic scattering cross-sections $\sigma_{\text{K}}^{\text{oxygen}}$ and $\sigma_{\text{K}}^{\text{carbon}}$ of the oxygen and carbon K-shells, respectively, can be estimated with the help of the hydrogenic model (see, e.g., [90]). It has been found that the oxygen content saturates at a value of around 15%, which is in good agreement with the predicted value for the double OsO_4 reaction and clearly inconsistent with the 29% oxygen content predicted for the single OsO_4 reaction, suggesting that crosslinking is the dominant mechanism during the staining procedure.

Spatially resolved EELS investigations were used by Siangchaew and Libera [102] to measure the interfacial width between the phase-separated components PS and poly(2-vinylpyridene) (PVP) in a PS/PVP blend. The experiments were performed in a Philips CM20 FEG TEM/STEM equipped with a post-column Gatan model 666 PEELS detector. There was sufficient contrast to identify regions of interest in unstained specimens by taking images in the STEM mode by means of an annular dark-field detector. Position-resolved core-loss electron energy-loss spectra at the carbon K-edge (284 eV) and the nitrogen K-edge (401 eV) were acquired at high spatial resolution. The carbon/nitrogen ratio was determined in the same way as expressed for the oxygen/carbon ration in Eq. 3.7. By taking into account the effect of the finite probe size with a corresponding deconvolution, the profile of the oxygen/carbon ratio across a PS/PVP interface was found to yield an interfacial width of 3.5 nm, which agrees well with neutron studies on PS/PVP lamellar block copolymer interfaces.

3.9.3 Electron Spectroscopic Imaging

The selection of energy windows in different regions of the electron energy-loss spectrum results in different spectroscopic imaging modes. Values of the energy window δE of between 10 eV and 30 eV have typically been used for the ESI of polymeric materials. Zero-loss imaging does not yield element-specific information, but it is often applied to improve the image quality of bright-field images, as shown in [103,104] for example. Its aim is to remove all inelastically scattered electrons so that only the unscattered electrons of the primary beam and the elastically scattered electrons that pass through the objective aperture contribute to the image, and hence zero-loss imaging will enhance the contrast and improve the resolution by avoiding chromatic aberration. This is more serious for a thicker specimen and demonstrated in Fig. 3.7 by a stained 400-nm-thick acrylonitrile-butadiene-styrene (ABS) semi-thin section imaged in a LEO 912 TEM with a Ω-filter used in the global mode (a) and by zero-loss filtering (b).

Elemental mapping by core-loss imaging at ionisation edges of the spectrum is the most important analytical application of energy-filtered images at large energy losses. As the elemental signal is superimposed on a high background in the spectrum, the background should be subtracted from the energy-filtered image to obtain

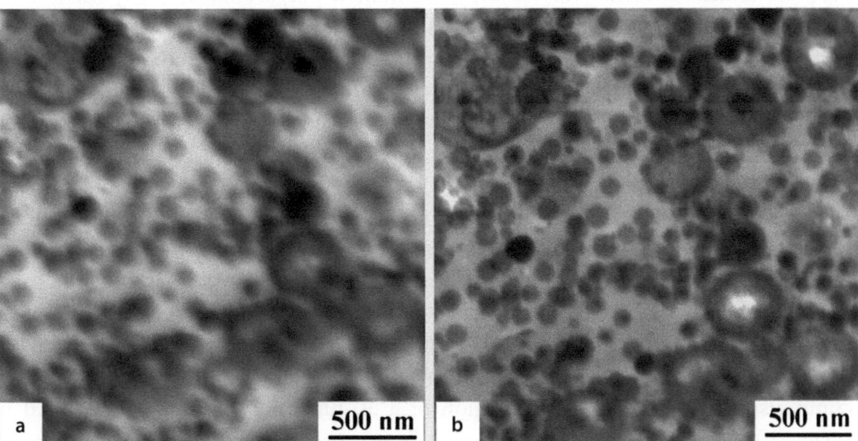

Fig. 3.7. TEM bright-field images of a stained 400-nm-thick ABS thin section produced by global mode (**a**) and zero-loss filtering (**b**). (From: Michler GH, Lebek W (2004) Ultramikrotomie in der Materialforschung. Carl Hanser Verlag, München. Reprinted with the permission of Carl Hanser Verlag)

a net elemental mapping. The standard procedure is to record, with a selected energy window δE, two images (A and B) below the edge and a third (C) beyond the edge of the ionisation energy of interest. The former two images are used to obtain a fourth image (D) by extrapolating to the same energy used in image C. Such an image extrapolation consists of calculating the grey-level value for each pixel, using the values measured from images A and B and the well-known exponential decay for the energy-loss spectrum background. Finally, the background image D is subtracted from the post-edge image C to obtain the digital difference image, revealing a map of the net elemental signal.

Elemental mapping is a very useful mode for identifying individual phases in heterogeneous polymers. However, due to the strong decay in intensity with increasing energy loss, core-loss imaging at ionisation edges and for large energy losses requires a relatively high intensity of the incident electron beam. Thus, the application of this technique to most polymeric materials has been limited by the rapid degradation of these materials in the electron beam, and the effects of sample drift can also cause insidious problems.

Therefore, it is often better to produce contrast between the phases in heterogeneous polymers via so-called structure-sensitive imaging [89]. This can be realised by taking a micrograph with an energy window δE in the spectral range between about 150 eV and 283 eV, i.e. close to but not reaching the carbon K-edge, where the scattering due to carbon atoms is at its minimum and a lot of other elements show increased intensity from the tails of their edges. This results in a dark-field-like structure-sensitive image with a resolution and a sensitivity that are superior to those of elemental mapping.

Successful applications of elemental mapping and structure-sensitive imaging of polymeric materials for identifying phases in phase-separated block copolymers [105–113], blends [107, 110], multicomponent polymer networks [110, 114], latex particles and latex films [110, 115–119], segmented polyurethanes [120] silicone oil modified ABS [121], polymeric nanocomposites [110, 122], electrically conducting polymers [123] and polymers for light-emitting diodes [124,125] have been reported.

Results of an investigation of a poly (ferrocenyldimethylsilane-*block*-styrene) block copolymer (PFS/PS block copolymer) by means of ESI are shown in Fig. 3.8. The experiments were carried out in an LEO 912 analytical TEM with an integrated Ω-filter at an accelerating voltage of 120 kV. Ultrathin sections of thickness 50 nm were cut at room temperature with the aid of a Leica UCT ultramicrotome from a solution-cast PFS–PS block copolymer film. Owing to the slow evaporation of the solvent and subsequent annealing at 190 °C, phase-separation results in a morphology of well-ordered PSF cylinders in a PS matrix if the PS volume fraction of the diblock copolymer is $\Phi_{PS} = 60\%$. The initial part of the energy-loss spectrum of the PSF–PS block copolymer is shown in image 3.8(a), and for the purposes of comparison the same spectrum region of a 50-nm-thick PS film is presented in image (b). An energy window of $\delta E = 30$ eV was used to record the zero-loss image (c) and the electron spectroscopic image at $\Delta E = 230$ eV (d) of the block copolymer. The zero-loss image is a bright-field image with mass-thickness contrast due

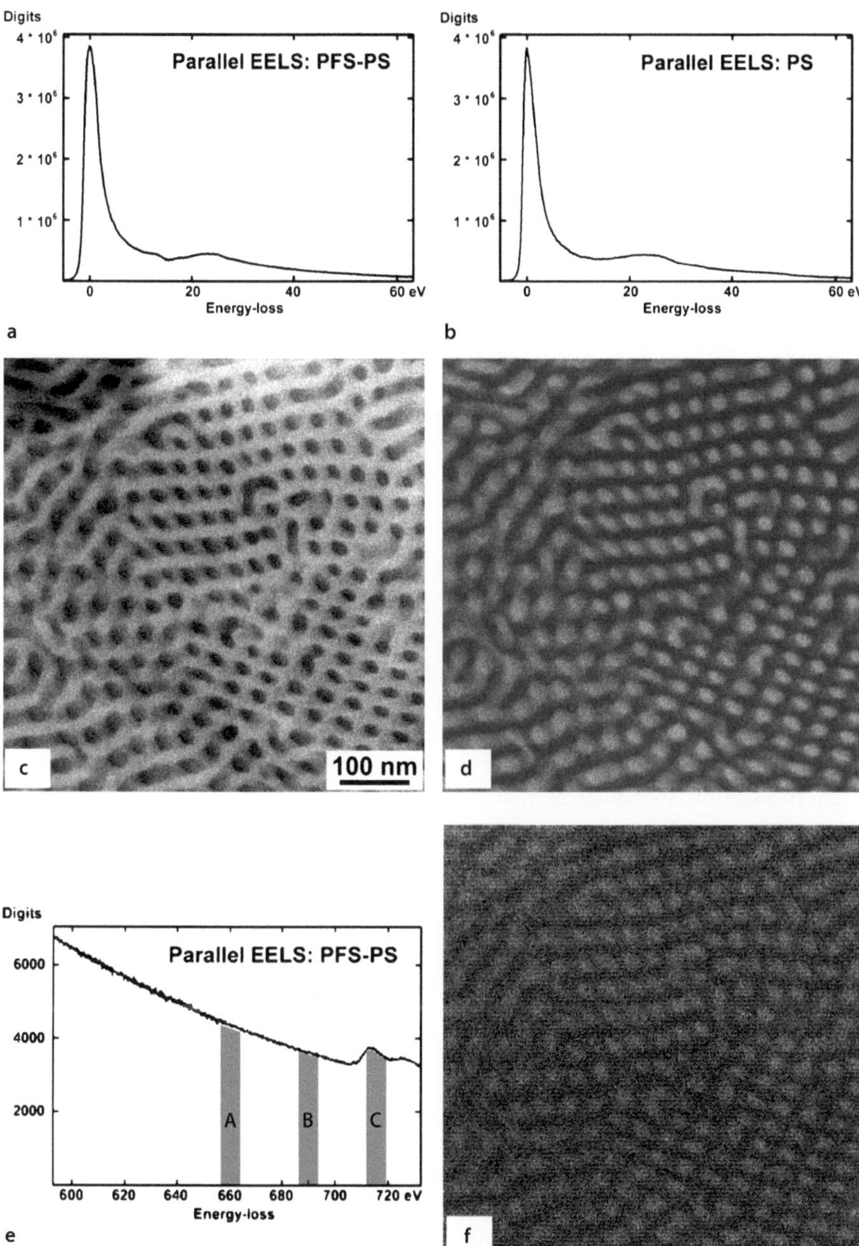

Fig. 3.8. Initial parts of electron energy-loss spectra of 50-nm-thick ultrathin sections of a PFS–PS block copolymer (**a**) and a PS homopolymer (**b**); EFTEM investigation of the PFS–PS block copolymer: zero-loss image (**c**), electron spectroscopic image at 230 eV energy loss (**d**) and elemental mapping (**f**) by means of the iron $L_{2,3}$ edge shown in the corresponding part of the energy-loss spectrum (**e**)

to the density differences between the PS and PSF phases of the block copolymer. While the repeat unit of PS consists of hydrogen and carbon atoms, i.e. of light elements only, that of PSF also contains one iron atom and one silicon atom. Thus, because of their increased elastic scattering, the PSF cylinders appear dark in image (c). Although, the iron $M_{2,3}$ edge and the silicon $L_{2,3}$ edge are not visible in the spectrum of the block copolymer, the spectroscopic image (d) recorded at an energy loss of 230 eV clearly reveals the block copolymer's morphology via structure-sensitive contrast. Furthermore, it has been found that, by recording a series of electron spectroscopic images at successively increasing energy losses, the bright-field contrast vanishes at an energy-loss of about 50 eV and a transition to the dark-field-like contrast takes place. The optimum dark-field contrast with inelastically scattered electrons was obtained for the spectral region between $\Delta E = 150$ eV and $\Delta E = 260$ eV, while the contrast disappeared beyond the carbon K-edge, as expected. Image (e) shows the area of the energy-loss spectrum of the block copolymer at around 700 eV, where the iron $L_{2,3}$ edge is detected. Additionally, the three windows A, B and C that were used to obtain the mapping of the net iron signal shown in image (f) have been marked. The mapping can be used to identify the morphological structures of the block copolymer, although the image quality of the structure-sensitive image (d) is superior to that of the mapping, as mentioned above.

The continuous change in contrast obtained by switching the energy window between $\Delta E = 0$ eV and $\Delta E = 280$ eV can also be used for contrast tuning. This method is of special interest for imaging more thickly stained sections. It allows one to change the contrast of strongly stained structures relative to weakly or unstained areas of a section, so that all of the structures can be seen with appropriate contrast in one micrograph. A corresponding example of a copolymer of PE and polypropylene (PP) stained with RuO_4 is described in [89]. While the boundaries between PE and PP cannot be clearly distinguished in either the unfiltered image or at $\Delta E = 50$ eV, maximum contrast and good separation of the phases was obtained at $\Delta E = 200$ eV.

3.10 Electron Tomography

3.10.1 Introduction

Valuable information about the structure and the chemistry of materials in the micron and nanometre range can be obtained by applying different investigation modes of transmission electron microscopy. However, since TEM micrographs provide only two-dimensional (2-D) projections of the three-dimensional (3-D) object under investigation, the interpretation of the material's structure can be ambiguous. This is why electron tomography was developed (see, e.g., [126–128]); it provides a means to reconstruct the 3-D structures of objects from a series of images taken at regular tilt intervals and usually recorded digitally on slow-scan CCD cameras. Whilst electron tomography has been used in the biological sciences for over 30 years (see, e.g., [126, 127]), it has only recently been applied to materials science [128–130], and has proven useful for providing 3-D information about a variety of polymeric structures, e.g. microphase-separated structures in block copolymers [131–140] and blends [135,137,141,142], self-assembly of rodcoil copolymer nanostructures [143] and material distributions in polymer nanocomposites [144,145].

3.10.2 Data Acquisition, Image Alignment and Reconstruction

The acquisition of the tilt series is carried out by sequentially tilting the specimen about a single axis, usually from one extreme tilt to the other. As the goniometer of the TEM must provide a high enough tilt range of typically ±60–70° or greater to minimise artifacts in the reconstruction process, TEMs with large-gap pole pieces are used for electron tomography investigations. Usually, a tilt series is acquired with equal angular increments of about 1° or 2°. On the other hand, an optimisation of the tilt increments has proposed by Saxon et al. [146] in order to improve the overall resolution for a given electron dose.

In principle, the acquisition of a tilt series can be performed manually. However, automated data acquisition and analysis procedures [147, 148] have been developed for dose-efficient data collection and improved image processing. The next generation of automation routines [149–152] based on computer-controlled precalibrated goniometers with highly reproducible movements have been adapted to TEMs and provide a user-friendly platform for all processing steps, including acquisition, alignment, reconstruction and analysis, resulting in high-throughput electron tomography.

When tilting the sample, image shifts and defocus changes can occur. The best possible resolution in the final 3-D electron tomogram can only be achieved when the tilt series is aligned very accurately. Therefore, the individual images need to be shifted onto a common tilt axis, and alignment is needed to remove any residual shifts of the individual images with respect to each other, requiring spatial and rotational shift and the correction of scan or lens distortions. Generally, the alignment of the individual image is carried out by cross-correlating the images [153] or by tracking fiducial markers within the images [154]. Combinations of both techniques have also been used. The former is usually a relatively uncomplicated and fast procedure, which however cannot correct for rotation or magnification changes between the individual images and is generally best-suited to nonshrinking specimens, the projected geometries of which do not change radically as a function of tilt [150]. An accurate marker-free alignment based on cross-correlation techniques with simultaneous geometry determination and reconstruction of tilt series has been reported recently [152]. Least-squares tracking of easily recognisable features in the images throughout the tilt series have been used as an alternative alignment technique. Usually, nanogold clusters deposited onto the specimen surface serve as markers, since most materials under investigation do not intrinsically contain fine globular particles which can be used as fiducial markers. Alignment data collected by means of fiducial marker tracking also contain information on changes in magnification and image rotation, and improved resolution of the reconstructed tomogram is obtained due to a more accurately aligned image stack.

After alignment of the data series, the 3-D reconstruction must be carried out. There are different reconstruction techniques in use. The most widely used technique is weighted back-projection (WBP), which is described in detail in [155, 156]. Weighting by a frequency filter back-projection involves projecting each 2-D image into a 3-D reconstruction space back along its original tilt angle. The superposition of the back-projected images results in the 3-D reconstruction of the specimen. The WKB technique is limited by noise and the loss of both high and low frequency information. Therefore, the application of iterative reconstruction techniques such as the algebraic reconstruction technique (ART) [157] and the simultaneous iterative reconstruction technique (SIRT) [158], which rely on iteratively optimising the 3-D tomograms, is preferable, in particular for the 3-D reconstruction of noisy and/or undersampled datasets [128, 150]. Special computer software packages such as IMOD [159, 160] have been developed which can be used as tools for analysing and viewing the reconstructed 3-D image data.

3.10.3 Resolution of Reconstructed Data

The resolution of the 3-D reconstruction of the specimen is significantly influenced by data collection. For the single-axis tilt geometry considered so far, the resolution parallel to the tilt axis is equal to the resolution of the recorded 2-D images. However, the resolution in the perpendicular direction is controlled by the number of projections acquired N. Assuming that the N projections cover the total tilt range of $\pm 90°$, the resolution in those directions is inversely proportional to N.

In practice, the tilt range is limited and cannot be extended to $\pm 90°$. Due to this missing information, the resolution in the direction parallel to the electron-optical axis, i.e. parallel to the incident electron beam, is degraded by a corresponding elongation factor. This problem can be better understood by considering the relationship between a projection in real space and the Fourier space. The "central slice theorem" upon which computerised tomography relies states that the Fourier transformation of a projection at a given angle is a central section at the same angle through the Fourier transform of that object. As the unsampled volume in the Fourier space becomes wedge-shaped, the cause of the degraded tomogram due to the limited tilt range is called the "missing wedge". A more detailed discussion of this problem has been given in [130, 140]. Here, only the conclusion is noted: to obtain maximum 3-D information, as many projections as possible should be acquired over a tilt range that is as wide as possible.

However, problems arising from a limited tilt range for a single-axis tilt series can be solved by applying dual-axis electron tomography [140, 161]. By using an additional second tilt series with its tilt axis perpendicular to that of the first series, the missing volume in the Fourier space can be significantly decreased by changing the "missing wedge" to a "missing pyramid", resulting in a correspondingly improved reconstruction if both individual tomograms have been properly combined. Thus, the results reported in [140] reveal that, in each individual reconstruction, the only cylindrical nanodomains of the block copolymer that were reproduced were the ones that were properly oriented with respect to the tilt axis used, and also that complementary information is obtained by using two tilt series with orthogonal tilt axes. Combining the two sets of reconstructed data in Fourier space, on the other hand, results in an improved reconstruction where all of the cylindrical nanodomains are successfully captured, irrespectively of their orientations. Moreover, a significant enhancement of the image quality was observed in the 3-D reconstruction obtained from dual-axis tomography compared to that obtained from single-axis tomography.

3.10.4 Application of Electron Tomography

The application of electron tomography relies on the assumption that the TEM operates as a projection tool that produces an image intensity distribution which fulfils the projection requirement. In principle, a monotonically varying function would be acceptable for a successful tomographic reconstruction [162]. Therefore, the mass-thickness contrasts of TEM bright-field images of biological and polymer samples fulfil the projection requirement very well. However, for the majority of specimens encountered in materials science (particularly crystalline materials), the TEM bright-field contrast depends on diffraction conditions, is caused by phase relations and does not show the required monotonic relationship to the amount of material through which the electron beam passes. To overcome this problem, special investigation techniques such as HAADF-STEM, HAADF-TEM and EFTEM are usually used when electron tomography needs to be applied in materials science (see, e.g., [128, 130]). Details of these new variants of electron tomography need not be described here, as polymer investigations are generally based on bright-field tomography. Nevertheless, the electron tomography of polymers takes advantage of advanced techniques; in particular, applications where EFTEM tomography is used to improve the quality of bright-field images by zero-loss imaging have been reported [136, 139, 140].

Although electron tomography has proven useful for revealing 3-D information about structures in polymeric materials [131–145], it should be noted that its application to polymers is mainly limited by the sensitivity of most of polymeric materials to beam damage, as the sample is extensively exposed to the electron beam during the acquisition of a tilt series, even when low-dose techniques are used.

References

1. Michler GH (1993) Appl Spectrosc Rev 28:327
2. Michler GH (1984) Ultramicroscopy 15:81
3. Dorset DL (1989) Electron diffraction from crystalline polymers. In: Booth C, Price C (eds) Polymer characterization (Vol. 1 of Allen G, Bevington JC (eds) Comprehensive polymer science). Pergamon, Oxford, p 651
4. Dorset DL (1991) Ultramicroscopy 38:23
5. Dorset DL (2003) Rep Prog Phys 66:305
6. Voigt-Martin IG (1996) Acta Polym 47:311
7. Martin DC, Thomas EL (1995) Polymer 36:1743
8. Kumar S, Adams WW (1990) Polymer 31:15

9. Martin DC, Chen J, Yang J, Drummy LF, Kübel C (2005) J Polym Sci Polym Phys 43:1749
10. Reimer L (1993) Transmission electron microscopy: physics of image formation and microanalysis, 3rd edn. Springer, Berlin
11. Kobayashi K, Ohara M (1966) Voltage dependence of radiation damage to polymer specimens. In: Uyeda R (ed) Electron Microscopy 1966 (Proc 6th Int Congr for Electron Microscopy). Maruzen Company Ltd., Tokyo, p 579
12. Richardson MJ, Thomas K (1972) Aspects of HVEM of polymers. In: Electron Microscopy 1972 (Proc 6th Eur Congr on Electron Microscopy). The Institute of Physics, Bristol, p 562
13. Salih SM, Cosslett VE (1974) Some factors influencing radiation damage in organic substances. In: Sanders JV, Goodchild DJ (eds) Proceedings of the 8th International Congress on Electron Microscopy, vol 2. Australian Acad Sci, Canberra, Australia, p 670
14. Thomas EL, Ast DG (1974) Polymer 15:37
15. Martinez JP, Locatelli D, Balladore JL, Trinquier J (1982) Ultramicroscopy 8:437
16. Ohno T, Sengoku M, Arii T (2002) Micron 33:403
17. Tsuji M, Kohjiya S (1995) Prog Polym Sci 20:259
18. Tosaka M, Hamada N, Tsuji M, Kohjiya S, Ogawa T, Isoda S, Kobayashi T (1997) Macromolecules 30:4132
19. Tosaka M, Tsuji M, Kohjiya S, Cartier L, Lotz B (1999) Macromolecules 32:4905
20. Tosaka M, Kamijo T, Tsuji M, Kohjiya S, Ogawa T, Isoda S, Kobayashi T (2000) Macromolecules 33:9666
21. Tsuji M (1989) Electron microscopy. In: Booth C, Price C (eds) Polymer characterization (Vol. 1 of Allen G, Bevington JC (eds) Comprehensive polymer science). Pergamon, Oxford, p 785
22. Plummer CJG, Kausch H-H (1996) Polym Bull 37:393
23. Kübel C, González-Ronda L, Drummy LF, Martin DC (2000) J Phys Org Chem 13:816
24. Drummy LF, Voigt-Martin I, Martin DC (2001) Macromolecules 34:7416
25. Petermann J, Gleiter H (1975) Phil Mag 31:929
26. Roche EJ, Thomas EL (1981) Polymer 22:333
27. Handlin DL Jr, Thomas EL (1983) Macromolecules 16:1514
28. Handlin DL Jr, Thomas EL (1984) J Mater Sci Lett 3:137
29. Tsuji M, Fujita M, Shimizu T, Kohjiya S (2001) Macromolecules 34:4827
30. Simon P, Rameshwar A, Lichte H, Michler GH, Langela M (2005) J Appl Polym Sci 96:1573
31. Danev R, Nagayama K (2001) Ultramicroscopy 88:243
32. Nagayama K, Danev R (2003) Microsc Anal 17:13 (EU)
33. Willasch D (1975) Optik 44:17
34. Tosaka M, Danev R, Nagayama K (2005) Macromolecules 38:7884
35. Danev R, Okawara H, Usuda N, Kametani K, Nagayama K (2002) J Biol Phys 28:627
36. Tonomura A (1999) Electron holography, 2nd rev edn. Springer, Berlin
37. Voelkl E, Allard LF, Joy DC (1999) Introduction to electron holography. Plenum, New York
38. Lichte H, Lehmann M (2002) Adv Imaging Electron Phys 123:225
39. Gabor D (1948) Nature 161:777
40. Gabor D (1949) Proc R Soc Lond Ser A 179:454
41. Binh VT, Semet V, Garcia N (1995) Ultramicroscopy 58:307
42. Chou T-M, Libera M, Gauthier M (2003) Polymer 44:3037
43. Chou T-M, Libera M (2003) Ultramicroscopy 94:31
44. Frost BG, Voelkl E (1999) Mater Char 42:221
45. Simon P, Huhle R, Lehmann M, Lichte H, Mönter D, Bieber T, Reschetilowski W, Adhikari R, Michler GH (2002) Chem Mater 14:1505
46. Simon P, Lichte H, Drechsel J, Formanek P, Graff A, Wahl R, Mertig M, Adhikari R, Michler GH (2003) Adv Mater 15:1475
47. Delong A, Hladil K, Kolařik V (1994) Microsc Anal 8:13(EU)
48. Wilska AP (1960) Low voltage electron microscope. In: Houwink AL, Spit BJ (eds) Proceedings of the European Regional Conference on Electron Microscopy, vol 1. De Nederlandse Vereniging voor Electronenmicroscopie, Delft, p 105
49. Wilska AP (1970) Low voltage electron microscope with energy filtering and amplitude contrast. In: Favard P (ed) Microscopie électronique 1970, vol 1 (Proc 7th Int Congr on Electron Microscopy). Société Française de Microscopie Électronique, Grenoble, p 149
50. Heinemann K, Möllenstedt G (1967/1968) Optik 26:11
51. Stolz H, Möllenstedt G (1971) Optik 33:35

52. Delong A (1992) Low voltage TEM. In: Ríos A, Arias JM, Megías- Megías L, López-Galindo A (eds) Electron microscopy 92, vol 1 (Proc 10th Eur Congr on Electron Microscopy). Secretariado de Publicaciones de la Universidad de Granada, Granada, p 79

53. Delong A, Kolařik V, Martin DC (1998) Low voltage transmission electron microscope LVEM-5. In: Benavides HAC, Yacamán MJ (ed) Electron microscopy 1998, vol 1(Proc 14th Int Congr on Electron Microscopy). Institute of Physics Publishing, Bristol, p 463

54. Lednický F, Coufalová E, Hromádková J, Delong A, Kolaøik (2000) Polymer 41:4909

55. Lednický F, Hromádková J, Pientka Z (2001) Polymer 42: 4329

56. Lednický F, Pientka Z, Hromádková J (2003) J Macromol Sci B 42:1039

57. Drummy LF, Yang J, Martin DC (2004) Ultramicroscopy 99:247

58. Michler GH (1986) Progress in polymer investigations by high voltage electron microscopy. In: Sedlacek (ed) Morphology of polymers. Walter de Gruyter & Co, Berlin New York, p 749

59. Michler GH (1987) Morphology of polymers. In: Bethge H, Heydenreich J (eds) Electron microscopy in solid state physics (Mater Sci Monogr 40). Elsevier, Amsterdam, p 386

60. Michler GH (1992) Kunststoff-Mikromechanik: Morphologie, Deformations- und Bruchmechanismen von Polymerwerkstoffen. Carl Hanser Verlag, München

61. Michler GH (2001) Crazing in amorphous polymers: Formation of fibrillated crazes near the glass-transition temperature. In: Grellmann W, Seidler S (eds) Deformation and fracture behaviour of polymers. Springer, Berlin, p 193

62. Michler GH (2005) Micromechanical mechanisms of toughness enhancement in nanostructured amorphous and semicrystalline polymers. In: Michler GH, Baltá-Calleja FJ (eds) Mechanical properties of polymers based on nanostructure and morphology. Taylor & Francis, Boca Raton, FL, p 379

63. Kim G-M, Michler GH, Rösch J, Mühlhaupt R (1998) Acta Polym 49:88

64. Laatsch J, Kim G-M, Michler GH, Arndt T, Süfke T (1998) Polym Adv Technol 9:716

65. Ivankova EM, Adhikari R, Michler GH, Weidisch R, Knoll K (2003) J Polym Sci Polym Phys 41:1157

66. Ivankova EM, Krumova M, Michler GH, Koets PP (2004) Colloid Polym Sci 282:203

67. Seydewitz V, Krumova M, Michler GH, Park JY, Kim SC (2005) Polymer 46:5608

68. Michler GH, Dietzsch (1982) Cryst Res Technol 17:1241

69. Golla U, Schindler B, Reimer L (1994) J Microsc 173:219

70. Cudby PEF, Gilbey BA (1995) Rubber Chem Technol 68:342

71. Crewe AV, Wall J, Welter LM (1968) J Appl Phys 39:5861

72. Cowley JM (1969) Appl Phys Lett 15:58

73. Zeitler E, Thomson MGR (1970) Optik 31:258,359

74. Crewe AV, Wall J (1970) Optik 30:461

75. Krause SJ, Allard LF, Bigelow WC (1978) STEM versus CTEM beam damage: paraffin single crystals. In: Sturgess JM (ed) Electron microscopy 1978, vol 1 (Proc 9th Int Congr on Electron Microscopy). Microscopical Society of Canada, Toronto, p 496

76. Batson PE, Dellby N, Krivanek OL (2002) Nature 418:617

77. Krivanek OL, Nellist PD, Dellby N, Murfitt MF, Szilagyi Z (2003) Ultramicroscopy 96:229

78. Handlin DL, MacKnight WJ, Thomas EL (1981) Macromolecules 14:795

79. Laurer JH, Winey KI (1998) Macromolecules 31:9106

80. Winey KI, Laurer JH, Kirkmeyer BP (2000) Macromolecules 33:507

81. Kirkmeyer BP, Weiss RA, Winey KI (2001) J Polym Sci Polym Phys 39:477

82. Kirkmeyer BP, Taubert A, Kim J-S, Winey KI (2002) Macromolecules 35:2648

83. Taubert A, Winey KI (2002) Macromolecules 35:7419

84. Kirkmeyer BP, Puetter RC, Yahil A, Winey KI (2003) J Polym Sci Polym Phys 41:319

85. Benetatos NM, Smith BW, Heiney PA, Winey KI (2005) Macromolecules 38:9251

86. Rose H, Plies E (1974) Optik 40:336

87. Krivanek OL, Gubbens AJ, Dellby N, Meyer CE (1992) Microsc Microanal Microstruct 3:187

88. Reimer L (ed) (1991) Energy-filtering transmission electron microscopy. In: Hawkes PW (ed) Advances in electronics and electron physics. Academic, Boston, MA

89. Reimer L (ed) (1995) Energy-filtering transmission electron microscopy (Springer Ser Opt Sci 71). Springer, Berlin

90. Egerton RF (1996) Electron energy-loss spectroscopy in the electron microscope, 2nd edn. Plenum, New York

91. Shindo D, Oikawa T (2002) Analytical electron microscopy for materials science. Springer, Tokyo

92. Ernst F, Rühle M (eds) (2003) High-resolution imaging and spectrometry of materials. Springer, Berlin

93. Ahn CC (ed) (2004) Transmission electron energy loss spectrometry in materials science and the EELS atlas, 2nd edn. Wiley-VCH, Weinheim
94. Varlot K, Martin JM, Quet C, Kihn Y (1997) Ultramicroscopy 68:123
95. Varlot K, Martin JM, Quet C, Kihn Y (1997) Macromol Symp 119:317
96. Varlot K, Martin JM, Quet C (1998) J Microsc 191:187
97. Varlot K, Martin JM, Gonbeau D, Quet C (1999) Polymer 40:5691
98. Varlot K, Martin JM, Quet C (2000) Micron 32:371
99. Hunt JA, Disko MM, Behal SK, Leapman RD (1995) Ultramicroscopy 58:55
100. Chou TM, Prayoonthong P, Aitouchen A, Libera M (2002) Polymer 43:2085
101. Ribbe AE, Bodycomb J, Hashimoto T (1999) Macromolecules 32:3154
102. Siangchaew K, Libera M (1999) Macromolecules 32:3051
103. Correa CA, Hage E Jr (1999) Polymer 40:2171
104. Correa CA, Bonse BC, Chinaglia CR, Hage E Jr, Pessan LA (2004) Polym Test 23:775
105. Kunz M, Möller M, Cantow H-J (1987) Makromol Chem Rapid Commun 8:401
106. Kunz M, Möller M, Heinrich U-R, Cantow H-J (1989) Makromol Chem Macromol Symp 23:57
107. Cantow H-J, Kunz M, Klotz S, Möller M (1989) Makromol Chem Macromol Symp 26:191
108. Du Chesne A, Lieser G, Wegner G (1994) Colloid Polym Sci 272:1329
109. Gerharz B, Du Chesne A, Lieser G, Fischer W, Cai WZ (1996) J Mater Sci 31:1053
110. Du Chesne A (1999) Macromol Chem Phys 200:1813
111. Tanaka Y, Hasegawa H, Hashimoto T, Ribbe A, Sugiyama K, Hirao A, Nakahama S (1999) Polym J 31:989
112. Ribbe AE, Hayashi M, Weber M, Hashimoto T (2000) Macromolecules 33:2786
113. Ribbe AE, Okumura A, Matsushige K, Hashimoto T (2001) Macromolecules 34:8239
114. Du Chesne A, Wenke K, Lieser G, Wenz G (1997) Acta Polym 48:142
115. Du Chesne A, Gerharz B, Lieser G (1997) Polymer Int 43:187
116. Amalvy JI, Asua JM, Leite CAP, Galembeck F (2001) Polymer 42:2479
117. Galembeck F, Leite CAP, da Silva MCVM, Keslarek AJ, Costa CARC, Teixeira-Neto E, Rippel MM, Braga M (2002) Macromol Symp 189:15
118. Rippel MM, Leite CAP, Galembeck F (2002) Anal Chem 74:2541
119. Rippel MM, Leite CAP, Lee L-T, Galembeck F (2005) Colloid Polym Sci 283:570
120. Eisenbach CD, Ribbe A, Günter C (1994) Macromol Rapid Commun 15:395
121. Heckmann W, McKee GE, Ramsteiner F (2004) Macromol Symp 214:85
122. Amalvy JI, Percy MJ, Armes SP, Leite CAP, Galembeck F (2005) Langmuir 21:1175
123. Lieser G, Schmid SC, Wegner G (1996) J Microsc 183:53
124. Lieser G, Oda M, Miteva T, Meisel A, Nothofer H-G, Scherf U (2000) Macromolecules 33:4490
125. Loos J, Yang X, Koetse MM, Sweelssen J, Schoo HFM, Veenstra SC, Grogger W, Kothleitner G, Hofer F (2005) J Appl Polym Sci 97:1001
126. Frank J (ed) (1992) Electron tomography: three-dimensional imaging with the transmission electron microscope. Plenum, New York
127. Frank J (1996) Three-dimensional electron microscopy of macromolecular assemblies. Academic, New York
128. Weyland M, Midgley PA (2004) Materials Today 7:32
129. Koster AJ, Ziese U, Verkleij AJ, Janssen AH, de Jong KP (2000) J Phys Chem B 104:9368
130. Midley PA, Weyland M (2003) Ultramicroscopy 96:413
131. Spontak RJ, Williams MC, Agard DA (1988) Polymer 29:387
132. Spontak RJ, Fung JC, Braunfeld MB, Sedat JW, Argard DA, Kane L, Smith SD, Satkowski MM, Ashraf A, Hajduk DA, Gruner SM (1996) Macromolecules 29:4494
133. Laurer JH, Hajduk DA, Fung JC, Sedat JW, Smith SD, Gruner SM, Agard DA, Spontak RJ (1997) Macromolecules 30: 3938
134. Jinnai H, Nishikawa Y, Spontak RJ, Smith RJ, Agard SD, Hashimoto T (2000) Phys Rev Lett 84:518
135. Jinnai H, Kajihara T, Watashiba H, Nishikawa Y, Spontak RJ, (2001) Phys Rev E 64:010803-1
136. Yamauchi K, Takahashi K, Hasgawa H, Iatrou H, Hadjichristidis N, Kaneko T, Nishikawa Y, Jinnai H, Matsui T, Nishioka H, Shimizu M, Furukawa H (2003) Macromolecules 36:6962
137. Jinnai H, Nishikawa Y, Ikehara T, Nishi T (2004) Adv Polym Sci 170:115
138. Takano A, Wada S, Sato S, Araki T, Hirahara K, Kazama T, Kawahara S, Isono Y, Ohno A, Tanaka N, Matsushita Y (2004) Macromolecules 37:9941
139. Kaneko T, Nishioka H, Nishi T, Jinnai H (2005) J Electron Microsc 54:437
140. Sugimori H, Nishi T, Jinnai H (2005) Macromolecules 38:10226
141. Sengupta P, Noordermeer JWM (2005) Macromol Rapid Commun 26:542

142. Jinnai H, Hasegawa H, Nishikawa Y, Sevink GJA, Braunfeld MB, Agard DA, Spontak RJ (2006) Macromol Rapid Commun 27:1424
143. Radzilowski LH, Carragher BO, Stupp SI (1997) Macromolecules 30:2110
144. Wilder EA, Braunfeld MB, Jinnai H, Hall CK, Agard DA, Spontak RJ (2003) J Phys Chem B 107:11633
145. Ikeda Y, Katoh A, Shimanuki J, Kohjiya S (2004) Macromol Rapid Commun 25:1186
146. Saxton WO, Baumeister W, Hahn M (1984) Ultramicroscopy 13:57
147. Dierksen K, Typke D, Hegerl R, Koster AJ, Baumeister W (1992) Ultramicroscopy 40:71
148. Fung JC, Liu W, de Ruijter WJ, Chen H, Abbey CK, Sedat JW, Agard DR (1996) J Struct Biol 116:181
149. Ziese U, Janssen AH, Murk J-L, Geerts WJC, Van der Krift T, Verkleij AJ, Koster AJ (2002) J Microsc 205:187
150. Schoenmakers R, Perquin R, Fliervoet T, Voorhout W, Schirmacher H (2005) Microsc Anal 19:13(EU)
151. Nickell S, Förster F, Linaroudis A, Del Net W, Beck F, Hegerl R, Baumeister W, Plitzko JM (2005) J Struct Biol 149:227
152. Winkler H, Taylor KA (2006) Ultramicroscopy 106:240
153. Frank J, McEwen BF (1992) Alignment by cross-correlation. In: Frank J (ed) Electron tomography: three-dimensional imaging with the transmission electron microscope. Plenum, New York
154. Lawrence MC (1992) Least-squares method of alignment using markers. In: Frank J (ed) Electron tomography: three-dimensional imaging with the transmission electron microscope. Plenum, New York
155. Rademacher M (1988) J Electron Microsc Tech 9:359
156. Rademacher M (1992) Weighted back-projection methods. In: Frank J (ed) Electron tomography: three-dimensional imaging with the transmission electron microscope. Plenum, New York
157. Gordon R, Bender R, Herman GT (1970) J Theor Biol 29:481
158. Gilbert P, (1972) J Theor Biol 36:105
159. Kremer JR, Mastronarde DN, McIntosh JR (1996) J Struct Biol 116:71
160. Boulder Lab For 3-D Electron Microscopy of Cells (2008) The IMOD home page. http://bio3d.colorado.edu/imod/index.html
161. Mastronarde DN (1997) J Struct Biol 120:343
162. Hawkes PW (1992) The electron microscope as a structure projector. In: Frank J (ed) Electron tomography: three-dimensional imaging with the transmission electron microscope. Plenum, New York

4 Scanning Electron Microscopy (SEM)

Today, scanning electron microscopy (SEM) is a versatile technique used in many industrial labs, as well as for research and development. Due to its high lateral resolution, its great depth of focus and its facility for X-ray microanalysis, SEM is often used in materials science – including polymer science – to elucidate the microscopic structure or to differentiate several phases from each other.

After a brief historic overview, this chapter explains the assembly and the mode of operation of SEM, which deviates from standard microscopes. This includes descriptions of the fundamentals of electron optics, the electron optical column, and the physical basics of electron–specimen interactions, which aid the understanding of contrast formation and charging effects. Because it is important to know the factors that influence X-ray microanalysis, a separate section about the origins of X-ray spectra and their interpretation has also been added. A discussion of environmental scanning electron microscopy (ESEM™) – a special development of SEM that is particularly useful when nonconducting or "wet" samples are to be examined – completes the chapter.

4.1 A Brief History of SEM

Parallel to the development of the transmission electron microscope (Sect. 2.1), Max Knoll had the idea of developing an electron microscope for investigating compact or bulk samples, and he demonstrated the basic principle for a scanning electron imaging device using two Braun-type cathode ray tubes in 1935 [1]. In both of these tubes, the electron beam is scanned by a pair of magnetic coils. In one of them, a plate with different surface layers was irradiated with a scanning beam, and the emitted secondary electrons that reached the anode were collected. The signal obtained was used to control the local brightness of the beam in the second tube. The surface layers of the irradiated plate could therefore be visualised by the second tube. This experiment was a proof of the material-specific dependence of secondary electron emission, but it was still not adequate for microscopic imaging.

Manfred von Ardenne developed an electron probe microscope in 1938 that used electron-optical lenses to focus the beam. This idea originated from the fact that resolution-limiting chromatic failures do not play a role in such a microscope [2–4]. He presented this instrument, in which electrostatic scanning plates were applied to a TEM, only a few years after Knoll's and Ruska's first TEM. This type of microscope is now known as a scanning transmission electron microscope (STEM). However, Ardenne's prototype fulfilled the minimum criteria for a scanning electron microscope (SEM) from today's point of view. The first scanning electron microscope of the usual type was described and developed by Zworykin et al. [5] in 1942 using three electrostatic lenses and electromagnetic coils between the second and third lenses.

Unfortunately there was no further development in the field of SEM in Germany, and Ardenne's instruments were lost in 1944 in an air raid during World War II. In the latter half of the 1940s, Charles Oatley continued research into SEM at Cambridge University, but it was still quite some time until the first commercial SEM was made available. One of his graduate students, McMullan, enhanced the instrument of Zworykin in order to achieve a lateral resolution of about 50 nm [6]. Smith [7] replaced the electrostatic lenses by electromagnetic ones and introduced a double deflection system to correct astigmatism. At the beginning of the 1960s, Everhardt and Thornley [8] improved the secondary electron detector by adding a light pipe between the scintillator and multiplier. All these improvements provided a great impetus for the development of the first commercial scanning electron microscope, which was named Stereoscan Mark 1 and sold by Cambridge Instruments in 1965.

Parallel to the Cambridge group, Coslett's group at the Cavendish Laboratory combined the X-ray analytical capability of Castaing's "microsonde electronique" (electron microprobe) [9] with an imaging facility and developed the scanning electron probe X-ray microanalyser [10]. From the middle of the 1960s until the present several improvements in electron generation and electron optics have been made that have enhanced the resolution power. At the end of the 1980s, computerised SEM entered the scene, which led to the introduction of the digital scanning generator for digital image recording and processing.

At that time a new type of SEM with a vacuum of some tens of mbar in the specimen chamber was developed. One reason for this development was the idea of directly preventing the charging of insulating samples; another reason was to make it possible to investigate humid samples. Because of the high-voltage flashovers of the Everhardt-Thornley detectors normally used, most instruments only allow imaging with backscattered electrons in the low vacuum mode. Danilatos provided a completely new principle, the "gaseous secondary electron detector" (GSED) [11]. Using this type of detector and subsequent developments from Danilatos [12–14], the newly formed ElectroScan Corporation produced the first "environmental scanning electron microscope" (ESEM™) in 1989, which was based on electron optics from Philips Eindhoven that had the ability to image with secondary electrons. After the acquisition of ElectroScan in 1996 by Philips/FEI, a new generation of ESEMs came onto the market. The special features of this instrument are explained in Sect. 4.6.

4.2 Fundamentals of Electron Optics and Signal Generation

4.2.1 Principle of SEM

The basic principle of the SEM [15–17] is based on Knoll's experiment [1] and von Ardenne's idea for a scanning probe transmission microscope [4].

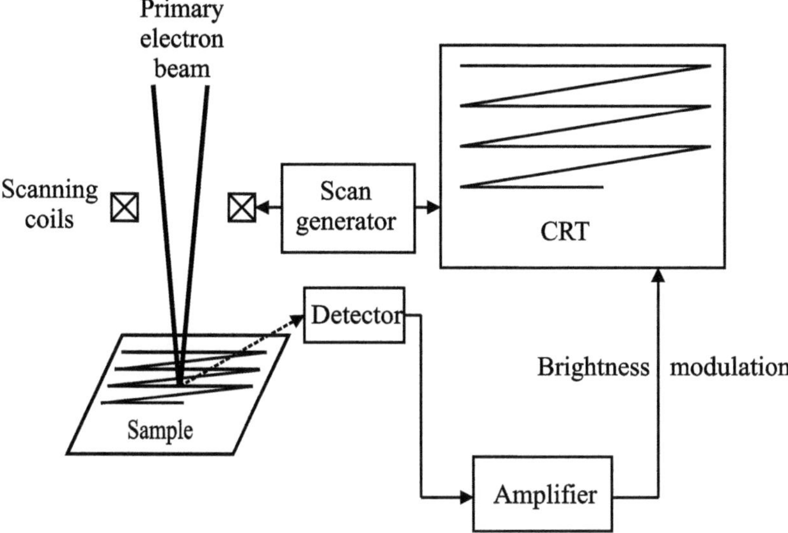

Fig. 4.1. Scheme showing the principle of SEM

In a similar process to the scanning of the electron beam in a cathode ray tube (CRT), the focussed electron beam scans line by line over the surface of the specimen in the evacuated microscope column and forms signals based on the interactions between the beam and the sample, which are electronically detected and amplified by suitable equipment. Originally, the response signal was displayed as a brightness modulation on a CRT where the electron beam is driven simultaneously to the beam in the column, as illustrated in Fig. 4.1. Nowadays, digital computer techniques have replaced the traditional CRTs. As the area of the displayed image remains unchanged, the magnification of the image is determined by the dimension of the scanned sample area (see Sect. 4.3.4). Generally, the resolution of the SEM image is determined both by the diameter of the electron probe focussed on the sample surface (see Sect. 4.2.2) and the interaction of the primary electrons (PE) with the sample (see Sect. 4.2.5).

4.2.2 The Lateral Resolution Power of SEM

Using the electron-optical system within the SEM column, the electron beam created by the electron gun (see Sect. 2.3.1) is reduced to a sufficient diameter to form a very fine focussed electron probe. The PE beam diameter in the electron gun depends on the type of cathode (respective gun type) and is formed by the first crossover of the electron trajectories (see Fig. 2.7d). The first crossover generated from the tungsten cathode must be scaled down to approximately a factor of 1000 to attain a reasonable resolution. Two or three (although sometimes more) electromagnetic lenses are needed to demagnify the beam diameter (see Fig. 4.2). Furthermore, the first lens or the first and second lenses (called condenser lenses) are used to vary the PE beam current.

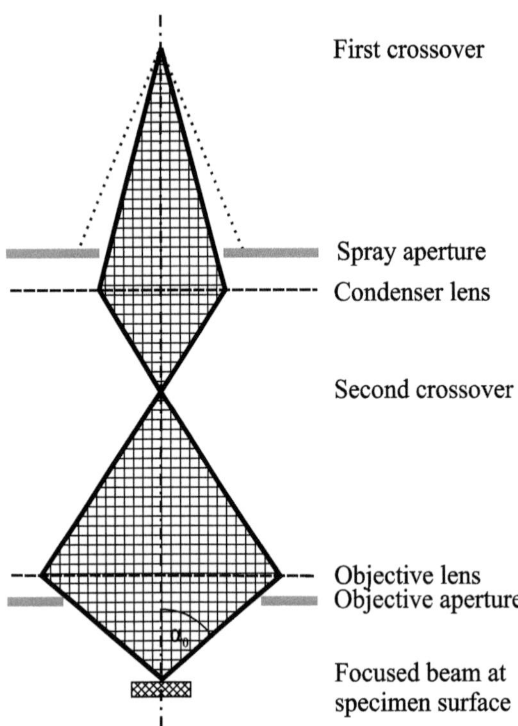

Fig. 4.2. Scheme of the electron-optical relationship in SEM

The minimal achievable diameter d_0 of the beam can be estimated with the help of the lens law (see Eqs. 2.8 and 2.9). Another quantity used to characterise the electron beam performance is the gun brightness R, defined by

$$R = \frac{j_c}{\pi \alpha_c^2} = \frac{j_o}{\pi \alpha_0^2} = const. \tag{4.1}$$

where j_c is the beam current density at crossover, α_c is the aperture angle at crossover, j_0 is the beam current density at the specimen surface and α_0 is the aperture angle at the specimen.

The aperture angle α_0 is determined by the ratio of the diameter of the last diaphragm r to the working distance L, which is the distance between the pole piece of the objective lens and the specimen:

$$\alpha_0 = \frac{r}{L}. \tag{4.2}$$

A homogeneous beam density within the beam diameter d_0 at the sample surface corresponds to the following beam current I_0:

$$I_0 = \frac{\pi}{4} d_0^2 j_0. \tag{4.3}$$

Combining Eqs. 4.1 and 4.3 results in

$$I_0 = \frac{\pi^2}{4} R d_0^2 \alpha_0^2 \tag{4.4}$$

and

$$d_0 = \left(\frac{4 I_0}{\pi^2 R}\right)^{1/2} \frac{1}{\alpha_0} = \frac{C_0}{\alpha_0}. \tag{4.5}$$

This equation is also valid when the homogeneous illumination is replaced with a Gaussian-distributed one. In this (more realistic) case, d_0 corresponds to the half-width of the full maximum (HWFM) of the distribution.

It can be seen from Eq. 4.5 that the diameter of beam is directly proportional to the beam current and indirectly proportional to gun brightness and aperture angle. In reality, the effective diameter of the beam d_b at the specimen is somewhat larger due to aberrations (see Sect. 2.2.3).

An acceptable signal-to-noise ratio in the secondary electron signal (see Sect. 4.3.3) usually requires a minimum beam current I_0 of between 10^{-12} A and 10^{-11} A. Figure 4.3 shows the dependence of d_b on the aperture angle and the beam current for a tungsten filament gun.

The figure also shows that there is an optimum value for the aperture angle as a function of the beam current. The gun brightness R of the Schottky emitter or field-emission guns (FEG) is higher than that of the tungsten or LaB_6 cathodes (see Table 2.4). According to Eq. 4.5, a much lower beam diameter is achieved at the same beam current with the FEG rather than thermionic guns. Thus, if an FEG is used, a resolution power of 2 nm or even less can be achieved.

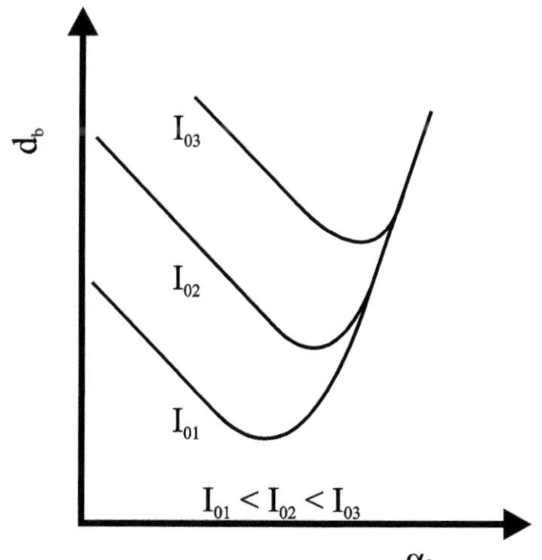

Fig. 4.3. Dependence of the effective beam diameter d_b (in the nm range) on the aperture angle α_0 (in the mrad range) and the beam current I_0 for a tungsten filament gun (schematic, after [15])

4.2.3 Comparison of Various Cathode Types

The types of cathodes (usually termed "guns") frequently used in an electron microscope have been already described (in Sect. 2.3.1 and Table 2.4). To select the most appropriate gun type, the operator should take into account first the minimum resolution power (Sect. 4.2.2) required for the investigations and secondly all of the other parameters of cathodes. Table 2.4 shows the main differences between tungsten and LaB_6 cathodes and the FEG. While the beam diameter increases with increasing probe current for the first two types, in the case of FEG there is a probe current interval where the beam diameter stays constant with increasing probe current. The most important feature of the gun for the user of an SEM is the available probe current of the emitters. For certain SEM techniques, like WDX analysis (see Sect. 4.5.6) or cathodoluminescence, thermionic emitters have been preferred in the past. Nowadays thermal FEG is a good compromise. This emitter type offers on the one hand a good resolution in the SE mode and on the other hand sufficient probe current for the application techniques mentioned before.

4.2.4 Depth of Focus

The depth of focus is one of the outstanding features of SEM. While the depth of focus of a light microscope for a magnification of about 200× is in the µm region, the depth of focus of the SEM at same magnification covers mm.

The depth of focus is a function of both the convergence angle α_0 of the electron beam and the magnification M (Fig. 4.4). Noting that $\tan \alpha_0 \approx \alpha_0$ at $\alpha_0 \ll 1$, the focus depth S is expressed by:

$$S = \frac{2r}{\alpha_0} = \frac{0.1\,\text{mm}}{\alpha_0 \cdot M} . \tag{4.6}$$

Here, $2r$ is the beam diameter which produces a corresponding spot on the display which can just about be resolved by the human eye (ca. 0.1 mm). Equation 4.6 reveals that the depth of focus decreases with increasing magnification and aperture angle α_0. The facet eye of a fly (see Fig. 4.5) demonstrates high depth focus in SEM imaging.

4.2.5 Interaction of Primary Electrons with Sample

Signal generation in SEM is a result of the interaction between the incident electron beam and a thin surface layer of the sample, which depends on the beam energy. The primary electrons are charged particles and so they interact strongly with the electrically charged particles of the atoms in the sample, i.e. both with negatively charged electron clouds and positively charged nuclei. The interaction is said to be *inelastic* if some of the energy of the primary electron is lost during the interaction. If no energy is lost the interaction is said to be *elastic*. The fundamentals and applications of these scattering processes are also described in Sects. 2.4.1 and 3.9.

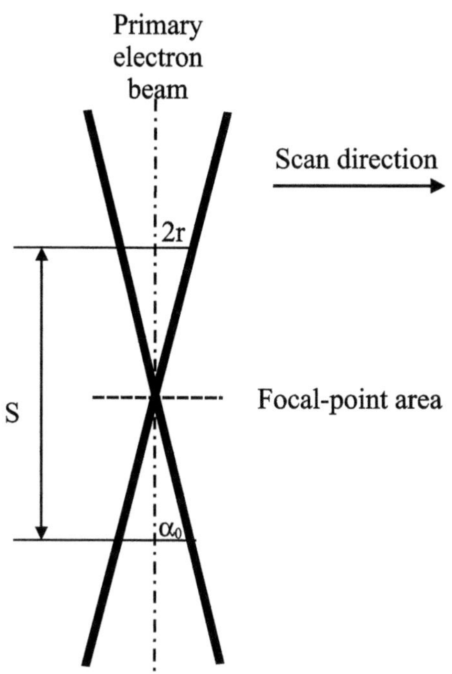

Fig. 4.4. Scheme of the derivation of depth of focus in SEM (after [15])

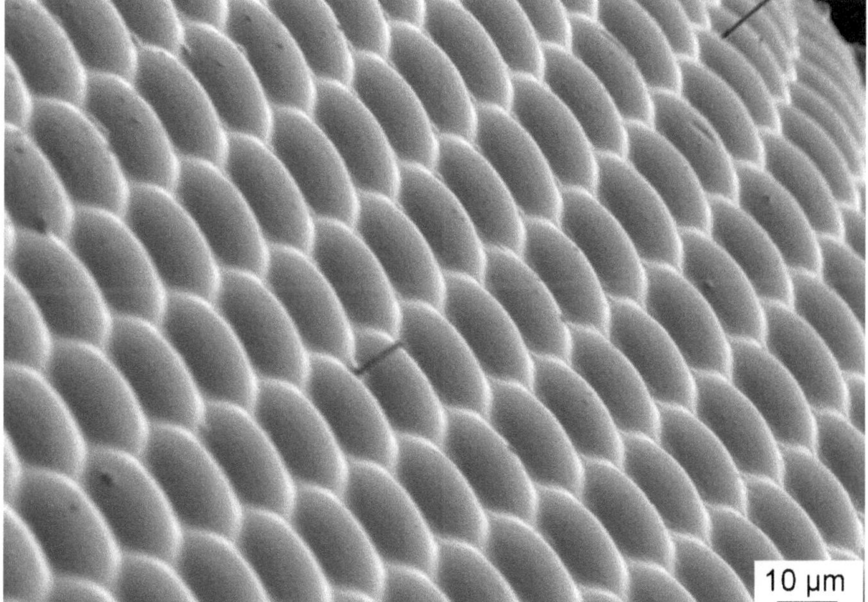

Fig. 4.5. SE micrograph of the facet eye of a fly showing a high depth of focus

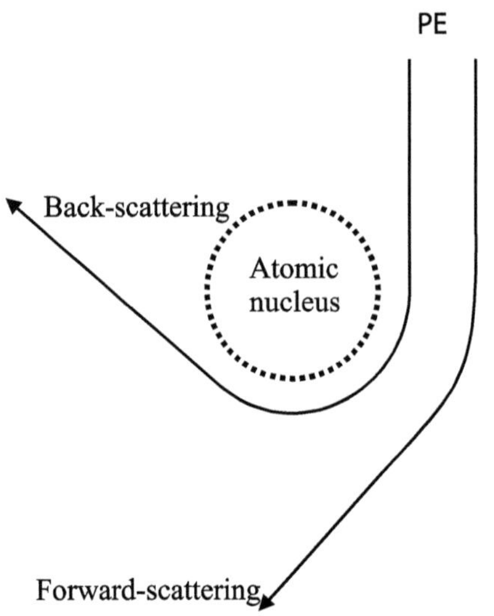

Fig. 4.6. Rutherford scattering of a primary electron in the Coulomb field of an atomic nucleus

PE/Nucleus Interaction

When electrons enter the electric field of an atomic nucleus, which can be partly screened by electron clouds, their paths are deflected. Generally, this so-called Rutherford scattering results in parabolic PE trajectories. The larger the atomic number of the atoms in the interaction region and the smaller the distance between the nucleus and the PE, the stronger the deflections of the latter and hence the more curved PE paths.

While contrast-forming interaction processes in a TEM are restricted to weak scattering events with small scattering angles (forward scattering), electron scattering in bulk samples in SEM also includes a high proportion of strong scattering events where the directions of the PE trajectories are significantly changed, with some even sent back along their original paths (back scattering), as illustrated in Fig. 4.6. While a negligible amount of energy is lost due to conversion into X-ray continuum radiation ("bremsstrahlung"; see Sect. 4.5.1), the interactions are assumed to be elastic. Electrons that undergo one or more Rutherford processes and leave the sample surface without notable energy loss are called backscattered electrons (BSE).

PE/Electron Cloud Interaction

PEs striking the sample surface may also interact with the electrons of atoms within the surface region. Owing to their equivalent masses and identical charges, the interaction is inelastic. If the energy of the incident electron is high enough, the valence electrons of the surface atoms can easily be released from the atoms (Fig. 4.7). These electrons are called secondary electrons (SE). Their kinetic energies are very low (only a few electron volts), just enough to surmount the work function of the sample. However, only SE originating from a very thin surface layer a few nm in thickness can contribute to the detectable signal, as all of the SEs generated in deeper regions of bulk samples will recombine. Generally, ionised or excited atoms will rapidly con-

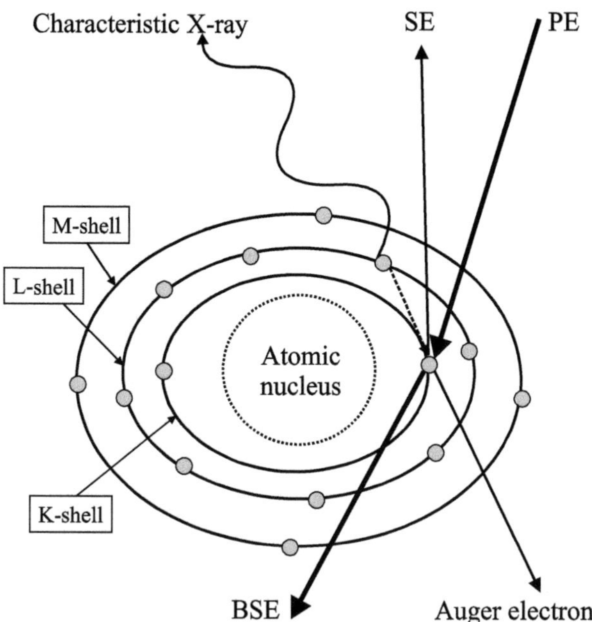

Fig. 4.7. Inelastic interaction between primary electrons and electrons in atomic shells

vert back to their initial states. These processes are accompanied by the emission of characteristic X-rays (see Sect. 4.5.1) or Auger electrons, which is the basis for analytical investigations by means of SEM.

4.3 The Instrumentation of SEM

4.3.1 The Column

Generation of the Focussed Electron Beam
As already described in Sect. 4.2.1, the SEM consists of a column (which is a unit containing the lens system that forms the finely focussed electron beam), a specimen chamber, detectors, as well as imaging and recording units (see Fig. 4.1).

The typical acceleration voltage range of SEM lies between some hundreds volts and 30 or 35 kV. PEs originating from the virtual source at the first crossover (in the electron gun) as a divergent beam pass through the anode aperture and enter the lens system of the SEM. Usually, a spray aperture is placed in the entrance plane of the lens system to block electrons moving along paths very far from the optical axis.

The lens system acts on the electron beam in two ways. On the one hand, it transfers the PEs from the gun crossover to the plane of the specimen surface, and in doing this it reduces the beam diameter considerably to form a very fine probe. On the other hand, the beam current must be controlled by the lens system. In principle, a system of just two lenses combined with an aperture diaphragm could be used to do this, as illustrated in Fig. 4.2. The first lens, called a condenser lens, that has a variable and relatively weak magnetic field (and a correspondingly large focal length) is mainly used to control the beam current, while the second one, called the objective lens, is a short-focal lens (with a strong magnetic field) which is primarily responsible for the demagnification of the beam diameter. However, the condenser lens is usually replaced by a system of condenser lenses that allow the beam current and beam diameter to be varied independently.

The aperture angle α_0 limiting diaphragm is usually placed within the pole piece of the objective lens (see Figs. 4.2 and 4.8). This aperture can usually be adjusted from outside and must be corrected after changing the high voltage and beam current. One should note that the middle points of the images remain unaltered during the focussing operation.

Scanning Unit and Stigmator

Two pairs of scanning coils to deflect the electron beam in the x- and y-directions are located within the objective lens (see Fig. 4.8). These coils are driven by two different sweep signals, where, as in a CRT, the slew rate of the pair of scanning coils for x-deflection is a multiple value of that for the other pair. The increase in the sweep rate over time influences the change in the magnetic field. Due to the change in the magnetic field in the coils, the electron beam will be deflected away from the optical axis and so the beam scans across the sample surface. In the process both of the sweep voltages range from a maximum negative value to the same positive value, so that the middle of the scanned area on the specimen surface occurs along the optical axis.

The higher the maximum sweep voltage, the stronger the magnetic field in the coils and the larger the maximum elongation at the sample. The size of the scanned area on sample surface can be changed by varying this maximum voltage. This procedure controls the magnification of the SEM (Sect. 4.3.4).

As mentioned in Sect. 2.2.3, astigmatism occurs in all electron lenses due to instrumental imperfections introduced during the manufacturing process. In SEM, this astigmatism also leads to a deformation of the ideal spherical cross-section of the electron beam. Due to the deflection of the beam during the scanning process, the beam spot presents an increased elliptical cross-section, in particular at the fringes of the frame. This elliptical distortion leads to loss of resolution and to reduced image

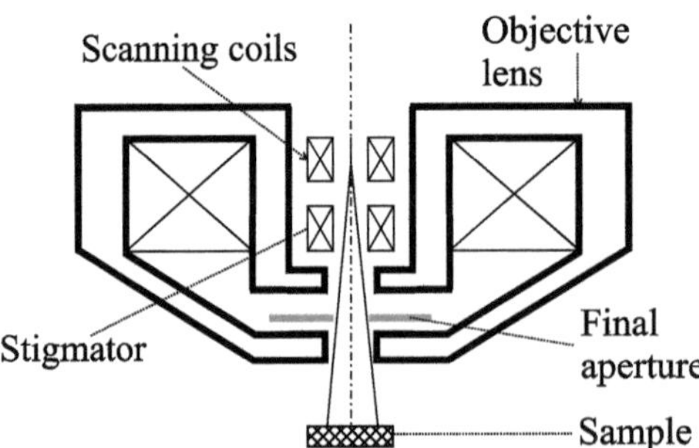

Fig. 4.8. Scheme of an objective lens with scanning coils and stigmator

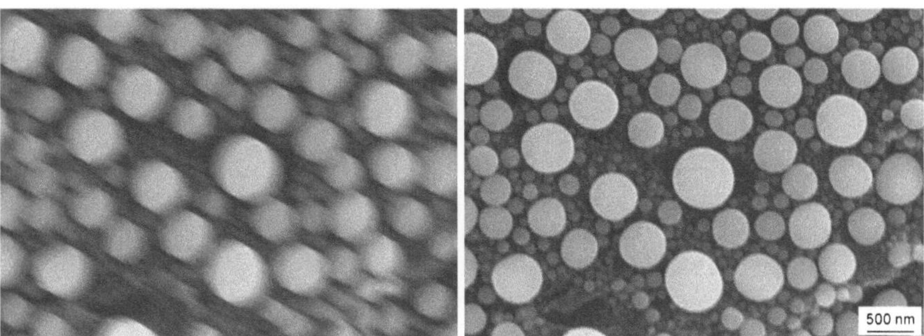

Fig. 4.9. SE image demonstrating the appearance of an elliptical focus due to astigmatism (*left*) and compensated image (*right*) (test sample Au on C)

sharpness. However, the so-called *stigmator* (see Fig. 4.8) can be used to compensate this distortion (Fig. 4.9).

4.3.2 Specimen Chamber and Goniometer

A specimen chamber with a goniometer unit, which enables sample movement to be defined, is attached to the microscopic column below the pole piece of the objective lens. Usually, the goniometer not only enables rotations and translations of the specimen in all (x, y and z) directions, but it also enables the specimen to be tilted. The maximum tilt angle is usually 90° towards the Everhardt-Thornley detector (see below) and 10–30° in the opposite direction. Modern goniometers allow specimen movements of more then 5 cm in the x- and y-directions and a few cm along the z-direction. For a constant deflection of the electron beam, the scanned surface area depends on the distance between the specimen surface and the scanning coils (theorem on intersecting lines), i.e. z-translation is very important when choosing the optimal working distance (WD). In general, WD is defined as the distance between the pole piece of the objective lens and the specimen surface. It should be noted that a higher resolution can be achieved at lower WD. Usually a corresponding decrease in the detected SE signal due to shadowing effects limits the reduction of WD in practice. The use of an optimal WD is of special importance for analytical investigations performed by X-ray microanalysis (Sect. 4.5).

Often the sample is transferred to the specimen support of the goniometer via an airlock; otherwise the chamber and the lower part of the column have to be vented. The specimen support of the goniometer is electrically insulated from the other parts of the microscope. Thus, absorbed electrons can be detected using a special amplifier or a warning signal can be generated using a suitable electronic unit when the specimen collides with the microscope.

It is also possible to incorporate special devices (heating or cooling units, tensile or bending modules, etc.) into the specimen chamber to carry out in situ tests of the sample (see Chap. 6).

4.3.3 Detectors

Secondary Electron Detector

An SEM is normally equipped with an Everhardt-Thornley detector [8] for imaging the sample surface via collected secondary electrons. This detector is a combination of a scintillator and a photomultiplier. Due to its low noise, a photomultiplier is favoured over other kinds of amplifiers used to amplify small currents in the range of nA to pA, which are typical of the emitted SE currents in SEM.

The Everhardt-Thornley detector is based on the following principle.

The secondary electrons that leave the sample possess energies of up to only 50 eV. These low-energy electrons can be collected with high efficiency by a grid electrode that is positively biased with a voltage of about 200 V. The collected SEs are subsequently accelerated toward the scintillator, which is covered with a thin aluminium layer and placed at a voltage of about +10 kV.

The accelerated SEs striking the scintillator possess sufficient energy to emit photons by converting their kinetic energies. The generated photons pass through a light pipe into the photomultiplier, where they cause the emission of photoelectrons. The latter are highly amplified by electron multiplication at the dynodes in the multiplier, and so finally an electronic signal that is proportional to the number of collected SEs is produced (Fig. 4.10).

If the bias voltage of the scintillator electrode switches off or if a negative bias is applied to the collector grid, only highly energetic BSEs can reach the scintillator, which leave the sample and head towards the detector. Due to geometric constraints, the signals from sample locations turned away from the detector do not contribute to the imaging; the resulting BSE image shows shadow phenomena. This mode is generally used as a simple method for BSE imaging. Another BSE detector with a higher detection efficiency is described below.

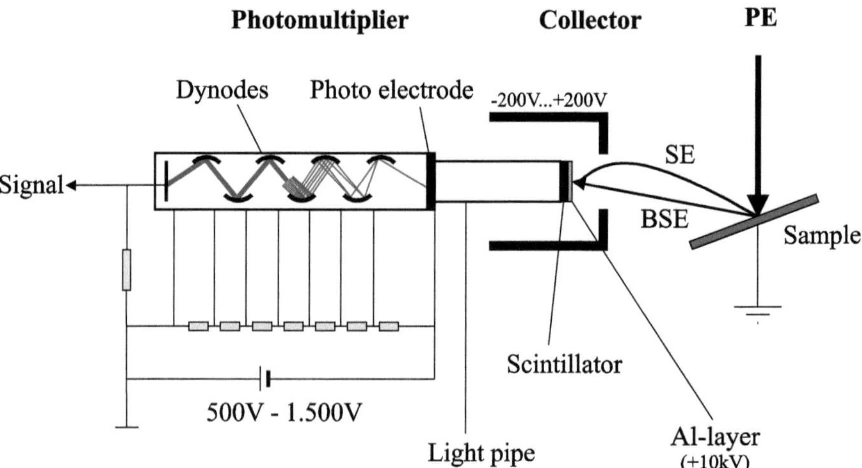

Fig. 4.10. Principle of the Everhardt–Thornley detector

The scintillator material consists of either a layer of fluorescent powder (placed on optically transparent platelets) or an yttrium-aluminium-garnet (YAG) crystal. The first type can degrade due to prolonged bombardment with high-energy electrons and hence need to be replaced at regular intervals.

Other Detectors (Including Solid State Backscattered Electron Detectors)

In addition to the type of detector discussed above, there are a large number of commercially available detectors that can be bought as complementary equipment for SEM. However, most of these are rarely used in polymer research. Only a few of them have found application in special electron microscopic tests, such as BSE detectors based on the electron voltaic effect in a solid state detector, and energy dispersive X-ray detectors, which will be discussed in Sect. 4.5.3.

The backscattered electron detector consists of two or four semiconductor diodes which are symmetrically arranged around the opening of the pole piece of the objective lens. In these diodes, the backscattered electrons generate electron-hole pairs in quantities that depend on the energy and the intensity of the electrons. The electron–hole pairs can be partially separated by the inner electric field at the p-n junction. By passing the p- and the n-type parts through an amplifier, an output signal can be obtained which is proportional to the backscattered electrons reaching the detector. By suitably coupling the signal outputs from different diodes, one can generate both topography-sensitive as well as material-specific signals (Sect. 4.4.2).

4.3.4 Signal Display and Magnification

The signals obtained by the detectors are simultaneously displayed on a monitor. Data representation can be achieved through either analogue signal processing on a CRT (as in the older generations of SEMs) or by digitisation after displaying the data with the aid of a computer. When the first option is used, the images should be obtained via photographic techniques. The digital technique offers the chance to obtain the sample information directly, as a digital image which can be saved, modified and transferred instantly in electronic form. Both of the imaging techniques have one feature in common: the surface scanning and the representation of the surface information on the display take place simultaneously, such that each specimen point corresponds geometrically to the identical location on the recorded SEM image.

The magnification in an SEM is the ratio of the lateral length of the image displayed or printed to that of the scanned area!

This definition makes it clear that the magnification always depends on how the information is presented. For example, an SEM micrograph of 10 cm × 6 cm with 20 000× magnification has double magnification as when the picture is presented in a 5 cm × 3 cm format. Therefore, it is usual to introduce a scale bar to an image calibrated by the microscope, which automatically appears in the SEM images. Otherwise a scale bar (also called a μ-marker) must be manually marked onto a micrograph.

The reasonable magnification of SEM (M_r) is determined by the resolving power of the resolving power of the human eye (m, ca. 0.1 mm) and the resolving power of the microscope (r). The latter depends on the signal used (e.g. SE, BSE, X-rays) and corresponds to the lateral diameter of the generation volume of the electron beam interaction product which contributes to the signal.

$$M_r = \frac{m}{r} \, .$$

(4.7)

The above statement means that the operator must calculate the upper limit of magnification using Eq. 4.7 before beginning the work. For instance, to compare an SE image with X-ray mapping, it is important to be aware that the SE image can be constructed from more pixels than X-ray mapping because the effective probe diameter for SE is on the order of 1–5 nm, whereas the X-ray probe diameter is in the order of several microns (μm), and also depends strongly on the acceleration voltage. Nevertheless, it makes sense to work with so-called empty magnifications, e.g. during X-ray point analysis. If one works with very high magnifications (for instance 500 000-fold conforms approximately to a point analysis), it is possible to observe the beam stability on the screen during the analysis.

4.4 Contrast Formation and Charging Effects

4.4.1 Secondary Electron Contrast

As already discussed in Sect. 4.2.5, PEs interact with the sample in an elastic or an inelastic manner. During the diffusion of the PEs, one or more of the processes discussed in the above section may take place. It is common to employ Monte Carlo simulation in order to obtain a statistical picture of the movement of the PEs. Figure 4.11 presents, for instance, typical path simulations for a light element (carbon) and a heavy metal (gold). For the lighter element and for high PE energies, the presence of a pear-shaped path distribution is typical. For heavy metals with lower PE energies, the distribution assumes a shape similar to that of a sectioned sphere. As a result, the penetration depth of the PEs for a lighter element is higher than that observed for a heavier one. SEs are generated during the diffusion of the PEs, but possess only a small amount of energy. Thus only the SEs produced by the surface layers can leave the surface and reach the detector. The SE efficiency and hence the signal-to-noise ratio of the image increases due to inelastic collisions close to the surface. The SE yield for elements with high atomic number Z is greater than for materials with low atomic numbers at identical PE energies, which means that the operator can reduce beam current at high Z.

The lateral resolution power of SE imaging is affected by other factors as well as the beam diameter. Besides the dependence on the electron optics, there is also a dependence on the interaction of the PEs with the sample, because SEs will be emitted from a larger region of the specimen. In particular, for light elements (such as those present in plastics), a considerable number of secondary electrons are produced far from the location of PE incidence, which ultimately deteriorate the lateral resolution. The resolution can be improved in such cases by coating a thin film of

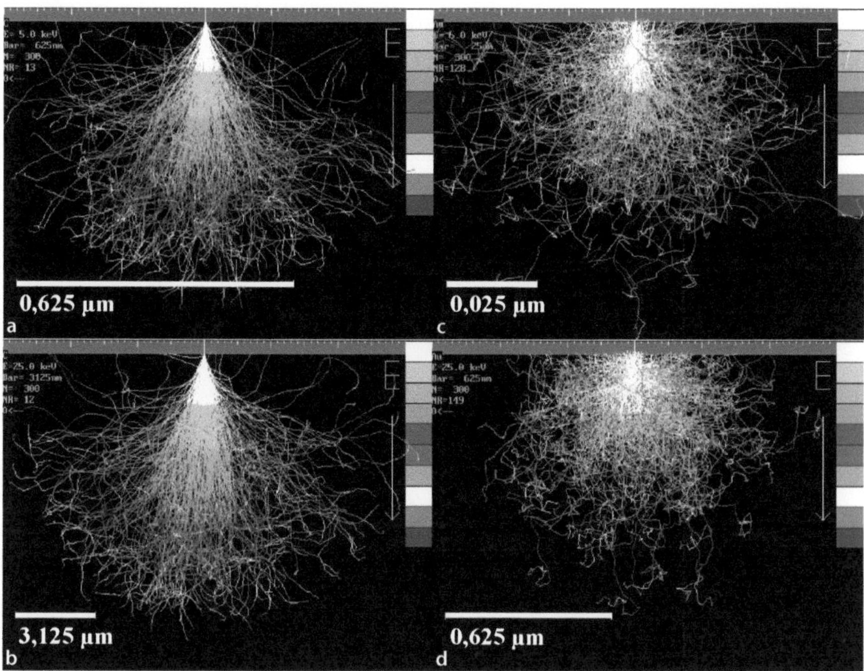

Fig. 4.11a–d. Monte Carlo simulations of the PE paths for carbon (C) and gold (Au) at different PE energies (E_P): **a** C, E_P = 5 keV; **b** C, E_P = 25 keV; **c** Au, E_P = 5 keV; **d** Au, E_P = 25 keV

heavy metal onto the sample surface. Such a coating also contributes to reducing the surface charging (see Sect. 4.4.3). The improvement results from the corresponding interactions of the PEs with the heavy metal layer. At the same time, more SEs are produced near the primary beam zone, which improves the signal-noise ratio.

The details of the SE emission determine the contrast in a SE image. Generally, good contrast is generated when the atomic number (and therefore the density) of the constituents differ significantly (see Fig. 4.12).

In polymeric materials, this kind of contrast (i.e. which depends on the atomic number) cannot be expected. The contrast formation in the SE mode is mainly determined by the local inclination of the sample surface with respect to the incident beam. This phenomenon, which is particularly apparent at surface edges (the so-called "edge effect") can be explained by the correlation between the surface and the interaction volume of the PE, as represented in Fig. 4.13. If the sample surface is not ideally flat, the interaction volume of the PE electron can pass through the side of the step, and from that location an additional number of secondary electrons can leave the sample surface, resulting in a higher SE signal. The maximum contrast enhancement occurs when the average penetration depth of the PE corresponds to the height of a step. Therefore, it can be deduced that the contrast in SEM imaging is determined not only by the atomic numbers of the elements but also the energy of the PEs selected for the SEM operation. Thus, to optimise the contrast, it is advisable not

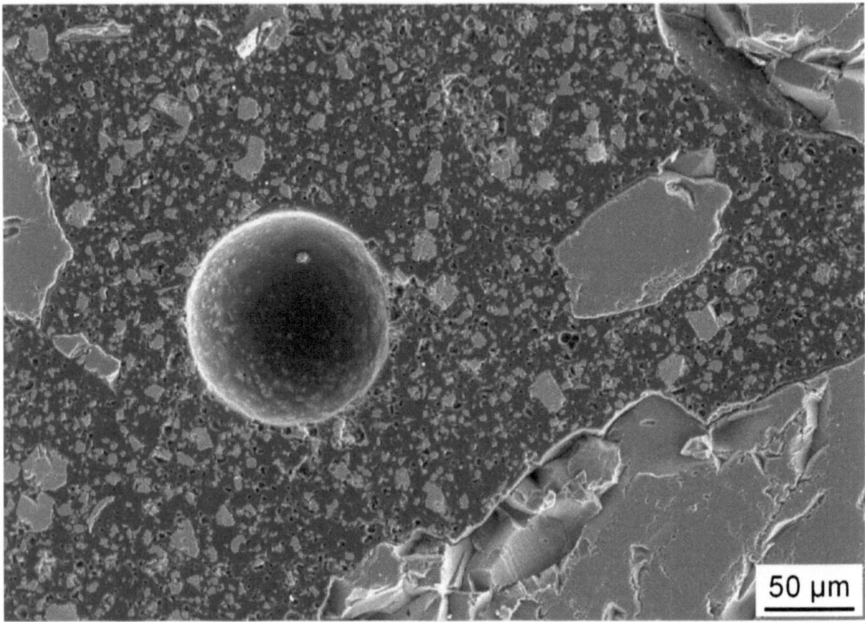

Fig. 4.12. Atomic number contrast in SE mode of a polymeric concrete sample consisting of epoxy, quartz (SiO_2, large particles), calcite ($CaCO_3$, small particles) and spherical pores

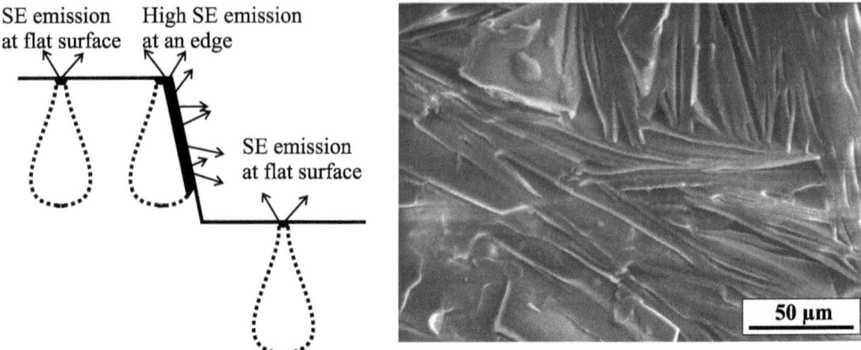

Fig. 4.13. Contrast formation in secondary electron mode due to surface relief (*right*: SE micrograph of paraffin crystals, 12 keV)

only to work at a fixed acceleration voltage, but to vary it in order to match it to the sample.

As well as a positive edge effect, a negative one can also occur; this is caused by the shadowing of SEs by the edge, and it results in a decreased SE signal.

4.4.2 Contrast of Backscattered Electrons (Solid State Detector)

The emission of the BSE can also be explained by the simulation scheme presented in Fig. 4.11. Due to the strong interactions of the PEs and the resulting lower penetration depths for elements with higher atomic numbers, these elements offer a greater probability of BSE emission. On the other hand, only BSEs that reach the detector can contribute to the signal.

The influences of the atomic numbers of the atoms within the region of interest and the surface topography on the signal detected by the individual segments (A and B) of a split BSE detector are illustrated in Fig. 4.14. By forming a sum signal (A+B) and a difference signal (A−B) of the signals A and B measured by the detector segments, the influences of the kind of material and the surface topography can be distinguished. Compositional contrasts are produced by the sum signal (A+B), while the difference signal (A−B) results from shadowing effects of the surface relief and therefore provides an impression of the sample topography.

Indeed, one should note that the efficiency of the BSEs is much lower than that of the SEs. Additionally, the noise levels of the detection and amplification systems significantly worsen the image quality when working with small beam currents in the BSE mode. In particular, the noise affects the difference signal (topography signal) of the split detector.

Furthermore, it should be noted that BSEs from a surface region corresponding to the lateral extension of the interaction volume actually contribute to the BSE signal detected, while in the case of SE detection the signal mainly results from SEs emitted from the small area where the PEs meet the sample surface.

4.4.3 Charging Effect

Both the secondary electron yield (ratio of the SEs emitted to the PEs that strike the sample) and the backscattering coefficient (the ratio of the BSEs emitted to the PEs that strike the sample) depend on the PE energy as well as the atomic number of the material and the angle of incidence of the PE [15]. Therefore, the charge supplied to the sample by incident PEs will generally differ from the charge release caused by the emission of SEs and BSEs. In most cases for bulk specimens, at PE energies of between several hundred eV and about 2 keV, the number of electrons that leave the sample exceeds the number of incident PEs, while at PE energies in the usual range (5–25 keV) the opposite is true. Consequently, the charging of insulating samples occurs (Fig. 4.15) apart from when the PE energy falls within a small range around 2 keV, where dynamic charge compensation takes place. For electrically conductive materials, the charge difference is compensated for by connecting the sample to the ground potential. To avoid the surface charging of insulating samples, their surfaces are usually coated with conducting layers of gold (Au) or carbon (C). However, it is important that the layer has sufficient contact with the specimen support held at ground potential.

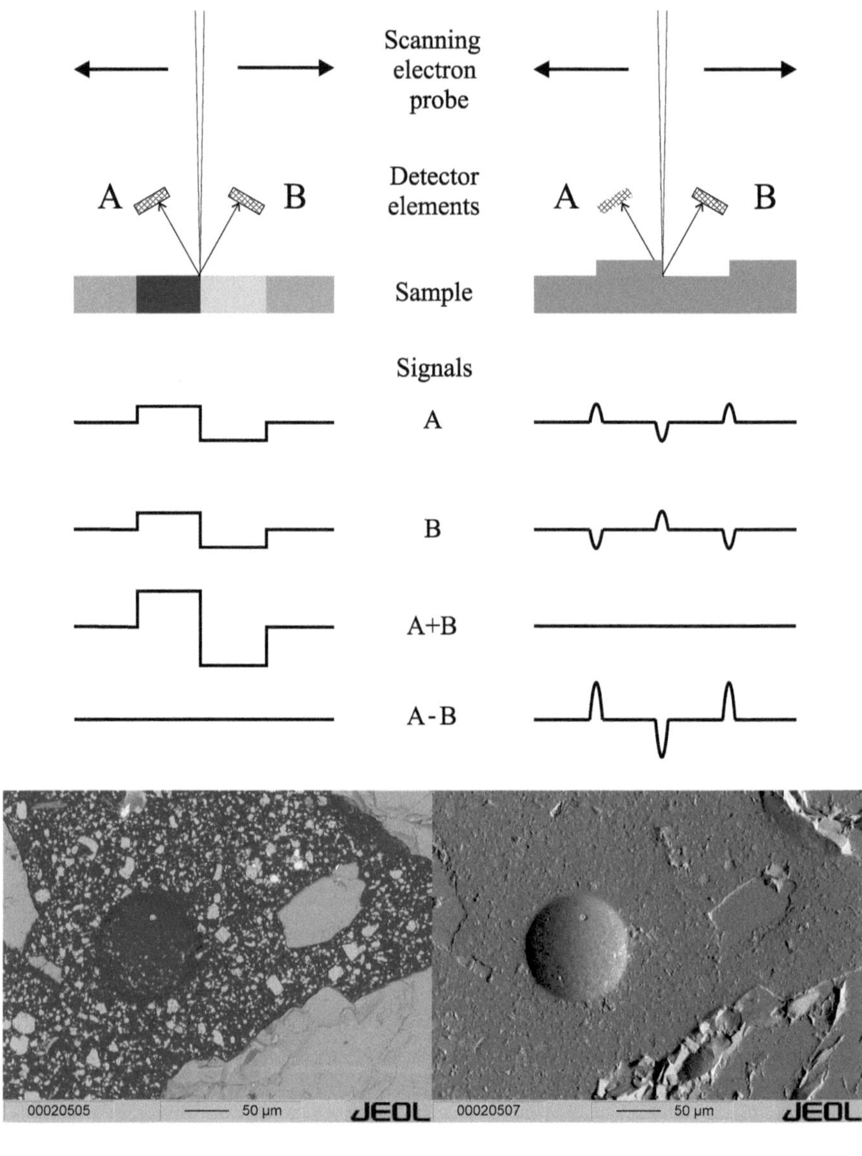

Fig. 4.14.   Demonstration of composition and topography contrast in BSE mode using two symmetric solid state detectors A and B (sample as in Fig. 4.12)

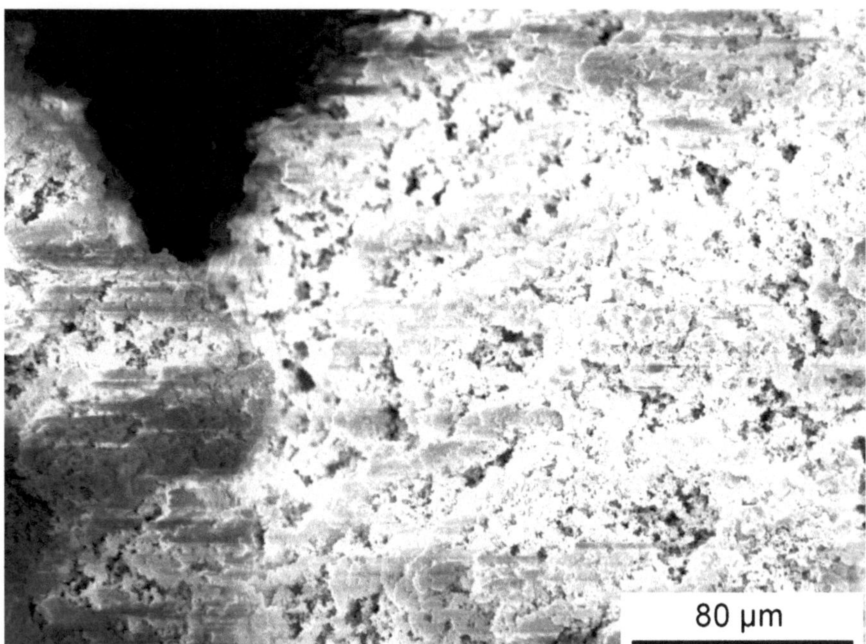

Fig. 4.15. SE image showing charging effect (porous UHMWPE reactor grains, 10 keV)

4.5 X-Ray Microanalysis

4.5.1 Physical Fundamentals of the Generation of X-Rays

The imaging of surface structures (topography) or compositional variations on the surface is sometimes insufficient to fully characterise the specimen. If, for instance, inorganic particles are embedded in a polymer matrix, X-ray microanalysis [16–19] can be useful for accurately determining the nature of these particles. So-called "characteristic" X-rays are generated by inelastic interactions of incident primary electrons with the orbital electrons of specimen atoms, as already mentioned in Sect. 4.2.5.

Orbitals, which represent the spatial probability distributions of electrons, are defined by quantum numbers. According to Pauli's exclusion principle, only one electron can have a given set of quantum numbers. The principal quantum number is the main influence on the binding energy and the distance between the nucleus and the orbital electron. The inner shells, designated K, L, M, etc., correspond to principal quantum numbers of 1, 2, 3 . . . , respectively. The other quantum numbers have a relatively small effect on the energy and cause the shells (other than the K shell) to be split into subshells.

If a localised electron is knocked out of an atom due to its interaction with a PE, the atom enters an excited high-energy state. At some later time, the empty electron orbital will be filled and the atom will relax, releasing the excess energy as a secondary effect either through the emission of a photon or alternatively through the emission of an Auger electron (see Fig. 4.7). If the vacant electron state is an outer state, then the excess energy will be small and the emitted photon will be in the visible wavelength range (cathodoluminescence). If, however, the vacant orbital is an inner one, the amount of energy released is greater, giving rise to the emission of an X-ray photon. The atomic states that are relevant to characteristic X-ray production can be represented as horizontal lines on an energy diagram, as shown in Fig. 4.16. The energy of the emitted X-ray photon is equal to the difference between the two excited energy states (the original one of the vacant orbital and the final one of the orbital from which the electron jumps). However, it should be noted that not all transitions are allowed by the rules of quantum theory. The X-rays generated by transitions from any higher energy levels to lower K, L, or M shells are denoted K, L and M radiations, respectively. In the usual system of line nomenclature, the Greek letters α, β and γ refer to groups of lines with similar wavelengths, in order of decreasing intensity, while numerical subscripts distinguish the lines within each group, also in order of decreasing intensity. For instance, a transition from the L3 subshell to the K shell results in a $K_{\alpha 1}$ X-ray photon, while a $K_{\beta 1}$ X-ray photon is emitted by the transition from the M3 level to the K level. Further, it should be noted that the transition probability that leads to X-ray emission decreases with increasing distance between the energy levels taking part in the transition. As a result, for example, the intensity of the K_α X-ray peak is always higher than the corresponding K_β emission peak.

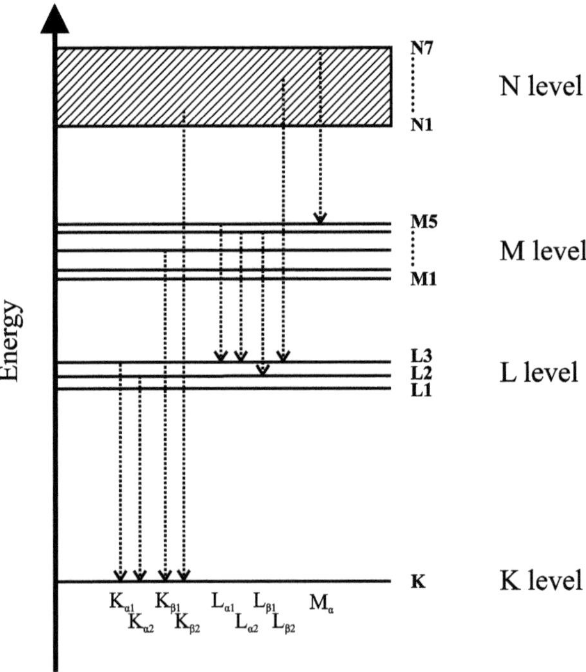

Fig. 4.16. Energy levels and possible electron transitions for the emission of characteristic X-rays (after [15, 18])

Table 4.1. Characteristic X-ray peaks of elements occurring in polymeric materials [20]

Element	K_α peak	$K_{\beta 1}$ peak	$L_{\alpha 1}$ peak	$L_{\beta 1}$ peak	M_α peak
C	0.277 keV				
N	0.392 keV				
O	0.525 keV				
F	0.677 keV				
Si	1.739 keV	1.829 keV			
Ca	3.690 keV	4.012 keV	0.341 keV		
Au			9.712 keV	11.440 keV	2.120 keV
Ti			10.267 keV	12.211 keV	2.268 keV

The energy of the characteristic radiation within a given series of lines varies monotonically with atomic number, i.e. the emitted radiation is element-specific. Therefore, for a qualitative chemical analysis, if the energy of a given K, L or M line is measured, the atomic number of the element that produces that line can be determined. Frequently observed characteristic X-ray emission peaks of elements commonly present in polymers are listed in Table 4.1.

Generally, the ionisation energy needed to excite an atom before the emission of a corresponding X-ray photon will take place exceeds the energy of the released photon. In practice, a rule of thumb is that optimal ionisation conditions are found when the energy of the incident PEs is about 2.5 times higher than that of the X-rays of interest.

X-ray emission in SEM takes place in the whole interaction volume of PE with the sample (see Fig. 4.11). Therefore, the lateral resolution of X-ray microanalysis is comparable to that of the BSE signals (see Sect. 4.4.2).

As well as the characteristic (i.e. element-specific) X-rays generated, nonspecific radiation (called "bremsstrahlung" or X-ray continuum radiation) is emitted over the same energy range. The latter results from Coulomb interactions of PEs along their paths through inhomogeneous electric fields of sample atoms. Bremsstrahlung produces a continuous spectrum with the primary electron energy as the upper limit (this corresponds to the lowest value λ_{min} in terms of wavelength; see Sect. 4.5.6). The contribution from bremsstrahlung must be removed from the measured X-ray spectrum to attain element-specific information. The superpositions of characteristic peaks and continuum radiation on the X-ray spectrum are shown in Fig. 4.17.

4.5.2 X-Ray Microanalysis Techniques

Two different experimental methods for X-ray microanalysis have been developed depending on the measuring technique applied:

- Energy dispersive X-ray microanalysis (EDX)
- Wavelength dispersive X-ray microanalysis (WDX).

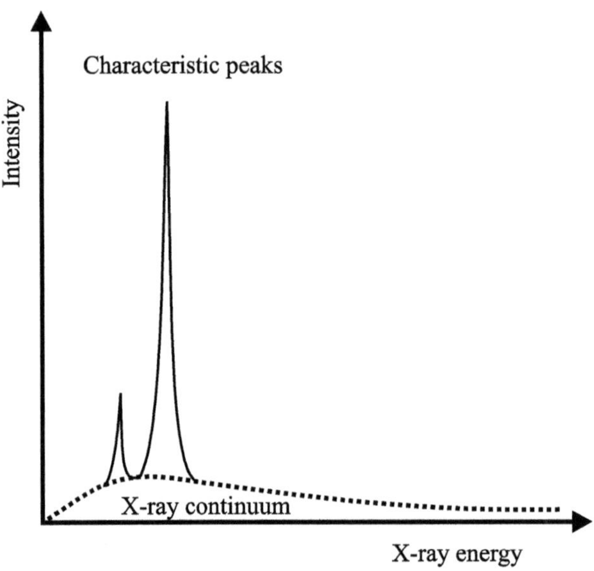

Fig. 4.17. Scheme of a typical X-ray spectrum. *Dotted line*: X-ray continuum, *solid lines*: characteristic peaks

As the names suggest, in each case the emitted X-ray will be analysed either as a function of the energy of the emitted radiation (EDX) or its wavelength (WDX). Both of these techniques are discussed briefly in the following sections.

4.5.3 Detector for EDX Analysis

EDX analysis [18–21] means energy dispersive spectroscopy of the X-rays emitted from a sample during electron irradiation. The detection technique resembles that of BSE solid state detector. X-rays penetrating the semiconductor detector are absorbed and generate electron–hole pairs. The formation of such a pair in a silicon semiconductor requires an energy of approximately 3.8 eV. Thus the number of pairs n_{eh} formed by the total absorption of the energy E_{ph} of one X-ray photon can be expressed by:

$$n_{eh} = \frac{E_{ph}}{3.8\,\mathrm{eV}}. \tag{4.8}$$

In order to determine the number of electron–hole pairs by measuring the corresponding current pulse, it is necessary to separate the pairs using an electric field in order to stop them from immediately recombining. To achieve this, an extended (Si(Li)) region in the initially n-doped silicon detector crystal is formed using drifted lithium. The charge carriers generated in this region have sufficient lifetimes due to the dominance of intrinsic p-i-n conduction. Furthermore, the electric field required to separate electron–hole pairs in this region is produced by applying a voltage between the outer gold contacts of the detector in such a way that the p-n junction formed at the interface between the drifted lithium region and the adjacent n-doped region is reversibly biased (see Fig. 4.18). The "separated" charges can be amplified in a field effect transistor (FET).

To stabilise lithium within the detector crystal and suppress the thermal noise of the amplifier, the detector is usually cooled with liquid nitrogen. A very thin window transparent to X-rays separates the evacuated detection unit (inside a stainless steel probe) from the atmosphere within the SEM specimen chamber, and thus prevents the cooled detector from becoming contaminated. Windows of older systems consisted mainly of beryllium, whereas in modern systems a polymer-based ultrathin window (UTW) supported by a silicon grid is used.

A sufficiently small time window is used for the basic detection cycle, during which all of the charges produced by this photon are integrated and processed to determine the energy of an individual detected X-ray photon. Using a multichannel analyser (MCA), the signal pulse is analysed based on its pulse height (which is proportional to the photon energy) and recorded as a count in the corresponding energy channel where all of these counts are accumulated. When the contents of the MCA channels are read out, an energy spectrum that shows how the accumulated counts of the channels (ordinate of the graph) depend on the corresponding photon energy (abscissa) is obtained, as demonstrated in Figure 4.19 for the spectrum of calcite.

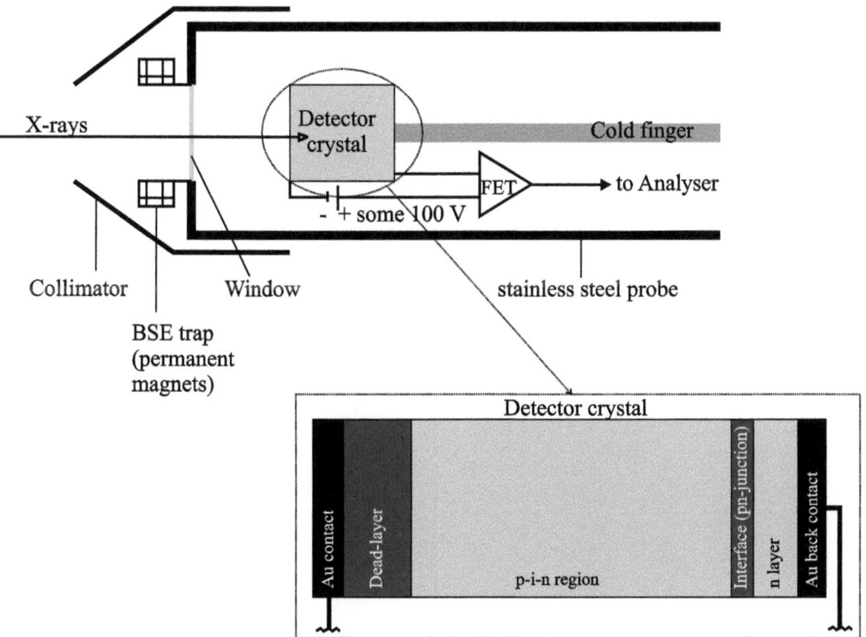

Fig. 4.18. Scheme showing the working principle of an energy dispersive X-ray detector

In order to properly evaluate the resulting X-ray spectrum, different detector specific influences should be taken into account:

1. The window and the gold layer absorb low-energy X-rays. If a beryllium window is used, only a spectrum beginning with the K_α rays for sodium (Na) can be recorded. A polymer-based UTW enables the detection of photon energies down to the K_α radiation of boron (B); however, absorption effects are significant in this low-energy region. Below 1 keV, the transparency of UTW is energy-dependent, which might lead to "virtual" peak shifts and has to be taken into account for quantitative considerations.

2. Between the gold electrode and the p-i-n region is a dead layer in which the electron–hole pairs can immediately recombine. X-ray photons with penetration depths that do not exceed the thickness of the gold contact and the dead layer cannot contribute to the signal.

3. If the X-ray energy of a photon is high enough that it can penetrate into the n-conducting region of the detector, electron–hole pairs that form in this part of detector do not contribute to the signal either. Therefore, the energy determined is too low, resulting in a loss of intensity at the actual energy of the photon and increased intensity at lower (falsely determined) energy.

4. The value of 3.8 eV mentioned as the energy needed to produce an electron–hole pair in Si(Li) is only an average value. Actually, there are a wide range of energies, so different numbers of electron–hole pairs can be generated for the same X-ray energy. This causes the peaks in the MCA to broaden, and this deviation increases with increasing X-ray energy. Peak broadening can lead to the overlapping of neighbouring peaks.

   The measured full width at half maximum (FWHM) of the K_α radiation of manganese (Mn) is usually used as the energy resolution when evaluating the quality of a detector. Good Si(Li) detectors possess energy resolutions of smaller than 130 eV.

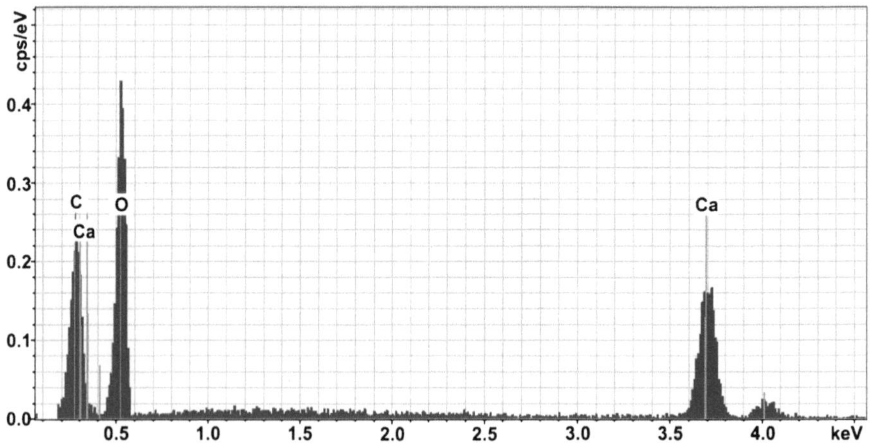

Fig. 4.19. EDX spectrum taken from a small particle (calcite, $CaCO_3$) in Fig. 4.12

5. X-ray absorption in the detector does not exclusively result in the formation of electron–hole pairs; it can also cause fluorescence emission from silicon (escape peak).

Consequently, peak heights are strongly influenced by measuring conditions and will not directly yield quantitative information on local concentrations of elements.

EDX detectors with take-off angles of between 35° and 45° are usually used. To increase the detection efficiency, the acceptance angle of the detector should be as high as possible, i.e. the detector must be placed very close to the sample. Furthermore, we must take into consideration that the X-rays to be detected should not be blocked by preceding surface structures on the sample. Therefore, the EDX analysis of a sample with a rough topography can be difficult.

Compared to integral element analysis methods, the sensitivity of EDX analysis is relatively low; however, it provides the ability to carry out a microanalysis within an interaction volume that is in the micron or even submicron range. A limit of detectability (in terms of concentration) of about 0.1 wt% holds for elements of medium and high atomic numbers, while elements with low atomic numbers need a concentration of more than 1 wt% to be detected.

4.5.4 Quantitative EDX Analysis

Quantitative EDX analysis [18–23] plays only a secondary role in polymer research. The reason for this is that the polymers are generally composed of the light elements carbon and hydrogen, with higher atomic number elements being present, if at all, only in very small amounts (generally below the detection limit; Sect. 4.5.3). Also, for polymers containing additives, the actual excitation volume of the additives is usually too small. The material in and around the excitation volume must be homogeneous. This requirement is usually not fulfilled by heterogeneous polymers containing relatively small particles. Additionally, a very flat surface is essential for a quantitative analysis. Therefore, only a short description of quantitative microanalysis will be presented here. As well as detector-specific properties, one should take also into account the X-ray emission in the excitation volume and the path of the photons in the material.

Initially, a qualitative analysis must be carried out. As absorption and fluorescence take place in the detector itself, the silicon escape peak should be subtracted from the spectrum. The spectrum obtained after this step is subjected to peak separation (separation of overlapping peaks) and then the peaks are fitted with Gaussian functions. Then the X-ray continuum for all of the elements present is calculated and will be subtracted. Finally, the intensity under each Gaussian peak is integrated. This results in intensity values for each peak. Two methods can then be applied: K ratios or ZAF correction.

K Ratios

If the intensity values of the peaks of standard elements recorded under equal conditions are known, then quantitative analysis can be conducted relatively easily. To do this, one constructs the so-called K ratio, which is the ratio of the measured peak intensity to the intensity of the standard sample peak. The result is, however, only a first approximation, because the real ratio corresponding to the absorption and fluorescence is not taken into account in this procedure.

ZAF Correction

Correction factors corresponding to the atomic number (Z), absorption (A) and fluorescence (F) are introduced in order to achieve an improved quantitative analysis. The Z term contains both the backscatter coefficients and the stopping power needed for the generation of an X-ray continuum.

X-ray photons emitted in the interaction volume can be absorbed (A term) along their paths through the specimen. On the other hand, absorbed radiation can also initiate the emission of fluorescence radiation with lower energy (F term).

Intensity values obtained and processed in this way can be used to conduct a quantitative analysis that considers all parameters. Comparisons with real standard samples (analysis with standards) or with virtual standards (standardless analysis) are made, whereas proportions of known elements that are not measurable can be calculated by a subtractive method.

If the sample thickness is smaller than the excitation volume for the X-ray radiation (as is the case for ultrathin sections), this fact should also be taken into account.

Manufacturers also deliver microanalysis systems with corresponding software that also take into account instrumental parameters for quantitative evaluations. During individual steps correction and calculation steps, it is also possible to carry out manipulations in an interactive way that minimises errors. In particular, when the specimen is not prepared as required for systematic studies (i.e. flat surface, homogeneous material, etc.), one should check the results for plausibility (e.g. the stoichiometry, see the discrepancy in quantitative results in Fig. 4.20 for SiO_2).

4.5.5 X-Ray Mapping

The SEM provides, beside the integral analysis discussed above (Sects. 4.5.1 and 4.5.2), the ability to determine the elemental distribution (elemental mapping) over a sample surface. To do this, in an EDX system one or more peaks from an interesting element are selected from the spectrum by so-called regions of interest (ROI). While scanning the electron beam, it is possible to record the X-ray counts of the ROI correlating to points on the specimen surface. By accumulating X-ray counts per pixel, it is possible to create an elemental distribution image (X-ray mapping, see Fig. 4.21).

4.5.6 Wavelength Dispersive X-Ray Microanalysis (WDX)

WDX microanalysis [16, 22] was the original method used for elemental analysis by electron-induced X-ray emission. However, due to the time-consuming nature of this method, this technique is currently employed only when high spectral resolution is required (5–20 eV), or element concentrations of less than 0.1wt% need to be measured.

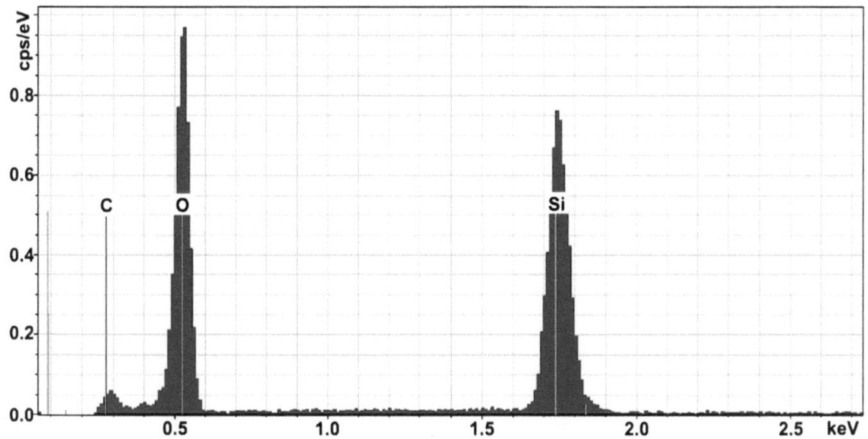

Quantitative EDX-analysis via ZAF correction

element	series	[weight-%]	[norm. weight-%]	[norm. at.-%]	error in %
silicon	K-series	37,390073	36,06903212	24,32263978	1,586829093
oxygen	K-series	66,27246187	63,93096788	75,67736022	8,579591982
	sum:	103,6625349	100	100	

Fig. 4.20. EDX spectrum taken from a large particle (quartz, SiO_2) in Fig. 4.12 and its quantitative (ZAF) results; C peak results from coating used to prevent charging

The relationship between the energy E and the wavelength λ of an X-ray photon is given by Planck's relation:

$$E = h\nu = h\frac{c}{\lambda} \tag{4.9}$$

where ν denotes the frequency of radiation, $h = 6.6260755 \times 10^{-34}$ Js is Planck's constant, and c is the velocity of light in vacuum.

For WDX analysis, the diffraction of X-rays by a crystal lattice is used to discriminate X-rays according to photon wavelength. Diffraction is an interference phenomenon, where intensity maxima only occur when the path differences between the waves diffracted in the same direction correspond to integer values n of the wavelength λ. Consequently, diffraction takes the form of a reflection of the incident beam at lattice planes, as illustrated in Fig. 4.22. The relation between the glancing angle θ, the inter-planar distance d of the lattice planes and the wavelength λ is given by Bragg's law:

$$n\lambda = 2d \sin \theta . \tag{4.10}$$

Thus, the incident X-ray beam can be monochromatized by using an appropriately selected inter-planar distance d, and the wavelength of the monochromatized beam can be changed by varying the glancing angle.

The WDX monochromator system comprises a crystal with a suitable value of d, an X-ray detector (a gas proportional counter) and a slit in front of it. The positions

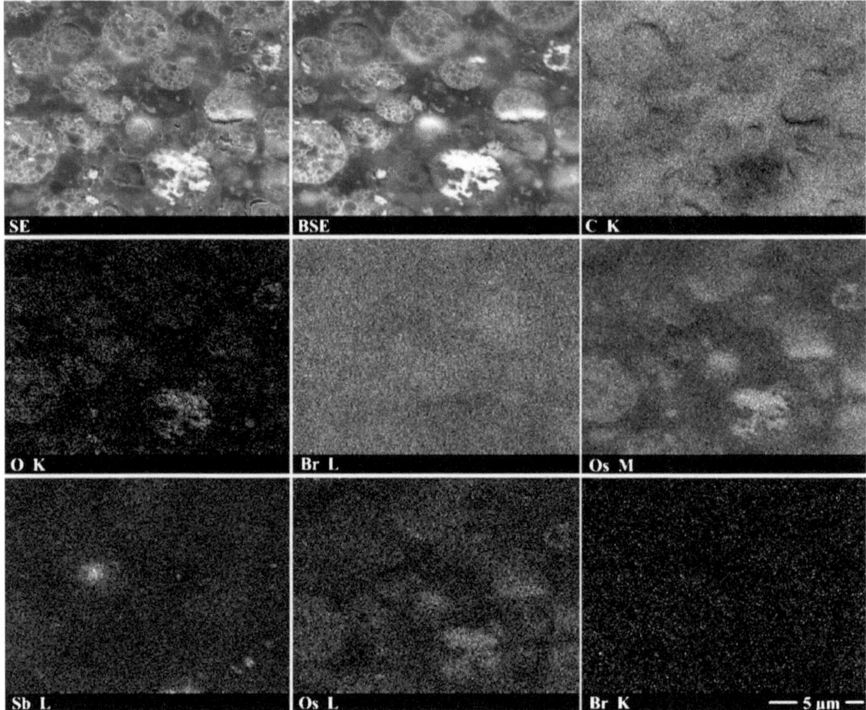

Fig. 4.21. SE image, BSE image and X-ray mapping of C, O, Br, Os, Sb (cut surface for TEM preparation of flame-retarded HIPS with OsO_4 staining)

of the sample, crystal and detector are arranged in such a way that the glancing angle of the radiation incident on the crystal and that of the radiation diffracted towards the detector slit are identical (see Fig. 4.23). This condition is fulfilled only when the sample, the crystal and the detector are positioned on a circle called a Rowland circle. When the arrangement shown in Fig. 4.23 is used, the detector measures the X-ray intensity of the wavelength λ filtered according to Eq. 4.10 (usually for $n = 1$) from the radiation striking the monochromator crystal, where the wavelength selected is

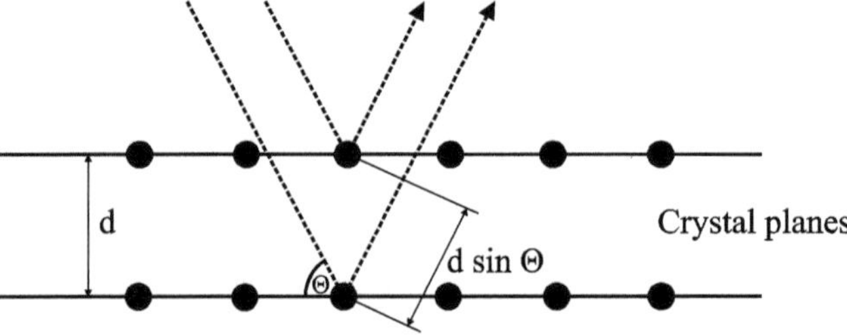

Fig. 4.22. Interference scheme for X-rays diffracted on crystal planes (Bragg's law)

determined by the glancing angle used. Therefore, to record a wavelength-dispersive spectrum, the sample position is kept fixed and the glancing angle is varied by shifting both the monochromator crystal and the detector along the Rowland circle to the position needed.

In order to make use of not only the central part but an extended region of the crystal in order to monochromatize the incident radiation, the crystal is concavely bent to a radius of curvature which is twice that of the Rowland circle and then ground in such a way that its surface has a radius of curvature equal to that of the Rowland circle. Such a spectrometer is known as Johnsson fully focussing spectrometer.

When only one individual crystal (i.e. only one value of d) is used, the measurable wavelength range is limited for technical reasons. Therefore, WDX systems are often equipped with several monochromator crystals with different d values, as shown in Table 4.2.

The measuring conditions for WDX analysis require the exact positioning of the region of interest on the sample surface at the point of intersection of the electron optical axis with the Rowland circle. The adjustment is usually carried out with the help of an optical microscope attached to the SEM.

Compared to EDX microanalysis, WDX analysis possesses a relatively poor signal-to-noise ratio and requires a higher electron beam current. This method is therefore not suited to use as a routine analytical technique for polymers due to the risk of beam damage. When it is used for quantitative analysis, the effects described in Sect. 4.5.4 should be taken into account.

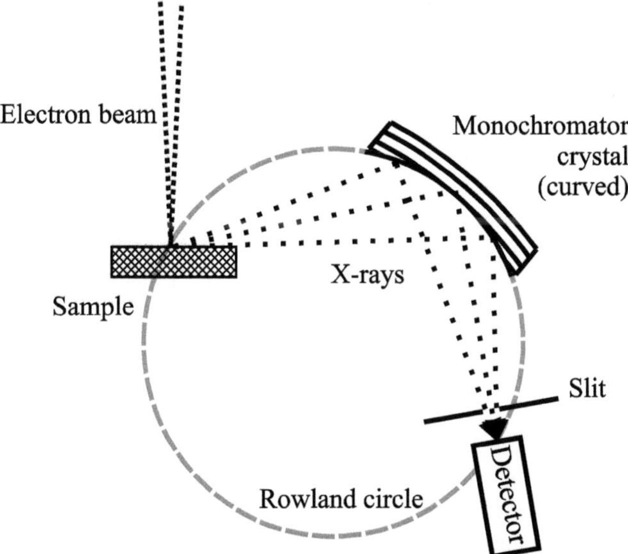

Fig. 4.23. Scheme of a Johnsson fully focussing wavelength dispersive spectrometer: specimen, crystal and detector are arranged on the Rowland circle

Table 4.2. Different types of monochromator crystals commonly used in WDX spectrometers

Crystal	$2d$ spacing of specific planes used (nm)	Region of detectable wavelength (nm)	Minimal detectable atomic number
LiF	0.40	0.08–0.4	19
α-Quartz	0.67	0.12–0.6	15
Pentanerythritol (PET)	0.87	0.15–0.8	13
Rubidium acid phthalate (RAP)	2.61	0.5–2.4	8
Potassium acid phthalate (KAP)	2.66	0.5–2.4	8
Pb stearate	10.4	2.0–9.0	5

4.6 Environmental Scanning Electron Microscope (ESEM™)

4.6.1 Low-Vacuum SEM and ESEM™

The environmental scanning electron microscope (abbreviated to ESEM) is a modified SEM that offers new applications and advantages over the conventional SEM, as described in more detail in Sects. 4.6.2 and 4.6.3.

Used in this context, the term "environmental" refers to the possibility of imaging wet samples such as biological samples using the SEM.

The ESEM was developed in the 1980s. At that time, many SEM manufacturers introduced scanning microscopes that had low vacuums of about 10^{-1} mbar in their specimen chambers, which were achieved using a special pumping system. In these low-vacuum SEMs, charging effects can be strongly reduced and hence isolated samples can be imaged without coatings.

Unfortunately, an Everhardt-Thornley detector (see Sect. 4.3.3) cannot be used in a low-vacuum SEM, as electrical breakdown at the high-voltage part of the detector cannot be avoided. Therefore, most producers of low-vacuum SEMs only have used BSEs to image the sample surface, which of course leads to limitations on the lateral resolution, as described in Sect. 4.4.2. Danilatos and coworkers developed and patented [24] a relatively simple SE detection system for a vacuum exceeding 1 mbar, which found application in a microscope called an ESEM [25]. The first ESEMs were constructed and commercialised by the company ElectroScan. Today, "ESEM" is a trademark of FEI.

4.6.2 Avoiding Charging

As described in Sect. 4.4.3, incident PEs usually cause negative charging of the surfaces of electrically nonconducting samples. Under low-vacuum conditions, a lot of residual gas molecules are ionised by the primary electron beam. The resulting positive ions are attracted by the negatively charged sample surface, and so the surface charge is compensated for by the impacting ions. Therefore, the surface charging of insulating samples is avoided or strongly reduced in a low-vacuum SEM. The presence of an additional electric field, as used in a "gaseous secondary electron detector" (see Sect. 4.6.4) can further enhance this effect, because more ions are formed by cascade ionisation, which are then available for neutralisation of the surface charge.

4.6.3 The Wet Mode

The ability of an ESEM to investigate wet samples is of great practical importance. Specimens do need not to be dried before the investigation, and they are not dried during the investigation either, as the pressure in the specimen chamber of the ESEM is comparable with the partial vapour pressure of water at room temperature. By using a differential pumping system [26], which involves introducing some diaphragms to separate regions with different levels of vacuum along the electron-optical axis from one another, an ESEM can work with a relatively high pressure in the specimen chamber while a high vacuum or even an ultrahigh vacuum is utilised in the microscope column and the electron gun. The maximum pressure achieved in the specimen chamber depends on the diameter selected for the last diaphragm of the pressure system. A smaller one must be selected to maintain a high vacuum in the column at increased pressure in the specimen chamber. A small diaphragm, however, affects the field of vision, especially at lower magnifications and lower working distances.

In the so-called "wet mode", the previously evacuated sample chamber is vented to a pressure of some mbar of water vapour, so that, to a good approximation, only water molecules are present in the chamber. Depending on the microscope type, the maximum achievable pressure can range from less than 1 mbar to 15 mbar.

Figure 4.24 shows the phase diagram of water close to its triple point. One can see that the transition from the liquid state to the gas phase at room temperature occurs at a pressure of 28 mbar. Therefore, if the water vapour pressure inside the sample chamber is 28 mbar and if the sample contains water, there will be a thermodynamic equilibrium. Under these conditions, the sample does not lose water by boiling and

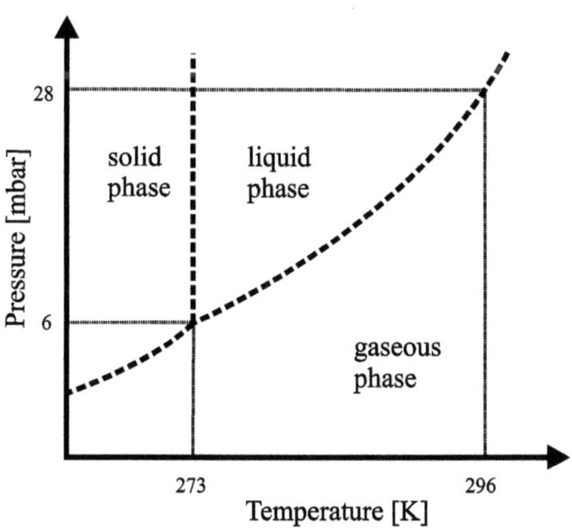

Fig. 4.24. Scheme of the phase diagram of water in the range usable in ESEM

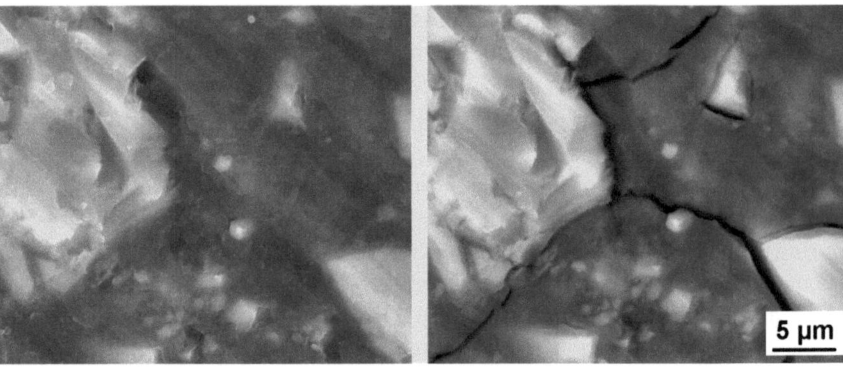

Fig. 4.25. ESEM (GSED) micrographs of a hydrated dental composite. *Left*: ambient conditions 7 mbar and 277 K. *Right*: after drying at lower pressure (3 mbar, 277 K)

a film of water is not deposited onto the sample surface. Upon increasing the pressure, the specimen will be covered with a film of water, while decreasing the pressure causes the sample to dry out. When working with an ESEM equipped with a FEG, one cannot create a pressure of 28 mbar, but one can shift the working conditions along the liquid/gas phase transition towards the lower pressure region by cooling the sample, e.g. by means of Peltier cooling.

The wet mode is the standard working mode of the ESEM. Water vapour can be produced without difficulty using distilled water. The water in the atmosphere also forms enough ions to eliminate surface charging. Therefore, this mode is ideal for water-containing samples as well as for imaging nonconducting polymeric materials without coatings. It is also a highly advantageous approach to use for in situ investigations of polymers. Besides micromechanical tests, special investigations involving variations in temperature, pressure or the use of a particular gas atmosphere can be carried out using this technique (see Chap. 6, Fig. 6.1). The reactions of the sample with gas molecules and the adhesion properties of its surface can also be comfortably studied using ESEM.

Figure 4.25 shows the effect of the surrounding atmosphere on SE imaging in the case of a hydrated dental composite sample. When imaging in wet mode (with the surrounding atmosphere at close to the partial pressure of saturated water vapour), sample drying can be reduced and the formation of cracks is avoided (left). When imaging at lower pressures, vacuum water is removed from the sample, so that cracks appear at the interface to the glass ceramic particles (right).

If it is impossible to use the water vapour pressure for practical reasons, the ESEM can also be operated under other gases. If, for example, nitrogen gas is used, charging effects can also be suppressed and SE imaging can be achieved using the GSED (see below).

4.6.4 The Gaseous Secondary Electron Detector (GSED)

The patented GSED [24] uses cascade ionisation of the residual gas molecules to amplify the secondary electron signal. Figure 4.26 shows the principle of this detector schematically. An isolated, positively biased electrode with a central hole, which the primary electron beam can pass through, is placed directly under the last diaphragm of the pumping system. If the operator intends to use a BSE detector at the same time, SE detection using a collection electrode that is not centrally arranged is also possible.

Secondary electrons that leave the sample surface are accelerated towards the positively biased electrode and ionise residual gas molecules along their paths to the collector. The free electrons thus generated are also forced to travel towards the collector electrode and can trigger new ionisation events. These cascade processes cause the number of originally released SEs to be multiplied enormously, resulting in a measurable current at the collector electrode, which can also be amplified electronically. The final current signal is proportional to the number of SEs emitted per unit time, and it can therefore be used to image the sample. However, it should be noted that this SE signal is also influenced by the number of residual gas molecules available for ionisation (i.e. the pressure in the specimen chamber), the ionisation probability (i.e. the type of gas molecules present), the distance and the voltage between the sample and the collector electrode.

A large working distance and a high pressure are favourable for high signal amplification. However, because of the widening of the primary electron beam due to collisions with residual gas molecules (skirt effect), highly resolved imaging (particularly at low PE energies) requires the use of a small working distance and/or a low pressure.

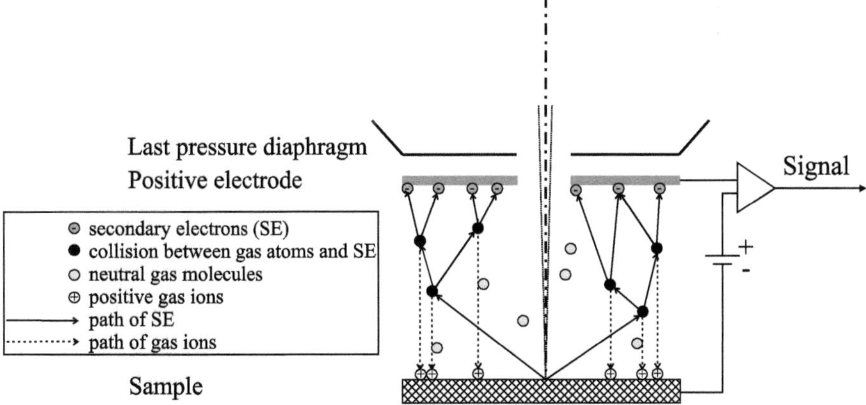

Fig. 4.26. Scheme showing the principle of the gaseous secondary electron detector (GSED)

References

1. Knoll M (1935) Z Techn Physik 16(11):467
2. Ardenne Mv (1938): Z Phys 109(9-10):553
3. Ardenne Mv (1938) Z Techn Phys 19:407
4. Ardenne Mv (1940): Elektronen-Übermikroskopie (in German). Springer, Berlin
5. Zworykin VK, Hiller J, Snyder RL (1942) ASTM Bull 117:15
6. McMullan D (1952) PhD thesis, University of Cambridge, Cambridge
7. Smith KCA (1956) PhD thesis, University of Cambridge, Cambridge
8. Everhardt TE, Thornley RFM (1960) J Sci Inst 37:246
9. Castaing R (1951) Application des sondes électroniques à une méthode d'analyse ponctuelle chimique et cristallographique. PhD thesis, University of Paris, Paris
10. Coslett VE, Duncumb P (1956). In: Sjöstrand FS, Rhodin J (eds) Proc Stockholm 11th Conf Electron Microsc. Almqvist and Wiksell, Stockholm, p 12
11. Danilatos GD (1983) Micron Microsc Acta 14(4):307
12. Danilatos GD (1985) Scanning 7:26
13. Danilatos GD (1986) In: Bailey GD (ed) Proc 44th Annual Meeting EMSA. San Francisco Press, San Francisco, CA, p 630
14. Danilatos GD (1986) In: Bailey GD (ed.) Proc 44th Annual Meeting EMSA. San Francisco Press, San Francisco, CA, p 632
15. Reimer L (1998) Scanning electron microscopy: Physics of image formation and microanalysis, 2nd edn. Springer, Berlin
16. Lee RE (1993) Scanning electron microscopy and X-ray microanalysis. PTR Prentice Hall, Englewood Cliffs, NJ
17. Goldstein JI, Newbury DE, Echlin P, Joy DC, Fiori C, Lifshin E (1981) Scanning electron microscopy and X-ray microanalysis. Plenum, New York
18. Reed SJB (1993) Electron microprobe analysis, 2nd edn. Cambridge Univ. Press, Cambridge
19. Chandler JA (1987) X-Ray microanalysis in the electron microscope. In: Glauert AM (ed) Practical methods in electron microscopy, vol 5, part II. Elsevier, Amsterdam
20. Friel JJ (1995) X-Ray and image analysis in electron microscopy. Princeton GammaTech Inc., Princeton, NJ
21. Russ JC (1984) Fundamentals of energy-dispersive X-ray analysis. Butterworths, London
22. Heinrich KFJ, Newbury DE (1991) Electron probe quantitation. Plenum, New York
23. Scott VD, Love G, Reed SJB (1995) Quantitative electron probe microanalysis, 2nd edn. Ellis Horwood, New York
24. Mancuso JF, Maxwell WB, Danilatos GD (1988) US Patent 4 785 182
25. Knowles WR, Schultz WG, Armstrong AE (1994) US Patent 5 362 964
26. Danilatos GD, Lewis GC (1989) US Patent 4 823 0062

5 Atomic Force Microscopy

Atomic force microscopy, also referred to as scanning force microscopy, is obviously not a technique based on electron microscopy. However, it has been common practice in the last few years to supplement electron microscopy with atomic force microscopy, as the latter enables the straightforward surface characterisation of polymers and provides additional insight into the structure and properties of homopolymers, blends and composites. Therefore, a brief survey of the fundamentals and relevant applications of this technique to polymer research is presented in this chapter. Because of its special importance for polymer investigations, tapping mode atomic force microscopy will be described in detail, while other modes of operation will be introduced more briefly.

5.1 Introduction

In preceding chapters, different methods and techniques of electron microscopy that are employed for the study of polymeric materials were described. While atomic force microscopy (AFM), also referred to as scanning force microscopy (SFM), is obviously not a technique that is based on electron microscopy, it has been common practice in the last few years to supplement electron microscopy with AFM, as the latter enables the straightforward surface characterisation of polymers and provides additional insights into the structure and properties of homopolymers, blends and composites. Therefore, a brief survey of the fundamentals and relevant applications of this technique to polymer research is presented in this chapter. For a detailed account of the fundamentals and applications of AFM techniques to polymers, the readers should consult the more concise recent reviews of the topic [1–5].

AFM belongs to the family of scanning probe microscopes (SPM), in which solid surfaces are scanned by extremely sharp mechanical probes. In an SPM technique, highly localised tip–sample interactions are measured as a function of position. Different types of SPMs are based on different kinds of interactions; the major types of SPM include: the scanning tunnelling microscope (STM), which measures electronic tunnelling current; the atomic force microscope (AFM), which measures interaction forces; and the scanning near-field optical microscope (SNOM), which measures local optical properties by exploiting the evanescent field, i.e. near-field effects. These SPM techniques allow the characterisation of a wide range of properties (structural, optical, mechanical, magnetic and electrical) of solid surfaces in different environments (vacuum, liquid or ambient conditions) and at different temperatures. Due to their ability to offer nanoscale resolution and versatile applicability, SPM techniques have proven to be indispensable part of modern nanoscience and technology.

A major revolution in the advancement of surface analytical tools occurred when the first STM was introduced in 1982 by Binnig and coworkers [6]. STM enabled the first ever three-dimensional imaging of solid surfaces with atomic resolution. Binnig and Rohrer were awarded the Nobel Prize in Physics in 1986 in recognition of their outstanding contributions.

STM is based on the electron tunnelling effect, which occurs when the distance between two conductors is close enough that the probability of charge transition via tunnelling through the potential barrier results in a measurable current. Generally, this is applicable for distances of smaller than 10 nm.

The tunnelling current measured between the probe and the sample increases exponentially as the distance decreases. Therefore, it can be used as an ideal means for controlling the distance between the probe and the sample surface over the nanometre and subnanometre ranges. Obviously, STM imaging is limited to surfaces that are electrically conductive to some extent. Therefore, there was a need to develop an SPM that could be employed for both conducting and insulating surfaces. Indeed, the invention of the AFM by Binnig [7], and its introduction by Binnig et al. [8] in 1986 enabled surface images of conductors and insulators to be obtained with atomic resolution by utilising very small interaction forces between the apex of the tip and the sample surface to control the distance between them.

The most basic function of AFM is to provide high-resolution imaging of the surface relief of a specimen between lateral scales of a few nanometres to about hundred micrometres, as demonstrated by the examples in Fig. 5.1. Figure 5.1a presents the contact-mode AFM height image of isotactic polypropylene (iPP). The sample was prepared by annealing a polymer film cast from its xylene solution. One can see the typical surface texture of the polymer, where different spherulites are clearly separated at their boundaries. Figure 5.1b presents a tapping-mode height image of an iPP surface area with a boundary between two spherulites, thus revealing the detailed morphology of α-type iPP with its cross-hatched texture (at the top of the image) and β-type iPP with its parallel arrangement of lamellae (at the bottom). In Fig. 5.1c, a sample area of only 600 nm × 600 nm is imaged to show the morphology of the polystyrene-*block*-polybutadiene-*block*-polystyrene block copolymer (SBS triblock copolymer) with symmetric polystyrene (PS) end blocks and the total PS volume fraction of 0.75%. A Fourier filter was applied to the tapping mode height image to reduce the noise, and so the image, which is of a high quality, shows the periodic pattern of the dark appearing, cylindrical butadiene domains in the PS matrix. That the AFM can be employed to image the structures of crystal surfaces down to atomic resolution is illustrated by the contact-mode height image of freshly cleaved mica (Fig. 5.1d). The periodic arrangement of the struc-

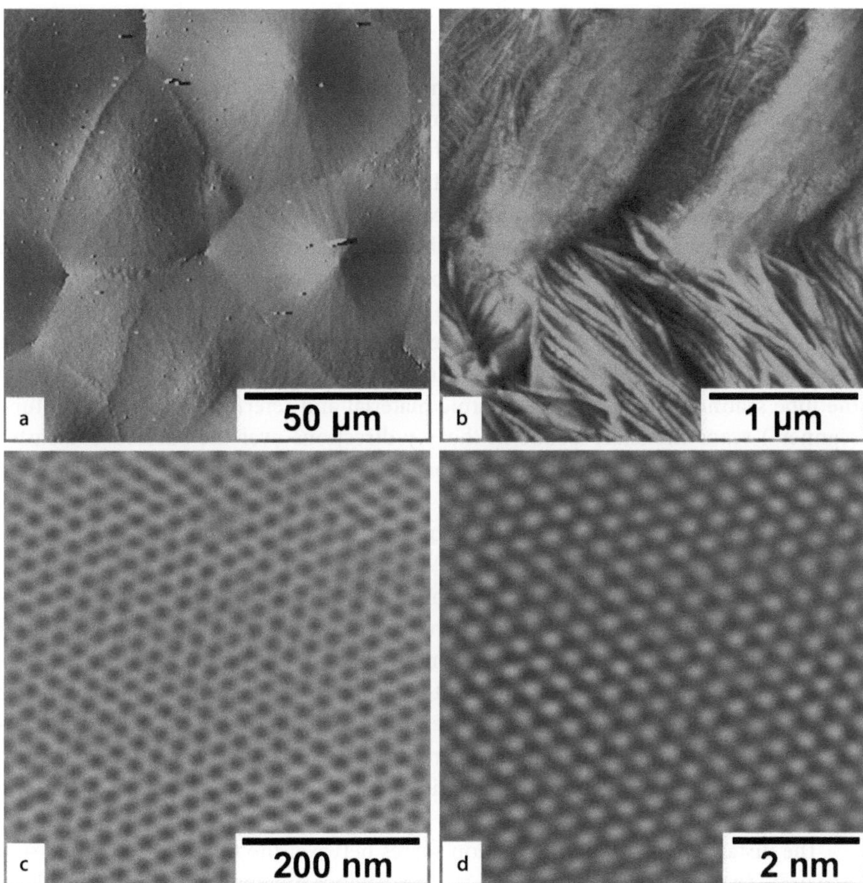

Fig. 5.1a–d. AFM images of different samples illustrating the ability of the technique to image structural details at different length scales: **a** contact-mode height image showing the distribution of spherulites in an annealed, solution cast film of isotactic polypropylene (iPP); **b** tapping-mode height image of an iPP sample showing the interface region between a α-type spherulite (*upper part of image*) and a β-type spherulite (*lower part of image*); **c** tapping-mode height image of a symmetric SBS triblock copolymer after image processing by a Fourier filter, showing the periodic pattern of dark-appearing cylindrical butadiene domains in the polystyrene matrix; **d** contact-mode height image of a freshly cleaved mica surface after image processing by a Fourier filter, showing the periodic pattern of the lattice

tures (with a period of about 0.5 nm) corresponds to the atomic lattice of the mica surface.

By applying special investigation modes, AFM can additionally be used to measure normal and lateral forces, adhesion, friction, elastic/plastic mechanical properties (such as indentation hardness and modulus of elasticity), electrical and magnetic properties in relation to the local position on the sample surface.

5.2 Methodical and Instrumental Fundamentals

Figure 5.2a shows the main part of a commercial AFM (a "Dimension 3000" from Digital Instruments Inc., a subsidiary of Veeco Instruments Inc., Santa Barbara, CA, USA) that has been widely used for polymer investigations in an ambient environment, and the corresponding sketch (b) illustrates the principle of its operation. The instrumentation shown is typical of AFMs which are designed to be used for the investigation of large samples. The sample, here mounted on a motorised xy-stage, is not connected to the parts of the equipment necessary for the operation of the AFM, and the force sensor (a cantilever with a sharp tip at its free end) mounted on the cantilever holder is fixed at the piezo-driven xyz-scanner. While the sample is held in position (which can be adjusted via the motorised xy-stage), the tube scanner (made of piezoceramic material and referred to as the "PZT tube scanner") scans the sharp probe in the x,y-directions over the surface in a raster pattern and controls the local distance between the tip and the sample surface via the very precise scanner motion in the z-direction. The vertical scanner motion (z-direction) is driven by a feedback loop that is linked to the sensed interaction force.

Unlike large-sample AFMs, small-sample AFMs (such as the widely used "Multimode" from Digital Instruments Inc.) that are primarily meant for high-resolution atomic- or molecular-scale imaging of sample surfaces hold the position of the cantilever holder fixed, and the implemented probe and PZT tube xyz-scanner are adjustably positioned below the cantilever. The small and lightweight sample (generally no larger than 10 mm × 10 mm) is mounted directly onto the scanner. During scanning, the cantilever base remains stationary and the sample is scanned under the sensing tip.

The distance-dependent interaction forces between the tip and the sample surface provide the physical basis for AFM. The distance dependence is, however, more complicated than the monotonic function with a strong rate of decay of the tunnelling current utilised in STM. Due to the superposition of long-range attractive forces and short-range forces, which change with decreasing distance from being attractive to being repulsive, the total interaction force is nonmonotonic and thus does not share the helpful characteristics of the tunnelling current. Consequently, when the tip is brought in close to the sample surface, at first it senses an increasing attractive force, which reaches a maximum value and then decreases. This attraction finally changes to a repulsion that strongly increases with decreasing tip–sample distance. Interatomic forces with one or several atoms in contact are 20–40 or 50–100 pN, respectively [9]. Thus, atomic resolution with an AFM working in contact mode is only possible with a sharp tip positioned on a flexible cantilever at a net repulsive force of 100 pN or lower [10].

The key component in an AFM is the sensor used to measure the force on the tip due to its interaction with the specimen. AFM probes consist of a force-sensing cantilever with a very sharp tip integrated at the lower side of its free end. The apex of the tip has to be as sharp as possible. The fixed end of the cantilever meets a (usually) prismatic block (the cantilever base), which is large enough to be

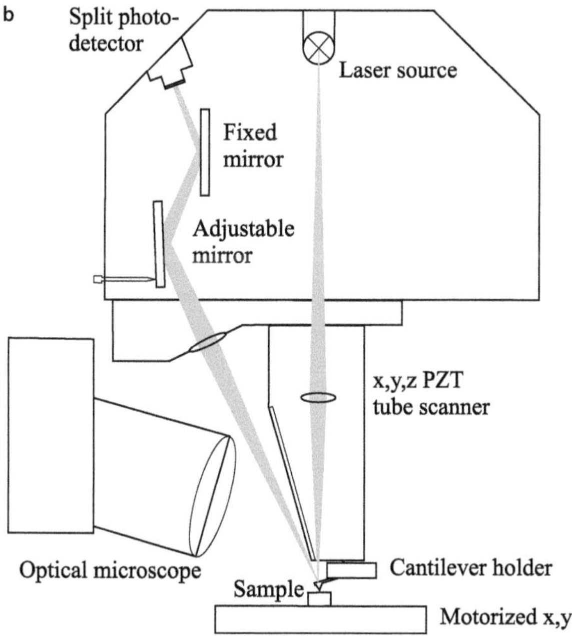

Fig. 5.2a,b. Main part of the "Dimension 3000", a large-sample AFM (**a**), and principle of its operation (**b**)

handled with tweezers. A cantilever with a correspondingly low spring constant is needed to sense small forces (0.1 nN or lower), but at the same time a high resonant frequency (about 10–100 kHz) is required in order to minimise the sensitivity of the AFM to vibrational noise from its surroundings, which is in the range of few Hz to 100 Hz. Commercially available cantilevers are batch-fabricated from silicon or silicon nitride by applying photolithographic and etching techniques. Rectangular single-crystal silicon cantilevers with a pyramidal tip are most commonly used. Tips with a radius of curvature of 5–50 nm are commonly available. Variations in the cantilever length (100–500 µm) and thickness (1–4 µm) result in cantilever spring constants of 0.01–50 $\mathrm{Nm^{-1}}$. Thus, a cantilever with an appropriate force constant for the special requirements of the specific operation mode performed is utilised.

The force on the tip due to its interaction with the sample is sensed by detecting the deflection of the lever. Taking into account the spring constant of the lever, it is obvious that the cantilever deflection will be very small (displacements are smaller than 0.1 nm). Therefore, in the pioneering work of Binnig et al. [8], the small cantilever deflection was sensed via the tunnelling current between the cantilever and a tip placed in the close vicinity of the rear side of the cantilever. A few other detection systems, including capacitance detection, piezoresistive detection and optical detection by means of optical interferometry, optical polarization, laser diode feedback and laser beam deflection, have also been used since then; these are reviewed in e.g. [9]. Here, only the optical beam deflection system will be described. This has a large working distance, is reliable and insensitive to distance changes, and is capable of measuring both normal forces (by sensing the bending of the cantilever) and lateral forces (by sensing the angular changes caused by cantilever torsion). Therefore, this is the detection method most commonly used in commercial SPMs. The sketch shown in Fig. 5.2b illustrates the principle of its operation. The laser beam source is adjusted in such a way that the vertical path of the focussed laser beam strikes the rear side of the cantilever near the free end of the latter, where the beam is reflected. As the cantilever is tilted downward at about 10° with respect to the horizontal plane, the reflected beam is separated from the primary beam and it can be deflected by means of an adjustable mirror system so that it meets a split-diode photodetector with four quadrants (also called a position-sensitive detector) in an appropriate manner. Perfect adjustment, which is carried out in the detached position of the cantilever, results in vanishing values of the difference signals (A + B) – (C + D) and (A + C) – (B + D), where A, B, C and D are the photointensities as measured by the individual quadrants of the split photodetector. Prior to scanning the surface, the cantilever is changed from the detached position to the measuring position in close vicinity to the sample surface. In this position, the difference signal from the top and bottom photodiodes, i.e. (A + B) – (C + D), provides the signal which senses the normal interaction force via the corresponding deflection of the laser beam caused by the cantilever bending. If the scan is initiated with a preselected deflection signal (i.e. with a preselected interaction force), topographical features will cause corresponding local changes in the deflection signal. Only small sample areas with an extremely low roughness (e.g. atomically flat cleaved surfaces, as shown in Fig. 5.1c) can be scanned to image the surface relief by using the deflection signal directly ("deflection image"). This mode of operation is commonly referred to as "constant height mode". The use of this mode to scan sample areas of high roughness will inescapably result in the cantilever becoming damaged. Therefore, the so-called "constant force mode" is the one most commonly used for topographical imaging. In this mode of operation, the normal force applied is kept constant during scanning by means of a feedback circuit. The preselected deflection signal is used as the reference value (the so-called set-point value) of the feedback loop. Topographical structures give rise to local changes in the measured deflection signal, which are interpreted as "error signals" by the input of the feedback loop. As a response, the system instantly compensates by changing the tip–sample distance by an appropriate amount. The output voltage of the feedback loop, which is proportional to the error signal, therefore controls the z-motion of the xyz-scanner. This output voltage signal is thus a linear response to the actual height variations on the sample surface. Therefore, when it has been calibrated by investigating a specimen with a known surface profile, it can be used as "height image" signal that shows the surface relief of the scanned area in a quantitative manner.

5.3 Modes of Operation

AFM techniques can be divided up based on whether dynamic (i.e. the AFM makes use of an additional probe oscillation) or static operational modes of the AFM are used.

Common "contact mode" AFM and also the imaging of lateral forces in a "lateral force microscope" (LFM) take advantage of static operation modes, whereas "non-contact" AFM and "tapping mode" techniques are categorised as dynamic operation modes.

Because of its special importance for polymer investigations, tapping-mode atomic force microscopy (TMAFM) will be described in more detail in this section, while other modes of operation will be more briefly introduced.

5.3.1 Contact Mode

In this mode, lateral scanning of the probe over the sample surface (as described in the previous section) is initiated when the vertical bending of the cantilever reaches the set-point value (chosen by the operator) that corresponds to a repulsive tip–sample interaction. To understand the reason for this assessment of the set-point value, it is useful to consider tip–sample forces in more detail using a so-called "force–distance curve". When a sawtooth voltage is applied to the z-control of the scanner, the force–distance curve describes the interaction force at a particular x,y-position based on the cantilever deflection signal as a function of the position of the scanner in the z-direction over a cycle. This cycle includes the tip's approach to the sample

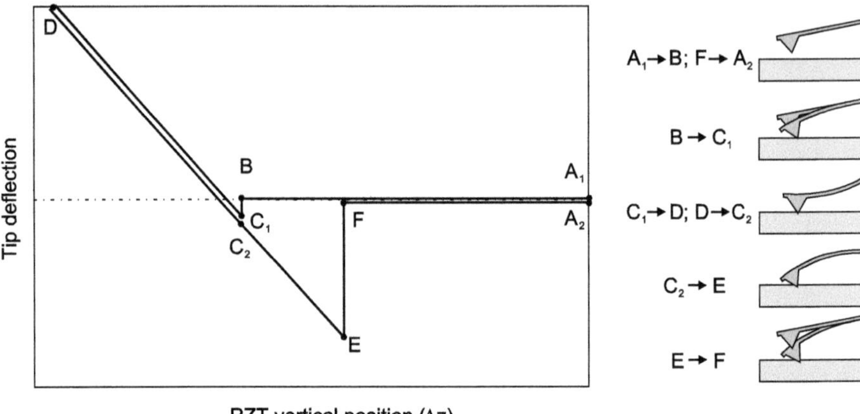

Fig. 5.3. Typical force–distance curve. A_1 and A_2 are the start and end points of the cycle, respectively; $A_1 \rightarrow B$ approach; $B \rightarrow C_1$ pull down, contact at C_1; $C_1 \rightarrow D$, $D \rightarrow C_2$ indentation region; $C_2 \rightarrow E$ adhesion region, $E \rightarrow F$ disengagement; $F \rightarrow A_1$ end of the cycle

surface and its retreat from it. Figure 5.3 shows the various features of the curve. The force measurement starts and ends with the sample being far away from the tip, at the rest position of the cantilever (points A_1 and A_2). As the extension of the z-piezo increases, the tip approaches the sample surface. As long as the tip is far from the sample surface ($A_1 \rightarrow B$), the cantilever shows no deflection, as indicated by the flat portion of the curve. However, as the tip approaches the sample to within a few nanometres (point B), the tip experiences an attractive force which causes the cantilever to bend downwards. If the attractive force exceeds the withdrawing force of the cantilever, the cantilever is pulled towards the sample and contact occurs at point C_1 of the plot. This will generally take place when a very flexible cantilever with a low spring constant (required for high sensitivity) is used. With the tip in contact, further extension of the z-piezo causes the cantilever to bend upwards, reflecting repulsive tip–sample forces. This is represented by the sloped portion of the curve ($C_1 \rightarrow D$, $D \rightarrow C_2$). As the z-piezo retracts, the tip is stuck to the sample by the capillary force, i.e. it goes beyond the zero deflection (flat) line into the adhesive region of the graph ($C_2 \rightarrow E$). The action of the capillary force results in a hysteresis of the force–distance curve. Capillary forces are mainly caused by a layer of liquid contamination that is present on samples in air. Therefore, a pull-off force (the difference in force between E and F) is needed to disengage the tip at point E of the graph. The pull-off force can be used as a measure of adhesion between the tip and the sample. Therefore, in the so-called "force volume mode", force–distance curves are collected at a large number of points in the area of interest, and the data at any particular force level is presented as a map [5]. When force data at strong repulsive force levels are used, surface maps of mechanical properties can be obtained, and maps of adhesive properties of the sample are provided by force data in the disengagement region. Similarly, the topography, local stiffness and adhesion can be simultaneously mapped by the "pulsed force mode (PFM)" [11]. In this mode, a complete force–distance cycle is carried out during scanning at a repetition rate of between 100 Hz and 2 kHz through sinusoidal modulation of the z-piezo. The special PFM electronics used avoid the need to completely digitise the curve and extract only the important features. The slope in the repulsive region is analysed for the local stiffness. Local adhesion is deduced from the value of the critical force where the tip and sample contact are detached, and the topography is obtained from the feedback control.

If scanning is carried out at set-point deflection, which corresponds to the attractive force near the pull-off point, imaging can be achieved with the lowest force for a given probe. However, stable imaging under these conditions is difficult due to possible tip disengagement, and the capillary effect limits the minimal force. The unwanted influence of the capillary force can be avoided by placing the sample and the measuring probe under liquid. Applications of AFM to biological systems often take advantage of this aspect. However, imaging under liquid has its own limitations. Therefore, contact-mode AFM is usually carried out with at set-point force that corresponds to a small repulsive interaction represented by the sloped portion of the force–distance curve near to the points C_1, C_2.

Topographical images with a vertical resolution of less than 0.1 nm (as low as 0.01 nm) and a lateral resolution of about 0.2 nm have been obtained with a SFM operated in the contact mode [9]. Although high-resolution contact-mode AFM is carried out at a normal force that is as small as possible, such a small repulsive force is within the range of chemical bond energies and is sufficient to wipe away atoms. This fact explains why defect-free atomic resolution has mostly been observed with AFMs working in contact mode. Therefore, measurements utilising attractive interactions in the (dynamic) noncontact imaging mode may be desirable for imaging with atomic resolution.

Normal and lateral forces between the tip and sample in standard contact-mode AFM are of tens and hundreds of nanonewtons. On the one hand, it is important to take into account that those forces cause damage to the surfaces of soft samples during such investigations. Therefore, its applicability to polymers may be limited. On the other hand, a tip with a normal load acts as a small indenter on the surface, and so its slide at a defined velocity over the sample surface during scanning is influenced by friction effects. As the lateral tip–sample forces can be used as a measure of friction, variations in them are recorded in a lateral force microscope (LFM) or friction force microscope (FFM) by sensing angular changes of the cantilever via the difference signal $(A + C) - (B + D)$ of the left hand and right hand sets of quadrants of the slit photodetector. In the so-called "friction mode", scanning is preferably carried out in such a way that the scan lines are orthogonal to the long axis of the cantilever beam in order to increase the torsional signal. Provided that the components have widely separated mechanical properties, the measurement of frictional force can be used to perform compositional mapping of heterogeneous polymers such as blends and composites. However, sufficient smoothness of the sample surface is a prerequisite for the meaningful FFM imaging of heterogeneous polymers.

5.3.2 Force Modulation Mode

Like the pulsed forced mode, the "force modulation mode" was introduced to extend contact-mode investigations by simultaneously mapping mechanical properties. For this reason, the probe or sample assembly is scanned with a small vertical (z-direction) oscillation (modulation) that is significantly faster than the scan rate. The force on the sample is modulated about the set-point scanning force such that the average force on the sample is equivalent to that in simple contact mode. Early designs added a modulation signal to the z-piezo of the scanner to induce the vertical oscillation [12]. In second-generation systems for force modulation, the probe is made to oscillate vertically by means of an additional piezo-actuator at its resonant frequency (5–8 kHz). This actuator is positioned at the cantilever base, i.e. at the fixed end of the cantilever. Therefore, when scanning in contact mode, the cantilever and its base also moves up and down with a small modulation amplitude that is induced by the piezo-actuator. The photodetector sensing the cantilever deflection collects force modulation data (fast-changing amplitude signal of deflection) and topographical information (slow-changing averaged signal of deflection) simultaneously. The amplitude response depends on the local stiffness of the sample, as the tip bounces from a stiff region on the sample and the amplitude is large, whereas the deflection amplitude is small at a soft sample region due to possible tip indentation. Therefore, the local heterogeneity in the mechanical properties of multicomponent polymers can be imaged by the amplitude signal of the force modulation technique, thus allowing surface compositional mapping.

Figure 5.4 shows height and amplitude images of "core–shell" rubber particles of modified poly(methyl metacrylate) (PMMA) collected during force modulation

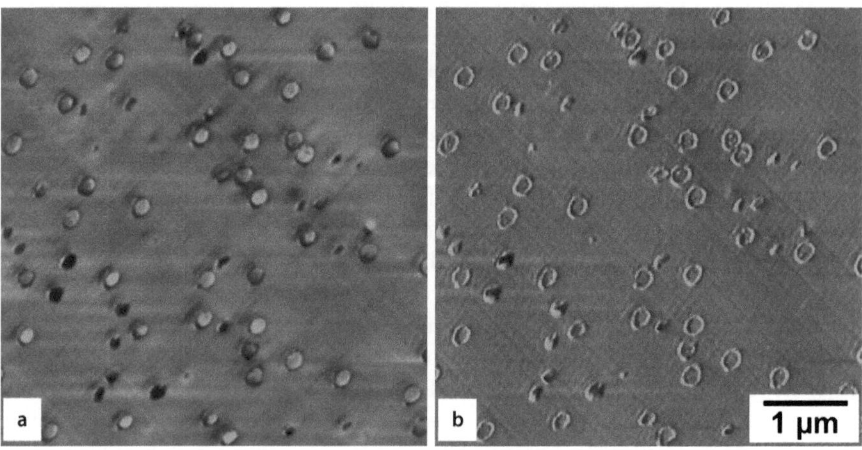

Fig. 5.4. AFM height (**a**) and amplitude (**b**) images for rubber-toughened poly(methyl-methacrylate) recorded in the force modulation mode; note how the soft shells of the dispersed particles appear brighter in the amplitude image

imaging. The particular method of force modulation used for this investigation was slightly different from the one described above. Utilising special scanning modes (so-called "interleaved scanning mode" and "lift mode") controlled by the Nanoscope IIIa controller from Digital Instrument Inc., each scan line was scanned twice. The surface topography along the scanned line was determined during the first ("main") scan. Contact-mode AFM or alternatively TMAFM can be used to do this. The measured surface relief was stored as a "height" signal pixel per pixel along the scanned line. The second ("interleaved") scan of the line was carried out under force modulation conditions. Based on the stored surface relief data, the tip approached the sample to an operator-adjustable value (corresponding to a repulsive set-point force) and additionally oscillated in the z-direction with a small amplitude, which was also adjustable. The amplitude of the response was collected as described above; however, due to the electronics of the controller, the brighter the image structure in the "amplitude" image the smaller the measured amplitude. Therefore, in Fig. 5.4b the soft rubber shells of the particles appear brighter than their stiff surroundings (the hard cores of the particles and the PMMA matrix) due to tip indentation and a correspondingly smaller deflection amplitude.

In spite of the ability of the force modulation technique to image heterogeneous polymers based on local differences in mechanical properties, the technique is sample-unfriendly because it is still possible to damage the surface.

5.3.3 Dynamic Operational Modes

In the dynamic mode of operation, the cantilever is excited such that it vibrates at or near its resonant frequency. Under the influence of tip–sample forces, the resonant frequency and consequently also the amplitude and the phase of the cantilever vibration will change and serve as measurement parameters. This is the basis for so-called "dynamic" AFM, which has been reviewed recently, e.g. in [13]. Two modes of operation dominate the application of dynamic AFM: amplitude modulation (AM) and frequency modulation (FM). In the amplitude modulation mode, the actuator is driven by a fixed amplitude at a fixed frequency f_0. When the tip approaches the sample surface, the interaction force between the tip and the sample causes changes in the amplitude and in the phase of the vibrated cantilever. These changes can be used as the feedback signal. While the AM mode was initially introduced as a noncontact mode, it was subsequently successfully used at a closer distance range that involved encountering attractive and repulsive interaction forces during each vibration cycle. This mode of operation, called the "tapping mode" (TM) or the "intermittent mode", has been widely used for AFM investigations of polymers under ambient conditions. Therefore, it will be considered in more detail in the next section. As the change in amplitude occurs on a timescale $\tau_{AM} \approx 2\,Q f_0^{-1}$ in AM mode, the scan must be performed at a correspondingly slow speed if the investigation is carried out with a high quality factor Q of the vibration in order to reduce noise.

In the FM mode, introduced by Albrecht et al. for analysing magnetic forces [14], a cantilever with a high Q factor is driven to oscillate at its eigenfrequency by performing positive feedback with an electronic circuit that keeps the amplitude of the cantilever vibration constant. The change in the eigenfrequency due to tip–sample interaction occurs on a timescale $\tau_{FM} \approx f_0^{-1}$, and so the benefits of a high Q factor and high-speed scanning are combined in this mode of operation. This combination results in improved resolution, and when applied to noncontact AFM performed under ultrahigh vacuum conditions with a very small minimal distance between the tip and the sample, it provides the means to achieve true atomic resolution in AFM investigations. Although noncontact AFM in the FM mode is one of the most powerful AFM methods due to its enormous potential for modern nanoscience and technology, its applications to polymer research have been rather limited. The need for vacuum conditions is undoubtedly one reason that this method is rarely demanded by polymer researchers, particularly in the case of routine applications. Therefore, a more detailed description of this technique is beyond the scope of this section, and interested readers should consult concise recent reviews on the topic, e.g. [15, 16].

5.3.4 Tapping Mode

The principle of TMAFM and the symbols used in the following discussion are illustrated in Fig. 5.5. The cantilever with the probing tip is forced by a driven actuator at the cantilever base to oscillate with certain amplitude A_0 of free vibration that is typically at or near its resonant frequency f_0. The cantilever is then brought close to the specimen and made to tap the surface with a certain reduced set-point amplitude A_{sp}.

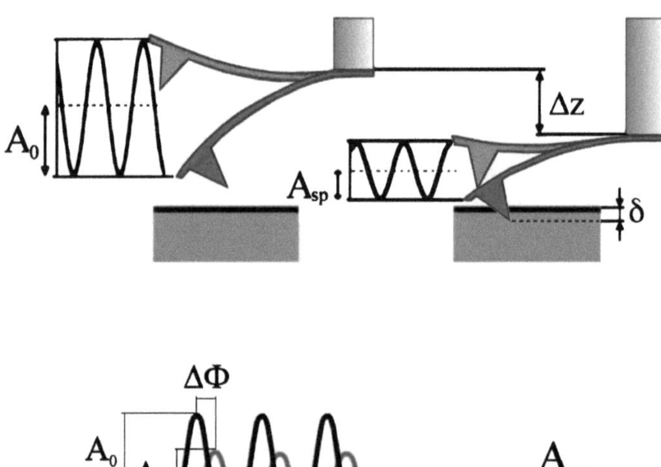

Fig. 5.5. Illustration of the principle of TMAFM

During each vibration cycle the tip taps the sample for only a very short time. Thus, lateral forces and surface damage are avoided as much as possible during scanning. Therefore, this gentle AFM mode is the preferred method for investigating soft materials like polymers. The probe–sample interaction also results in a shift of the resonant frequency and in a phase shift $\Delta\Phi$ of the vibration with respect to that of the freely oscillating cantilever. The resonant frequency and the phase shift are sensitive measures of the forces acting on the probe. Attractive forces acting on the AFM probe cause a negative shift in its resonance frequency, while repulsive ones lead to a positive shift. In the usual TMAFM imaging scheme, the specimen surface is scanned at a set-point amplitude A_{sp} that is kept constant by a feedback loop. During the scan the vertical displacements Δz needed to keep the amplitude constant are displayed as a "height" image and the locally varying phase shift $\Delta\Phi$ is displayed as a "phase" image. In principle, the height image should reflect the sample topography, while phase images show morphological structures of heterogeneous polymers. However, unlike contact-mode AFM, where force measurement by means of the deflection signal is straightforward and hence the formation of contrast in the scanned area is well-defined, the tip–sample interaction in TMAFM is rather complex. Therefore, the contrasts of the height and phase images strongly depend on experimental conditions. Factors that significantly affect the height and phase images in TMAFM of multicomponent polymers are the cantilever force constant, the tip shape, the amplitude A_0 of free vibration and, in particular, the set-point amplitude ratio $r_{sp} = A_{sp}/A_0$, as described in e.g. [5, 17–21].

Results from systematic TMAFM studies of a flat polymer blend surface, where height and phase images were simultaneously recorded at varied set-point amplitude ratios ($r_{sp} = A_{sp}/A_0$ ranging from 0.95 to 0.1), are shown in Fig. 5.6. Using a Dimension 3000 AFM and a commercial NCL silicon cantilever from Nanosensors with a resonant frequency of 165 kHz and a cantilever spring constant of about 40 N/m, the investigations were performed under ambient conditions by driving the cantilever with an amplitude of $A_0 \approx 40$ nm at its resonant frequency. The material used in this study was a blend consisting of 25 wt% high-density polyethylene (HDPE) and 75 wt% ethylene/1-octene copolymer (EOC). Both of the components, HDPE 53050E (density 0.952 g/cm^3) and EOC AFFINITY* EG 8150 (density 0.868 g/cm^3), were commercial products from the Dow Chemical Company. Due to the phase separation of the components, the blend shows a matrix with a morphology resembling that of the pure EOC. The formation of regions with ordered crystalline lamellae, as found in the pure HDPE, is hindered in the blend; instead, heaps and bundles of crystalline lamellae are embedded in the matrix. Nearly the same specimen area is recorded in the series of TMAFM height (a,c,e) and phase (b,d,f) images at set-point ratios r_{sp} of 0.95 (a,b), 0.5 (c,d) and 0.1 (e,f). Due to the variation of r_{sp}, small changes in contrast are observed in the height images (a,c,e). Phase image (b), recorded at light tapping with $r_{sp} = 0.95$, shows no contrast. However, lamellar structures and matrix regions can be clearly distinguished in the phase images (d) and (f), recorded at harder tapping with set-point amplitude ratios of 0.5

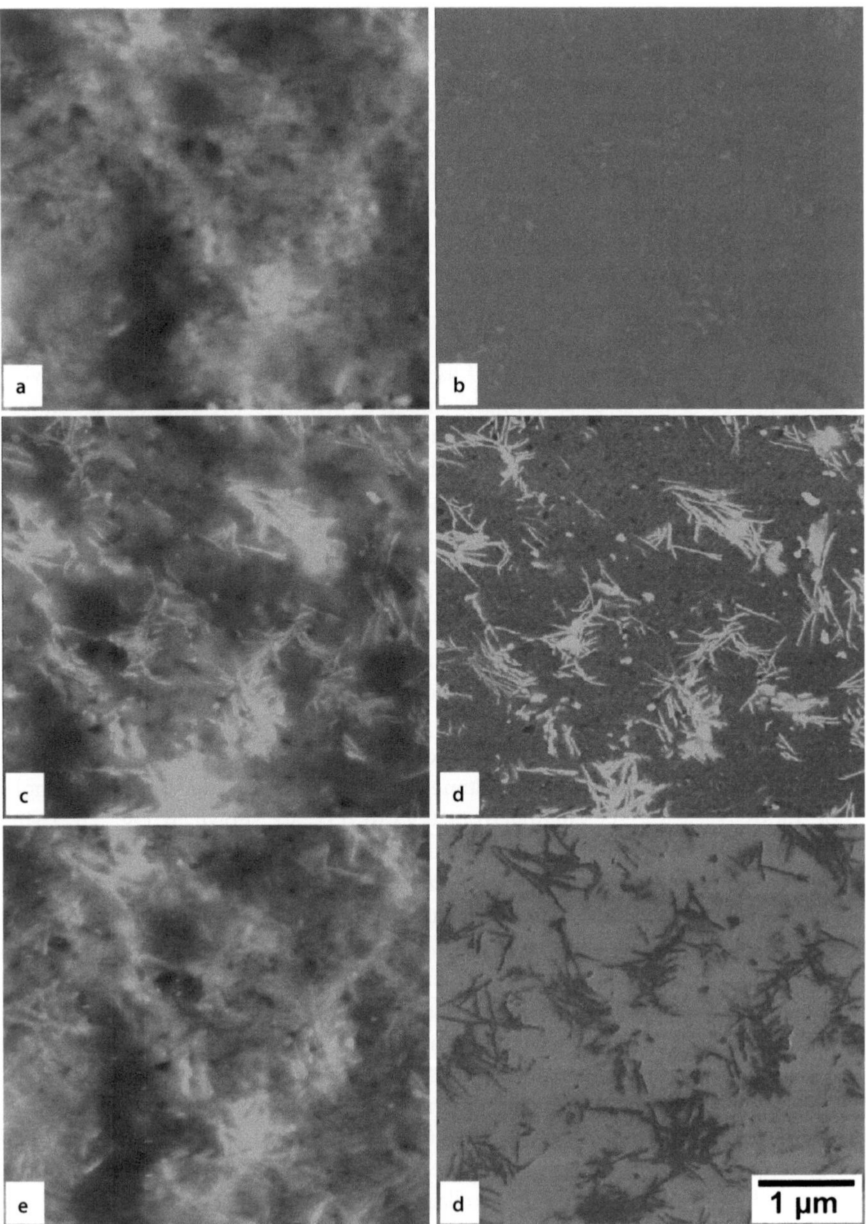

Fig. 5.6. TMAFM height (**a,c,e**) and phase (**b,d,f**) images of the same area of the HDPE/EOC blend recorded at three different amplitude set-point ratios r_{sp} of 0.95 (**a,b**), 0.5 (**c,d**) and 0.1 (**e,f**). The contrast covers height variations in the 120 nm range (**a,c,e**) and phase shift variations in the 100° range (**b,d,f**)

and 0.1, respectively. On the other hand, the contrast inversion illustrates the problems involved in interpreting the contrast of TMAFM images, as discussed in detail in e.g. [18, 21, 22].

To examine the dynamics of the tip–sample interaction that cause the image contrast, additional TMAFM experiments where the lateral position of the tip is fixed and the amplitude signal and the phase shift $\Delta\Phi$ of the tapping cantilever are measured as a function of the varied tip–sample distance Δz can be carried out. These TMAFM experiments resemble those performed in order to record force–distance curves, which are important for contact-mode AFM. Figure 5.7a shows amplitude and phase shift versus distance curves recorded for various regions of the blend sample: in a soft matrix area of the blend (dotted lines) and at a specimen site where a bundle of crystalline lamellae was located (solid lines). These curves generally show the following typical features. When the drop in amplitude is small, a negative phase shift indicates that the overall tip–sample interactions are attractive. As the amplitude drops further, repulsive interactions become dominant, as seen by the switch in the phase shift from negative to positive. Finally, the phase shift drops to zero when the amplitude is reduced to zero and the vibration has ceased. Taking into account the initial amplitude and the correlation of the phase shift $\Delta\Phi$ at an arbitrary point on the abscissa Δz with the corresponding reduced amplitude of the amplitude versus distance curve, the $\Delta\Phi$ versus Δz curves were transformed into the $\Delta\Phi$ versus r_{sp} plots, as shown in Fig. 5.7b. The differences between the two plots presented in Fig. 5.7b recorded for the lamellar and the matrix regions of the blend, respectively, directly correlate with the contrast of the phase images at different r_{sp} values. At light tapping with r_{sp} close to 1, both plots almost coincide, and so phase shift contrast is not observable under this condition, as shown by Fig. 5.6b. In agreement with Fig. 5.6d, the greatest difference between the plots in Fig. 5.7b at about $r_{sp} = 0.5$ causes a maximum phase shift contrast. Lamellae appear bright under this condition, as the corresponding curve in Fig. 5.7b exceeds that of the matrix region. A crossover of both curves takes place at about $r_{sp} = 0.2$, where the phase shift contrast vanishes again. A further decrease in r_{sp} results in a contrast inversion of the phase images, as shown in Fig. 5.6f.

The interpretation of contrast in height images is more complicated. It takes advantage of an evaluation of the tip indentation δ in soft materials. Due to indentation δ, amplitude versus distance curves recorded on polymer samples deviate from the straight line that describes the drop in amplitude when the sample is hard, such as a silicon wafer. In the latter case, the amplitude drops linearly with the vertical distance and vanishes completely when the sample surface coincides with the cantilever baseline. Figure 5.8a shows the amplitude versus distance curves of Fig. 5.7a together with the corresponding curve obtained for a silicon wafer (thin solid line denoted "Si"). As described in detail in, e.g., [22, 23], the indentation δ is estimated at an arbitrary point of the amplitude versus distance curve for the polymer by calculating the difference between it and the corresponding reduced amplitude of the hard silicon wafer. As the slope of the Si amplitude curve is $45°$, the difference in either the vertical direction or in the horizontal direction can be

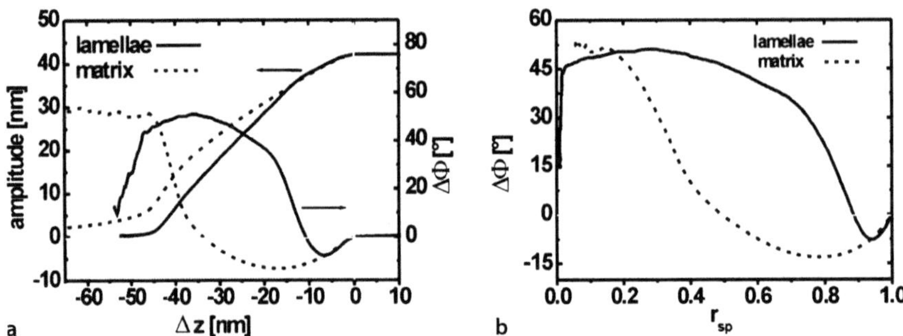

Fig. 5.7. **a** Amplitude versus distance Δz and phase shift $\Delta\Phi$ versus distance Δz curves of lamellar (*solid lines*) and matrix (*dotted lines*) regions of the blend. **b** Phase shift $\Delta\Phi$ versus r_{sp} plots of lamellar (*solid lines*) and matrix (*dotted lines*) regions of the blend, as estimated from a transformation (see text) of the $\Delta\Phi(\Delta z)$ of **a**. (Reprinted from [22] with the permission of Elsevier)

used to obtain δ, as illustrated in Fig. 5.8a. The estimation of δ as the difference between the ordinate values of the amplitude curves for the polymer and Si corresponds to the experiment used to record amplitude–distance curves, where the drop in amplitude with Δz is weaker in soft materials, due to indentation, than it is for hard materials, where no indentation takes place. On the other hand, the indentation δ expressed as the difference in abscissa values measured at the same reduced amplitude in the two amplitude versus distance curves corresponds to the variations in δ monitored in TMAFM height images, where the set-point amplitude is kept constant by changing Δz via the feedback loop. Figure 5.8a also shows the estimated indentation curves, where solid and dotted lines again correspond to lamellar and matrix regions of the blend, respectively. Using the procedure applied to phase shift versus distance curves, a useful transformation into plots of

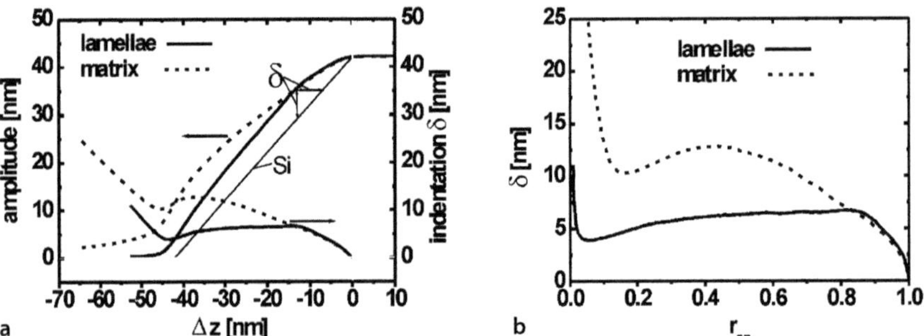

Fig. 5.8. **a** Amplitude reference curve of a silicon wafer (SI), amplitude versus distance Δz curves of Fig. 5.7a and estimated indentation δ versus distance Δz curves of lamellar (*solid lines*) and matrix (*dotted lines*) regions of the blend. The $\delta(\Delta z)$ curves were estimated using the ordinate or abscissa differences marked by δ (see text). **b** Indentation δ versus r_{sp} plots of lamellar (*solid lines*) and matrix (*dotted lines*) regions of the blend, as estimated from a transformation of the $\delta(\Delta z)$ of **a**. (Reprinted from [22] with the permission of Elsevier)

δ versus r_{sp} (as presented in Fig. 5.8b) was carried out. At light tapping with r_{sp} smaller than about 0.85, both plots are similar, and so height images monitored under this condition, such as Fig. 5.6a, show the actual surface topography. When harder tapping ($r_{sp} < 0.8$) is used, the indentations in the lamellar and matrix regions of the blend differ significantly, and so height images recorded under this condition (Fig. 5.6c,e) show a superposition of topographical and morphological information.

In agreement with results reported in [5], these investigations reveal that the profile in the height image only presents the true surface topography when r_{sp} is close to 1. This image regime is known as "light tapping". However, a much harder tapping with $r_{sp} \approx 0.5$ is necessary to observe maximum phase shift contrast between stiff and soft regions of the blend. Thus, when TMAFM is used in the usual manner a single scan cannot yield optimum topography and morphology information. Therefore, interleaved scanning was suggested in [22] as a way to simultaneously record a height image at light tapping, with r_{sp} close to 1 in the main scan, and a phase image with maximum contrast at correspondingly harder tapping in the interleaved scan.

Further examples of quantitative contrast interpretations of TMAFM height and phase images by means of results of amplitude/phase versus distance curves have been reported for different blends [5, 18, 21] and block copolymers [24–26].

5.4 Typical and Special AFM Applications

It was mentioned at the beginning of this chapter that the application of AFM complements electron microscopic investigations of polymeric materials. Comparable results can be achieved by applying TEM and TMAFM in the usual way, as revealed by pairs of corresponding micrographs for the following examples: (1) arrangement of lamellar structures in α- and β-spherulites of isotactic polypropylene (Fig. 16.4); (2) lamellar morphology of a styrene/butadiene block copolymer (Fig. 11.4); phase-separated HDPE/VLDPE blend with elastomeric particles embedded in a semicrystalline matrix (Fig. 17.5); Kraton SEBS triblock copolymer with cylindrical polystyrene domains (Fig. 19.3). The examples provided here also make it clear that sample preparation varies depending on the morphological structures that are to be visualised by TEM and TMAFM. TEM investigations usually take advantage of samples which have been selectively stained, e.g. by OsO_4 or RuO_4 as described in detail in Sect. 13.3, while hard and soft structural entities directly cause image contrast in TMAFM phase images if the differences in local stiffness are sufficient.

In addition to the routine application of the AFM to characterise the morphologies of polymeric materials, the method has been extended to study many specific polymer properties. In such cases, special equipment is often attached to the AFM. Thus, due to the advantage of being able to visualise the sample morphology without chemically treating the sample, AFM is preferably used to investigate the local deformation behaviour in the micron and nanometre ranges by means of in situ tensile tests in an AFM equipped with an attached tensile module [27–29]. The corresponding results for a phase-separated HDPE/VLDPE blend are demonstrated in

Figs. 6.10, 17.10 and 17.11, while Fig. 6.13 shows a series of TMAFM images of a carbon nanotube-filled ethylene/octene copolymer deformed step-by-step to different strains.

A heating stage is another very helpful AFM attachment. Using this combination, Hobbs et al. [30] succeeded in a series of experiments in which processes such as crystallisation, crystal thickening and crystal deformation were followed in situ and in real time, providing significant new insights into long-standing problems in polymer science. When this technique was used to investigate the crystallisation of polyethylene shish kebab crystals in real time with nanometre resolution, images of the extended chain backbone and the overgrowth and subsequent interdigitation of lamellae were obtained [31].

Due to the gentle surface scanning associated with TMAFM investigations, they are also useful for studying changes in local surface regions caused by sample treatment at high resolution. Thus, by eroding the specimen step-by-step and imaging the same surface area via TMAFM after each step, Magerle [32] expanded the AFM technique from surface imaging to volume imaging. By combining the series of TMAFM images in a similar way to computed tomography, this kind of "nanotomography" was used to reconstruct the 3-D distributions of polystyrene and polybutadiene in a poly(styrene-*block*-butadiene-*block*-styrene) [32] and the crystalline regions in an elastomeric polypropylene [33].

Utilising the results of the TMAFM investigations described in the previous section and demonstrated in Figs. 5.6–5.8, the same HDPE/EOC blend was used to study the structural changes induced by different surface treatments. By applying interleaved scanning, TMAFM investigations of exactly the same specimen area were carried out before and after several surface treatments in order to evaluate the influence of the surface treatment in a direct way. The investigated surface treatments include chemical etching with a permanganic etchant, electron beam irradiation by scanning the surface area of interest in an environmental scanning electron microscope (ESEM), and plasma etching in an oxygen atmosphere. The results of these investigations are demonstrated by the corresponding series of TMAFM micrographs shown in Figs. 5.9–5.11. In each series, the first row of micrographs shows the original untreated surface area while corresponding micrographs of the following rows reveal the modification of the sample due to the surface treatment in the area of interest. For light tapping, the height images in the first column of each series show the actual surface topography of the sample, while at harder tapping the blend morphology is revealed in the TMAFM phase images in the second column due to changes in the local stiffness of the sample.

Starting from a very flat sample surface (micrographs a, b), Fig. 5.9a,c,e (height images) illustrate the formation of surface relief with raised crystalline lamellae due to etching with a permanganic etchant (200 ml sulfuric acid + 20 ml water + 2.2 g potassium permanganate) for 15 s (micrographs c, d) and 30 s (micrographs e, f). A selectively etched surface can be a helpful alternative for visualising the morphology of heterogeneous polymers if local stiffness differences are insufficient to cause contrast in the TMAFM phase image.

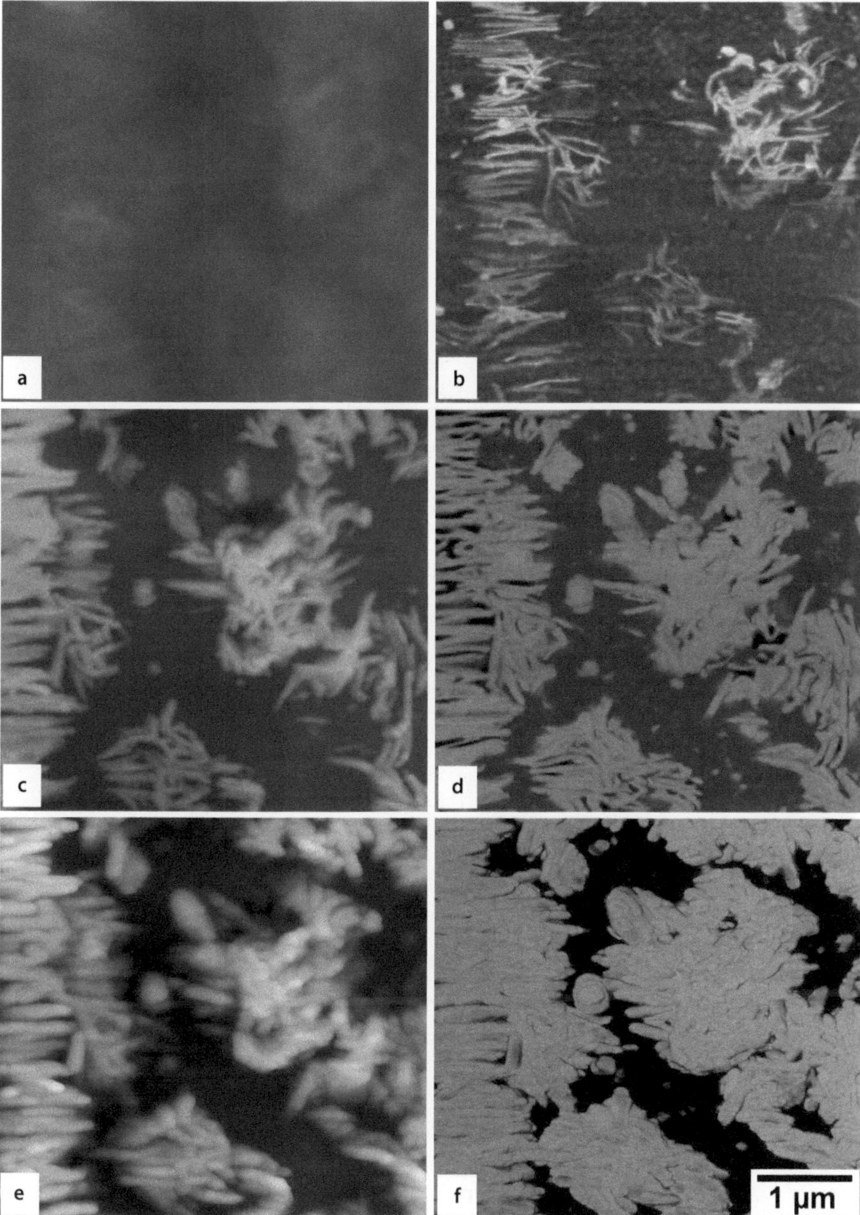

Fig. 5.9. Series of TMAFM height (**a,c,e**) and phase (**b,d,f**) images of the same HDPE/EOC blend region before (**a,b**) and after permanganic etching for 15 s (**c,d**) and 30 s (**e,f**)

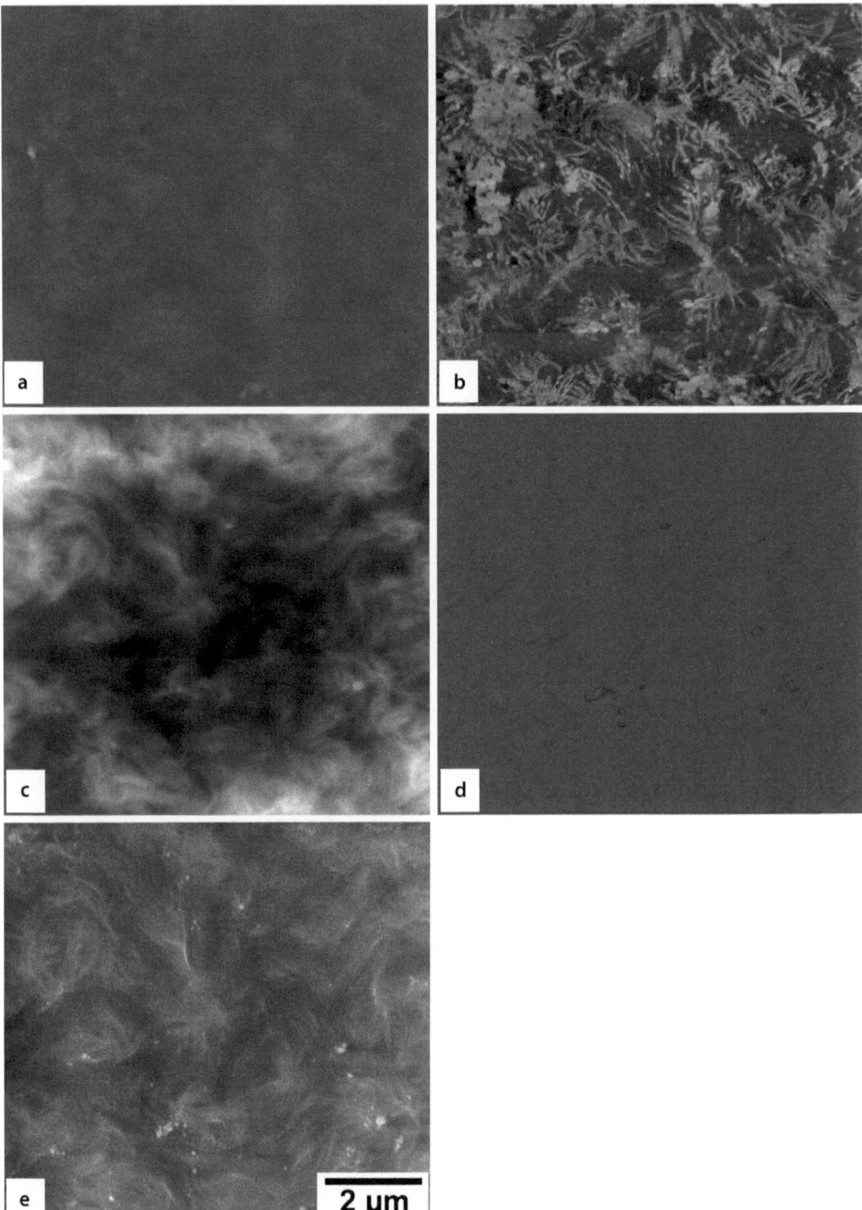

Fig. 5.10. Series of TMAFM height (**a,c**) and phase (**b,d**) images of the same HDPE/EOC blend region before (**a,b**) and after (**c,d**) ESEM inspection; secondary electron image of the ESEM inspection (**e**)

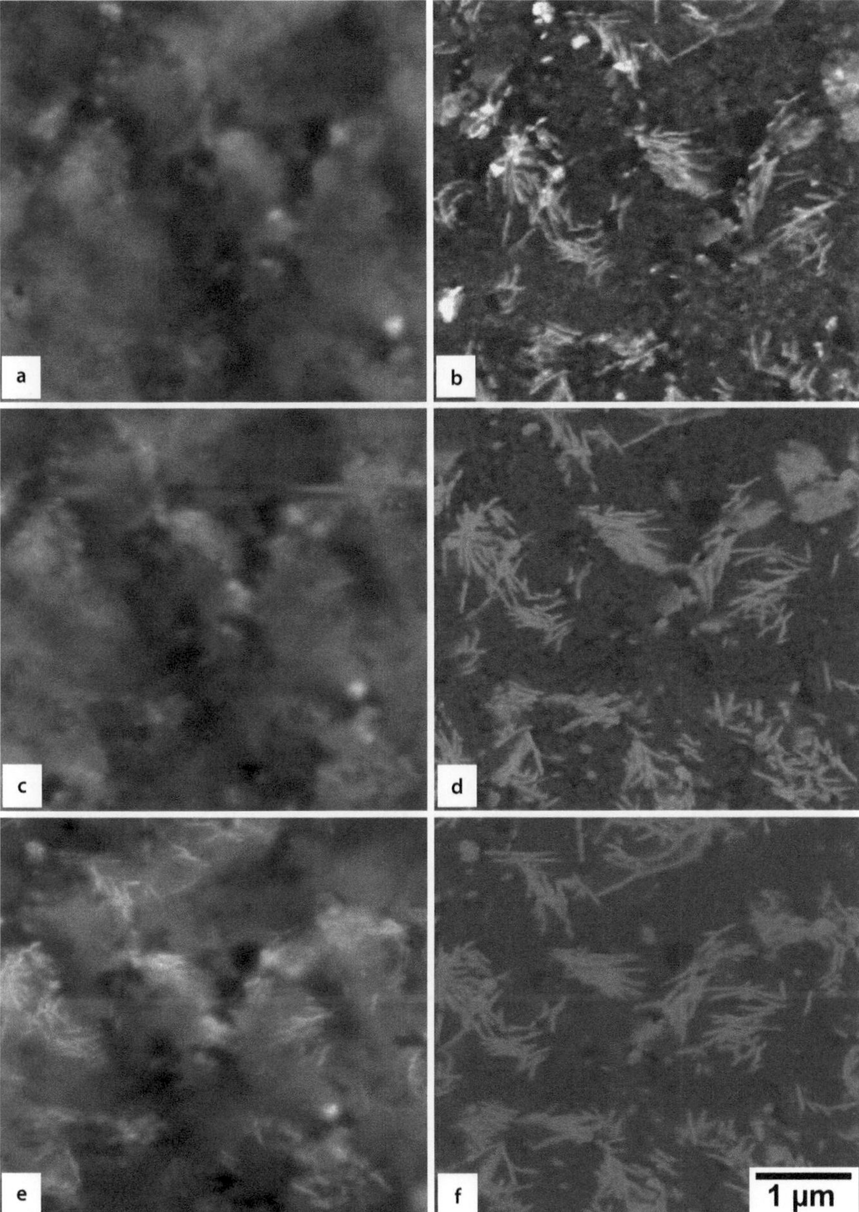

Fig. 5.11. Series of TMAFM height (**a,c,e**) and phase (**b,d,f**) images of the same HDPE/EOC blend region before (**a,b**) and after oxygen plasma treatment for 10 s (**c,d**) and 40 s (**e,f**)

In Fig. 5.10, the radiation damage caused by the impact of a 12-keV electron beam of low intensity on a selected sample area while scanning at a magnification of 20 000× in an ESEM-FEG XL30 environmental scanning electron microscope (Philips Electron Optics) for about half a minute is highlighted by the TMAFM images taken before (a, b) and after (c, d) the ESEM inspection. The area of interest in the ESEM was scanned at 9.6 s per image and recorded as an AVI video consisting of 31 individual images of size 800 × 600 pixels. While the original sample did not show any surface structure, the second image of the video showed the formation of surface relief due to the appearance of corresponding image contrast (Fig. 5.10e). The analysis of the individual images of the video, performed by measuring the distances between distinctive image structures, revealed a central shrinking of the scanned area. The relative changes in lateral distances reach values of about 0.96 in the fifth image and 0.925 in the last image of the video. The shrinking observed, which corresponds to the deep central cavity recorded by the TMAFM height image (c), may be caused by electron beam-induced chain scissions and the volatilisation of small chain segments. On the other hand, a comparison of TMAFM phase images recorded before (b) and after (d) the ESEM investigation reveals that the contrast between the crystalline lamellae and their surroundings vanishes upon electron beam illumination. This equalisation of local stiffness is clear evidence that crosslinking has occurred. The reduction of contrast in TMAFM phase images of the blend due to crosslinking was also observed in corresponding experiments where TMAFM was applied to evaluate the influence of irradiation with gamma-rays and the changes caused by argon plasma treatment.

A useful sample treatment leads to the results presented in Fig. 5.11. This series of micrographs shows the original blend area of interest and the minor changes caused by oxygen plasma. The treatment was carried out in a PLS 500P plasma reactor by means of microwave plasma source (ECR-MW; 246 GHz; 300 W) at a working pressure of 6.5×10^{-3} mbar and an oxygen flow rate of 50 $cm^3 min^{-1}$. The effects of selective surface etching can be detected in the height image (e), where a few raised crystalline lamellae appear after surface treatment for 40 s. Due to cross-linking in the EOC component, a small decrease in contrast is observed in the corresponding TMAFM phase image (f). This affects the sample for only 10 s, during which time the selective surface etching is suppressed and surface structures remain nearly unchanged, according to a comparison of the height images (a) and (c). On the other hand, the duration of the treatment is sufficient to clean the surface, and so unlike the original surface (b), both crystalline lamellae and matrix appear with uniform contrast in the TMAFM phase image (d). This example illustrates that a short sample treatment in oxygen plasma is useful since it cleans the surface by pushing away contaminated surface layers.

References

1. Magonov SN, Whangbo M-H (1996) Surface analysis with STEM and AFM. VCH, Weinheim
2. Magonov SN, Reneker D (1997) Ann Rev Mater Sci 27:175
3. Tsukruk VV (1997) Rubber Chem Technol 70:430
4. Jandt KD (1998) Mater Sci Eng R21:221

5. Magonov SN (2000) Atomic force microscopy in analysis of polymers. In: Meyers RA (ed) Encyclopedia of analytical chemistry. Wiley, Chichester, UK, p 7432

6. Binnig G, Rohrer H, Gerber C, Weibel E (1982) Phys Rev Lett 49:57

7. Binnig G (1986) Atomic force microscope and method for imaging surfaces with atomic resolution. US Patent 4,724,318

8. Binnig G, Quate C, Gerber C (1986) Phys Rev Lett 56:930

9. Bhushan B, Marti O (2004) Scanning probe microscopy: Principle of operation, instrumentation, and probes. In: Bhushan B (ed) Springer handbook of nanotechnology. Springer, Berlin, p 325

10. Ohnesorge F, Binnig G (1993) Science 260:1451

11. Rosa A, Hild S, Marti O (1997) Meas Sci Technol 8:1

12. Maivald P, Butt HJ, Gould SAC, Prater CB, Drake B, Gurley JA, Elings VB, Hansma PK (1991) Nanotechnology 2:103

13. Schirmeister A, Anczykowski B, Fuchs H (2004) Dynamic force microscopy. In: Bhushan B (ed) Springer handbook of nanotechnology. Springer, Berlin, p 449

14. Albrecht TR, Grutter P, Horne HK, Rugar D (1991) J Appl Phys 69:668

15. Morita S, Giessibl FJ, Sugawara Y, Hosoi H, Mukasa K, Sasahara A, Onishi H (2004) Noncontact atomic force microscopy and its related topics. In: Bhushan B (ed) Springer handbook of nanotechnology. Springer, Berlin, p 385

16. Giessibl FJ (2005) Materials Today 8:32

17. Brandsch R, Bar G, Whangbo M-H (1997) Langmuir 13:6349

18. Bar G, Thomann Y, Brandsch R, Cantow H-J, Whangbo M-H (1997) Langmuir 13:3807

19. Whangbo M-H, Bar G, Brandsch R (1998) Appl Phys A 66:S1267

20. Bar G, Brandsch R, Whangbo M-H (1999) Surf Sci Lett 422:L192

21. Bar G, Ganter M, Brandsch R, Delineau L, Whangbo M-H (2000) Langmuir 16:5702

22. Godehardt R, Lebek W, Adhikari R, Rosenthal R, Martin C, Frangov S, Michler GH (2004) Eur Polym J 40:917

23. Bar G, Delineau L, Brandsch R, Bruch M, Whangbo M-H (1999) Appl Phys Lett 75:4198

24. Kopp-Marsaudon S, Leclere Ph, Dubourg F, Lazzaroni R, Aime JP (2000) Langmuir16:8432

25. Knoll A, Magerle R, Krausch G (2001) Macromolecules 34:4159

26. Leclere Ph, Dubourg F, Kopp-Marsaudon S, Bredas JL, Lazzaroni R, Aime JP (2002) Appl Surf Sci 188:524

27. Godehardt R, Rudolph S, Lebek W, Goerlitz S, Adhikari R, Allert E, Giesemann J, Michler GH (1999) J Macromol Sci Phys B38:817

28. Michler GH, Godehardt R (2000) Cryst Res Technol 35 :863

29. Godehardt R, Lebek W, Michler GH (2001) Morphology and micro-mechanics of phase-separated polyethylene blends. In: Grellmann W, Seidler S (eds) Deformation and fracture behaviour of polymers. Springer, Berlin, p 267

30. Hobbs JK, Winkel AK, McMaster TJ, Humphris ADL, Baker AA, Blakely S, Aissaoui M, Miles MJ (2001) Macomol Symp 167:1

31. Hobbs JK, Humphris ADL, Miles MJ (2001) Macromolecules 34:5508

32. Magerle R (2000) Phys Rev Lett 85:2749

33. Rehse N, Marr S, Scherdel S, Magerle R (2005) Adv Mater 17:2203

6 In Situ Microscopy

Each type of electron microscope is suitable not only for studying the morphologies of polymers but also for visualising changes in these materials under different and varying conditions. Such experimental tests are commonly known as in situ techniques. Beside the effects of electron irradiation and different ambient atmospheres, micromechanical in situ tests are discussed here in some detail. Some examples are presented after describing the technical equipment used for in situ investigations, including miniaturised deformation devices for common electron microscopes, which enable tensile tests of ultra- and semi-thin specimens to be performed at low or high temperatures. SEM and (particularly) ESEM are very effective ways to study deformation, crack propagation and fracture processes. The TEM and HVTEM techniques permit higher resolution to be achieved and can be used to characterise effects at the micro- as well as the nanoscale. AFM can be applied to monitor micromechanical deformation processes without the limiting factors associated with EM, such as vacuum and electron irradiation damage.

6.1 Overview

The great diversity of polymeric materials reflects their enormous structural variety from the molecular up to the macroscopic levels, which ultimately determines their physical properties. Therefore, it is fundamentally important for the development of materials with improved properties (e.g. mechanical, thermal, electrical, etc.) to gain an understanding of the relationships between morphology and these properties under different application conditions. Each type of electron microscope is suitable for studying not only the morphologies of polymers but also the changes in these materials under different and varying conditions. Such experimental tests are commonly known as *in situ techniques*. The term "in situ" is Latin, and means "in its place". There are – depending on the type of microscope and the space available in its specimen chamber – several types of in situ experiments that can be performed:

- mechanical deformation tests
- heating or cooling
- electron irradiation
- application of electric or magnetic fields
- application of different ambient atmospheres.

Fig. 6.1a,b. In situ experiment performed in an ESEM (at 5 kV) to investigate the influence of the surrounding gas atmosphere on cellulose fibres. **a**: in a dry atmosphere (1.9 Torr N_2); **b**: after 1 min of swelling in an H_2O vapour atmosphere

The main application of the in situ microscopy of polymers is the investigation of micromechanical processes [1], which is discussed in more detail in Chap. 6.2. Heating and cooling are often coupled with mechanical tests in order to study low- or high-temperature mechanical mechanisms. Additionally, the melting and crystallisation of semicrystalline polymers can be followed in situ, especially using AFM because this avoids the need for any electron irradiation [2]. However, the effects caused by electron irradiation can actually aid the morphological analysis of polymers. One result of electron irradiation is mass loss, which can differ significantly from one polymer to another. In Chap. 17, Fig. 17.13 shows a PVC/SAN blend in which the mass loss of the PVC phase is much greater that for the SAN phase. The dark PVC phase at the beginning of electron irradiation becomes much brighter after intense irradiation. Therefore, this irradiation-induced change in contrast can help us to identify different polymer phases. Another example is the improved contrast of spherulites in semicrystalline polymers; *irradiation-induced contrast enhancement* like this is discussed in detail in Chap. 13 (see, e.g., Figs. 13.10–13.12). Studies of the effects of electric or magnetic fields do not play a crucial role in polymer research, since most polymers are electrical insulators and are even nonmagnetic. The influences of different ambient atmospheres can be studied using a special environmental (gas or liquid) cell for high-voltage electron microscopes (see Chap. 3), in the chamber of an environmental SEM (ESEM, see Chap. 4) or in the AFM (see Chap. 5). Figure 6.1 shows an example where the shape of a sample changes when moisture is introduced into the atmosphere. The cellulose fibres are shrunken and close together in a dry atmosphere (Fig. 6.1, left) but they expand in a moist atmosphere (right). In the following, in situ techniques for studying micromechanical processes of deformation and failure using different microscopes are discussed.

6.2 Micromechanical In Situ Tests

6.2.1 Technical Equipment

Over the last few decades, significant advances have been made in the visualisation of failure mechanisms in materials upon loading. Significant improvements in measuring methods, loading apparatus design, and sample preparation have resulted in the application of in situ tensile tests, where the material is deformed by applying a tensile force while its deformation is observed using a microscope. A number of in situ techniques for scanning electron microscopes (SEM), transmission electron microscopes (TEM) and atomic force microscopes (AFM) have been developed to date. One of the most widespread is the in situ tensile test performed within an electron microscope, which is used to take sequences of pictures of deformation processes in real time. This test not only provides us with information about the "macroscopic" properties of the material, but also a deeper understanding of its deformation behaviour at the micro- and nanoscopic scales. In situ tools allow the local morphology and the deformation in a material that is subjected to local mechanical stresses to be mapped out. Load/deformation fields in the vicinity of crack tips in materials can be used to characterise and validate the various localised fracture mechanics models. In situ tensile tests on materials provide dynamic rather than static observations of the nucleation and propagation of cracks upon the application of a load. There are several in situ techniques for the investigation of micromechanical deformation processes [1, 3–11]. The appropriate in situ technique and tensile stage to use depends on the geometry, thickness, stress state and electron beam sensitivity of the samples. The data and specifications for several commercial tensile stages are listed and summarised in Table 6.1, and the corresponding apparatuses are demonstrated in Fig. 6.2.

The tensile stages for SEM, ESEM and AFM can deform relatively thick specimens as well as semi-thin sections. Only semi-thin or ultrathin samples can be used for tests in HVTEM and TEM.

Table 6.1. Data for commercially available in situ deformation devices (see also Fig. 6.2)

	SEM	ESEM / AFM	HEM	TEM
Device	Tensile module B156	Tensile and compression module 1000 N	Tensile stage	Single tilt cooling holder Model 671
Manufacturer	Oxford Instruments	Kammrath & Weiss GmbH	JEOL	Gatan
Sample thickness	0.5 μm to 0.5 mm	0.5 μm to 1 mm	100 nm to 5 μm	100 nm to 500 nm
Temperature range	−180 °C to +200 °C	Room temperature	Room temperature to +200 °C	−180 °C to +120 °C
Resolution	>10 nm	ESEM: >5 nm AFM: >1 nm	>1 nm	>0.5 nm

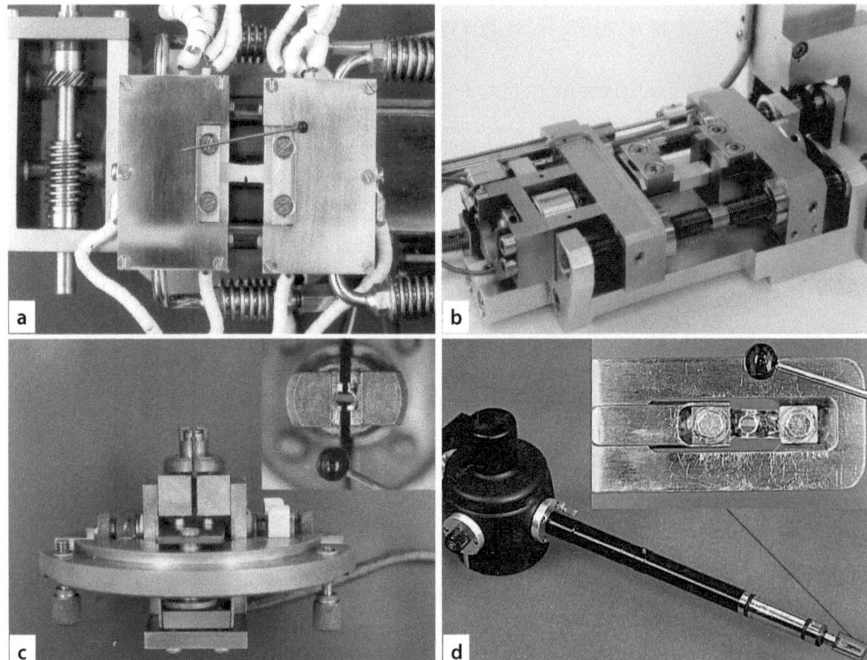

Fig. 6.2. Commercially available in situ deformation devices (**a**) for SEM, (**b**) for ESEM and AFM, (**c**) for HVTEM, and (**d**) for TEM (see also Table 6.1)

6.2.2 In Situ Microscopy in (E)SEM

One of the most applicable in situ microscopic techniques is performed in a SEM. The tensile device for a SEM is shown in Fig. 6.2a. Using this tensile stage, one can readily study crack nucleation and propagation in detail by stretching and/or compressing the relatively thick samples inside a SEM. The preparation of samples is comparatively easy since a conventional microtome can be used. Figure 6.3 shows a typical example of an in situ deformation test where crazes are formed in a semi-thin section of PS [1].

However, in most cases this technique is difficult to apply to polymeric materials because of the presence of charging electrons during in situ tensile tests. To overcome this limitation, a promising new technique is available, called environmental SEM (ESEM, see Chap. 4). ESEM allows the examination of specimens surrounded by a gaseous environment. This means that the specimen does not need to be coated with a conductive material. Many types of samples like greases, adhesives, liquids, foods, gels, and other biological spices can be examined using this technique. One relevant example of the use of this technique is shown in Fig. 6.4. Semi-thin sections with a thickness of about 1 µm from a PE nanocomposite modified with spherical SiO_2 filler particles (which have an average diameter in the range of 250 nm) were microtomed from the bulk at $-80\,^{\circ}\mathrm{C}$ using a Leica Ultracut E ultramicrotome. In this nanocomposite, the severe formation of oversized agglomerates is observed, which

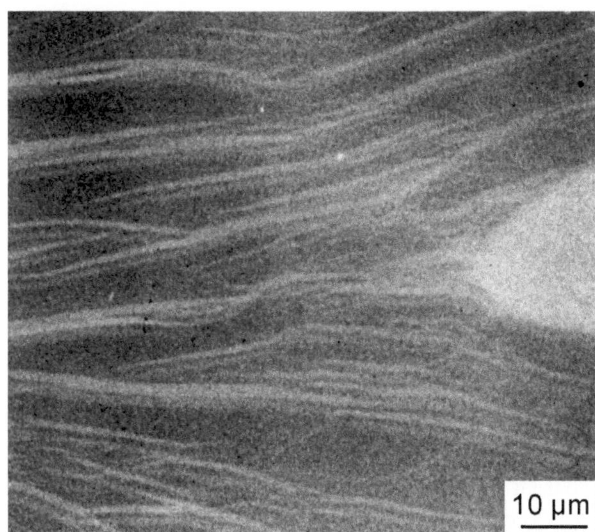

Fig. 6.3. Crazes in a deformed PS specimen 10 μm thick (SEM micrograph, deformation direction vertical). (From [1], with the permission of Hanser)

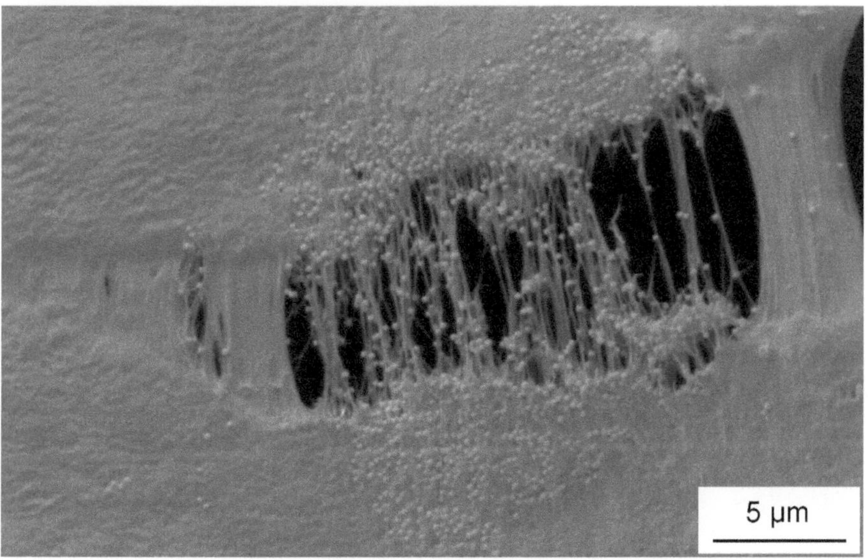

Fig. 6.4. Deformation structure of an agglomerate in a PE composite with 7 wt% SiO_2 (ESEM micrograph, deformation direction vertical)

is a phenomenon that is commonly observed when working with nanofillers. These agglomerates, in general, result in a degradation of mechanical properties because of the premature failure of the material during the early stages of deformation (see Chaps. 21 and 22). In Fig. 6.4, a large agglomerate is moderately elongated in the tensile direction and fibrils are formed inside the deformed agglomerate, mostly at its equatorial region [12].

An in situ deformation test in SEM of the step-wise development of craze-like deformation bands in a HDPE composite filled with 28 wt% Al_2O_3 particles is presented in Chap. 21 (see Fig. 21.7).

6.2.3 In Situ Microscopy in (HV)TEM

Due to the increasing demand for higher resolution and magnification in order to investigate micromechanical deformation processes, the in situ TEM technique is in great demand and is still undergoing development. In general, small-scale deformation stages can be used to characterise the strengths of micro- and nanoscale reinforcements, the adhesive bond strengths of small structural features, as well as the kinetics of crack initiation and propagation. A representative device for this technique is shown in Fig. 6.2d, for which ultrathin or not too thick semi-thin sections are usually used to allow the transmission of electrons. In other words, the materials should be "electron transparent", which strongly depends on the atomic weight of the material, the accelerating voltage of the microscope and the intended experimental conditions. The sample thicknesses typically required for this technique lie in the range 100–500 nm, which can be achieved using a well-equipped ultramicrotome. As an example, the deformation structure observed from an impact-modified SAN with PBA core-shell particles (with an average particle diameter of about 300 nm) is demonstrated in Fig. 6.5. Specimens about 0.3 µm in thickness were sectioned from the bulk at −110 °C using a Leica Ultracut E ultramicrotome, and then strained within the TEM while recording the deformation structures using a low-dose technique [5].

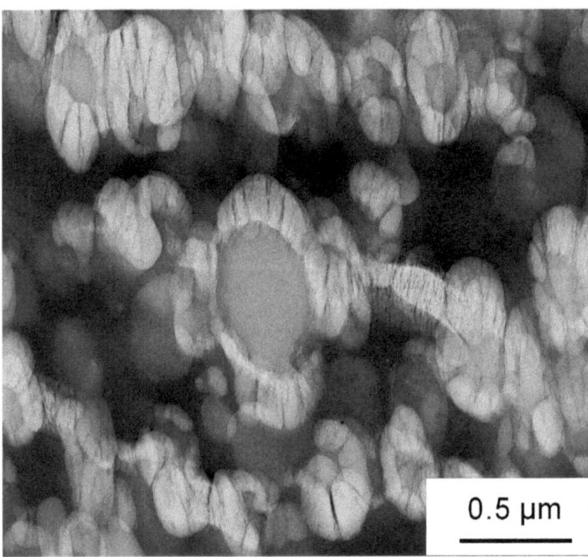

Fig. 6.5. Deformation structure of core–shell modifier particles in a SAN/PBA blend at room temperature (TEM micrograph, deformation direction vertical)

The deformation structure shows the formation of microvoids inside the modifier particles in combination with the fibrillation of the PBA shells while most of the cores persist during the tensile loading of the specimen [13].

Micromechanical properties at low and high temperatures can be investigated using straining devices with cooling and heating facilities. Such a cooling/heating/straining holder (shown in Fig. 6.2d) makes it possible to deform thin sections at temperatures of between −180 °C and +120 °C at constant strain rates. Figure 6.6 compares micrographs of thin sections of a SAN/PBA blend subjected to in situ tensile tests performed at −20, 23 and 60 °C and a constant strain rate of 0.05 s^{-1} using a straining device in a 200 kV TEM [13]. The micrographs show obvious changes in deformation behaviour with temperature. At −20 °C (Fig. 6.6a), the main mechanisms are crazing of the SAN and fibrillation of the elastomeric shells of the PBA particles around their PMMA cores (compare Fig. 6.5). The bright crazes show up clearly against the dark background of the undeformed matrix and the paler grey of the moderately stretched modifier particles. At 23 °C (Fig. 6.6b), above the T_g of the rubbery phase, nearly all of the particles cavitate with fibrillation of the PBA shell, as shown in Fig. 6.5. At 60 °C (Fig. 6.6c), there is very little evidence of crazing, and the matrix deforms almost exclusively by shear yielding in a ductile manner.

As mentioned above, the requirement for ultrathin sections for in situ tensile tests with conventional TEM restricts its applicability. The higher the electron energy used, the thicker the specimen that can be proved. As a consequence, high-voltage electron microscopy (HVTEM) permits samples with thicknesses on the order of 1–5 μm, which may reveal bulk-like behaviour, to be studied. The tensile stage for HVTEM is shown in Fig. 6.2c. As a typical example of the use of this technique, Fig. 6.7 shows a sequence of three micrographs taken at low magnifications during an in situ deformation test of an ∼0.5-μm-thick section of PS, where the initiation of crazes between artificial cracks is shown (Fig. 6.7b). The crazes appear as bright lines that are up to ∼100 μm long and possess a lower density because of their internal structure of microvoids and fibrils. After a small amount of additional deformation, the largest crazes break in a very brittle manner (brittle fracture, Fig. 6.7c). The fibrillated structure of the crazes is visible at larger magnifications in Chap. 15 (see Figs. 15.6 and 15.8) and the crack propagation inside the crazes is shown in Fig. 15.10.

Another example is provided by a PP blend. This blend is composed of polypropylene (PP) as matrix and polyamide 6 (PA) as the inclusion, which is surrounded by maleic anhydride-grafted polystyrene-*block*-poly(ethene-co-1-butene)-*block*-polystyrene (SEBS-g-MA) as compatibilizer [14]. The 1.5-μm-thick specimens were sectioned from the bulk at −80 °C using a Leica Ultracut E ultramicrotome, and then strained within HVTEM using an accelerating voltage of 1 MV. Figure 6.8 shows a typical deformation structure of this material. During the deformation process, the shell consisting of SEBS-g-MA is first stretched in the tensile direction. Once the strain of the shell has reached a certain critical value, fibrils form at the interface between the PA inclusion and the matrix. These fibrils appear at the polar regions of the modifier particles and are aligned in the direction of applied stress. This kind of micromechanism is attributed to the relatively strong phase adhesion between the PA inclusion and the matrix [14].

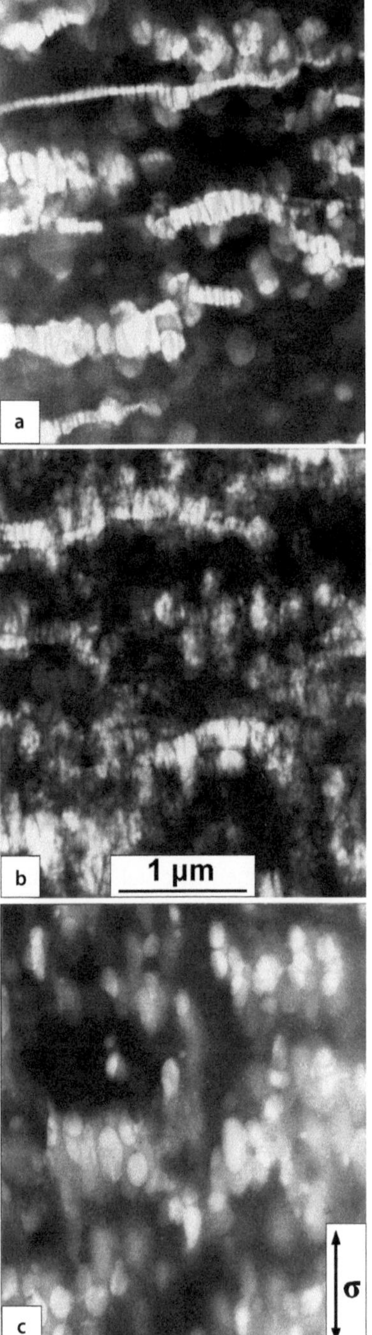

Fig. 6.6a–c. Deformation structures in a SAN/PBA blend at various temperatures and at a constant strain rate of 0.05 s^{-1}: **a** –20 °C; **b** +23 °C; **c** +60 °C. (Cryoultramicrotomed thin sections, TEM micrographs, deformation direction vertical, see *arrow*; from [15], with the permission of Chapman & Hall)

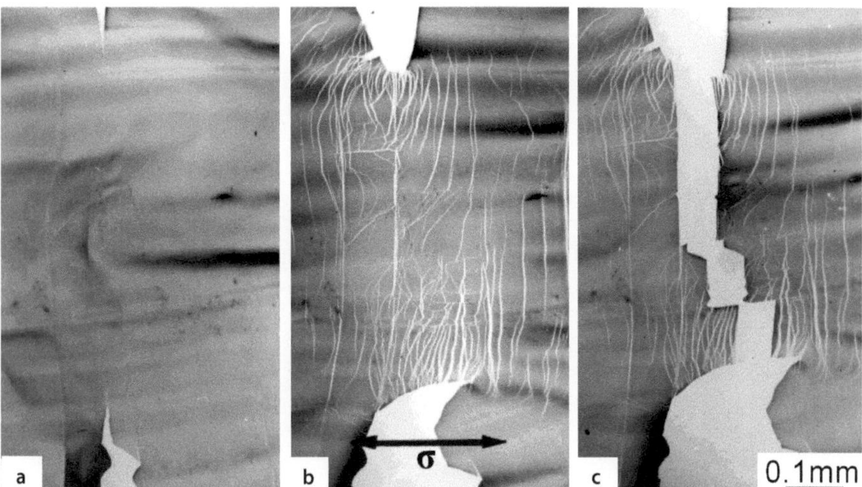

Fig. 6.7. In situ deformation of a 0.5 µm thick PS section in HVTEM (**a**) with artificial cracks before straining; (**b**) after deformation of about 10% with crazes; (**c**) after crack propagation and brittle fracture (for deformation direction see *arrow*; from [1], with the permission of Hanser)

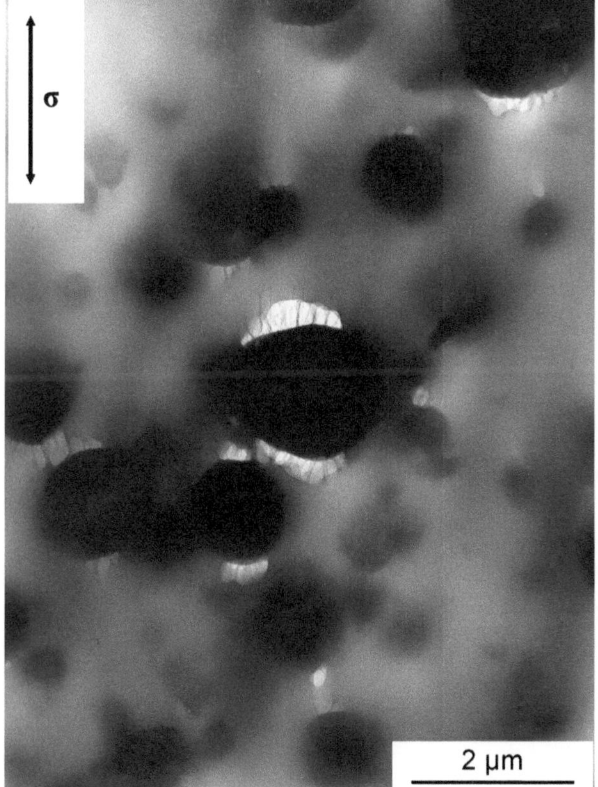

Fig. 6.8. Deformation structure of a PP/PA/SEBS-g-MA blend (HVTEM micrograph, the *arrow* indicates the deformation direction)

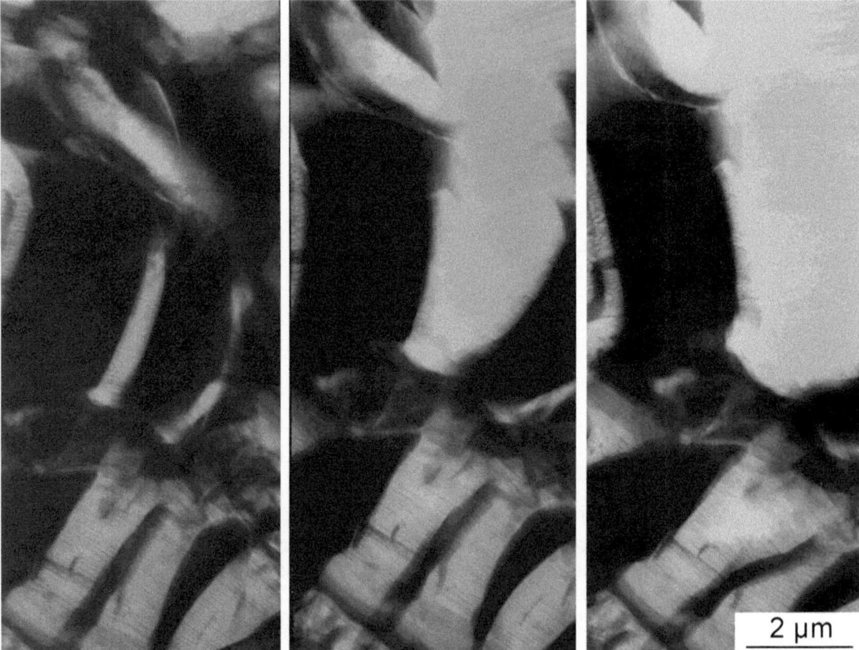

Fig. 6.9. In situ deformation with crack propagation and crack stop effect in a thin section of HIPS. (HVTEM, deformation direction horizontal)

Figure 6.9 shows crack propagation in a thin section of HIPS during in situ deformation in HVTEM. In a deformed area with crazes, a crack runs from the top, through a craze and is stopped in the soft rubber particle in the centre of the picture (the micrograph in the middle). After increasing the load, the craze below the large rubber particle rupture (right micrograph) and the crack continues to propagate through the rubber particle and the craze. This illustrates how the crack stop mechanism of the rubber particles is a precondition for high toughness in rubber-toughened polymers (see Chapter 18).

6.3 In Situ Microscopy in AFM

Up to now, we have established that AFM is a powerful technique for investigating the topographies as well as surface morphologies of materials. The advantages of working with this technique compared with electron microscopy can be listed as follows:

- no vacuum is necessary
- no sample damage occurs due to electron irradiation during experiments
- sample preparation is relatively easy, since pretreatments such as staining, etching, etc., are not required.

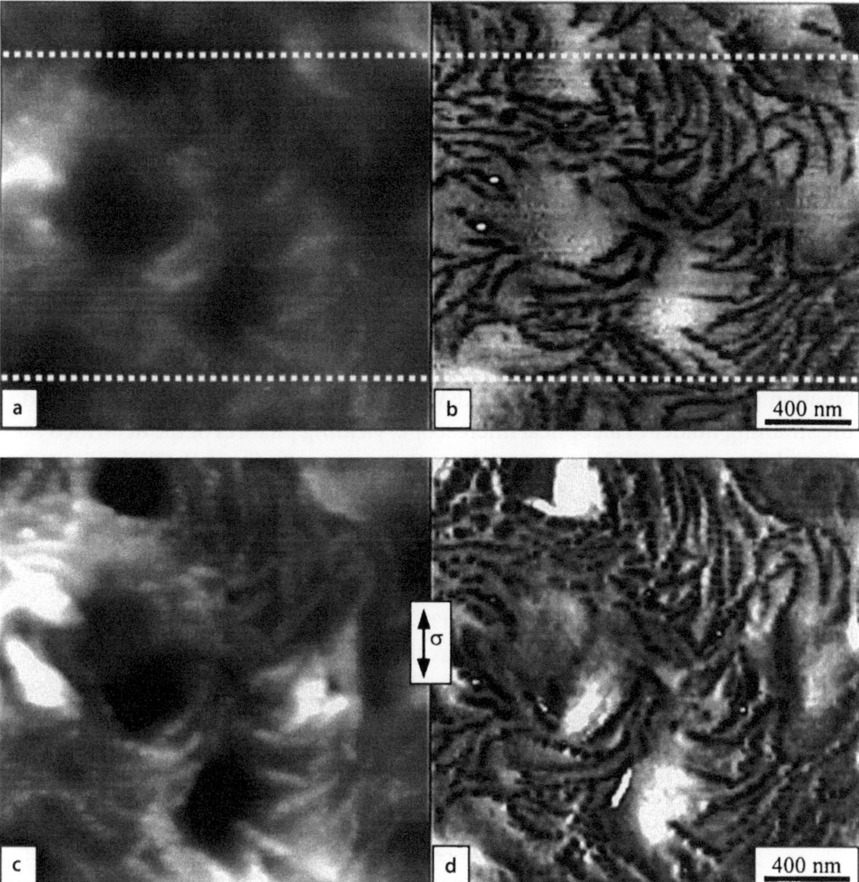

Fig. 6.10a–d. In situ deformation test performed in AFM in tapping mode: **a,c** height images and **b,d** phase signal images; **a,b** before deformation and **c,d** after deformation

It was also recently shown that AFM can be used to monitor micromechanical deformation processes in real time (e.g. in situ) upon the external loading of materials. As an example, a blend of high-density polyethylene/1-hexene copolymer with a weight fraction of 80/20 is shown in Fig. 6.10. Specimens of size 40 μm × 8 mm × 4 mm were prepared from the bulk material at room temperature with a Leica RM2165 microtome, and annealed at 150 °C in order to flatten the rough surfaces caused by microtoming. In situ deformation tests were carried out in an AFM (Digital Instruments, D 3000 atomic force microscope equipped with a Nanoscope IIIa controller) using a special device (see Fig. 6.2b). The micrographs in Fig. 6.10 illustrate typical results from the AFM in situ tensile test. On the left hand side, height signal images show the surface topography, and the sample's morphology is visible in phase signal images on the right hand side. A comparison of the micrographs shows that the area of 2 μm × 1.48 μm between the dotted lines in the micrographs in the first row is almost homogeneously strained to 2 μm × 2 μm in the tensile stress direction (marked by an arrow). This corresponds to a mean strain of the area of about 35%. In addition, the strong deformation of the elastomeric particles is accompanied by a very heterogeneous deformation of the surrounding semicrystalline matrix, and the alignment of crystalline lamellae into the direction of applied load is directly observed as the stress increases [15]. An additional improvement in the contrast between the lamellae and the amorphous parts can be achieved by image processing, as illustrated in the images of Fig. 6.10b,d and in Fig. 17.11b,d.

The ability to perform and observe in situ deformation tests at the same sample position at large magnification – as illustrated in Fig. 6.10 – is a great advantage of AFM. However, it is difficult to get an overview of the sample during deformation at low magnifications, which is easy to do in ESEM or HVTEM. In this case a combination of optical microscopy and AFM is helpful. A small sample is deformed in the straining holder of an optical microscope, yielding pictures of the deformed sample at different stages of the stress–elongation curve; see Fig. 6.11. The utilisation of a sample with an evaporated pattern of silver makes it possible to observe changes in the local deformation using the distances between the silver dots; see Fig. 6.12. The silver dots also allow exactly the same sample area to be found using the AFM at larger magnifications. Figure 6.13 shows a thin section of a soft, very low density polyethylene (VLDPE) filled with 10 wt% carbon nanotubes (CNT). Using the silver dots, the same place can be found as the elongation increases. The sequence of micrographs in Fig. 6.13 demonstrates the local deformation behaviour of the stiffer

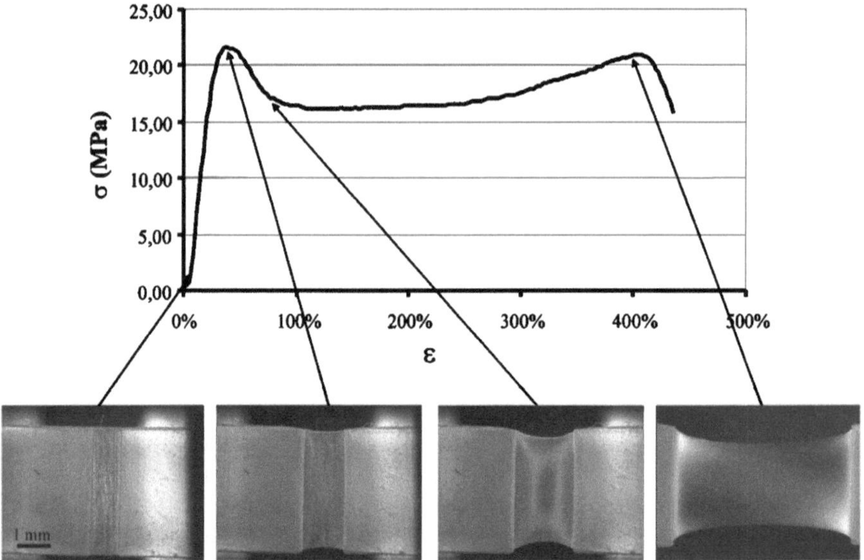

Fig. 6.11. Demonstration of an in situ deformation test where pictures were taken of a deformed miniaturised sample at different stages of the stress–elongation curve

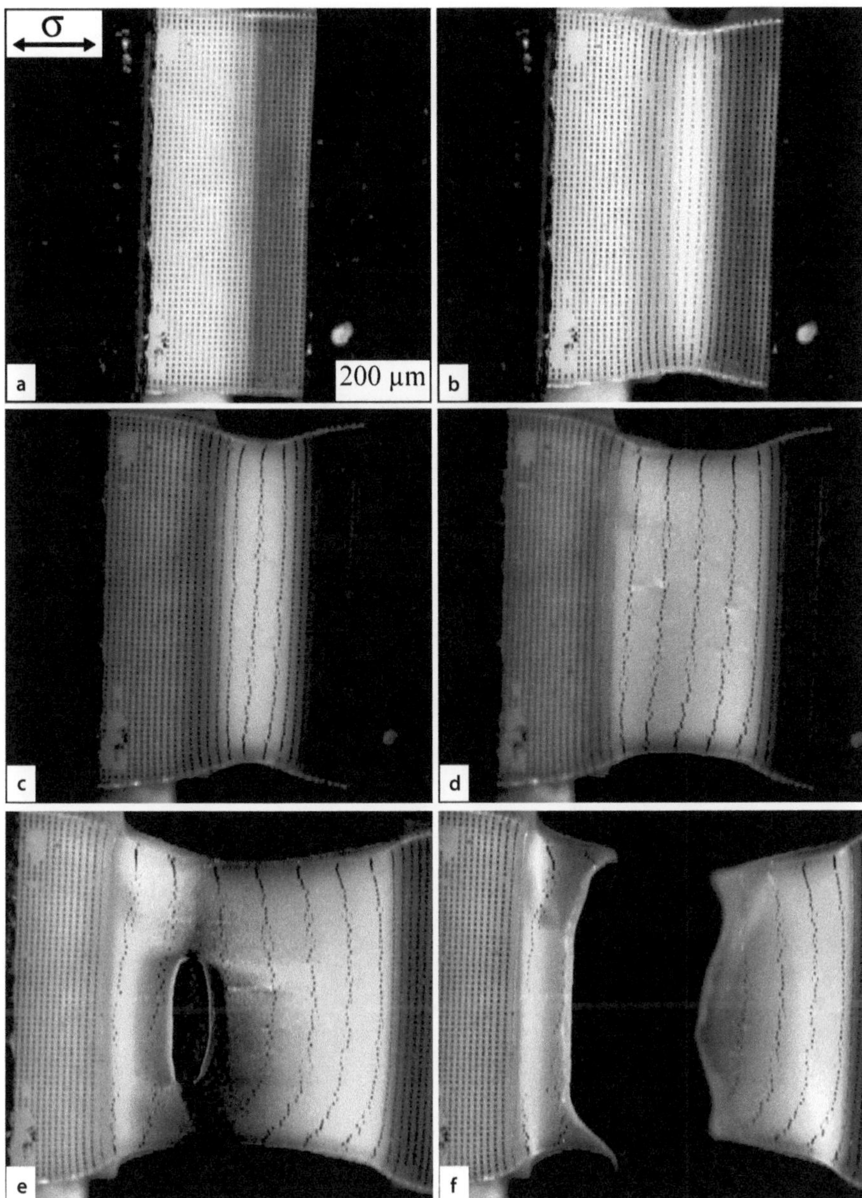

Fig. 6.12a–f. Optical images of a miniaturised sample with an evaporated pattern of silver, recorded during in situ deformation test

carbon nanotubes in the soft polymeric matrix. Some nanotubes that were originally perpendicular or oblique to the loading direction are twisted into the tension direction.

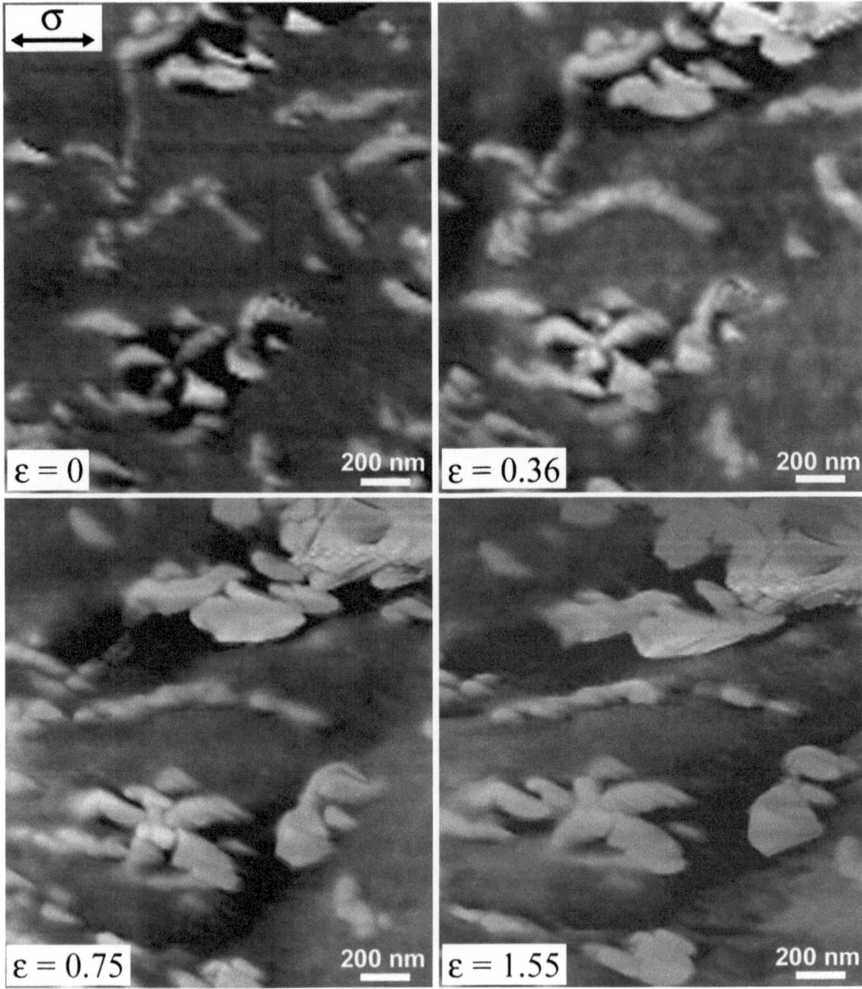

Fig. 6.13. Series of AFM images of a carbon nanotube-filled polyethylene sample deformed step-by-step to different strains (0%, 36%, 75%, 155%)

References

1. Michler GH (1992) Kunststoff-Mikromechanik: Morphologie, Deformations- und Bruchmechanismen. Carl Hanser, München
2. Pearce R, Vancso GJ (1998) J Polym Sci Polym Phys 36:2643
3. Michler GH (1993) Appl Spectrosc Rev 28:327
4. Michler GH (1995) Phys Status Solidi A 150:185
5. Michler GH (1995) Trend Polym Sci 3:124
6. Michler GH (1996) In: Salamone JC (ed) Polymer materials encyclopedia, vol 3 (D–E). CRC Press, Boca Raton, FL
7. Michler GH (1998) Polym Adv Technol 9:812
8. Kim GM, Michler GH (1998) Polymer 39:5689, 5699
9. Michler GH (1999) J Macromol Sci Phys B38:787

10. Michler GH (2001) J Macromol Sci Phys B40:277
11. Michler GH (2005) In: Michler GH, Baltá-Calleja FJ (eds) Mechanical properties of polymers based on nanostructure and morphology. Taylor & Francis, Boca Raton, FL, Ch 10, pp 379–432
12. Kim GM, Lee DH (2001) J Appl Polym Sci 82:785
13. Starke JU, Godehardt R, Michler GH, Bucknall CB (1997) J Mater Sci 31:1855
14. Kim GM, Michler GH, Rösch J, Mühlhaupt R (1998) Acta Polym 49:88
15. Michler GH, Godehardt R (2000) Cryst Res Technol 35:863

7 Image Processing and Image Analysis

Today, the use of computer-assisted methods for the analysis and evaluation of microscopic images is inevitable. The initial task of image processing is to enhance the quality of digital images for further analysis. This optimisation comprises the use of greyscale, contrast, shading correction, specific filtering methods (e.g. sharpness, high pass, low pass, etc.), as well as arithmetic operations (e.g. addition, multiplication, logic operation). This chapter also reviews methods that are used to quantitatively determine specific image information, such as relative composition, particle size, interparticle distance, intensity profile, etc. Special image analysis procedures for the determination of periodicities in micrographs (such as the orientations of structural details, long periods, domain thicknesses, etc.) involve the use of Fourier analysis. The chapter also describes the application of stereoscopic imaging to show the topography of the sample surface, e.g. in scanning electron microscopy or optical microscopy.

7.1 Overview

The microstructures of modern materials are becoming more and more complex and nonuniform. Therefore, it is very important to evaluate microscopy images in detail, objectively and quantitatively. Today, the use of computer-assisted methods for this analysis and evaluation is inevitable.

Image processing and analysis aims to modify an image (such as an electron micrograph) using specific hardware and software in such a way that the quality of the image is improved and the quantification of structural details in the image is enabled [1–11].

A digitised image, which can be obtained through various means, is a prerequisite for image processing. Modern transmission electron microscopes equipped with a CCD camera or image plates as well as scanning electron microscopes and atomic force microscopes allow the direct recording of digital images of the objects investigated [6,8,12,13]. In older electron microscopes the images were recorded on films or negatives. These analogue images can be converted into digital images using a digital camera or scanner.

Generally speaking, a "picture" is a reproduction of a structure in an analogous form. A digital image is a picture that has been created or converted into a discrete

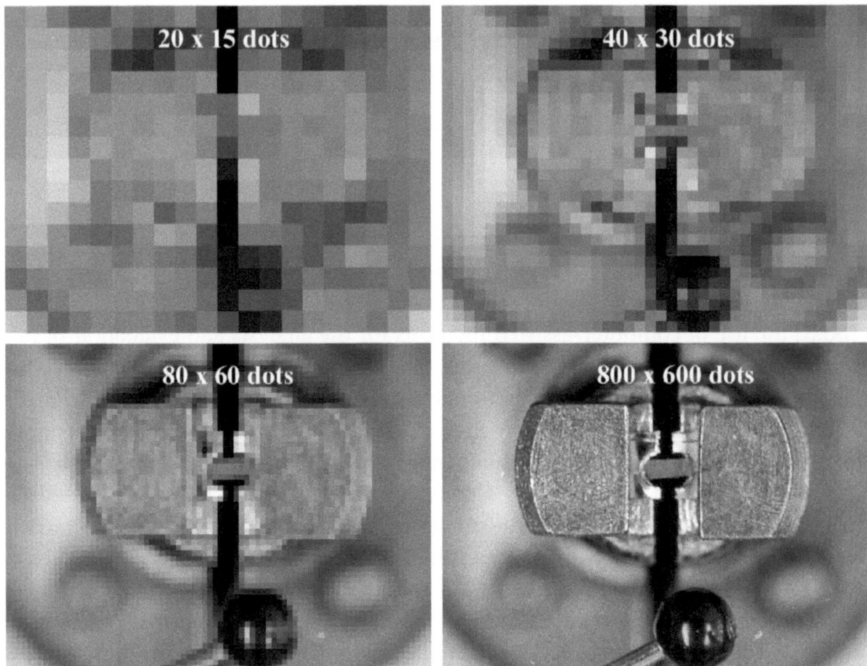

Fig. 7.1. Dependence of the perceptibility of the structural details of an object on the number of pixels; the object is a tensile module for HVEM

form. The crucial factors associated with the quality of a digital image are the resolution of the image (i.e. the number of image points or pixels, see Fig. 7.1) as well as the bit depth used for each pixel (e.g. every pixel of an eight-bit greyscale image corresponds to 256 points on the grey scale). The number of image points (pixels) that can be included in the image is determined by the technique used to take the photograph.

The greyscale value is typically between 8 and 24 bits in length. Images can be saved in different formats; one can differentiate between pixel and vector formats as well as between compressed and uncompressed ones. In microscopy, the uncompressed pixel-based TIFF format is generally used, which contains all the image information required for further processing and analysis. A description of other conventional image formats, such as those used in photography or computer-aided design, is beyond the scope of this chapter.

7.2 Image Processing

The quality of the recorded image depends on the parameters of the microscope (such as the beam intensity in TEM) and camera parameters (such as illumination time, contrast enhancement, image integration, etc.). Processing the digitised picture

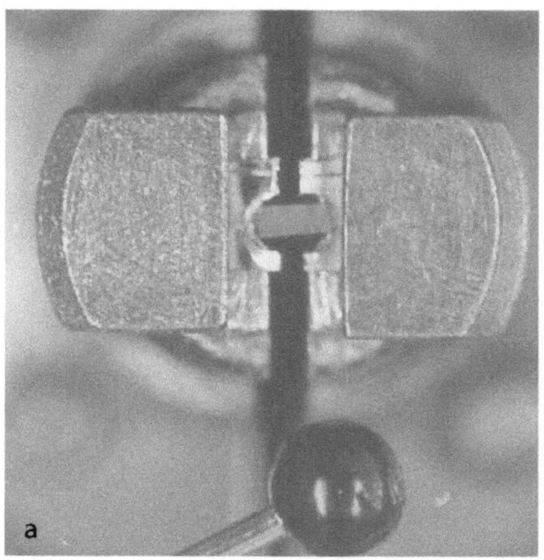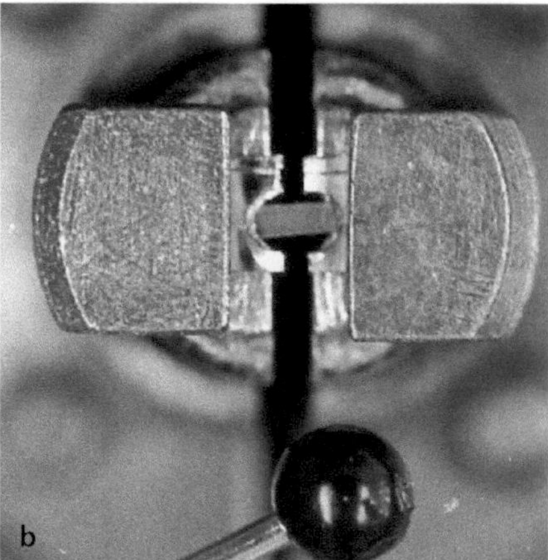

Fig. 7.2a,b. Example of the optimisation of greyscale values; the object is a HVEM tensile module: **a** original image; **b** optimised image

enhances the quality of the image (such as greyscale, contrast, etc.), and the resulting image is hence optimised for further analysis. This optimisation comprises various operations, such as shading correction (equalisation of an inhomogeneous image background), specific filtering (e.g. sharpness, high pass, low pass, etc.) as well as arithmetic operations (e.g. addition, multiplication, logic operation).

The full range of values available for adjusting pixel intensity often remains unused. This means that images will appear weak in contrast, or an image will seem too bright or too dark. An example of the optimisation of a greyscale image is shown in Fig. 7.2.

Using various filter operations, it is possible to highlight or attenuate specific structural details. After applying a low-pass filter (e.g. a Gaussian filter) to an image it will appear less sharp, with the edges of the grey areas becoming blurred and noise diminishing. In contrast, a high-pass filter will makes the image appear sharper, highlighting the finer structural details (such as edges) and removing the homogeneous regions of an image (see Fig. 7.3).

7.3 Image Analysis

There is a proverb that is often used by microscopists to brag about the value of their work: "A picture says a thousand words". However, from the point of view of modern information theory, one would ask how much information these thousand words (the picture) actually provide. It is more important to extract valuable information about the materials from the microscopy images. Microscopists have always extracted such information about the materials by interpreting micrographs. This form of qualitative study is subjective in many cases. In the industrial world, the microstructures of samples are often very complex and nonuniform, so a large number of images have

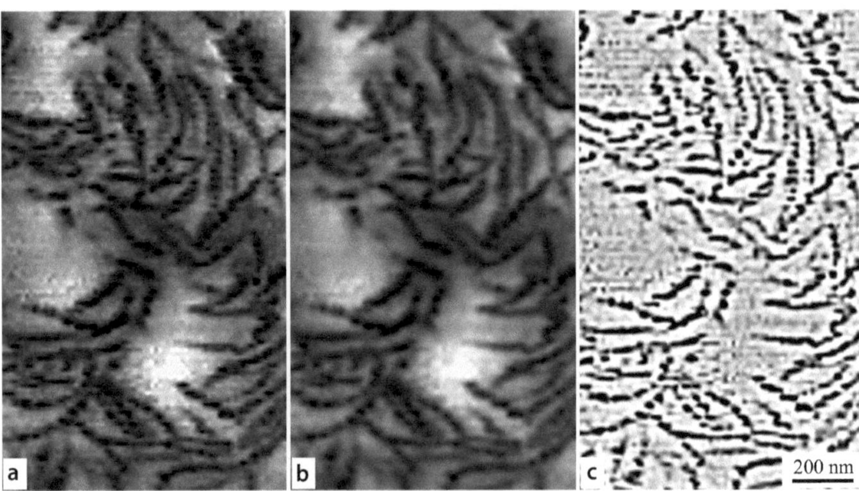

Fig. 7.3a–c. Example of filter operations leading to changes in image details: **a** AFM phase image of a P(E-co-H) blend; **b** after low-pass filter; **c** after high-pass filter

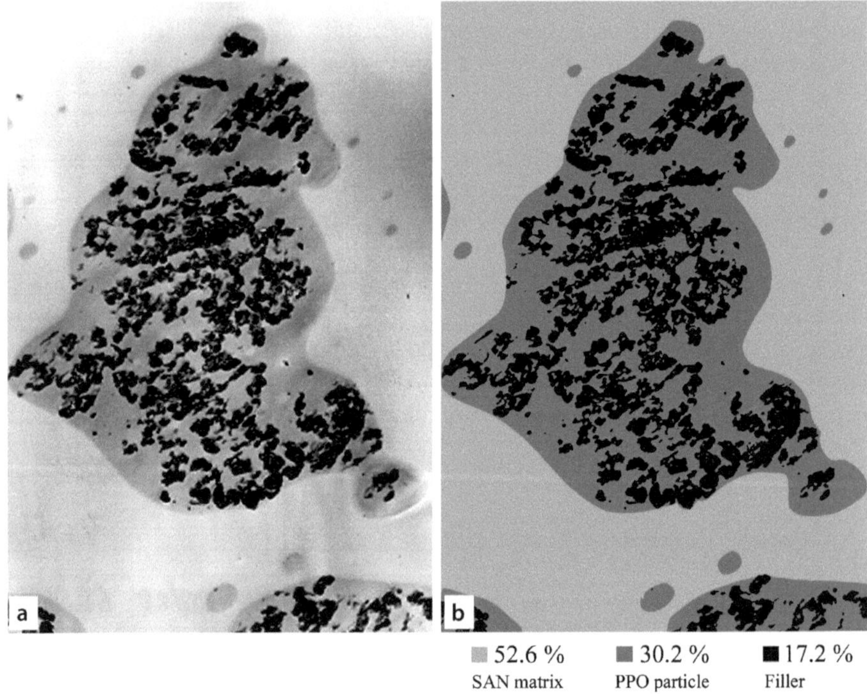

Fig. 7.4. **a** TEM micrograph of a filled SAN/PPO blend; **b** determining the volume fractions of the constituents by assigning defined greyscale values to individual components

to be analysed. The subsequent image analysis serves to quantitatively determine the specific information of interest contained by the processed images, such as the relative composition, particle size, interparticle distance, intensity profile, etc. Automatic evaluation of the image information is generally possible provided that the measured image information can be separated from the background (greyscale separation and binarization). However, many images (in particular TEM micrographs) only permit an interactive evaluation, i.e. where the image details are marked using a computer mouse.

The fundamental possibilities offered by image processing for the quantitative evaluation of structural details are clarified by the following two examples. Figure 7.4 illustrates how one can determine the volume fraction of the components in a SAN/PPO polymer blend. By defining the greyscale regions of the image under consideration (which practically reflect the amounts of the respective constituents), and calibrating the image (i.e. based on the total number of pixels), one can easily calculate the relative compositions of different phases in multicomponent systems. In Fig. 7.5 a direct (automatic) determination of the diameter of PS latex particles is presented using a special image analysis algorithm. The hardest task in automatic image analysis is to separate the individual latex particles, as there is no clearly visible boundary between them. This special analysis was performed using the image processing and analysis software "Analysis" (from Soft Imaging System, Münster, Germany).

7.4 Fourier Transformation

Special image analysis procedures for determining the periodicities in micrographs (such as the orientation of structures, long periods, domain thicknesses, etc.) involve the use of Fourier analysis, which is based on so-called "fast Fourier transformation (FFT)" [1, 14, 15]. During the Fourier transformation of an image, the periodic components of the image are determined and presented as a function of the frequency of corresponding periodicity. The evaluation of the Fourier image enables the preferred orientation direction of the structures in the image to be determined. By evaluating the reflexes of the Fourier image followed by inverse Fourier transformation, one can also quantify the size of the structural periodicity. Figure 7.6 presents the procedure for determining the lamellar periodicity in a β-iPP. In the Fourier image (see the greyscale profile in Fig. 7.6b, the regular lamellar spacing is represented by the reflexes. The average lamellar spacing can be calculated directly by evaluating each peak (Fig. 7.6b) of the reflexes.

Using this technique, additional details can be revealed for long-period distributions. Figure 7.7a shows an electron micrograph of a shish-kebab structure in a HDPE. Direct optical measurement of the long period of the oriented lamellae shown in the micrographs revealed values in the range of 10–30 nm. Laser light diffraction patterns (Fig. 7.7b) from the negatives of the electron micrographs show two maxima. Fourier transformation of the electron micrograph (Fig. 7.7d) gives two maxima for long periods, $L_1 = 36$ nm and $L_2 = 60$ nm. Long periods of the

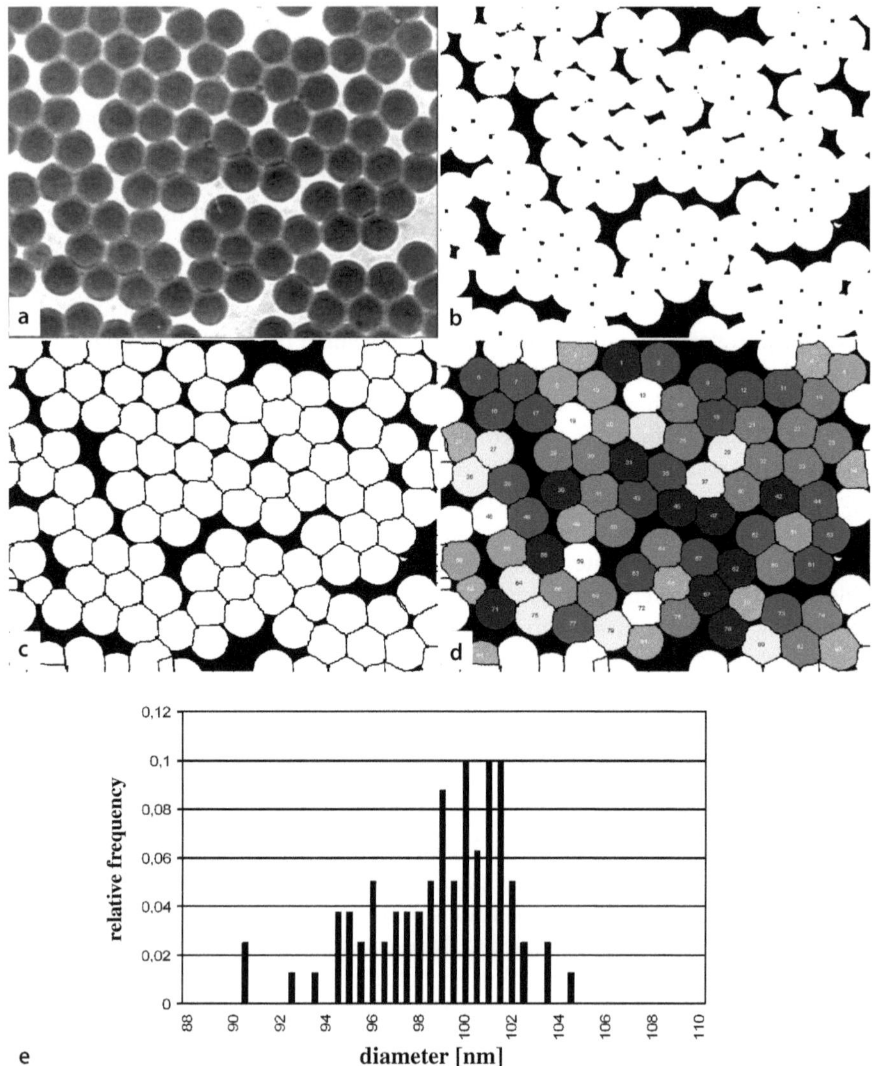

Fig. 7.5a–e. Automatic determination of the diameter of PS latex particles: **a** original image; **b** binarised image; **c** separation of particles using special software; **d** particle size analysis and assignment in a predefined category; **e** frequency distribution of particle size

same length were also detected by SAXS measurements [16]. This means that scattering techniques (laser light, SAXS) and Fourier transformation yields long periods between lamellae of around 30 nm with clear contrast (laser light) and electron density (SAXS), whereas direct inspection of the micrographs also yields smaller long periods between thinner, not so strongly stained lamellae (in the range of 10 nm).

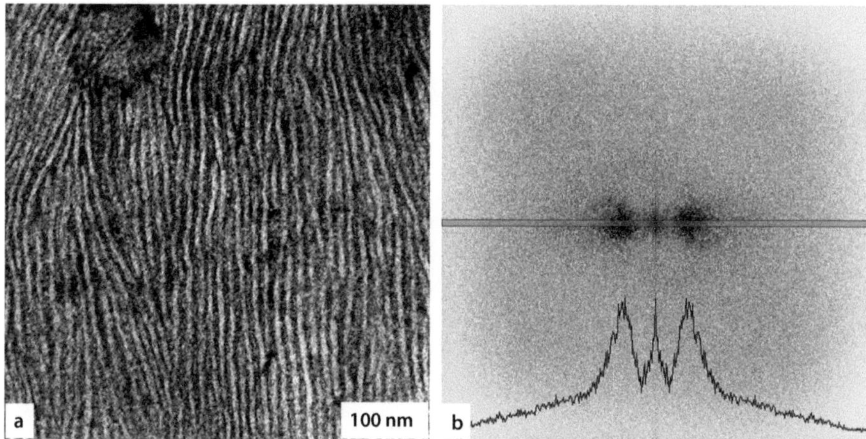

Fig. 7.6a,b. Determination of lamellar spacing in a β-iPP: **a** TEM micrograph of a stained ultrathin section (PP lamellae white, amorphous parts dark); **b** Fourier image with greyscale profile (line plot)

By applying specific filters, it is possible to enhance (or even eliminate) different reflexes of the Fourier image so that the back-transformation results in a so-called "corrected" image. This procedure modifies the image information somewhat, but the images gain clarity, and present clearly noticeable structural details (see Fig. 7.8 with a PB/PS block copolymer as an example).

Using Fourier transformation, it is possible to correct different kinds of errors in electron micrographs (such as preparation-induced striations of sections, folding, etc.) or to eliminate artefacts that do not belong to the morphology of the investigated material (see Fig. 7.9).

7.5 Stereoscopic Imaging

Stereoscopic imaging is particularly suited to the illustration of surfaces in scanning electron microscopy or optical microscopy for example. Any stereoscopic imaging is based on the concept that the human brain computes spatial information from the different images that an object creates in the retinas of the left and right eyes [17,18]. The difference between the images in the left and right retinas is caused by the different observing angles of the eyes. This is why a human is capable of seeing in three dimensions and estimating spatial distances.

A stereoscopic image pair consists of two images that display the same object before and after a tilt by a certain stereo angle ω. Stereo angles of about 3–6° are generally used. The optimum tilting angle depends on the microscopic magnification and the roughness of the sample surface. Special software creates stereo images, for example in certain colour combinations (e.g. red–green or red–blue), which produce a three-dimensional impression if viewed through coloured glasses.

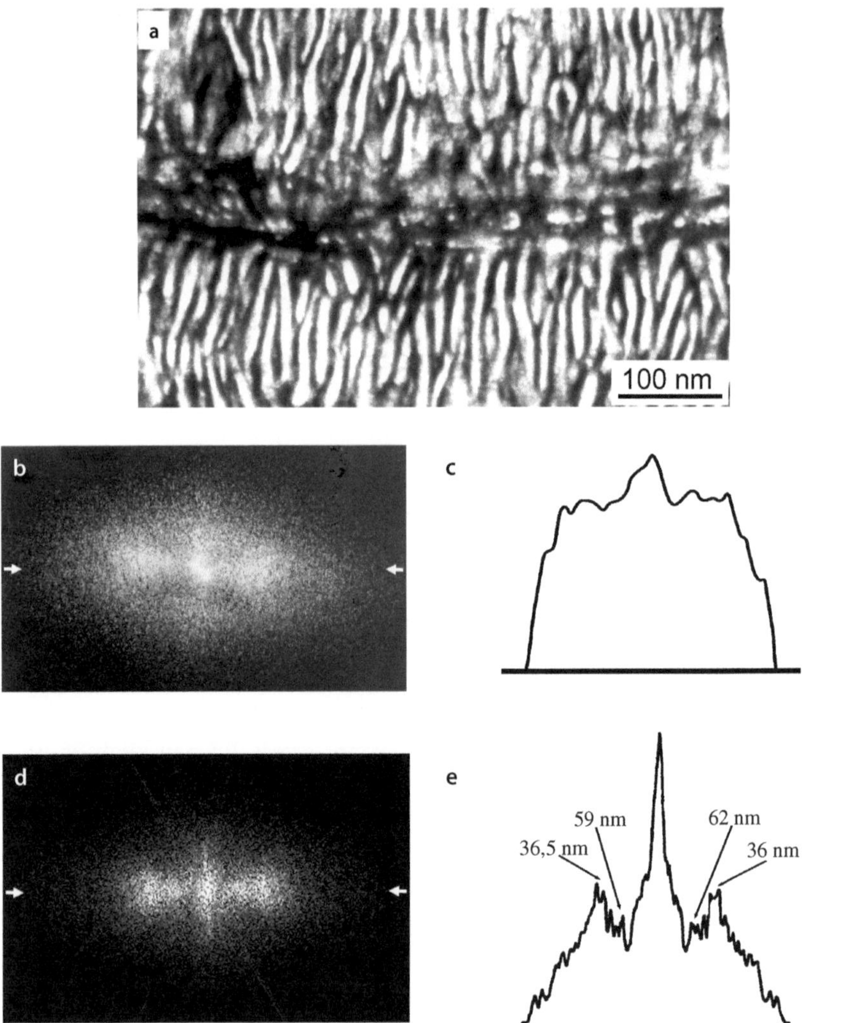

Fig. 7.7a–e. Determination of long period of HDPE: **a** TEM micrograph of shish-kebab structure; **b** laser light diffraction pattern from negatives of electron micrographs obtained at lower magnification; **c** microdensitometer curve of **(b)** between arrows; **d** FFT spectrum of **(a)**; **e** greyscale value distribution of **(d)** between arrows

The heights of image points can be computed based on two stereo images with a defined stereo angle ω between them. Due to the tilt, individual points on the object are usually at different positions in the images of the stereoscopic pair. In the initial position, the tilt axis must be parallel to the vertical borders of the images. The object is then tilted by $-\omega/2$ and the first image is taken. Then the object is tilted by $+\omega/2$ from the initial position and the second image is obtained. The height h of any point resulting from its x-shift x_{sh} between

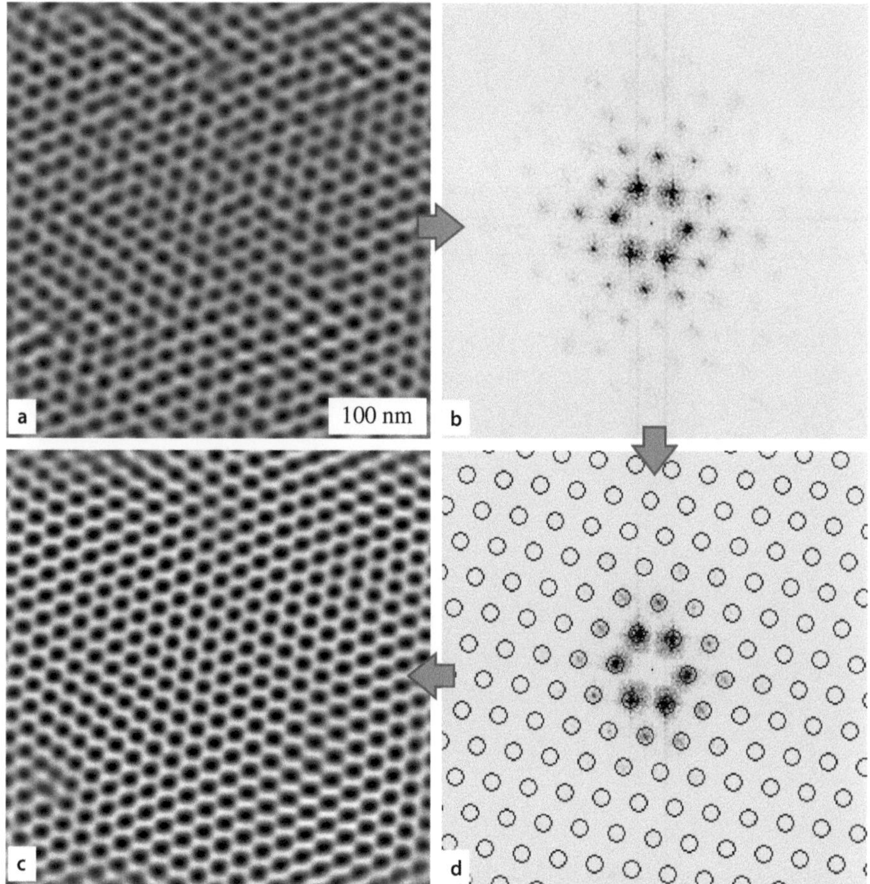

Fig. 7.8a–d. Optimum presentation of hexagonal arrangement of PB cylinders in a PS/PB block copolymer: **a** original AFM image (PB cylinders appear dark); **b** FFT image of **(a)**; **c** inverse-transformed image after filtering; **d** modified FFT image with lattice filter

the two images is:

$$h = \frac{x_{\text{sh}}}{\sin \omega} \, . \tag{7.1}$$

When the height is measured, one image from the stereoscopic pair will be used as the reference image and the other as the search image. The height of a point is calculated by considering the x-shift between the reference point and the corresponding search point. Usually the individual image points have different heights and so they are shifted to different degrees and the search image is more or less distorted. When a human observer compares the two image patterns, he will use the terms "the same pattern" or "nearly the same pattern" to describe the degree of congruence. The degree of coincidence or correlation between the reference pattern and the distorted

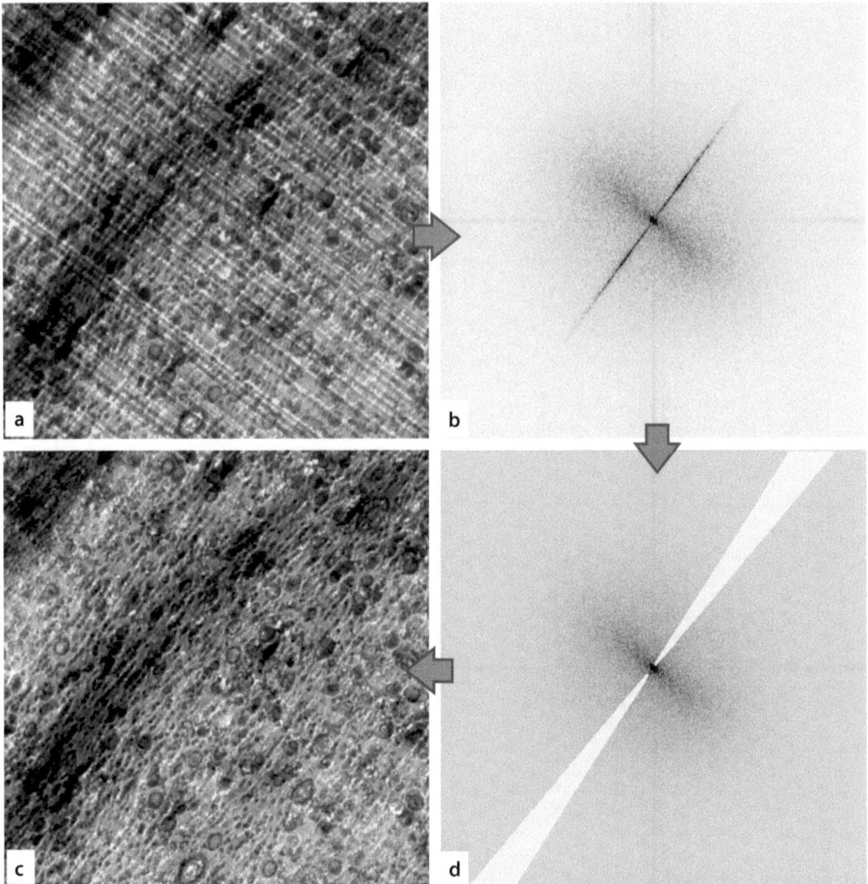

Fig. 7.9a–d. Elimination of preparation (ultrathin sectioning)-induced striations from a TEM image of an ABS material due to Fourier transformation and filtering: **a** original image; **b** FFT image; **c** inverse-transformed image after filtering; **d** filtered FFT image

and redetected pattern in the search image can be described by a mathematical magnitude that ranges from 0 to 1. The closer the correlation value is to 1, the better the congruence between both patterns.

There is no generally valid relation between the correlation and the accuracy of the calculated height. If the reference pattern is extremely distorted due to the tilt, a bad correlation should be expected but the resulting height might still be correct. On the other hand, if there are no clear structures in the images, the correlation might be excellent but the height is completely wrong, because the reference pattern cannot be detected unambiguously.

Three-dimensional electron tomography is a technique in which 3-D information on a specimen is obtained by tilting the specimen in a transmission electron microscope and acquiring projection images at many different viewing angles. Subsequent reconstruction in the computer reveals spatial information on the specimen under consideration. As tomography is increasingly being used as a routine research technique in microscopy, it has become necessary to automate and streamline the entire process of 3-D reconstruction, from sample preparation to visualisation and analysis of the resulting 3-D data cubes [19–22]. When attempting to obtain high-quality 3-D sample data using electron tomography, there are several aspects that are of major importance, some of which are associated with the TEM during the acquisition of the data, while others come into play during the subsequent reconstruction process (see also Chap. 2).

References

1. Castleman KR (1996) Digital image processing. Prentice-Hall, Upper Saddle River, NJ
2. Sawyer LC, Grubb DT (1996) Polymer microscopy, 2nd edn. Chapman and Hall, London
3. Kirkland EJ (1998) Advanced computing in electron microscopy. Springer, New York
4. Pitas I (2000) Digital image processing algorithms and applications. Wiley, New York
5. Rourdeyhimi B (2000) Imaging and image analysis applications for plastics. William Andrew, Norwich, UK
6. Zhang XF (2001) Progress in transmission electron microscopy 1. Springer, Berlin
7. Russ JC (2002) The image processing handbook. CRC Press, Boca Raton, FL
8. Yamanaka CR, Zhigang RL, Li L, Khigang RL (2003) Industrial application of electron microscopy. Marcel Dekker, New York
9. Jahne B, Jahne J (2004) Practical handbook on image processing for scientific and technical applications. CRC Press, Boca Raton, FL
10. Jähne B (2005) Digital image processing: Concepts, algorithms, and scientific application. Springer, Berlin
11. Umbaugh SE (2005) Computer imaging: Digital image analysis and processing. Taylor & Francis, Boca Raton, FL
12. Hader DP (2001) Image analysis. CRC Press, Boca Raton, FL
13. De Graef M (2003) Introduction to conventional transmission electron microscopy. Cambridge University Press, Cambridge
14. Graham D, Barrett AN (1997) Knowledge-based image processing systems. Springer, Berlin
15. Wojnar L, Wojnar W (1999) Image analysis: Applications in materials engineering. CRC Press, Boca Raton, FL
16. Michler GH, Naumann I, Baltá Calleja FJ, Ania F (1997) Acta Polym 48:36
17. Russ JC, Dehoff RT (2000) Practical stereology. Kluwer Academic/Plenum, New York
18. Staniforth M, Goldstein J, Newbury DE, Lyman CE, Echlin P, Lifshin E, Sawyer LC Michael JR, Joy DC (2002) Scanning electron microscopy and X-ray microanalysis. Kluwer Academic/Plenum, New York
19. Frank J (1992) Electron tomography: Three-dimensional imaging with the transmission electron microscope. Springer, New York
20. Yaroslavsky L (2003) Digital holography and digital image processing: Principles, methods, algorithms. Kluwer Academic, Boston, MA
21. Schoenmakers RHM, Perquin RA, Fliervoet TF, Voorhout W, Schirmacher H (2005) Microsc Anal 96:13
22. Fliervoet T (2005) Imag Microsc 2:45

Part II

Preparation Techniques

8 Problems Associated with the Electron Microscopy of Polymers

This introductory chapter to Part II – which deals with preparation techniques – summarises the main problems associated with the investigation of polymers by electron microscopy. The irradiation sensitivity of polymers can be reduced by taking precautions with the instrumentation and the manner of operation. It should also be noted that the irradiation sensitivity of polymers can be utilised for special contrast development effects. The problem of the low contrast between structural details of polymers can be overcome through the use of chemical staining, physical effects or surface etching. The third problem is the preparation of ultrathin specimens from bulk polymers, which is successfully solved by applying (cryo)ultramicrotomy. Some additional methods are mentioned, and the applicabilities of various methods for studying the morphologies of several classes of polymers are summarized. All of these methods are described in detail in subsequent chapters in Part II.

8.1 Overview

The polymers that are studied by electron microscopy can take various shapes, forms and sizes. Examples include powders obtained directly from macromolecular synthesis and granules from compaction or a first extrusion step, and injection and extrusion-moulded test parts in the centimetre range or large pieces for different applications. In general, direct investigations of polymers by electron microscopy involve three problems:

1. The usual preparation techniques applied to inorganic samples, particularly for TEM investigations, cannot be applied, and the preparation of ultrathin specimens from bulk polymers is often difficult
2. Polymers (since they are organic substances) are particularly sensitive to electron beam irradiation
3. The contrast between structural details is often very low because polymers usually consist of the same light elements (C, H, O, and others) that interact only weakly with the electron beam (see Sects. 2.4 and 4.2.5).

Several preparation and investigation techniques have been developed to overcome these difficulties. These are reviewed in this chapter, and the main steps are discussed in detail in subsequent chapters in Part II.

8.2 Irradiation Sensitivity of Polymers

The effects of irradiation on polymers have been discussed in many papers and several reviews (e.g. [1–5]). The primary effects of the interaction of electrons with organic matter are inelastic scattering processes, which yield ionisation and break chemical bonds. Secondary effects are mainly chain scission or crosslinking, mass loss, fading of crystallinity, heat generation, and charging-up. The sensitivity to irradiation decreases with increasing carbon content in the polymeric samples; in other words in the sequence: PTFE, PVC, PMMA, PC, PE, PS [4]. Irradiation processes usually proceed very quickly during irradiation in EM, which means that investigations of polymers can often only be performed on badly damaged molecules. These damaged specimens are frequently well suited for investigating the supramolecular structure (the morphology) because the morphology is often unaltered by molecular processes. This generally holds for amorphous polymers, since image formation is based on differences in mass thickness (e.g. between different polymer phases, particles, domains). For semicrystalline polymers, however, the crystalline structure should also be investigated using diffraction contrast, which disappears upon molecular damage. The damage to the specimen can be reduced by taking precautions with the instrumentation and the manner of operation:

1. If possible, "low-dose" techniques should be used [6], which means focussing on one place on the sample and taking micrographs of a different (i.e. previously nonirradiated) place. Low-dose techniques involve performing adjustments to parameters such as the magnification, matching the dosage to the sensitivity of the viewing medium, and focussing on an area of the specimen adjacent to the feature. The beam is then moved with the beam deflection coils, and the shutter is closed to reduce the irradiation of the site adjacent to the feature. The beam is directed onto the feature only when it is being recorded.
2. The presence of evaporated thin carbon layers on both sides of the specimen improves the conductivity and can reduce specimen movements and the volatilisation of molecular fragments [7, 8]. The loss of contrast is small, and the increase in dosage has been shown to be as much as tenfold for organic crystals [9].
3. The sensitivity of the photographic material can be enhanced by using special emulsions with a higher silver halide content or special development techniques [10, 11], or the use of electron image intensifiers with coupled image recording can be helpful.
4. The application of a higher accelerating voltage for the electrons will lead to a reduction of the inelastic scattering cross-section. The corresponding enhancement of the radiation resistance upon changing from 100 to 200 kV is about 50%, and it increases about threefold upon changing from 100 kV to 1 MV [12–14]. (However, it should also be remembered that the exposure of the photographic film depends on ionisation process, and so the exposure time also increases).
5. In a HVTEM with an accelerating voltage of 1000 kV or more, the high energies of the electrons allow the use of highly sensitive photographic films, such as special X-ray films with a double coating of thick layers of emulsion and a higher

silver halide content [15]. Such films cannot be used in a usual TEM because of the lower penetration power of the 100–200 kV electrons.

6. The use of a scanning transmission electron microscope (STEM) instead of a fixed-beam microscope can reduce the damage somewhat [16].

7. By cooling the specimens to low temperatures (cryomicroscopy), the mobility of the polymer molecules and all secondary processes (e.g. mass loss, amorphisation rate of crystals, crosslinking) can be reduced. Cooling specimens to the temperature of liquid nitrogen only has a small effect [1, 2], but cooling to liquid helium temperatures yields a significant increase in the polymer lifetime [8, 17, 18]; see also Sect. 3.3.2. However, it is still quite difficult to use liquid helium in a special cryomicroscope, and this rules it out as a routine operation for polymer samples.

On the other hand, it is also possible to make use of the irradiation sensitivities of polymeric materials for special effects of contrast development:

– In the case of polymer blends, the different sensitivities of the polymeric components can result in a stronger mass loss in one component than in another, yielding a the development of contrast at the start of irradiation in the EM. An example is shown in Fig. 17.13.

– In the case of semicrystalline polymers, the primary irradiation processes (rupture of chemical bonds, ionisation, etc.) are the same, but the secondary processes of crosslinking may be stronger in the amorphous regions than in the crystalline ones. This gives rise to some other effects, and the result is the development of contrast between the amorphous and crystalline parts; see Sect. 13.3.1 with Figs. 13.10–13.12.

As mentioned above, the irradiation-induced changes (at the macromolecular level) often do not impede morphological investigations performed at the supramolecular level. Moreover, the morphologies of specimens can be stabilised by applying chemical fixation and staining treatments – essentially by crosslinking the macromolecules and incorporating atoms of heavy elements – which is often necessary when preparing ultrathin sections for TEM or surfaces for SEM investigations. Another common and serious problem is poor performance during in situ deformation tests, although this problem can be overcome by implementing several precautions; see Chap. 12.

The total avoidance of any irradiation-induced effects is only possible if replicas of the materials of interest are investigated (see Sect. 9.5) or if an AFM is used.

8.3 Low Contrast of Polymers

A second problem associated with the direct investigation of polymers by transmission electron microscopy (TEM) is the low contrast between structural details. However, there are several methods that can be used to enhance the contrast:

1. The chemical staining of details can be achieved through the selective incorporation of heavy elements. Different structural details (lamellae, amorphous regions, interfaces, regions of different molecular packing densities or different free volumes, several polymer phases, and others) possess different reactivities to staining media. Also, different staining media are applicable to different materials (e.g. osmium tetroxide, ruthenium tetroxide, phosphoric-tungsten acid, chlorosulfonic acid).

2. In addition to the commonly used method of chemical staining, there are some physical methods that enhance the contrast between structural details. One method is based on the abovementioned different irradiation sensitivities of different polymeric components. Irradiation-induced changes can also enhance the contrast of semicrystalline polymers; for instance the structure and lamellar arrangements inside spherulites. Another physical method involves developing the contrast between different polymeric components by mechanically straining the sample (this is known as "straining-induced contrast enhancement").

3. Structures at surfaces can be "developed" by chemical or physical etching methods. The contrast of structural details can also be enhanced by evaporating metal atoms at small angles (shadowing).

These chemical and physical methods of contrast enhancement will be discussed in detail in Chap. 13, and the methods of structural development at surfaces are explored in Sect. 9.3.

8.4 Methods of Investigating the Morphologies of Polymers

8.4.1 Powders, Particles, Fibres

Powders, other small particles or fibres can be investigated "as is", i.e. without any additional treatment. After mounting the material on a sample holder, its shape, size, and surface structure can be studied directly in SEM or ESEM. Dispersions of very fine particles or fibres can be mounted on a (C-coated) grid for direct TEM inspection. Embedding and ultramicrotomy of the particles is usually needed to reveal the internal particle structure in TEM (see Chap. 10).

8.4.2 Bulk Polymers

The sizes of the supramolecular structures or morphological features of bulk polymers range from the smallest details that are just above the molecular level (smaller than 1 nm) up to structures that are larger than 100 μm – a range of more than five orders of magnitude. Several different preparation techniques and microscopic methods are available for studying structure. However, there is a general methodology for structure determination that should be applied to all microscopic investigations. The various steps included in preparation and investigation methods, as well as the corresponding results obtained, are schematically illustrated in Fig. 8.1.

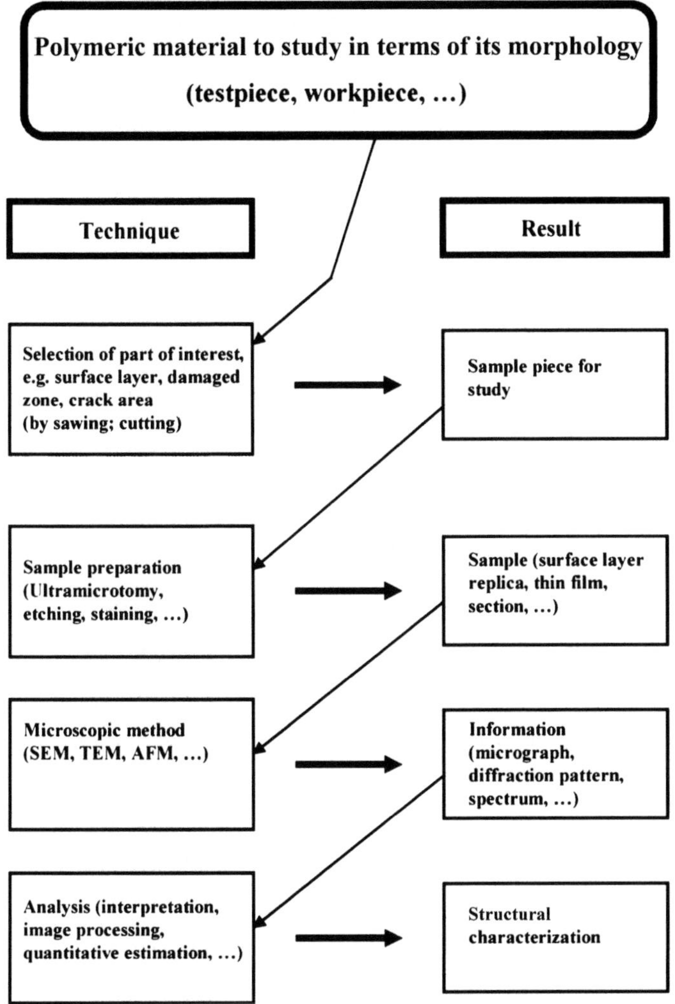

Fig. 8.1. Illustration of the steps involved in sample preparation and investigation methods, and the corresponding results from them during structure determination

Preparation methods for investigating the morphologies of bulk polymers differ from the techniques used for inorganic materials. The usual technique applied to inorganics – chemical or electrolytical thinning – is not applicable to polymers because they swell in solvents and therefore cannot be thinned continuously. Three methods for investigating the morphologies of polymers are generally available (Fig. 8.2):

1. The preparation of special surfaces (e.g. brittle fracture surfaces, smooth and selectively etched surfaces) that yield information on the internal structure of the material. These surfaces are investigated by means of replicas in the TEM or directly in the SEM or AFM.

2. The preparation of thin sections by ultramicrotomy, generally after special fixation and staining procedures have been performed. Investigations are carried out by conventional TEM, HVTEM or AFM.
3. The preparation of special thin films [solution-cast films, focussed ion beam (FIB) sections] with an additional staining treatment (as used for ultramicrotomy) and then studying by TEM or AFM.

The methods used to prepare surfaces are covered in the next chapter, while those used to prepare ultrathin or semi-thin sections are discussed in Chaps. 10 and 12 and the preparation of special thin films is reviewed in Chap. 11.

These three groups of preparation techniques are generally applicable to various polymers; however, some are more convenient for some groups of polymers while others are better for other polymer groups. A rough summary of the applicability of the preparation techniques to various types of polymers is provided in Table 8.1.

Whether or not these different preparation and investigation methods are successful depends on the kind of sample under examination and its structural details, and they do not work to the same extent for all morphological types. Therefore, to perform a full structural determination it is often necessary to apply several techniques. This approach is illustrated for semicrystalline polymers in Fig. 14.9. Additionally, different techniques enable different maximum resolutions in EM. Figure 13.17 shows resolution that can be obtained for semicrystalline, lamellar polyethylene when it is prepared with different techniques and contrast enhancements.

The structures observed in electron micrographs will usually be estimated qualitatively or semiquantitatively in order to derive information about shapes, distributions, arrangements or types of structural components. Quantitative information about structural details, such as the diameters of particles, the thicknesses of lamellae, or others, are usually determined directly from the micrographs using a PC. The automatic analysis of structural details is covered in the discussion of image processing in Chap. 7.

8.5 Methods for Studying Micromechanical Processes

Due to the variety of different structural details that can occur in polymers, there are also a wide variety of micromechanical processes that can appear under load. These include changes to individual macromolecular segments (on a nanometer scale), localised plastic yielding in the form of crazing or shear bands (at the micrometre scale), up to crack propagation and macroscopic fracture (at the millimetre scale); see Fig. 12.1. Therefore, different techniques for studying these processes are required, which are discussed in detail in Chap. 12.

There are additional advantages of using micromechanical testing for morphology determination:

– Components with different mechanical properties can show enhanced contrast due to their different deformabilities (e.g. soft, rubber-like particles in a stiffer

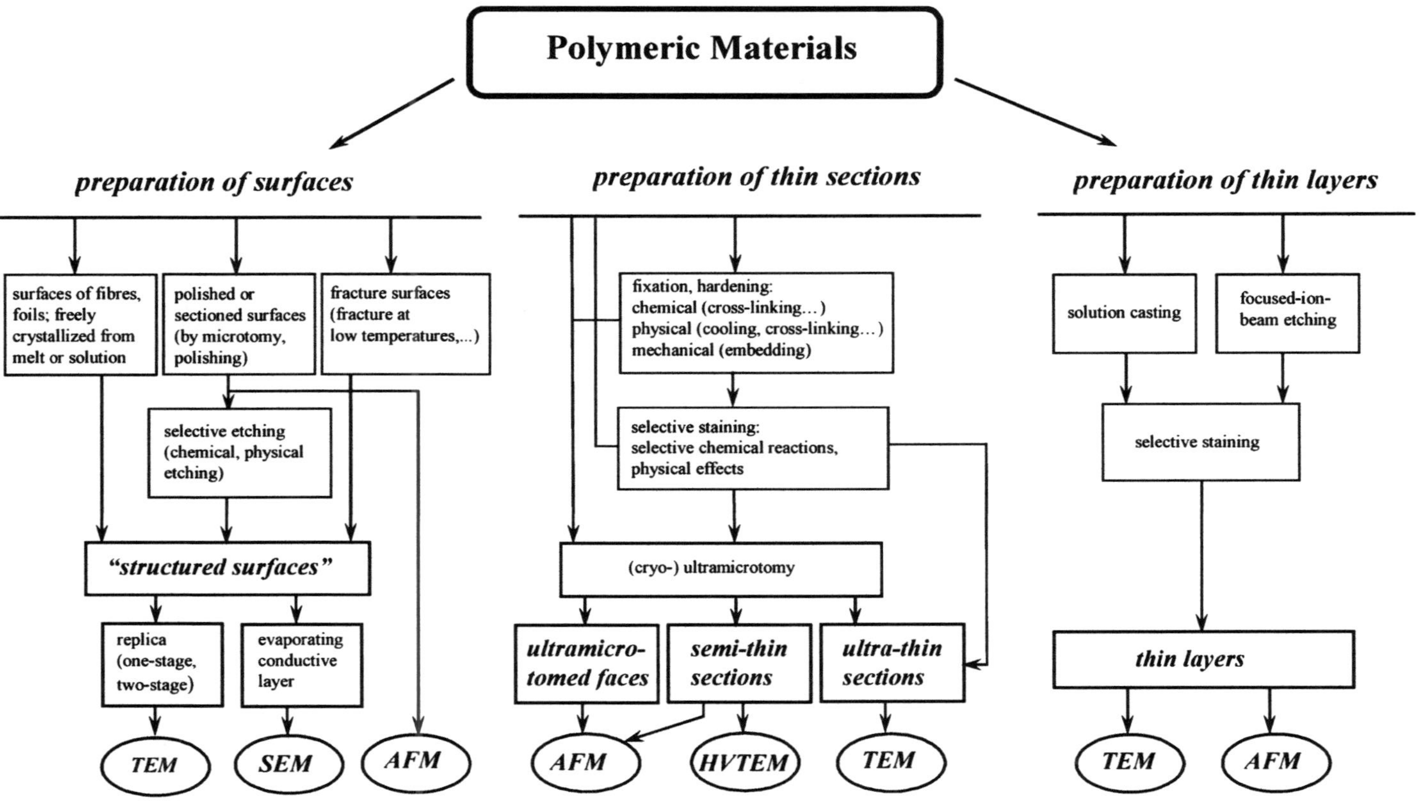

Fig. 8.2. Overview of the preparation techniques and electron microscopic methods often used to investigate the morphologies of bulk polymers

Table 8.1. Summary of the specimen preparation techniques used for EM investigations of the morphologies of different classes of polymers

Preparation technique and EM method	Amorphous polymers	Semicrystalline polymers	Polymer blends, Block copolymers	Composites
1. Preparation of "structure developed surfaces" and study of (single-stage or two-stage) replicas by TEM, studying directly by ESEM, AFM or after coating by SEM (see Chap. 9)				
a) Surfaces "as is" if they show a structure, e.g. after free crystallisation of films or foils from melt or solution		x		
b) Smooth (polished or sectioned) surfaces and selective etching (chemically using solvents or physically using ion beams or glow discharge)	x	x	x	x
c) Fractured surfaces from low-temperature fracture, brittle fracture		x	x	x
d) Surfaces from "melt fracture" (high-temperature fracture, soft matrix fracture)			x	x
2. Preparation of ultra- or semi-thin sections by (cryo)ultramicrotomy after chemical or physical fixation and study by TEM, STEM, HVTEM or AFM (see Chaps. 10, 13)				
a) Sufficient density differences exist for contrast between structural details			x	x
b) Enhancement of contrast by chemical treatment (with one chemical agent, by combined attack of several media, or after physical activation)	x	x	x	
c) Enhancement of contrast via physical effects				
– irradiation-induced contrast enhancement		x		
– straining-induced contrast enhancement	x		x	
d) Use of diffraction contrast, preparation of thin sections with cryoultramicrotomy		x		
3. Preparation of special thin films with additional staining procedure (see Chap. 11)				
a) Solution-cast films	x	x	x	
b) Thin sections cut with focussed ion beam (FIB)				x

matrix). This effect is termed "straining-induced contrast enhancement" (see Sect. 13.3.2).

– The strengths of interfacial layers in heterogeneous polymers can be checked through straining. In particular, a low interfacial strength will result in interfacial rupture (phase decohesion, unbonding, see Fig. 21.6) under strain. This can allow, for instance, the effects of using surface treatments or coupling media to be evaluated.

References

1. Grubb DT (1974) J Mater Sci 9:1715
2. Reimer L (1975) Review of the radiation damage problem of organic specimens. In: Siegel BM, Beaman DR (eds) Electron microscopy. Wiley, New York, p 231
3. Glaeser RM (1975) Radiation damage and biological electron microscopy. In: Siegel BM, Beanan DR (eds) Electron microscopy. Wiley, New York, p 205
4. Misell DL (ed) (1977) Developments in electron microscopy and analysis (Conf Ser 36). Institute of Physics, Bristol, UK
5. Michler GH (1993) Appl Spectrosc Rev 28:327
6. Fischer EW (1976) Proc 4th Int Conf Phys Non-Cryst Solids, Clausthal-Zellerfeld, Germany, 13–17 Sept 1976, p 34
7. Baier P et al. (1980) Proc 7th Eur Congr Electron Microscopy, Vol 2, The Hague, The Netherlands, 24–30 Aug 1980, p 638
8. Dubochet J, Knapek E, Dietrich I (1981) Ultramicroscopy 6:77
9. Fryer JR, Holland F (1983) Ultramicroscopy 11:67
10. Boudet A, Kubin LP (1978) Proc 9th Int Congr Electron Microscopy, Vol 1, Toronto, Canada, 1–9 Aug 1978, p 498
11. Parsons DF, Marko M, King MV (1980) J Microsc 118:127
12. Thomas EL, Humphreys, CJ, Duff WR, Grubb DT (1970) Radiat Effects 3:89
13. Thomas EL, Ast DG (1974) Polymer 15:37
14. Martinez JP et al. (1982) Ultramicroscopy 8:437
15. Michler GH, Dietzsch Ch (1982) Cryst Res Technol 17:1241
16. Krause SJ, Allard LF, Bigelow WC (1978) Proc 9th Int Congr Electron Microscopy, Vol 1, Toronto, Canada, 1–9 Aug 1978, p 496
17. Glaeser RM, Taylor KA (1978) J Microsc 112:127
18. Dorset DL, Zemlin F (1985) Ultramicroscopy 17:229

9 Preparation of Surfaces

This chapter covers various preparation steps, techniques and results concerning the imaging of surfaces of polymers. Firstly, examples of the direct imaging of surfaces of polymeric samples like powders, fibres, polymer latices, etc. that do not require chemical or physical pretreatment are mentioned. Secondly, chemical etching procedures like permanganic etching and physical etching techniques like plasma or ion etching are described, and the results of these procedures are discussed for several groups of polymers like polyolefins, block copolymers, and others. Thirdly, the preparation of fracture surfaces is explained and illustrated, including cryofracture at low temperatures and the special method of soft matrix fracture or melt fracture at elevated temperatures. Finally, examples of the preparation of surface replicas that can be imaged in the TEM are presented.

9.1 Overview

Figure 8.2 in the preceding chapter showed that the preparation and investigation of surfaces is one of the ways in which the morphologies of polymers can be studied. The external surfaces of polymeric solids only contain any information on the internal morphology in some cases, because they are usually strongly modified by the processing conditions, e.g. pressure or injection moulding, extrusion, or others. Therefore, special "structured surfaces" from the interior must be prepared, using different etching techniques or fracture processes. These different techniques are discussed in the following sections.

9.2 Surfaces in Their Natural Form

Surfaces only reveal a polymeric morphology in some cases, such as for foils or fibres that were freely crystallised from the melt or solution. Another example is powder obtained directly from the crystallisation process, where the surface morphology is of great research interest (see the UHMWPE powders in Fig. 9.4).

In the simplest cases, these "as is" surfaces can be mounted onto a sample holder and studied directly by ESEM or, after evaporating a thin conductive layer, by typical SEM. To avoid agglomeration in the case of fine powders, the particles should be well separated in a liquid suspension or through ultrasound treatment. Particles that are small enough, i.e. that have diameters that are smaller than about 200 nm, can be dispersed onto a carbon-coated grid for TEM inspection. Latex particles, for instance, can be advantageously sprayed onto a grid from the dispersion in an ultrasound bath. Latex particles of hard polymers, e.g. PS particles, are stable enough to maintain their spherical shape, whereas rubber latex must be stabilised by chemical fixation before spraying (e.g. polybutadiene rubber particles can be fixed with OsO_4 or bromine). By the way, latex particles of PS with exactly the same molecular weight and therefore the same diameter can be prepared, and these can be used as a simple way to check the magnification of a TEM.

9.3 Smooth and Etched Surfaces

Smooth internal surfaces of bulk polymers can be prepared by polishing [1] or sectioning in an ultramicrotome (Chap. 10). Very smooth surfaces prepared by applying ultra- or cryoultramicrotomy of semicrystalline to heterogeneous polymers are suitable for AFM inspection. If the surfaces are not flat enough, they can be pressed against a really smooth surface, e.g. a mica cleavage surface at higher temperatures. Moreover, a parallel alignment of the sample surface in relation to the scanner is desirable for successful AFM imaging. Deviations from a very flat and well-aligned surface can lead to erroneous interpretations of AFM data (see Chap. 5).

Smooth surfaces of particle-filled polymers (composites) can be studied in SEM with backscattered electrons, i.e. in material contrast, revealing the size, shape and distribution of the filler particles. However, the preparation of brittle fracture surfaces (see below) gives better results. For polymer blends, one polymer phase can be chemically stained (e.g. rubber particles in high-impact polymers can be stained with osmium tetroxide), thus giving a sufficient material contrast in SEM (using backscattered electrons); one example is shown in Fig. 18.4.

In all other cases, the internal polymer structure must be "developed" on the surface via selective etching processes, which can be performed chemically with solvents, acids, and bases and physically using ion etching or glow discharge. Etching, in general, is a process that erodes material from the surface of a sample, changing surface properties like the topography, wettability, functionality, optical qualities, etc. In this context, etching is employed as an industrial process for material modification. On the other hand, chemical etching procedures have long been used to facilitate metallographic texture analyses. Etching for contrast enhancement in polymer microscopy includes the removal of one or more components (or "phases") of the polymer, polymer blend or composite by physical or chemical means. In most cases, the constituents of a heterogeneous polymer system are more or less sensitive to a certain treatment. The idea is to produce a surface topography that can be utilised for image contrast formation using secondary electrons in the SEM or ESEM. Etching procedures can be applied to original sample surfaces as well as to fracture surfaces or block faces smoothed by a microtome.

9.3.1 Chemical Etching

The most common etching method is the selective elimination of a material component through the application of selective solvents, strongly oxidising acids, basic compounds, or mixtures of these substances [2–4]. Excellent results are achieved, for instance, by applying a mixture of water, sulfuric acid, orthophosphoric acid and potassium permanganate to semicrystalline polyolefins [5,6]. Essentially, the etching effect is due to the higher sensitivity of the amorphous, less ordered portion of the material to the strong oxidising agent. Differences in the degradation rate produce a relief, creating "mountains" on the crystalline material; spherulites and stacks of crystalline lamellae become visible, as do fibrillar structures of oriented fibres [5–7]. In rubber-toughened polymers, rubber particles can be removed by the application of a proper etching agent. For instance, a mixture of sulfuric acid, orthophosphoric acid, water and chromic acid is used to remove the rubber phase [8]. For optimal results, it is necessary to find the appropriate etching agent, and one must balance (at least) the following parameters: composition and concentration of the etchant, temperature, and duration. The sample should be agitated vigorously using a magnetic stirrer or a shaker to remove debris from the surface and to avoid the redeposition of low molecular weight degradation products. Some authors use ultrasonic cleaners to achieve better etching results. In any case, subsequent rinsing and drying steps will be necessary. The following example will elucidate these processes.

This etching procedure, commonly denoted "permanganic etching", follows the recipes provided by Olley et al. that have been applied with great success to a number of polyolefins and polyolefin blends [5–7,9]. The somewhat simplified procedure described here is easy, fast and safe [10,11]. It can easily be modified to meet changing requirements when material parameters (e.g. crystallinity, crystalline modifications) are varied. For plate-pressed or injection-moulded standard iPP in the crystalline α-form, the procedure is as follows:

- Cut a sample of appropriate size from the region of interest.
- Produce a flat surface using a microtome. If possible, a cryomicrotome should be used to avoid plastic deformation of the polymer.
- Prepare the etchant: 5 ml of distilled water are placed into a flask that contains a Teflon-coated magnetic stirrer. Add 50 ml of concentrated sulfuric acid slowly. Add 0.55 g of potassium permanganate powder slowly under continuous stirring. This work must be done under a ventilated hood. Wear gloves and eye protection! Do not try to produce more of the etchant; it will degrade quite quickly when stored in the laboratory. There is a risk of explosion if the mixture is created in the wrong way. The proper etchant has a deep olive green colour. If it is used it becomes violet or brown.
- Place the sample in a test tube, add the etchant using a pipette, and shake it well for 20 min.
- Remove the sample and rinse it with distilled water. Put the sample in another (clean) test tube, add water, and place it in an ultrasonic bath for several minutes.
- Place the sample on filter paper and allow it to dry.

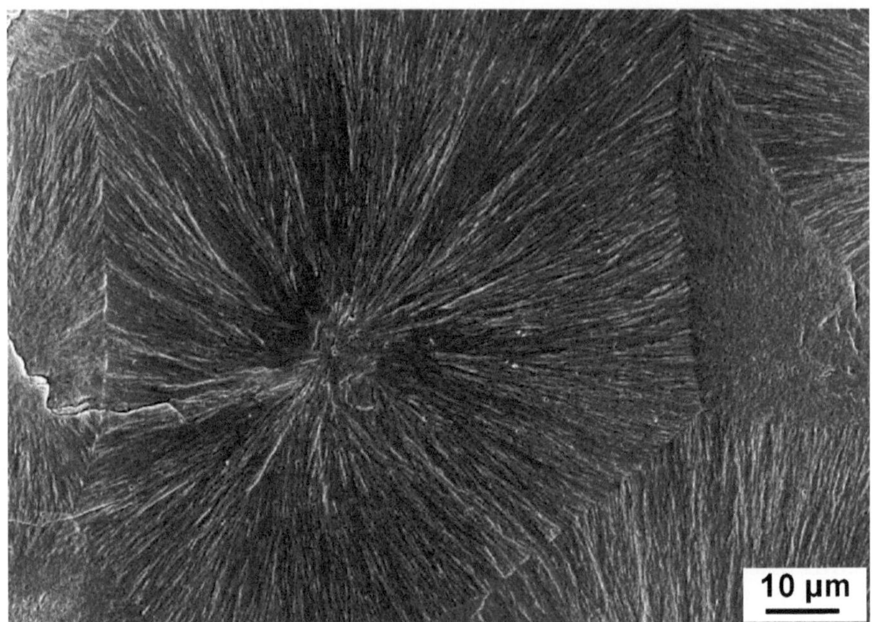

Fig. 9.1. Spherulitic morphology of α-iPP after permanganic etching. SEM image

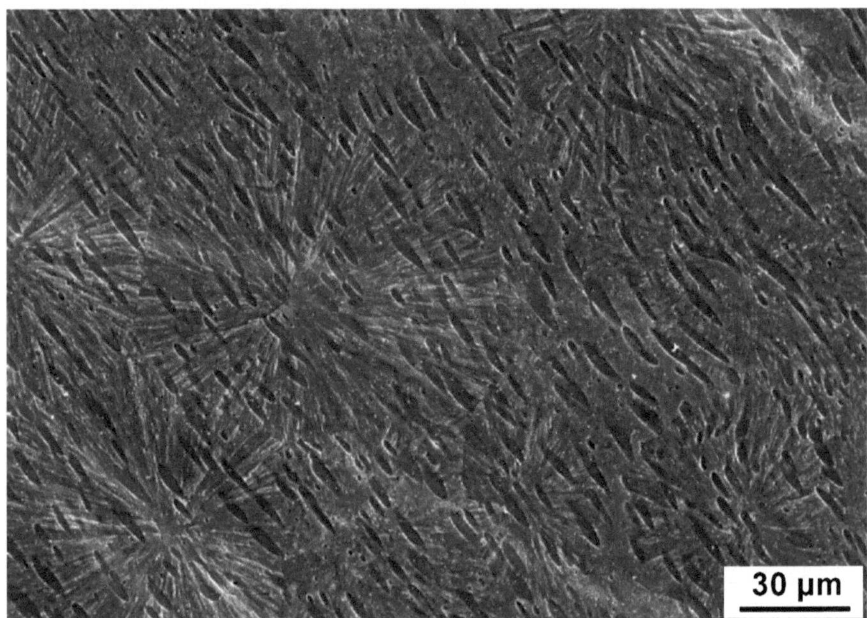

Fig. 9.2. Morphology of a blend of PP and PEO after permanganic etching. PEO phase is selectively removed, and the spherulitic morphology of PP is visible. SEM image

Of course, the rinsing procedure can also be modified. For instance, there can be an additional rinsing step using acetone. Alternatively, the procedure also can be applied to oriented fibres, cryofracture surfaces or free surfaces produced by solution-casting techniques.

Some examples of surfaces etched in the described manner are given in Figs. 9.1 to 9.3. In Figs. 9.1 and 9.2, the spherulitic morphology of α-iPP is transformed into a surface topography that is detectable using the secondary electron signal of the SEM. Moreover, the amorphous PEO phase of a PP/PEO blend (Fig. 9.2) is removed. More images of etched PP samples are reproduced in Chap. 16.

In Fig. 9.3 the same procedure is applied to a SBS block copolymer. The result is a nanoscopic surface topography arising from the typical phase nanostructure of the SBS.

The same etchant is used for syndiotactic PS (Chap. 16), ethylen-octen copolymers (Chap. 5, Fig. 5.9), and rubber-toughened PP (Chap. 18, Fig. 18.19), where only the etching times have been changed.

Table 9.1 provides an overview of selected polymers and etching agents that have been used successfully. Besides literature references, references for the examples that are discussed in this book are provided.

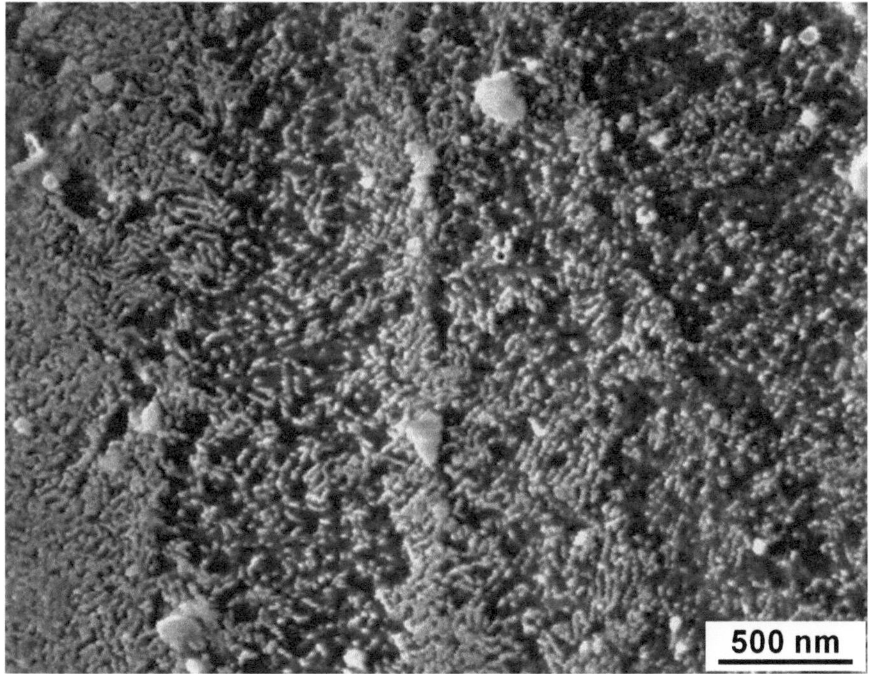

Fig. 9.3. Nanostructured surface of SBS block copolymer after permanganic etching. The polybutadiene phase is selectively removed. ESEM image

Table 9.1. Etching techniques for selected polymers

Polymer		Etching agent, action, result	Example in Chapter	References
Polyolefins				
	PE	– permanganic – solvent (xylol) amorphous phase removed; spherulitic texture, lamellar structure	16	[5–7, 12–16]
	PP	– permanganic amorphous phase removed; spherulitic texture, lamellar structure	16	
Copolymers of polyolefins	EOC	– permanganic amorphous matrix degraded, crystalline lamellae visible	5	[17]
	COC	– permanganic amorphous phase removed; spherulitic texture	16	
Blends of polyolefins		– permanganic removal of the amorphous or less crystalline phases	17, 18	
Polyesters				
	PHA	– methylamine – alcohols amorphous phase removed; spherulitic texture, lamellar structure	23	[18, 19]
	PLA	– acids amorphous phase removed; spherulitic texture	23	
	PET	– hydrolysis – nitric acid fibrillar structure of oriented fibres	9	[20, 21]
	PA	– solvents: heptane, benzol, xylol; amorphous phase removed; spherulitic texture, lamellar structure		[22, 23]
	PC	– diethylamine, ethanol amorphous phase removed; spherulitic texture, lamellar structure (if applicable)		[24]
	POM	– alcohols amorphous phase removed; spherulitic texture		[19]

Table 9.1. *(continued)*

Polymer		Etching agent, action, result	Example in Chapter	References
Blends				
	PP/PEO	– permanganic PEO phase removed amorphous phase of PP removed; spherulitic texture, lamellar structure	9	
	PET/PP	– permanganic PET phase removed amorphous phase of PP removed; spherulitic texture, lamellar structure	9	
	HIPS	– chromic acid – permanganic rubber phase removed		[8]
	ABS	– chromic acid – permanganic rubber phase removed		[8]
	PCL/PS			[25]
	PE/NR	– permanganic		[26]
Block copolymers				
	SBS	– permanganic butadiene phase removed	9	
Others				
	PVDF	– phosphorus pentoxide, – chromium trioxide – permanganic amorphous phase removed; spherulitic texture, lamellar structure		[27, 28]
	PEEK	– permanganic amorphous phase removed; spherulitic texture, lamellar structure	9	[29]
	sPS	– permanganic amorphous phase removed; spherulitic texture, lamellar structure	16	

9.3.2 Physical Etching

Another way to achieve topographic contrast on polymer surfaces is physical etching
using gaseous ions. This treatment usually takes place in a vacuum chamber where
argon ions are generated by cascade ionisation in a low-pressure vacuum
(0.1–2 mbar). This ion plasma procedure is similar to the process used in the sputter coater to create specimen surfaces conducive to SEM (Chap. 4). In contrast to
a sputter coater in a plasma etching device, here the specimen itself acts as the target. The plasma etching can be performed either in a directly coupled electric field
for anisotropic etching, e.g. at a certain angle to generate a well-defined structure on

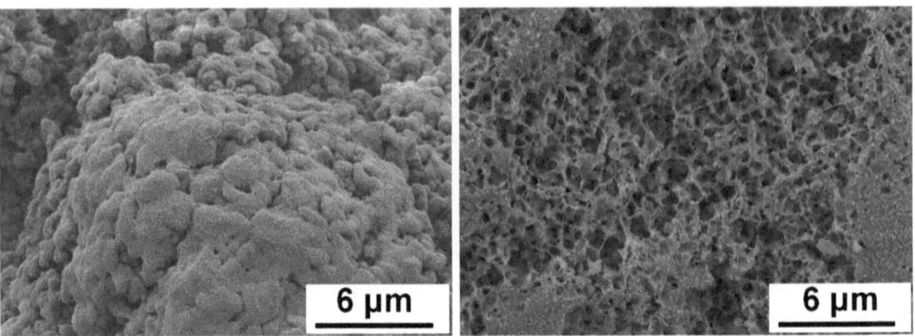

Fig. 9.4. SEM micrographs of high-frequency plasma-etched surfaces of UHMWPE powder particles. *Left*: etching by Ar ions; *right*: etching by O ions. Conditions: pressure 2.3×10^{-1} mbar, frequency 40 kHz , plasma power 40 W, etching time for argon ions 30 min and for oxygen ions 40 min

a specimen surface, or in a high-frequency electric field for isotropically etching the surface structures of the specimen. The latter is a very sparing and sensitive etching technique in comparison to the chemical one mentioned above.

On the other hand, a high-frequency plasma treatment can also be carried out by reactive ions like oxygen, chlorine or fluorine; among these oxygen is the most important gas in plasma processes. The application of oxygen [30–32] leads to oxidative reactions with the organic molecules of the polymer that generate O_2 radicals that are able to crack hydrocarbon chains and to oxidise them to CO_2 and H_2O.

Due to the fact that different polymeric structural details (crystalline and amorphous parts, different polymer phases, inorganic particles) have different etching rates, the result of plasma etching can be the elucidation of the structural details, depending on the etching time. However, the etched surfaces only represent near-surface structures during the initial stages of etching. At advanced stages, i.e. after longer etching times, special etch figures that are more influenced by the etching conditions often occur [33].

In Fig. 9.4 two examples of high-frequency plasma etching by argon and oxygen ions of a reactor powder of ultrahigh molecular weight polyethylene (UHMWPE) are demonstrated.

9.4 Fracture Surfaces

The simplest way to study multiphase, heterogeneous materials is to produce brittle fracture surfaces. This is usually done at low temperatures (e.g. at the temperature of liquid nitrogen) to avoid plastic deformation, which would hide the morphology. The fracture path occasionally follows structural details, e.g. spherulites in semicrystalline polymers, phase boundaries, or interfaces in polymer blends, so that they become visible in SEM. The application of the etching techniques discussed in Sects. 9.3.1 and 9.3.2 occasionally enhances structure visibility. Fracture surfaces of polymer composites often clearly reveal the size, shape and distribution of the inorganic filler particles in a polymer matrix. Figure 9.5 shows a particulate-filled PMMA

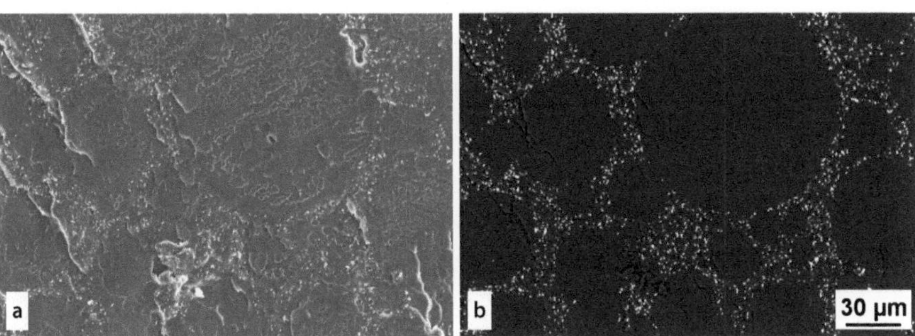

Fig. 9.5a,b. SEM images of the same area of a carbon-coated fracture surface of a particulate-filled PMMA, used as bone cement (see Sect. 23.3.3): **a** SE image highlighting the topography of the fracture surface; **b** BSE material contrast image showing the distribution of the inorganic filler in the PMMA matrix

used as bone cement (see Sect. 23.3.3) in SEM micrographs with secondary electrons (a) and with backscattered electrons, i.e. in material contrast (b). Using EDXA, the chemical nature of the filler particles is made detectable.

Brittle fracture paths do not always follow distinctive structures, particularly if there are only small differences in the mechanical stiffnesses of the individual components in the frozen state or if there are no clearly marked interfaces between components. Therefore, different accidental features are often created on the surfaces, which vary with the fracture process itself, e.g. crack velocity. These accidental features make the identification of real structures somewhat difficult. This problem can be partly overcome by using an unusual technique to produce a fracture surface at higher temperatures, e.g., in the melting range of the matrix polymer (called "melt fracture" or "soft matrix fracture" [34, 35]). Using this technique, hard filler particles in a matrix become more clearly detectable, as illustrated in Fig. 9.6. Another example of this unconventional way to prepare fracture surfaces is a carbon black modified rubber fractured at 150 °C, see Fig. 20.5. This unconventional "soft matrix fracture" is based on a specific property of polymer materials, namely the (up to three-decade) difference in the Young's modulus below and above the glass transition temperature. In this approach, the fracture surface must be obtained at a temperature at which the matrix is soft (the temperature must be sufficiently high above the glass transition temperature of the matrix) and the inclusions are hard (below the glass transition temperature of the particles) at the same time [35]. The temperature of fracture must be in a range ΔT where the matrix has a low modulus and the inclusions have a high modulus; see Fig. 9.7.

This difference in moduli can be obtained by varying the sample temperature. Besides the study of polymer composites, this procedure can also yield successful results for polymer blends, provided that a sample state can be reached in which the matrix is rubber-like while the particles are hard. Such an example was presented previously in [36] with an ABS polymer, where the soft rubber particles were hardened by osmium tetroxide treatment and the sample was then notched and fractured at 130 °C. This temperature is well above the glass transition temperature of the styrene–

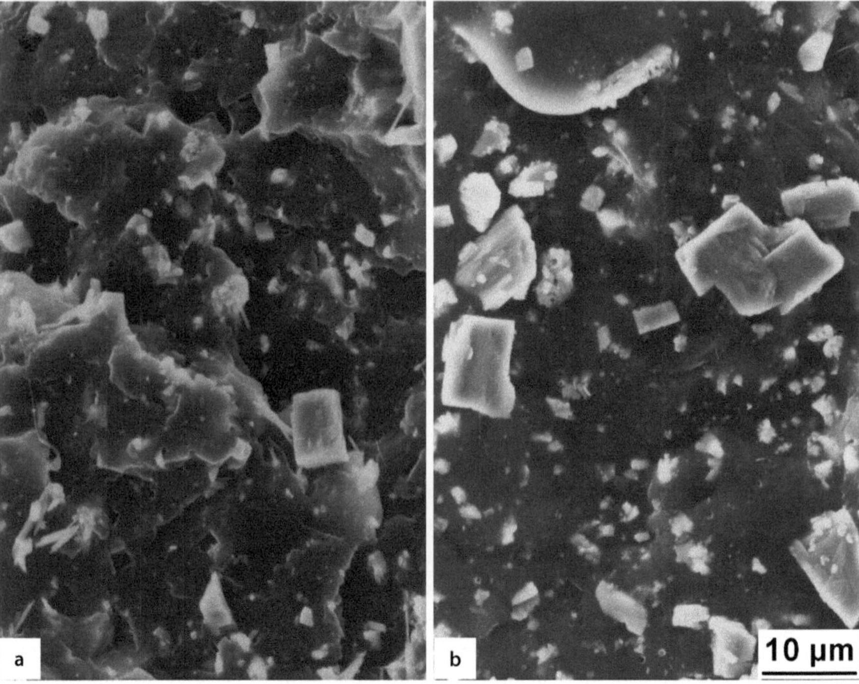

Fig. 9.6a,b. Visibility of inorganic filler particles ($CaSO_4$ and TiO_2) in a PP matrix on fracture surfaces in SEM, using different fracture processes: **a** usual brittle low-temperature fracture; **b** high-temperature fracture, prepared at 140 °C ("soft matrix fracture"). From [35], reproduced with the permission of Chapman and Hall

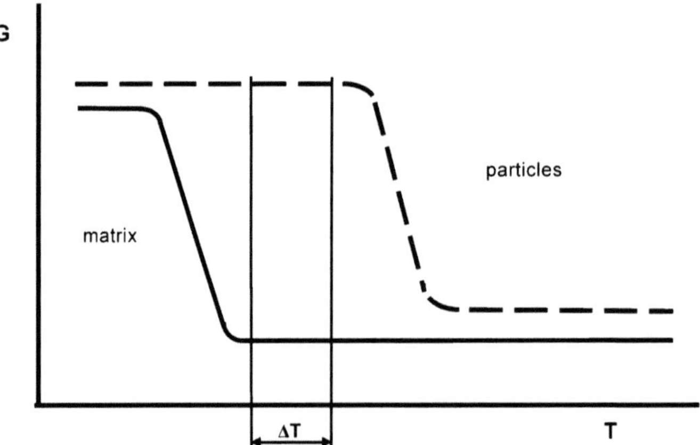

Fig. 9.7. A sketch of the moduli (G) conditions needed to apply the method using soft matrix fracture in the temperature range ΔT (after [35])

acrylonitrile matrix, but the stained particles remain hard, so that the conditions for the moduli (Fig. 9.7) are fulfilled. Please note that the same staining of rubber particles is used to visualise the particles on cut surfaces in SEM using material contrast with backscattered electrons (see Fig. 18.4) and in thin sections in TEM (see Figs. 18.2 and 18.3).

In general, methods for the preparation and electron microscopic investigation of surfaces (etched surfaces or fracture surfaces) are relatively easily and successfully performed if the polymers contain large structures or structurally clearly distinguishable parts. They are unsuited to revealing very fine details or for examining the surface at high resolution.

There are two other surface techniques, which are only interesting from a historical point of view:

- The method of decorating the surface with gold, i.e. the evaporation of Au under high vacuum onto substrates, is very useful for studying surface steps of monoatomic height on inorganic crystal surfaces [37]. However, there is no clear decoration effect on polymer surfaces. Another problem arises from the need to cover the gold particles with a carbon film and to remove film and particles from the substrate, which is very difficult for polymers (see below). There are reports of the decoration of ultrathin solution-cast films of PE and PA [38, 39] and of the decoration of polymer crystals with paraffin [40].
- Another rarely used technique with a highly variable success rate is called "surface rupture". When a matrix material is peeled off a polymer's surface, especially after activation by oxygen etching, a thin polymer layer can be torn from the polymer surface and directly studied by TEM [41].

9.5 Investigation of Surfaces

Polymer surfaces exhibiting structural development after etching or fracture are usually studied directly by SEM or ESEM. SEM investigation demands that the surface is covered with a very thin conductive layer, which is usually performed by evaporating carbon or, for a better contrast, carbon/gold under a high vacuum or by a sputtering process.

In the past, surfaces have been investigated by TEM using the replica technique. A "single-step replica" is produced by evaporating a thin carbon film onto the surface followed by platinum/carbon shadowing (see below) and partial dissolution of the matrix. This film with platinum structures on original surface edges gives an impression of the surface and can be investigated in the TEM. Since polymers usually swell strongly during dissolution, a thin evaporated carbon film will usually be destroyed. Therefore, a "two-step replica" can be used instead. In a first step, a primary replica is made of the surface by means of a thick, mechanically stable plastic foil (matrix). Next, after the foil has been mechanically removed from the surface, a secondary replica is made of the contact plane of the latter, e.g. by using a platinum/carbon shadow cast (see below). After dissolving the matrix foil, the C-film is investigated in the TEM. Matrix materials that are particularly well suited to this are the solvoplastic substances acetyl cellulose ("Bioden", soluble in methyl acetate) and cellulose acetobutyrate ("Triafol", soluble in acetone), which, after slight dissolution of their contact surfaces, are pressed as a thin foil onto the specimen to be replicated. Polymer solutions such as cellulose acetate in acetone and polymerising substances, or, as a special technique, an evaporated silver matrix reinforced by an electrolytically produced stable copper layer, are also suitable. Polymer residues that adhere to the matrix foil can be removed by activated oxygen plasma or by thermal decomposition in a high vacuum [42].

To enhance the contrast of the structures, shadowing is carried out in a vacuum evaporator using a heavy metal, such as gold, platinum or a platinum/palladium mixture. Of the numerous replica techniques developed for electron microscopy, the simultaneous evaporation of platinum/carbon film 20–50 nm thick [43] has proven to be the best. The evaporation of the metal is performed at an oblique angle to the surface (generally at an angle of 30 °C to 60 °C). The edges of the structure facing away from the source of the heavy metal will not be coated. The length of this uncoated area (or shadow) of carbon film depends on the evaporation angle and the height of the structure.

References

1. Linke U, Kopp WU (1980) Prakt Metallogr 17:479
2. Berndtsen N, Jansen FJ (1979) Präparation von Polymeren für die Licht- und Elektronenmikroskopie. Institut für Kunststoffverarbeitung an der RWTH, Aachen, p 19
3. Olley RH (1986) Sci Progr 70:17
4. Sawyer LC, Grubb DT (1996) Polymer microscopy, 2nd edn. Chapman and Hall, London, p 122
5. Olley RH, Hodge AM, Bassett DC (1979) J Polym Sci Polym Phys 17:627
6. Olley RH, Bassett DC (1982) Polymer 23:1707
7. Bassett DC, Olley RH (1984) Polymer 25:935
8. Bucknall CB, Drinkwater IC, Keast WE (1972) Polymer 13:115
9. Olley RH, Bassett DC, Hine PJ, Ward IM (1993) J Mater Sci 28:1107
10. Henning S, Michler GH, Ania F, Baltá Calleja FJ (2005) J Colloid Polym Sci 283:486
11. Henning S, Michler GH (2005) In: Michler GH, Baltá Calleja FJ (eds) Mechanical properties of polymers based on nanostructure and morphology. CRC Press, Boca Raton, FL, p 245
12. Mercx FPM, Benzina A, van Langeveld AD, Lemstra PJ (1993) J Mater Sci 28:753
13. Li J, Lee YW (1993) J Mater Sci 28:6496
14. Shahin MM, Olley RH, Blissett MJ (1999) J Polym Sci Polym Phys 37:2279
15. Zok F, Shinozaki DM (1994) J Mater Sci Lett 13:940
16. Reding FP, Walter ER (1959) J Polym Sci 38:141
17. Doshev P, Lohse G, Henning S, Krumova M, Heuvelsland A, Michler G, Radusch HJ (2006) J Appl Polym Sci 101(5):2825
18. Organ SJ, Barham PJ (1989) J Mater Sci Letters 8:621
19. Shahin MM, Olley RH (1999) J Polym Sci Polym Phys 40:124
20. Cagiao ME, Baltá Calleja FJ, Vanderdonckt C, Zachmann HG (1993) Polymer 34(10):2024
21. Yoshioka T, Okayama N, Okuwaki A (1998) Ind Eng Chem Res 37:336
22. Kallweit (1975) Progr Coll Polym Sci 57:164
23. Bartosiewics L (1974) J Polym Sci Polym Phys 12:1163
24. Kämpf G (1960) Kolloid-Zeitschrift 172(1):50
25. Shabana HM, Olley RH, Bassett DC, Jungnickel BJ (2000) Polymer 41:5513
26. Hudec I, Sain MM, Kozankova J (1991) Polym Test 10:387
27. Vaughan AS (1993) J Mater Sci 28:1805
28. Hsu TC, Geil PH (1989) J Mater Sci 24:1219
29. Olley RH, Basset DC, Blundell DJ (1986) Polymer 27:344
30. Spit BJ (1963) Polymer 4:109
31. Spit BJ (1967) Faserforsch Textiltech 18:161

32. Kämpf G (1975) Progr Colloid Polym Sci 57:249
33. Friedrich J, Gähde J (1980) Acta Polym 31:52
34. Lednicky F (1986) In: Sedlaèek B (ed) Morphology of polymers. Walter de Gruyter & Co, Berlin, p 541
35. Lednicky F, Michler GH (1990) J Mater Sci 25:4549
36. Lednicky F, Pelzbauer Z (1987) Polym Test 7:91
37. Bethge H, Heydenreich J (eds) (1987) Electron microscopy in solid state physics. Elsevier, Amsterdam
38. Bassett GA, Blundell DJ, Keller A (1967) J Macromolec Sci Physics B1:161
39. Spit BJ (1968) J Macromol Sci Phys B 2:45
40. Wittmann JC, Lotz B (1985) J Polym Sci Polym Phys 23:205
41. Preuß HHW (1975) Plaste und Kautschuk 22:958
42. Marichin VA, Mjasnikova LP (1970) Prib Tekh Eksp No 6:196
43. Bradley DE (1959) Brit J Appl Phys 10:198

10 Preparation of Thin Sections: (Cryo)ultramicrotomy and (Cryo)microtomy

Microtomy and ultramicrotomy are standard methods used for the preparation of ultrathin or semi-thin sections and flat surfaces of plastics, biological and biomedical objects. Brief discussions of the instrumentation required for these techniques, of working with a microtome and ultramicrotome, of specimen preparation (embedding, trimming, fixation), of sectioning (wet and dry sectioning, sectioning at room or cryo temperatures) and of modern trends in this specific field are presented in this chapter. In particular, the handling of the devices in practice and the influence of the sectioning parameters are described and illustrated using many examples from polymer research. In the final section, typical errors encountered during the different working steps and solutions to them are discussed in detail.

10.1 Overview

Ultramicrotomy (including cryoultramicrotomy) is a standard method for the preparation of ultrathin/semi-thin sections as well as very flat surfaces of plastics, biological and biomedical objects for various microscopic investigations. Improvements in preparation techniques over the last few decades have demonstrated that thin sections of different materials that are free from artefacts can be successfully prepared for electron microscopic investigations. Therefore, successful sectioning now depends primarily on the experience of the experimentalist rather than on the instrumentation used. In order to avoid sectioning-induced errors and to fully exploit the capability of an ultramicrotome, one must master the optimum specimen preparation and sectioning technique.

The traditional application of ultra- and cryoultramicrotomy involves the sectioning of soft materials such as the abovementioned biological or biomedical substances and polymers. Recent experiences demonstrate, however, that even hard materials can be successfully ultramicrotomed. In addition to soft metals like aluminium (Al) or copper (Cu), materials as hard as steel, ceramics and even very hard substances like sapphires and carbides have been successfully sectioned by ultramicrotomy. These developments show that the technique of ultramicrotomy can be expected to extend to other sectors of materials science, supplementing existing preparation techniques [1–13].

10.2 Instrumentation

Depending on the section thickness and sectioning temperature, the following instruments are necessary.

10.2.1 Microtomes

One generally prepares thin (>5 μm) and semi-thin (0.5–5 μm) sections at room temperature with the aid of a microtome. The sections and the flat surfaces of the blocks where the sections were cut can be inspected by means of SEM, AFM and HVTEM.

10.2.2 Ultramicrotomes

Using ultramicrotomes, semi-thin (thickness 0.2–3 μm) as well as ultrathin (thickness <0.2 μm) sections can be prepared for inspection by TEM and AFM. Depending on the mechanical properties of the materials (e.g. hardness), the sectioning can take place at room temperature without any sample preparation (in the case of hard materials), after chemical fixation, or under cryogenic conditions (see below). It should be noted that the semi-thin sections, after deformation outside or inside a microscope, are also used for the detailed study of the deformation mechanisms of the materials. As in the previous case, the flat surfaces remaining after ultramicrotomy can be studied by means of AFM and SEM.

10.2.3 Cryomicrotomes and Cryoultramicrotomes

Cryo devices are ideal equipment for sectioning soft materials without fixation or staining. Both the knife and the specimen are cooled by liquid nitrogen, whereby the sectioning temperature can be adjusted down to $-185\,°C$. These sections are mainly studied by means of TEM, while the flat surfaces can be scanned by AFM.

10.2.4 Knives

The knives used for sectioning are made of glass or diamond. Using a special glass-breaking device and good-quality glass strips, an expert can easily manufacture glass knives for broad application. Both types of knives have advantages and disadvantages. The selection of the type of the knife is determined by the price and also by the nature of material to be sectioned. For wet sectioning it is necessary to mount a truf or collecting tray onto the glass knife, or to use a diamond knife with a truf. Figure 10.1 shows the angles between the knife and the sample that are important for cutting.

Glass Knives
Glass knives are sharp-edged glass prisms manufactured from special glass strips free of reams (no window glass!). Strips 400 mm long and 25 mm wide with thicknesses of

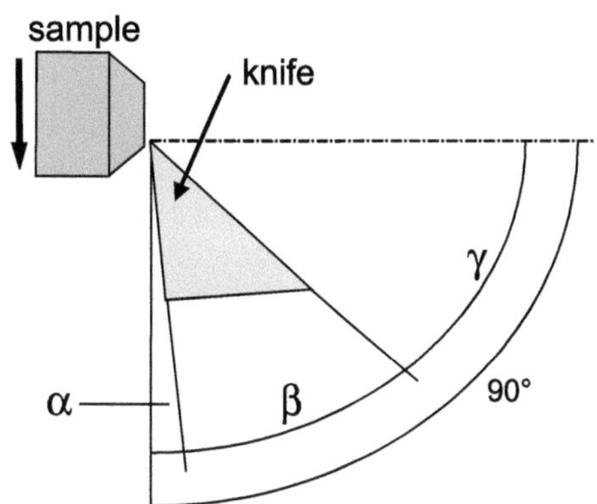

Fig. 10.1. Angle of inclination of knife during sectioning (α: clearance angle; β: knife angle, γ: rake angle; α + β: sectioning angle)

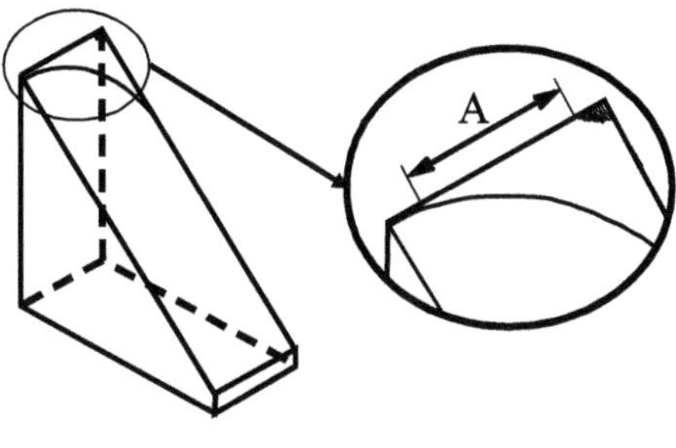

Fig. 10.2. Scheme showing the sharp edge of a glass knife that can be used for thin sectioning (A: useable cutting edge)

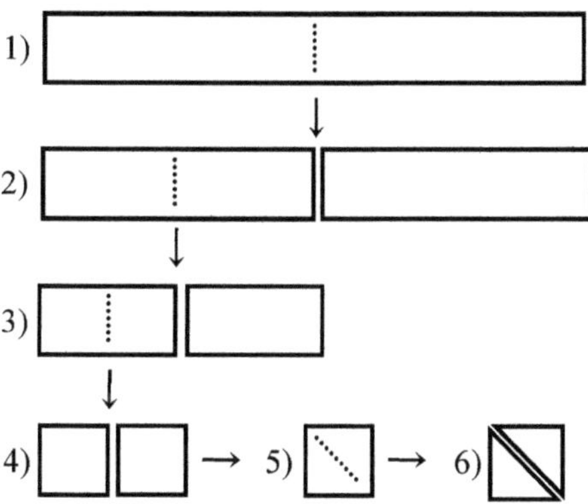

Fig. 10.3. Breaking process for glass knives ("balanced break technique")

1   – glass strip with score
2, 3, 4 – broken glass strips
5   – square with score
6   – two knives (broken square)

6.4 mm, 8 mm or 10 mm, respectively, are on the market. When selecting the thickness of a glass strip, it is important to take into consideration the requirements of the specimen (in terms of its nature and size) as well as the aim of the subsequent microscopic examination. The sharp edge of a glass knife that can be used for thin and ultrathin sectioning is shown in Fig. 10.2.

Special equipment (a knife maker) is necessary for the production of glass knives of high quality. The best (qualitatively) results are obtained with the so-called balanced break method. The balanced break technique (see Fig. 10.3) enables a standard 400-mm-long strip of glass to be broken continuously into equal halves. The resulting flat-sided squares are broken in the next step into two glass knives. Due to the crack propagation process along the diagonals of the squares, the cutting angle is larger than 45°.

Diamond Knives

Diamond knives are manufactured from naturally occurring diamonds. During the manufacturing procedure, the knife angle β (see Fig. 10.1) can be defined exactly to be 45° or 35° (smaller than in the case of glass knives). An important advantage of the smaller angle of inclination is the fact that one can significantly reduce the bending and contraction of the sections. The capabilities of 35° diamond knives have already been demonstrated, e.g. in two publications by Jésior [14,15].

The main advantages of diamond knives are as follows:

- Diamond has the greatest hardness of all materials, i.e. all materials can be sectioned.
- Diamond is very strong in the sectioning direction.
- Various inclinations are available, and smaller inclinations can be obtained than for glass knives.
- Long lifetimes are possible if carefully applied.

The disadvantage of diamond knives is their high price; however, this is compensated for over time by the high quality of the sections that they yield, which therefore lead to valuable scientific information from microscopic examinations.

10.2.5 Modern Trends

As mentioned above, ultramicrotomes have improved greatly over the last few decades, such that they now enable the quick and secure preparation of high-quality sections for microscopic examinations of different materials. There are some recent trends in ultramicrotomy, as summarised below.

Oscillating Knife

When sectioning using the microtome, a pressure is exerted on the sample along the direction of sample holder movement. In order to reduce this pressure, a new technique has been introduced into sectioning: the use of an oscillating knife. This new technique makes use of our practical experience that the back-and-forth movement of a knife reduces the pressure on the object being cut (bread, cheese, sausage …), which decreases the shrinkage of the slices. The company Diatome Ltd. (Biel, Switzerland) has recently developed a new diamond knife which is made to oscillate during sectioning by a piezoelectric element. Experiments have revealed the advantages of this knife at reducing the compression of polymeric materials [16–18].

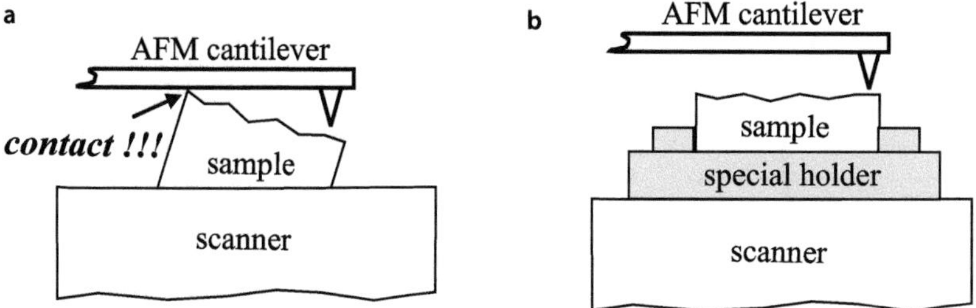

Fig. 10.4a,b. Transferring an ultramicrotomed flat surface from a microtome to an AFM sample holder: **a** normal transfer with danger of tilted fastening and wrong contact; **b** with the aid of special holder developed by the company Leica Microsystems

Special Sample Holders

Particularly for the study of ultramicrotomed surfaces of materials by AFM, a parallel alignment of the sample surface relative to the probe scanner is required. Recently, the company Leica Microsystems has developed special miniaturised sample holders that can be mounted into the ordinary sample holder of the ultramicrotome, so that thin sections can be prepared and employed for TEM and flat surfaces for AFM. The special holder insert can be transferred to the AFM to minimise problems while studying the morphology (see Fig. 10.4).

10.3 Working with a Microtome and an Ultramicrotome

An (ultra)microtome is a sensitive instrument for thin sectioning which is equipped with a knife made of glass or diamond. The microtome sample holder is moved towards the knife a given distance during every cycle. Figure 10.5 shows the principle of an (ultra)microtome.

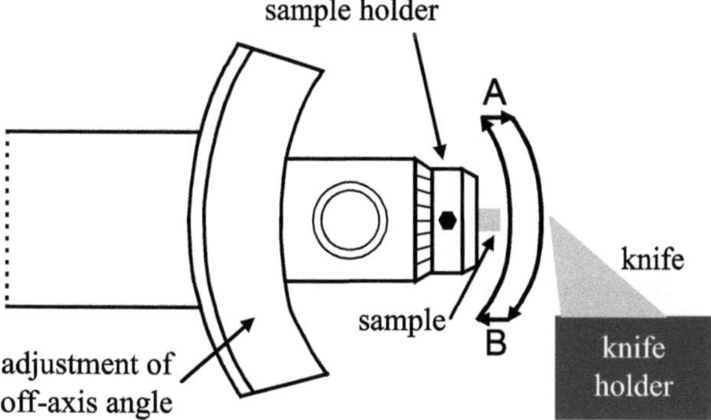

Fig. 10.5. Scheme showing the movement of the sample relative to the knife during the sectioning process (A: forward motion; A → B: cutting: B: backward motion; B → A: upward movement); "length (A)" = "length (B)" + "section thickness"

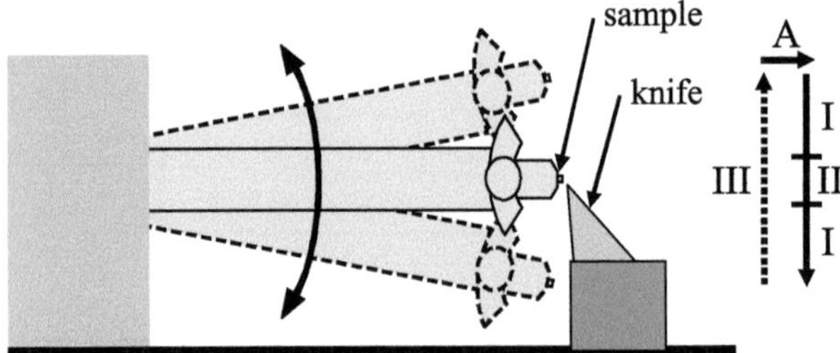

Fig. 10.6. Schematic representation of variation of speed of sample holder during sectioning: *I*, high speed downwards; *II*, specified sectioning speed (beginning and end of sectioning process); *III*, automatic backwards and upwards motion of sample holder until the starting point is reached (high speed); *A*, forward motion (selected section thickness) and beginning of new cycle

During the downward motion of the sample holder (A → B, cutting step), a section is produced which has a thickness that is specified by the forward motion (difference: A–B). The thickness of the resulting section is determined not only by the amount of forward motion but also by other parameters such as the type, quality and inclination of the knife as well as the properties of the material and the size of the surface being sectioned. Depending upon the material itself and the geometry of the sections being prepared, it is necessary to optimise the speed of sectioning, which can be adjusted across a wide range in an ultramicrotome. The backward motion of the sample and the sample holder takes place at another speed, which depends on the selected sectioning speed, in such a way that the time needed to perform the whole cycle remains constant (Fig. 10.6).

The sections are generally allowed to swim on a liquid placed on the knife truf, which are transferred to grids and are then ready for microscopic investigations. Modern ultramicrotomes are equipped with video microscopes and computers that are used to display and save sectioning parameters, respectively.

10.4 Specimen Preparation

The aim of specimen preparation is to obtain a piece of the sample that can be fixed to the microtome sample holder in order to be able to prepare good-quality sections. It is extremely important that the structure of the material remains unchanged at the various stages of specimen preparation. The first step is to decide which part of the specimen is of interest (surface or subsurface region, centre of the sample) and what information is needed. The embedding, trimming, chemical modification (e.g. staining) and ultramicrotomy steps are then performed.

10.4.1 Embedding

Sometimes the specimens cannot be mounted directly into the microtome sample holder, e.g. if they are very small samples, very thin films, powdered materials, or if there is the possibility that the sample could be deformed as it is fixed into the sample holder. In such cases the sample must be embedded into a stable medium that does not react with the sample. Other advantages of embedding include protection against the breakage of the specimen at its edge and against the fragmentation of sections of porous or extremely brittle materials. There are three media in general use: epoxy resins, polyester resins and methacrylates. The ease of sectioning and the stability of the medium under high vacuum and under the electron beam are important. The use of an embedding medium that has a similar hardness to the polymer is also very important for good sectioning. The recipe used to create each medium can be varied in order to change its hardness. One of the reasons for the widespread use of epoxies is the fact that they are most stable in the electron beam, whereas methacrylates are relatively unstable.

10.4.2 Specimen Trimming

Trimming means achieving the optimum size and geometry of the specimen prior to microtomy. This can be done by hand using a trimming blade under a stereomicroscope, or with the aid of a trimming device. The specimens can also be trimmed inside an ultramicrotome using a trimming blade. The trimmed surface, which is generally a rectangular area on top of a pyramid, should be ca. $0.5\,\text{mm} \times 0.5\,\text{mm}$ or smaller. For samples that pose difficulties during sectioning, it is possible to reduce the area down to $0.1\,\text{mm} \times 0.1\,\text{mm}$ or even smaller than this. It is important to take the following points into consideration for sectioning:

- The specimen should not be pointing out like a needle and it should not extend too far outside the sample holder
- Both the knife and the sample should be positioned stably and fixed.

In order to specify the direction or orientation of the sample, the surface area can be trimmed to a special shape, such as a trapezium or triangle (see Fig. 10.15).

10.4.3 Fixation and Staining

Soft polymeric materials, polymers containing soft components or organic substances (such as biological or biomedical materials) need to be fixed as well as hardened before undergoing sectioning. This hardening or fixation can be achieved through physical processes (i.e. γ-irradiation) or chemical reactions (i.e. by creating crosslinks by reacting the material with OsO_4 or RuO_4). Different techniques can be employed for the selective staining procedure often used in polymer research, either on the bulk material before sectioning or on the thin sections obtained after cutting. Additionally, the sample should be sectioned at very low temperatures, below its glass transition temperature (T_g). The process of fixation is coupled to staining, and these are treated in detail in Chap. 13.

10.5 Ultrathin Sectioning

10.5.1 Sectioning Parameters

After selecting a microtome type and mounting the specimen, the sectioning parameters must be specified; for example:

- Whether sectioning is done at room temperature or cryosectioning is used
- The temperature of specimen and knife during cryosectioning
- Wet or dry sectioning?
- Knife selection: glass or diamond knife?
- Preselection of the cutting region (region of lower speed) and the sectioning speed
- Section thickness (< 100 nm for organic substances and polymers and some tens of nanometres for metals)
- The positioning of the specimen and knife
- Whether to use an ioniser to discharge the section (so that sections stay in place and do not fly away).

10.5.2 Wet and Dry Sectioning Techniques

The collection of sections is performed either in the presence of a floating liquid (so-called "wet sectioning" using a special knife truf), or directly on the surface of the knife (so-called "dry sectioning", see Fig. 10.7). If the sections are prepared at room temperature, the wet method is favoured, while at cryogenic conditions the dry technique is generally used. However, wet sectioning is also possible at lower temperatures.

Distilled water is used for wet sectioning at room temperature. A mixture of DMSO and distilled water or (up to $-50\,°C$) even p.a. ethanol can be employed for wet sectioning at reduced temperatures. The advantages and disadvantages of both of the techniques are summarised in Table 10.1.

Using an eyelash attached to a wooden needle, the movement of sections away from the knife edge can be controlled, which reduces or even eliminates the deformation of thinner sections and the rolling of thicker sections (see Fig. 10.8).

The sectioning procedure also depends on the sample/knife response at the edge of the knife (see Fig. 10.9). For instance, if the sample is too soft, compression or

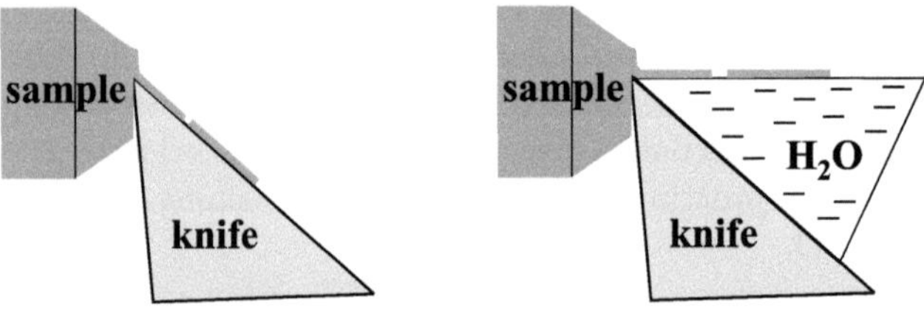

Fig. 10.7. Dry and wet sectioning

Table 10.1. Comparison of dry and wet techniques for section collection

	Dry sectioning	**Wet sectioning**
Pro	– Specimen and knife have identical temperatures – No effect of floating liquid	– Easy collection of sections – Less section compression – Low electrostatic charging – Knife-friendly
Contra	– Electrostatic charging – Difficulty in collecting sections – High compression	– Reaction of floating liquid with sample is possible – Sample and knife temperature may vary during cryosectioning

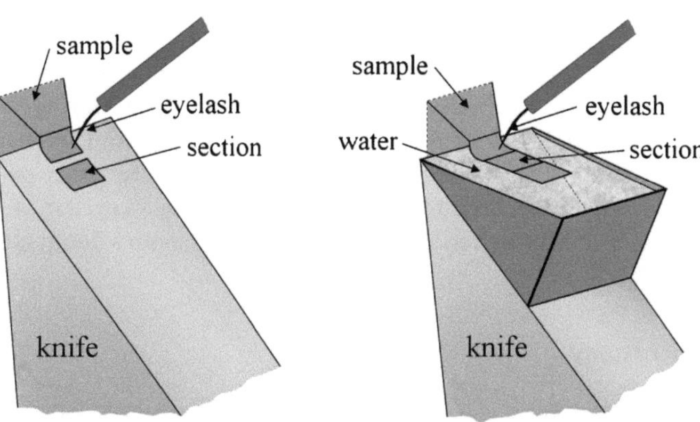

Fig. 10.8. Controlling the paths of the sections with the aid of an eyelash during dry and wet sectioning

shrinkage of the sections may dominate, while brittle materials may exhibit crack formation just ahead of the knife edge and the specimen may even break.

If the sample is very hard and tends to undergo brittle fracture, the sections may damage the knife edge, resulting in the blunting of the knife. In such cases, it is advisable to choose the smallest width possible (~10 μm) for microtomy in order to minimise damage to the knife. Fresh edges can then be moved into sectioning position to continue the sectioning of hard objects.

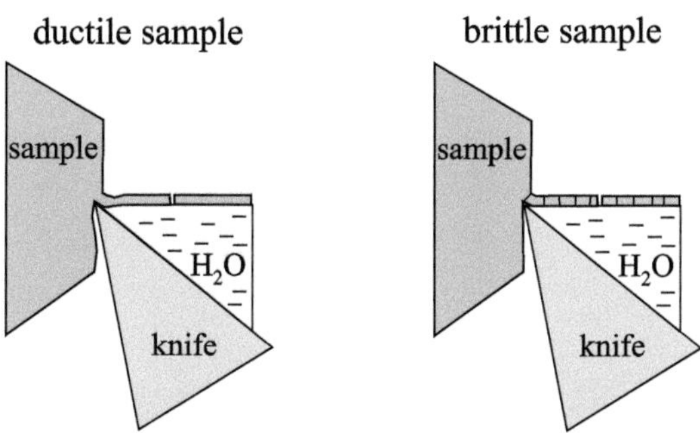

Fig. 10.9. Comparison between the sectioning of a soft (ductile) and a brittle material

10.5.3 Room Temperature Ultramicrotomy

Room temperature ultramicrotomy is only suitable for samples that are hard enough for successful sectioning. After embedding and trimming the specimen, sections 60–100 nm thick (for plastics) or 10–60 nm thick (for inorganic materials) can be produced using glass or diamond knives. If the specimen is not hard enough, a fixation procedure (for example, treating samples with heavy metal compounds in the case of heterogeneous polymers) must be performed prior to sectioning. Otherwise, the soft samples can also be sectioned at cryogenic temperatures.

Room temperature sectioning can be performed using floating liquid or under dry conditions (see Fig. 10.7). The liquid used is generally distilled water. Section thickness can easily be estimated from the interference colours of the sections (see Table 10.2).

Sections with a relatively large area and with thicknesses in the range of a few microns are needed for studying micromechanical processes that occur during deformation tests by means of HVTEM or AFM (see Chapter 12). To avoid sections rolling during the preparation of thick and large-area sections, one can either perform dry sectioning and help the sections to glide on the knife using an eyelash (see Fig. 10.8), or the sections can be allowed to slip from the knife edge onto a liquid film towards the truf.

10.5.4 Cryoultramicrotomy

Sectioning at cryogenic temperatures is advantageous if the materials are too soft and can be hardened by cooling. For instance, cooling polymers below their glass transition temperatures makes them hard and the internal structural details are frozen in. If cryoultramicrotomy is required, the instrument must be equipped with a cryo-compartment. The latter is cooled by means of liquid nitrogen and sectioning is generally carried out at temperatures of up to $-185\,^{\circ}\mathrm{C}$. Any type of knife can be used, including glass knives. There are also special cryo-diamond knives that are suitable for both dry and wet sectioning.

After trimming, the specimen is fixed in the sample holder and the knife is mounted into the microtome. Then it is necessary to wait until the temperatures chosen for the specimen and the knife are reached. It is important to note that the knife

Table 10.2. Relation between section thickness and interference colour during wet sectioning

Section thickness (nm)	Interference colour
0 to 50	Grey
50 to 70	Silvery
70 to 120	Golden
120 to 160	Copper-brown
160 to 200	Purple
200 to 240	Blue
> 240	Green

is not cooled as strongly as the specimen. Additionally, the cooling temperature is limited if the sectioning is done in the presence of floating liquid. For example, when p.a. ethanol is used, cooling can only be done down to $-55\,°C$. When DMSO/water mixtures are used, even lower temperatures are possible.

Next, sections with the required thickness can be prepared. The techniques and the selection of sectioning parameters that were described for room temperature are also valid in the case of cryoultramicrotomy. However, estimating section thickness via interference colour is not possible during cryosectioning. If the sections need to be studied by TEM under cryoconditions, a special cooling holder is required to transfer the sections from ultramicrotome to the TEM.

10.5.5 Collecting Sections on Grids

After they have been cut, the sections can be collected on a grid. The collection can be performed in different ways.

Section Collection After Wet Sectioning
First of all, each section or a row of sections are moved away from the knife edge with the aid of an eyelash (see Fig. 10.8) to avoid damaging the knife edge during subsequent operations.

The sections are then fished up (see Fig. 10.10) using different tricks:

– Placing the grid from the top onto the sections. This may, however, lead to folding and faults in the sections (case I in Fig. 10.10).

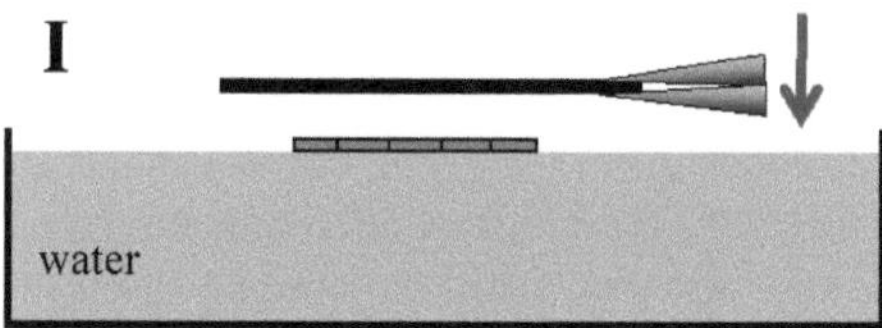

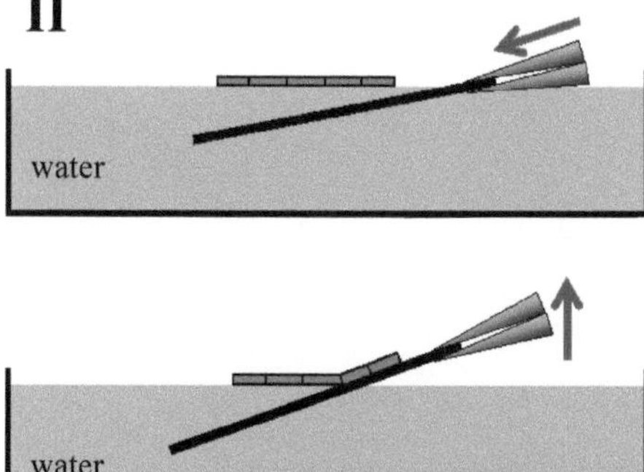

Fig. 10.10. Different possibilities for collecting sections from the knife truf of the ultramicrotome

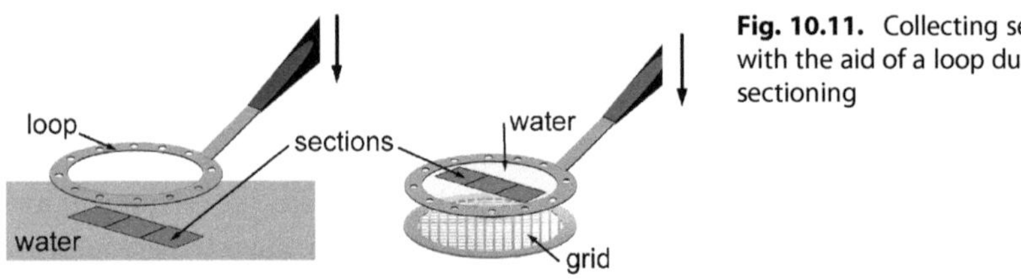

Fig. 10.11. Collecting sections with the aid of a loop during wet sectioning

 – The grid is carefully brought close to the sections in the water and then they are slowly pulled with the aid of an eyelash onto the grid (case II in Fig. 10.10). This is the recommended method.

The sections can be also be collected using a special loop (see Fig. 10.11). This accessory is brought down towards the sections until a water drop is formed. The water drop then contains the sections. After removing the loop from the ultramicrotome, the water drop is transferred onto the grid and the water is absorbed by filter paper.

Section Collection During Dry Sectioning
A loop with a drop of distilled water or sugar solution is usually used (see Fig. 10.12). In this case, the sections are collected directly with the aid of the "drops". The drop containing the sections is transferred onto a grid placed on a filter paper. The underlying filter paper soaks up the water and the sections remain intact on the grid surface. If a sugar solution is used to fish out the sections, the grid containing the sections should be washed thoroughly with distilled water in order to avoid the presence of artefacts on the grid.

The pros and cons of performing section collection with the aid of either distilled water or sugar solution are listed in Table 10.3.

A further method of collecting sections is to directly transfer them onto the grid. In the case of a well-formed series of sections, one can carefully transfer the sections by allowing them to slide over the grid (see Fig. 10.13).

Sections that are thicker and more stable can be easily fished out using an eyelash and transferred onto a grid. If the sections have wavy surfaces, they can be stabilised and flattened by fixing a second grid over the grid holding the sections (see Fig. 10.14). The sections can be inspected under an optical microscope to check their quality.

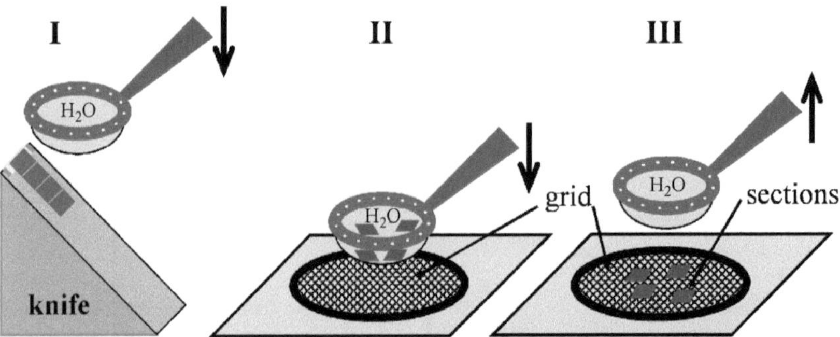

Fig. 10.12. Collecting sections during dry sectioning using a drop of water in a loop

Table 10.3. Advantages and disadvantages of using distilled water or sugar solution for to fish out sections during dry sectioning

	Sugar solution	**Distilled water**
Pro	– Can be used while sectioning at both room temperature and cryotemperatures	– Can be used during room-temperature sectioning
Contra	– Rapid working is essential, otherwise the drops become frozen	– Cannot be used at cryotemperatures
	– Sugar solution should be removed. The grids containing the sections are washed carefully with distilled water and then dried	

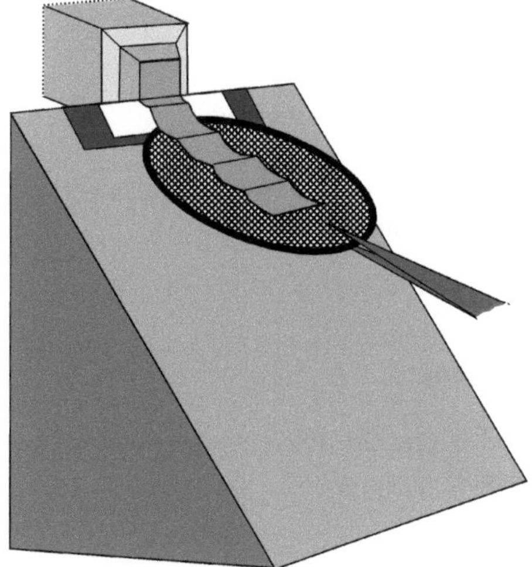

Fig. 10.13. Direct transfer of a series of sections onto a grid

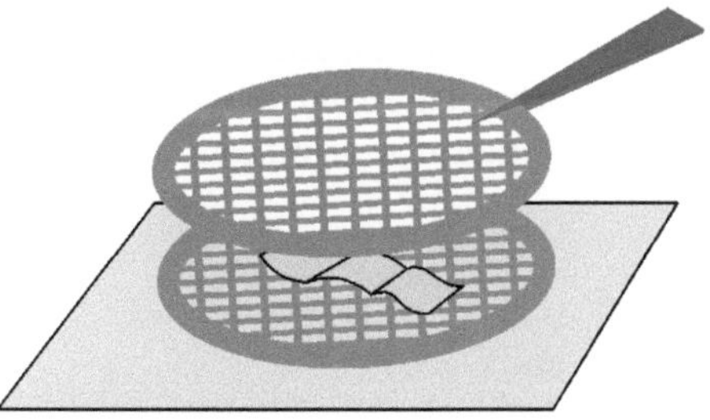

Fig. 10.14. Fixation of sections with the aid of a second grid

10.6 Problems, Errors and Solutions

10.6.1 Overview

Despite careful operation, errors can still occur during sample preparation and due to the wrong choice of sectioning technique [13,19]. Instrumental errors seldom take place. The most common errors encountered during sectioning are the following:

- Inappropriate embedding is used
- Insufficient fixation or hardening is employed
- The surface area is too large to be sectioned
- The specimen and/or knife is loose
- There are defects or damage on the knife edge
- An unsuitable section thickness is chosen
- The wrong sectioning parameters (speed, temperature, etc.) are used
- An improper method of sectioning is applied
- There is a lack of skill in transferring the sections onto the grid
- Impure solvents and dirty apparatus are used
- Inhomogeneous materials with hard and soft inclusions (such as polymers with glass fibres or inorganic fillers) are being investigated

Typical errors and their origins are listed in Table 10.4.

10.6.2 Typical Errors and Possible Solutions

In the following, typical errors encountered during the different working steps and solutions to them are discussed in more detail.

Embedding

- Check whether other fixation methods without embedding are possible
- Choose the correct embedding agent, especially the final hardness
- Check whether the composition and homogeneity of the embedding agent is correct (the deviation from the standard composition may significantly affect the section consistency).

Staining (See also Chap. 13)

- Select the most suitable agent (usually RuO_4 and OsO_4)
- Check the staining procedure (e.g. whether the block material or the section is stained)
- Optimise the time and temperature of the staining.

Trimming

- Trim the surface to be cut properly (a pyramid-like surface is suitable)
- Sectioning surface area should be 0.1 mm × 0.1 mm or smaller depending on the sample material

Table 10.4. Typical errors and their origins

Characteristics of the errors	Possible origin
Periodic waves, different thicknesses along the section	– Damaged or blunt knife edge – Improperly or loosely placed knife – Needle-like trimmed sample – Elastic sample – Insufficiently fixed sample or sample holder – Building vibrations or contact with the ultramicrotome during sectioning – Sample is too soft
Single or several nonperiodic waves in a section	– Extreme building vibrations – Unstably positioned ultramicrotome
Changing thicknesses along an ultrathin section	– Inhomogeneities (i.e. hard and soft phases) in the sectioning area – Wrong trimming
Variation of thickness from section to section	– Air movement and temperature fluctuations in the laboratory – Section thickness attempted is too low – Material is too soft
Strongly compressed sections	– Wrong knife angle – Wrongly adjusted free angle – Sample material is too soft
Folds in ultrathin sections	– Use of a bad knife (damaged edge) – Sectioning velocity is too high – Very large sectioning area or sections are too thick
Impurities in sections	– Impure water used as floating liquid – Insufficiently cleaned apparatus (such as glass envelope, tweezers, needles, etc.) – Impure staining solution, excessive staining
Folds in sections on the grid	– Wrong floating and transfer of the sections onto the grid – Sections are too large

–   Detection of the structural details of the samples (such as inclusions, orientations, deformation zones, etc.) in the EM is facilitated if the surface is trimmed to produce asymmetric sections (see Fig. 10.15).

The surface of the pyramidal tip of the specimen can be modified in such a way that the common problem of the adherence of the sections to the edge of the specimen can be minimised or even eliminated. This can be done by trimming the upper edge of the specimen into a roof-like shape (see Fig. 10.16). It is generally advisable to trim hard materials (such as metals, ceramics, or semiconductors) as small as possible and to section them along or oblique to the specimen edge, so that only a very small area of the knife edge can be damaged during the sectioning. Experience has shown that only a few sections of very hard substances can be produced by sectioning on a given knife area.

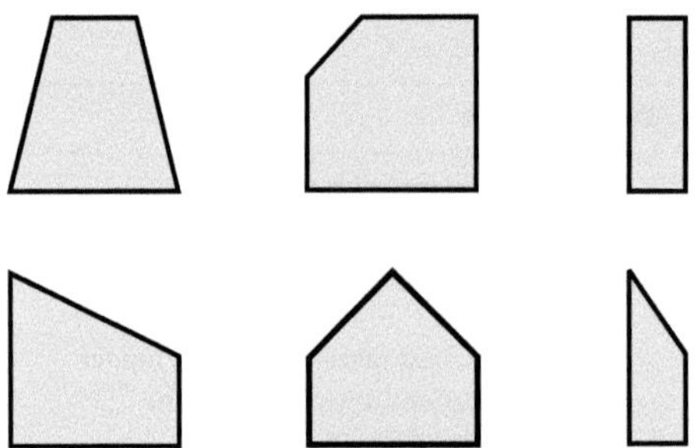

Fig. 10.15. Varying the shape of the trimmed surface to investigate structural details such as inclusions, orientations, deformation zones, etc.

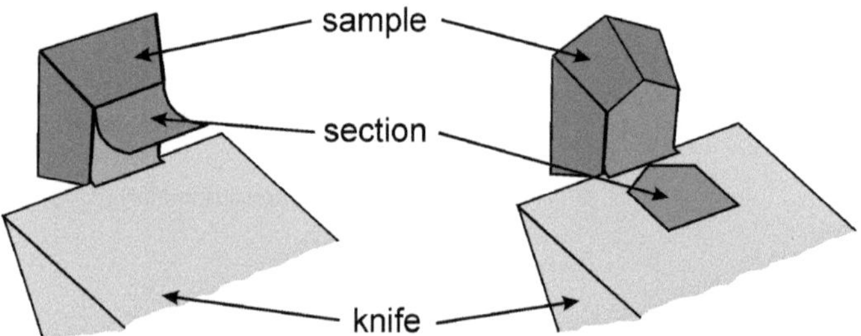

Fig. 10.16. Reducing the stickiness of sections by trimming the specimen into a roof-like shape (*right*)

Compression

During polymer sectioning, one always observes that the length of the sections is less than the length of the specimen area from which the sections are derived (compression). Even the sections of hard, stiff polymers like PS or PMMA can show compressions of up to 30% (see Fig. 10.17). This effect can be minimised by selecting suitable sectioning parameters, such as:

- Hardening the specimen if the specimen fixation and staining is not sufficient
- Trimming the cutting surface to make it as small as possible; one may also vary the geometry of the sectioning area
- Varying the sectioning angle and speed as well as the section thickness and temperature.

Knife Edge Defects

- Choose a suitable knife (glass or diamond knife) and suitable knife angles in the case of a diamond knife (35° or 45°)
- Avoid damaged edges. Do not bring foreign materials such as tweezers, needles, or fingers in contact with the knife edge. Ensure that there are no dust particles or fine ice particles on the knife edge. Defects on the knife may be pressed into the sample surface during sectioning, causing defects on the sections throughout the length of the section. The sections may even break into pieces (see Fig. 10.18).
- Clean the knife before and after sectioning.

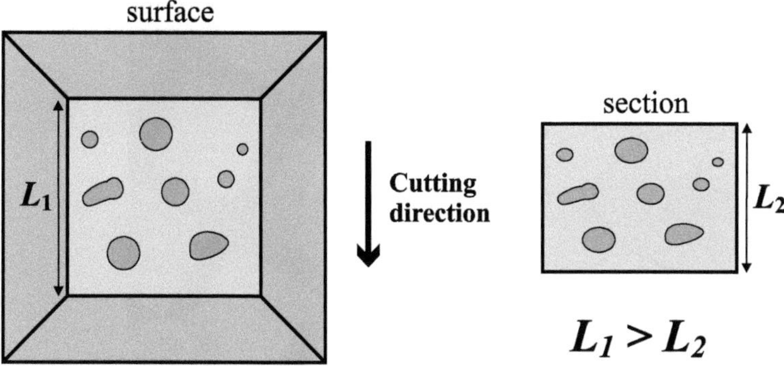

Fig. 10.17. Compression of sections along the sectioning direction

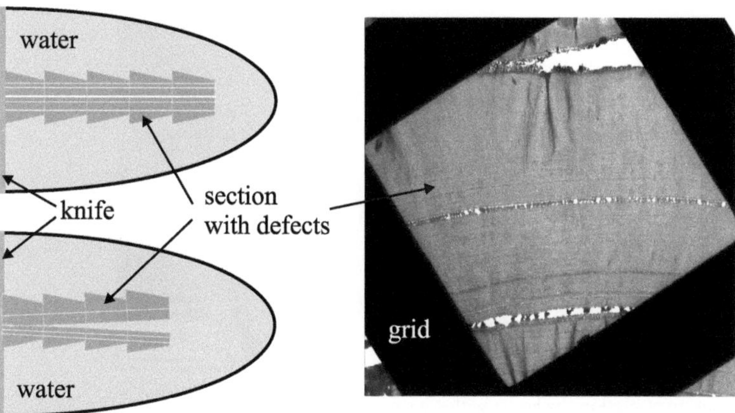

Fig. 10.18. Influence of a knife with defects on the quality of sections: schematic representation of defects in the sections and tearing along the section (*left*), TEM micrograph of an ultrathin section with defects (*right*)

Variable Section Thickness (Thickness Variation From One Section to Another)
Quite often one encounters the situation in which the thickness of the first section is higher and that of the second one is lower than the specified thickness (see Fig. 10.19). This situation can even result in a situation where one thicker section alternates with a missing section. These problems can be eliminated by performing the following actions:

– Changing the sectioning temperature
– Improving the hardness/staining of the sectioning object
– Altering the sectioning speed and the free angle.

Zones with Varying Thicknesses in a Single Section
The variation of thickness along a single section is generally connected to the properties of the material, sectioning parameters or the transfer of sections from knife to grid (see Fig. 10.20).

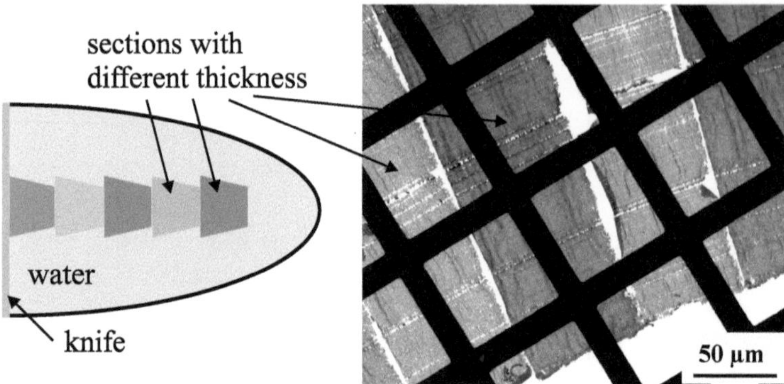

Fig. 10.19. Series of sections revealing variable thickness

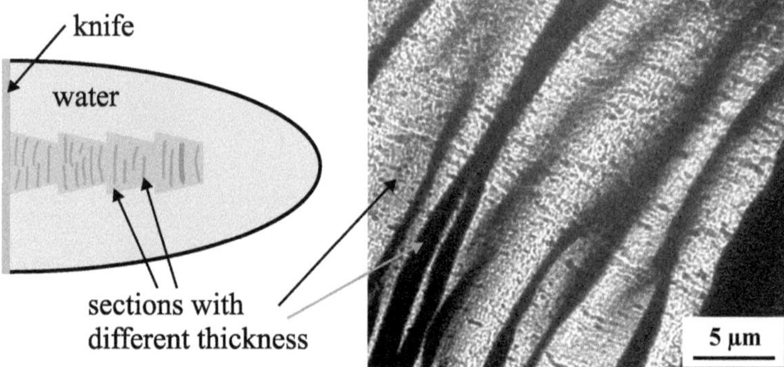

Fig. 10.20. Variation of section thickness within a single section

The following actions may help to reduce this negative effect:

– Equalise the different hard sample components (such as a soft matrix and hard inclusions, particles, fibres, or a brittle matrix with soft particles such as rubbery inclusions) by further hardening (sometimes by changing the staining agent: multiple staining) or varying the sectioning temperature
– Adjust the sectioning temperature as well as the speed.

Unsuitable Sectioning Parameters
If optimum cutting does not occur, a steady correction of the sectioning parameters
is necessary:

– Ensure that the specimen and the knife are suitably placed
– The specimen should neither be trimmed to a needle-like form nor protrude too
  much beyond the specimen holder
– Align the sample in such a way that the smaller specimen edge reaches the knife
  edge first

- Ensure the parallel alignment of the specimen and the knife edge and carefully select the cutting area
- Adjust (or vary) the sectioning speed and temperature.

Since there are so many problems that can appear during the preparation of thin sections, no single recipe can be suggested that can eliminate all of the difficulties. Thus, the general rule of microscopy holds in this situation too: "**be creative**".

References

1. Thompson-Russell KC, Edington JW (1977) Electron microscope specimen preparation techniques in materials science. Macmillan, London
2. Malis TF, Steele D (1990) Ultramicrotomy for materials science (Symp Proc). Materials Research Society, Pittsburgh, PA
3. Maniette Y (1990) J Mater Sci Lett 9:48
4. Reid N, Beesley JE (1991) Practical methods in electron microscopy, vol 13: Sectioning and cryosectioning for electron microscopy. Elsevier, Amsterdam
5. Cahn RW, Haasen P, Kramer EJ (1991) Materials science and technology/Characterization of materials, pt 1, vol 2A. Wiley-VCH, Weinheim
6. Gianvill SR (1995) Microsc Res Techniq 31:275
7. McMahon G, Malis T (1995) Microsc Res Techniq 31:267
8. Sitte H (1996) Scan Microsc Suppl 10:387
9. Sawyer LC, Grubb DT (1996) Polymer microscopy, 2nd edn. Chapman and Hall, London
10. Quintana C (1997) Micron 28:217
11. Schubert-Bischoff P, Krist T (1997) Microsc Microanal Proceed 359
12. Bozzola JJ, Russell LD (1999) Electron microscopy: Principles and techniques for biologists, 2nd edn. Jones and Bartlett, Boston, MA
13. Michler GH, Lebek W (2004) Ultramikrotomie in der Materialforschung. Hanser Verlag, München
14. Jésior JC (1986) J Ultrastruct Mol Struct Res 95:210
15. Jésior JC (1989) Scan Microsc Suppl 3:147
16. Al-Amoudi A, Dubochet J, Gnaegi H, Lüthi W, Studer D (2003) J Microsc 212:26
17. Studer D, Gnägi H (2000) J Microsc 197:94
18. Vastenhout JS, Gnaegi H (2002) Microsc Microanal 8:324
19. Sitte H (1981) Ultramikrotomie – Häufige Fehler und Probleme. Reichert-Jung Optische Werke AG, Wien

11 Special Preparation Techniques

This chapter focuses on alternative preparation methods to ultramicrotomy. Various methods for thin and ultrathin film preparation based on polymer solutions are presented. The examples that are given for specimens prepared by spin coating, dip coating and solution casting cover several classes of polymers, including block copolymers and others. The essential preparation steps and the advantages of the focussed ion beam (FIB) technique are also described. It is shown that this technique is particularly convenient for preparing ultrathin specimens of composite materials that allow TEM analyses of interfaces of different classes of materials in their most undisturbed states.

11.1 Preparation of Polymer Films from Solutions

11.1.1 Introduction

When new polymeric materials are produced in industrial or academic research laboratories, they are very often produced in small amounts, since either the components are expensive or the yields of laboratory syntheses are too small. Starting with only a few grams (or even less) of a powder, small granules or a solution of a polymeric material, the production of relatively large standard test pieces by the usual processing methods (e.g. extrusion, injection or compression moulding) is absolutely impossible. Smaller polymer samples can be prepared using special microprocessing tools (e.g. a miniaturised extruder, microinjection technique) which are convenient for preparation by ultramicrotomy, as described in the previous chapter. An alternative method is the preparation of semi- or ultrathin polymer films from polymer solutions. Depending on the microscopic techniques that are subsequently used to investigate the sample, the film thickness can be adjusted within a range of between a few tens of nanometres and several hundreds of micrometres. In many cases, such films allow broad characterisation of the polymeric materials, including examinations of morphology and micromechanical mechanisms. This method permits the chemical staining of the polymer films for contrast enhancement directly after preparation or after subsequent specimen treatments, such as annealing, isothermal crystallisation or deformation experiments. However, it is important to consider that films of polymeric materials prepared from the solution might exhibit a different morphology and modified micromechanical behaviour compared to the bulk material prepared from the melt by extrusion, injection moulding, or pressing. In polymer blends or block copolymers, for instance, the morphology is controlled by complex phase separation phenomena. Moreover, the morphology will depend on the solvent evaporation rate and subsequent annealing procedures. In semicrystalline polymers, the lamellar arrangement and the formation of spherulitic superstructures will also depend on film thickness. One must consider that for particulate-filled or rubber-modified systems, the film thickness must not be lower than the size of the heterophase (see Fig. 12.3). In particular, a rough rule of thumb for micromechanical deformation experiments is that the thin films must be of a thickness that is at least 5–10 times the size of the morphological unit that is of interest. Therefore, it is crucial to check whether it is possible to transfer the results from solution films to the bulk material.

11.1.2 Solution Behaviour of Polymers

The dissolution of a polymer in a solvent with a low molecular weight is determined by the polymer–solvent interactions [1]. The general solubility of a polymer in a solvent can be estimated by means of the solubility parameters. The best solvent of a polymer is the one that has a solubility parameter that is closest to the solubility parameter of the polymer. There is a generally accepted rule: a polymer should be dissolved in a chemically similar solvent. In Table 11.1, various solvents of mainly commercial polymers are listed. In addition to the use of one solvent, mixtures of different solvents or non-solvents of a polymer can be applied to dissolve the polymer.

Besides the constitution of the polymer, other properties (e.g. temperature, molecular weight) also have an influence on the solubility of the polymer. For example, increasing the molecular weight of a soluble polymer can yield a transition from a polymer solution to a gel-like state where the addition of solvent causes the polymer to swell rather than dissolve. Cross-linked polymers also only exhibit a gel-like state. Due to the wide variety of factors that influence the solubility of polymers, the process of dissolution may take few minutes to several days.

11.1.3 Spin-Coating

The spin-coating process is a well-established and very common technique used to produce thin and smooth films on flat substrates. It can be applied to many classes of polymers, and can yield a wide range of film thicknesses [2–4]. Compared to other methods, such as solution-casting and dip-coating, spin-coating is perhaps the most prominent procedure for obtaining uniform films of a specific thickness.

The spin-coating process of film formation, which leads to a liquid–solid phase transition, is complex and involves time-dependent changes in the properties of the applied polymer solution. There are distinct stages in the spin-coating process [5], as shown schematically in Fig. 11.1: deposition, spin-up, stable fluid outflow, and finally evaporation.

1. Deposition
The polymer solution is deposited on the disk-shaped substrate. The disk is stationary or rotates very slowly in this start phase of the spin-coating process. There are dif-

Table 11.1. Summary of some examples of polymers and commonly used solvents

Polymer	Abbreviation	Solvent
Polystyrene	PS	Toluene, chloroform, benzene, o-dichlorobenzene
Poly(ethylene oxide)	PEO	Toluene, distilled water
Poly(vinyl alcohol)	PVA	Distilled water
Poly(vinyl methyl ketone)	PVMK	Chloroform
Poly(ethylene terephthalate)	PET	Tetrachloroethylene, hexane
Poly(vinyl chloride)	PVC	N, N-dimethylacetamide (DMA), cyclohexanone, acetone/carbon disulfide
Poly(vinylidene fluoride)	PVF_2	Tetrahydrofuran (THF)
Poly(methyl methacrylate)	PMMA	Acetone, benzene, toluene
Polyethylene (high density)	HDPE	Above 80 °C: 1,2,4-trichlorobenzene, decalin, di-n-amyl ether, halogenated hydrocarbons, higher aliphatic esters and ketones, hydrocarbons, xylene
Polyethylene (low density)	LDPE	As for HDPE, but temperature 20–30 °C lower, depending on degree of branching
Poly(hydroxyl butyrate)	PHB	Chloroform (60 °C)
Atactic polypropylene	aPP	Toluene
Isotactic polypropylene	iPP	See polyethylene
Poly(cis-1, 4-butadiene)	PB	Toluene
Nylon 11	PA 11	Trifluoroacetic acid (TFA)
Polyacrylonitrile	PAN	N, N-dimethylacetamide (DMA)
Styrene/acrylonitrile copolymer	SAN	N, N-dimethylformamide (DMF), methyl ethyl ketone (MEK), acetone
Polycarbonate	PC	Chloroform, dichloromethane
Poly(p-phenylene ether sulfone)	PPES	N, N-dimethylacetamide (DMA)
Poly(ether ether ketone)	PEEK	Pentafluorophenol (>40 °C)
Polysulfone	PDF	Chlorobenzene
Poly(ether sulfone)	PES	Cyclohexanone
Poly(ether imide)	PEI	Pentafluorophenol (>40 °C)

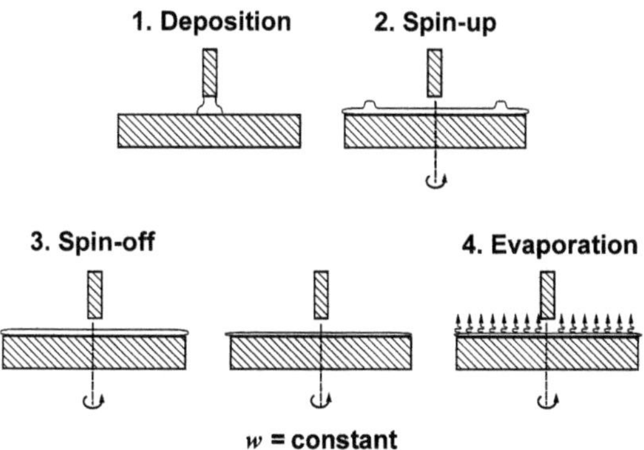

Fig. 11.1. Stages of the spin coating process: fluid dispensing, spin-up, spin-off (stable fluid outflow), evaporation (and finally evaporation-dominated drying)

ferent ways to cover the disk with the polymer solution. For example, the polymeric solution can be placed under a heavy rain that covers the whole disk, or the solution can be introduced as a bolus or a continuous stream at the centre of the disk. In the last two variants, centrifugal force drives the liquid from the centre outwards across the entire disk.

2. Spin-Up

After dispensing the fluid, the substrate is accelerated rapidly to the desired process rotation rate – usually several thousand revolutions per minute (rpm). By rotating the substrate at these high rotational speeds, the polymer solution flows radially due to centrifugal force, and covers the substrate completely. In this process phase, a high shear rate acts in the polymer solution. In the spin-up phase, which usually lasts for a few seconds, the solvent evaporates dramatically and the viscosity of the polymer solution increases.

3. Spin-Off

At the spin-off phase, the film of polymer solution thins due to convective stable outflow driven by centrifugal force, and the excess polymer solution is ejected from the edge of the substrate in the form of droplets, leaving a thin film. During the spin-off phase, the final rotation rate is commonly in the range 2000–8000 rpm.

4. Evaporation

In contrast to fluid dispensing, spin-up and fluid outflow, the evaporation-dominated drying step occurs simultaneously to the other process phases and also at the final phase of the spin-coating process. In the evaporation phase, the transition from a liquid to a solid polymeric layer takes place and the polymer film thins due to the evaporation of the remaining solvent. The spin-coating process is completed by evaporating all of the remaining solvent. The most important parameters affecting the film thickness are the spin speed, the volatility of the solvent, and the initial polymer concentration, i.e. the viscosity of the polymer solution [6].

11.1.4 Dip-Coating

Dip-coating is another coating technique that is widely used to form thin films on irregularly shaped and rigid metal, glass or polymer objects. Again, the polymer must be dissolved in an appropriate solvent. In some cases heating will be necessary. The optimum concentration should be between 5 and 10% (weight) and will depend on the actual polymer and the film thickness one wishes to achieve. Secondly, a clean substrate (e.g. a glass slide) is dipped into the solution and then pulled out slowly at a constant rate. To gain uniform films, a very slow motor can be used to pull the substrate from the solution. A possible setup is given in Fig. 11.2. Thirdly, the solvent is allowed to evaporate from the thin film. To estimate the film thickness, one must check the interference colour of the film using the same colour table that is used for ultramicrotome sections (Chap. 10).

It will undoubtedly take a number of trials to find the proper concentration. Fine-tuning can be done through addition or removal (i.e. evaporation) of the solvent.

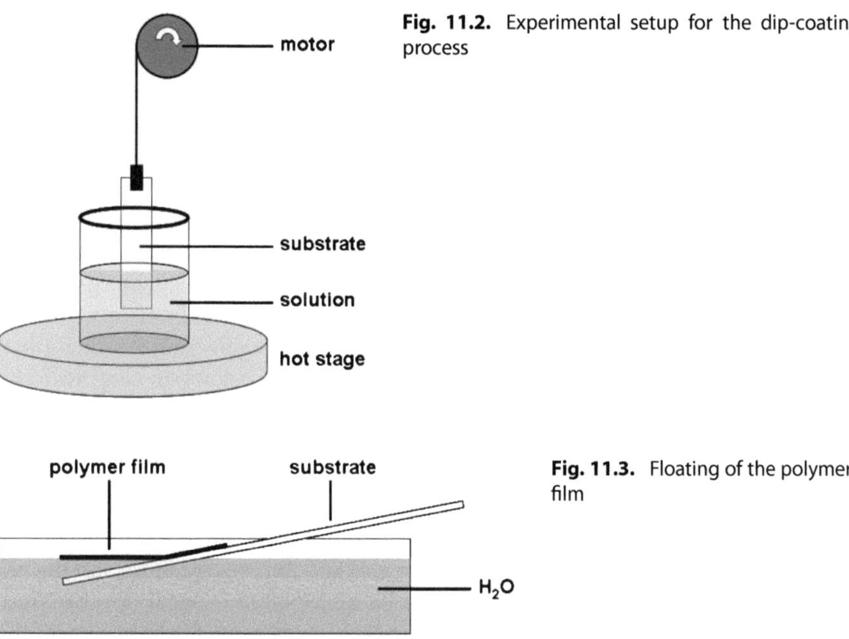

Fig. 11.2. Experimental setup for the dip-coating process

Fig. 11.3. Floating of the polymer film

The substrate is selected based on subsequent treatments. In general, glass is a good choice, but the polymer film might stick if subsequent annealing or recrystallisation steps are involved. In this case, freshly cleaved mica or silicon wafer material may be used.

The thin film can be removed from the substrate by floating on water, as depicted in Fig. 11.3. It is wise to define the required size of the thin film specimen by scribing with a scalpel or a blade beforehand. This leads to the production of a number of films that will cover the TEM meshes without overlap. The floating films can then be collected in the same way as described for ultrathin sections produced by ultramicrotomy (see Chap. 10). After drying, the films can be transferred directly to the TEM for investigation, although in most cases a subsequent staining procedure is inevitable, as it has been demonstrated for the chemical staining of ultrathin sections (Chap. 13).

11.1.5 Solution Casting

Solution casting is another quite elementary method for the preparation of polymer samples from solution. Consequently, this static preparation technique is very common. The process is performed in the following manner. After the fabrication of a polymer solution of adequate concentration, a defined amount of the solution is simply effused into an appropriate jar that must have a perfectly flat bottom (e.g. a Petri dish). Subsequently, the solvent is allowed to evaporate. The duration of the evaporation process is influenced by the environmental temperature or/and the environmental atmosphere. An atmosphere with a high concentration of the gaseous solvent results in a longer evaporation time. Thus, the evaporation rate can be controlled, for instance, by covering the dish with one or more layers of filter paper. Depending on the volume and the concentration of the solution, the evaporation process may take several days to weeks. The film thickness can be controlled by the area of the dish that is used and by the volume and the concentration of the solution. Depending on the type of polymer, the morphology of the film can strongly depend on the aforementioned parameters, especially on the evaporation rate and temperature. For example, a time-dependent sedimentation of particles (e.g. inorganic filler particles, including nanoparticles) can occur in particulate-filled systems; also, the morphology formation of block copolymers, which is driven by complex microphase separation processes, can be influenced dramatically (see below). In contrast to the ultrathin films produced by dip-coating, the films that are obtained by solution casting are too thick for direct observation in the TEM. In this case ultramicrotomy is required as an additional preparation step.

One variety of solution casting is to place a droplet of the solution onto a glass slide, freshly cleaved mica, or another suitable substrate. Again, the thickness of the resulting film can be estimated from the film area and the concentration and the volume of the solution. The resulting film can be subjected to thermal or other treatments. A 10–20 μm thick film on a glass slide is convenient for light optical microscopy, whereas the surface and – especially if a perfectly smooth substrate is chosen – the reverse side are often used for AFM studies.

11.1.6 Examples and Problems

Most of the polymeric samples created so far using solution techniques have been amorphous polymers. Here the main aim of this procedure is not to study any morphology, but to investigate micromechanical processes. Early results from imaging deformed ultra- and semi-thin films with thicknesses of between 60 and 500 nm via TEM (revealing the structure of crazes in detail) were published by Kramer, Donald and Plummer [7–9]. The deformation of the films was performed using the copper-grid method, described in detail in Sect. 12.2 (see also Figs. 12.10 and 12.11). Unlike the large number of publications that deal with amorphous polymers, there are only few papers about solution films of semicrystalline polymers [6,10–12].

From the number of publications, most of the amorphous polymers that have been examined have been self-assembled polymeric materials, like block copolymers. Because of there are a wide variety of microphase-separated structures with a huge range of different properties, lamellar block copolymers are often used (see Chap. 19). Investigations of block copolymers tend to focus on the influence of variations of solvent concentration, evaporation, evaporation rate or film thickness on the phase separation and morphology.

In contrast to the nonequilibrium morphologies that result from the extrusion or injection processing of block copolymers, it is possible to use the solution film preparation technique for the formation of equilibrium microstructures. Solution casting is the most suitable method for this because of the very slow rate of evaporation, which enables a very slow increase in the interactions between the different blocks of the copolymer as well as a sufficient large macromolecular chain mobility. Figure 11.4 shows examples of TEM (a) and AFM phase micrographs (b) of an equilibrium morphology of a styrene/butadiene block copolymer (the extruded morphology of this material is shown in Fig. 19.10 in Chap. 19). The solution-cast film of this styrene/butadiene block copolymer was prepared with toluene as solvent. The time required for the evaporation was two weeks. The solution-cast film was then held at a temperature of 120 °C under vacuum for two days. This procedure ensures that an equilibrium morphology is formed. In contrast to the highly oriented lamellae of the extruded sample, the lamellae of the solution-cast film appear as randomly oriented grains.

Another impressive example of the distinctive differences between equilibrium and processing-induced morphologies is the formation of the morphology of a polystyrene-polybutadiene-polystyrene (SBS) star block copolymer/homopolystyrene blend, which is characterised by a strong inclination to exhibit phase separation between the contrary phases at the equilibrium state (see Fig. 11.5). This pronounced inclination towards segregation leads to macrophase separation between the star block copolymer and the homopolystyrene in solution-cast films; see Fig. 11.5a. In contrast, processing the blend apparently increases the compatibility of the blend components. This results in the formation of a new "droplet-like" morphology in injection-moulded blends. Figure 11.5b shows this droplet-like microstructure, which – by the way – exhibits excellent mechanical properties [13].

The preparation of solution-cast films must be carried out carefully, especially the process of evaporating the solvent. The selection of the correct evaporation time requires much experience or a lot of pilot tests. If the preparation conditions are not optimum, preparation-induced defects can develop in the solution-cast films. In particular, solution-cast films of polymer blends and composites are very sensitive

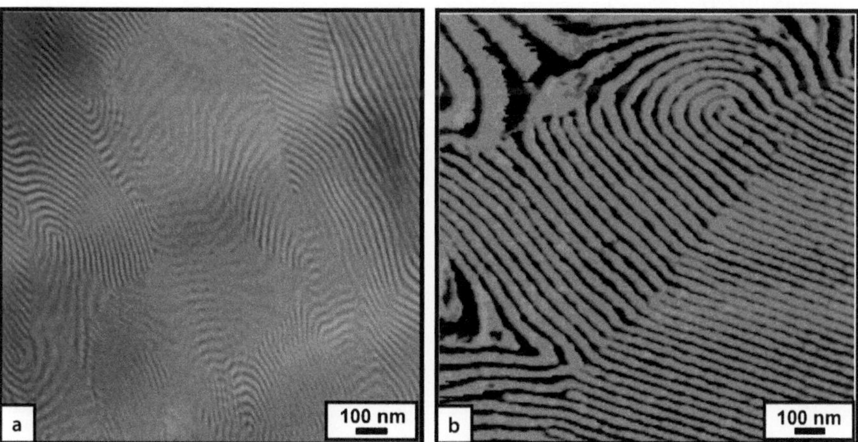

Fig. 11.4a,b. TEM (**a**) and AFM phase micrograph (**b**) of the lamellar equilibrium morphology of a styrene/butadiene block copolymer prepared by solution casting

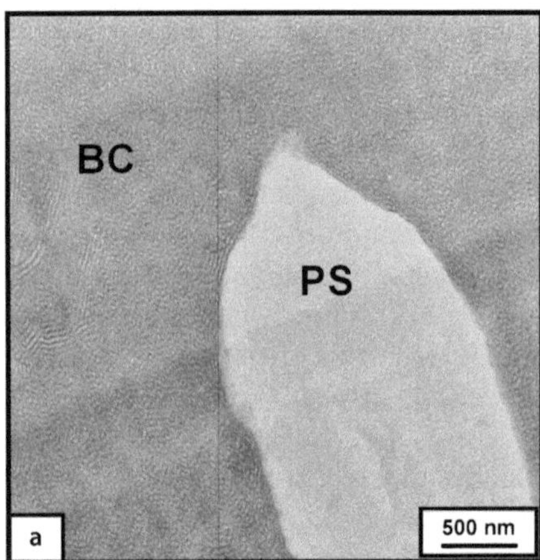
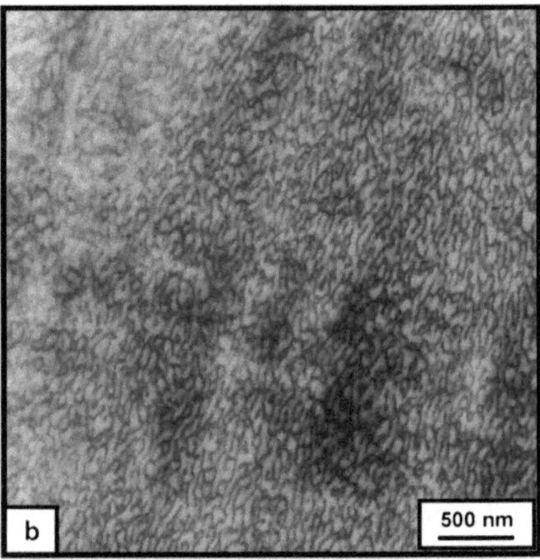

Fig. 11.5a,b. Morphology of a blend of a polystyrene-polybutadiene-polystyrene star block copolymer (BC) and homopolystyrene (PS): **a** macrophase separation in a solution-cast film; **b** apparently increased compatibility with a "droplet-like" morphology in an injection-moulded sample. Stained ultrathin sections, TEM

to preparation defects. Examples of two common defects of solution-cast films of styrene/butadiene block copolymer/homopolystyrene blends prepared with toluene as solvent are shown in Figs. 11.6 and 11.7. A typical defect in a solution-cast film is shown in the optical (a) and TEM (b) micrographs of Fig. 11.6. The solution-cast film is divided into two areas characterised by different morphologies. In the upper area, near the phase boundary between the polymeric film and the air (area I), a block copolymer matrix with macrophase-separated homopolystyrene inclusions is formed. The homopolystyrene phase appears bright in the optical and TEM micrographs.

In contrast to the upper area, the area close to the surface of the substrate (area II) shows an inverse morphology: the matrix is formed by the homopolystyrene and the macrophase-separated inclusions by the block copolymer.

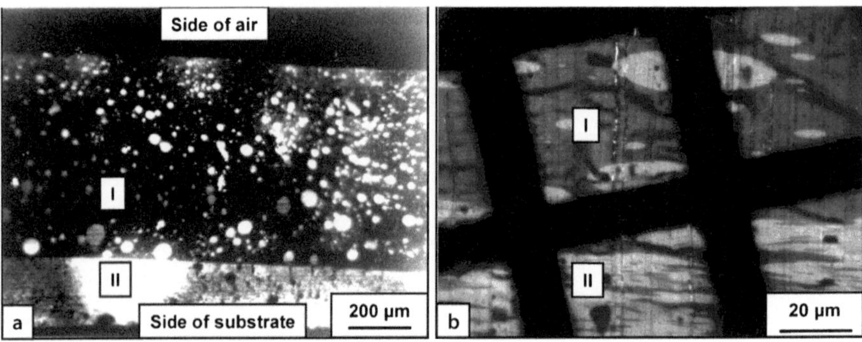

Fig. 11.6a,b. Example of the formation of two areas with different macrophase-separated morphologies of a poly(styrene-block-butadiene) block copolymer/homopolystyrene blend (*I*: block-copolymer-rich, *II*: homopolystyrene-rich): **a** light microscope micrograph; **b** TEM micrograph

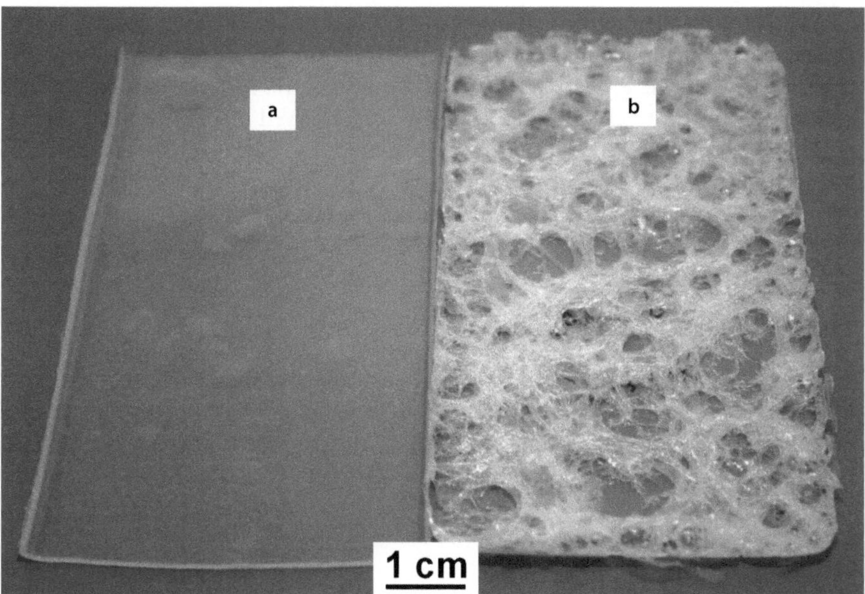

Fig. 11.7a,b. Comparison of the results of a successful solution casting preparation (**a**) and a typical mistake during the solution casting preparation (**b**) of styrene/butadiene block copolymer/homopolystyrene blends

Figure 11.7 shows the results of a successful solution-casting preparation (a) and a typical mistake (b) during such a preparation of styrene/butadiene block copolymer/homopolystyrene blends. After the slow evaporation of the solvent, the film of the blend with the low content of homopolystyrene is very smooth. If there is some solvent (toluene) in the film with the higher homopolystyrene content, longer annealing of the solution-cast film above the glass transition temperature of polystyrene causes the remaining toluene to develop into bubbles, destroying the film.

11.2 Preparation Using the Focussed Ion Beam Technique

11.2.1 Introduction

Besides the main TEM preparation techniques that are discussed in Chap. 10 and in Sect. 11.1, the FIB technique is a relatively new method that was originally established for inorganic materials but has also become interesting for polymers in some cases. The main advantage of this technique is that complex structures consisting of different classes of materials, such as layered materials, integrated circuits and composites, can be thinned to electron-transparent lamellae suitable for TEM investigations without delamination artefacts at the interfaces [14–20]. Moreover, the method allows samples to be prepared from a specific region of interest (e.g. a single transistor in a semiconductor chip) because the field of operation can be observed using a SEM that is coupled to the FIB instrument ("dual beam" or "cross beam"). Nevertheless, it has only limited applicability to polymers because many of them are sensitive to beam damage. PMMA and PMMA composites, for instance, are known to be extremely sensitive to electron irradiation when observed in TEM or SEM, and experiments where the FIB technique was applied to them were not successful. So far it has proven impossible to provide an overview of which polymer systems are suitable for FIB preparation procedures and which are not. Only a limited set of positive examples can be supplied [21–23].

11.2.2 Principle

In principle, the experimental setup of a FIB instrument is similar to that of the SEM. Instead of a focussed electron beam, a beam of gallium ions emitted from a liquid metal ion source and focussed by electrostatic lenses is used. The spot size is in the range of a few nanometres. The interactions of the ions with the sample are dramat-

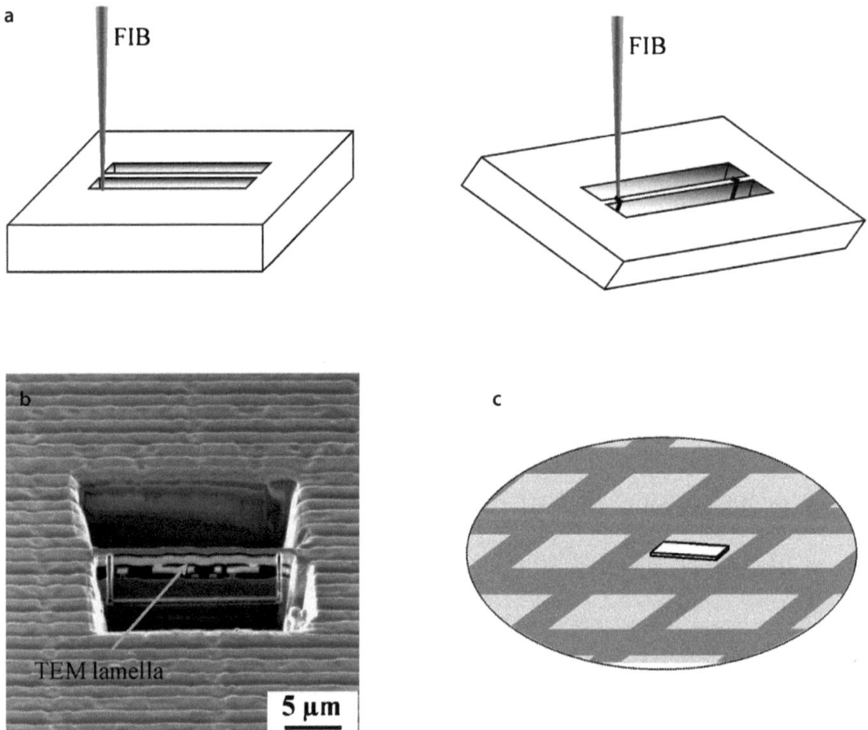

Fig. 11.8a–c. Typical steps in the preparation of ultrathin lamellae by the FIB technique: **a** etching of trenches on both sides of the region of interest, cutting of the thin ligaments, and transfer of the sample to a coated TEM mesh using a micromanipulator; **b** SEM image of a lamella thinned by the FIB technique (from the Fraunhoferinstitut für Werkstoffmechanik Halle); **c** lamella lying on the Formvar film of a TEM copper mesh

ically different from the interactions of electrons with the sample in SEM: the impinging beam of ions sputters atoms from the sample surface. This procedure is very similar in principle to etching with argon ions, as described in Sect. 9.3.2. Essentially, the focussed ion beam can be thought of as acting like a very narrow sandblast that cuts through the material. However, there are other interactions that may cause severe damage to the polymer in the vicinity of the beam. Polymers known to be sensitive to electron beam damage will probably also undergo degradation during FIB milling.

The principal preparation steps for FIB are depicted in Fig. 11.8. First, a thin lamella is produced by sputtering (cutting) trenches on both sides of an imaginary line (Fig. 11.8a). The lamella is then cut on both sides and the free-standing lamella is transferred to a TEM mesh using a micromanipulator. This last procedure is illustrated in Fig. 11.8b,c. Image (b) shows a thin lamella after the first two steps have been performed, and image (c) illustrates the placement of the lamella on a TEM mesh ready for investigation.

11.2.3 Examples

The FIB method has recently been used for the preparation of highly filled composites and nanocomposites, including mineralised biological tissues, such as bone, teeth, nacre, and others. The preparation of such materials by ultramicrotomy (Chap. 10) results in a great deal of abrasion of the knife, decreasing its lifetime significantly. Additionally, the combination of very hard fillers (e.g. ceramic nanoparticles, carbon nanotubes, etc.) with a relatively soft polymeric matrix may result in cutting artefacts at the interface between the components.

Fig. 11.9. **a** Lamella of compact bone prepared by FIB, lying on a transparent collodium film, low magnification EFTEM image; **b** EFTEM image at larger magnification; the periodic pattern is produced by a regular arrangement of hydroxy apatite crystals in the collagen matrix

Figure 11.9 shows a TEM micrograph of a FIB lamella cut from cortical bone. The bone material can be thought of as a biological nanocomposite consisting of a soft polymeric matrix (i.e. collagen) and hard inorganic nanoparticles (i.e. hydroxy apatite). Micrograph (b) shows fibrillar collagen with a periodic pattern of hydroxy apatite crystals. Although it was demonstrated that the preparation of ultrathin sections of such hard materials by ultramicrotomy is possible, the FIB method offers two main advantages. First, the sample is practically free of artefacts that are almost unavoidable when dry bone is microtomed, and second, the severe abrasion of the diamond knife due to hard nanoparticles straying onto the cutting edge is avoided.

References

1. Block DR (1999) Solvent and non solvent for polymer. In: Brandurp J, Immergu EH, Grulke EA (eds) Polymerhandbuch. Wiley-Interscience, New York
2. Emslie AG, Bonner FT, Peck LG (1958) J Appl Phys 29:858
3. Arcivos A, Shah MG, Petersen EE (1960) J Appl Phys 31:963
4. Meyerhofer D (1978) J Appl Phys 49:3393
5. Bornside D, Macosko CW, Scriven LE (1987) J Imaging Technol 13:122
6. Mellbring O, Kihlman OS, Krozer A, Lausmaa J, Hjertberg T (2001) Macromolecules 34:7496
7. Donald AM, Kramer EJ (1982) Polymer 23:461
8. Kramer EJ (1983) In: Kausch HH (ed) Crazing in polymers. Springer, Berlin
9. Plummer CJG (1995) Macromolecules 28:7157
10. Despotopoulou MM, Frank CW, Miller RD, Rabolt JF (1996) Macromolecules 29:5797
11. Kressler J, Wang C (1997) Langmuir 13:4407
12. Bartzak Z, Argon AS, Cohen RE, Kowalewski T (1999) Polymer 40:2367
13. Buschnakowski M, Adhikari R, Ilisch S, Seydewitz V, Goderadt R, Lebeck W, Michler GH, Knoll K, Schade C (2006) Macromol Symp 233:66
14. Stevie FA, Vartuli CB, Giannuzzi LA, Shofner TL, Brown SR, Rossie B, Hillion F, Mills RH, Antonell M, Irwin RB, Purcell BM (2001) Surf Interface Anal 31:345
15. Rubanov S, Munroe PR (2001) J Mater Sci Lett 20:1181
16. Kato T, Muroga T, Iijima Y, Saitoh T, Yamada Y, Izumi T, Shiohara Y, Hirayama T, Ikuhara Y (2004) J Electron Microsc 53(5):501
17. Mucha H, Kato T, Arai S, Saka H, Kuroda K, Wielage B (2005) J Electron Microsc 54(1):43
18. Abolhassani S, Gasser P (2006) J Microsc 223(1):73
19. Dehm G, Legros M, Heiland B (2006) J Mater Sci 41:4484
20. Obst M, Gasser P, Mavrocordatos D, Dittrich M (2005) Am Mineralogist 90:1270
21. Loos J, van Duren JKJ, Morrissey F, Janssen RAJ (2002) Polymer 43:7493
22. Virgilio N, Favis BD, Pépin MF, Desjardins P, L'Espérance G (2005) Macromolecules 38:2368
23. Luchnikov V, Stamm M, Akhmadaliev C, Bischoff L, Schmidt B (2006) J Micromech Microeng 16:1602

When polymers are placed under load, a wide variety of micromechanical processes may appear in them, including changes on the macromolecular level (on the nm scale), microvoid formation, crazing, shear band formation and yielding processes (on the μm scale), up to crack initiation, propagation and fracture (up to the mm scale). This chapter describes different techniques used to study these processes with various microscopes, including the investigation of fracture surfaces in the SEM and deformed samples by SEM, TEM or AFM. In particular, the preparation of semi-thin sections by (cryo)ultramicrotomy, mounting on a microtensile stage and investigation after deformation or in situ by TEM, ESEM or AFM are discussed.

12.1 Overview

Because of the many different structural details that can occur in polymers, there a wide variety of different micromechanical processes are possible in polymeric materials under load. These include changes in individual macromolecule segments (on a nm scale), microvoid formation, localised yielding processes, craze and shear band formation (on a μm scale), up to crack initiation, propagation and fracture (up to the mm scale); see Fig. 12.1. Therefore, very different techniques for studying these processes are required. Techniques which show both the morphology and micromechanical processes are of particular advantage, since they offer a very direct way to determine structure–property correlations. Electron microscopic methods can reveal morphology and micromechanical properties in the same sample area and, therefore, these methods can be used directly to study the influence of structural details on mechanical behaviour.

An overview of methods that have been successfully applied to this task is given in Fig. 12.2 [1]. Three main methods exist:

1. Investigation of fracture surfaces (from tensile or impact test, events that damage materials, etc.) directly in the SEM (or previously with TEM using replicas).
2. Bulk sample deformation (from the first elongations up to fracture) is followed by investigating surface changes via SEM (or after replication by TEM). Changes inside the bulk material are studied by preparing semi- or ultrathin sections using an ultramicrotome (occasionally after cooling or chemical staining) and investigating by TEM or AFM.

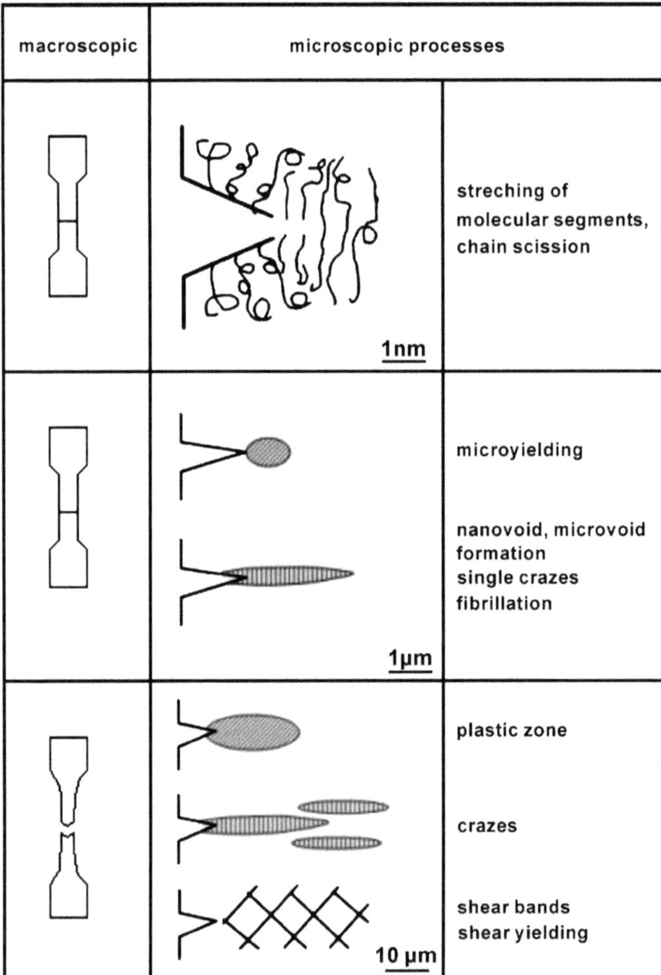

Fig. 12.1. Some basic types of localised plastic deformation in polymers, as occur in tensile bars (at the macroscopic level, on the *left*) and at the microscopic level

3. Deformation of thin films or semi-/ultrathin sections in a tensile device and investigation after deformation or in situ by TEM, HVTEM, ESEM or AFM.

Changing from method 1 to method 2 and 3 yields more and more details on the micromechanical processes and their dependence on the actual morphology. The analysis of fracture surfaces by SEM (method 1: microfractography) mainly yields information on the processes of crack initiation and crack propagation up to the final fracture. In particular, the influence of structural heterogeneities ("defects") on the initiation and propagation of cracks and the consequences of phase separation or low interfacial strength in polymer combinations can be studied.

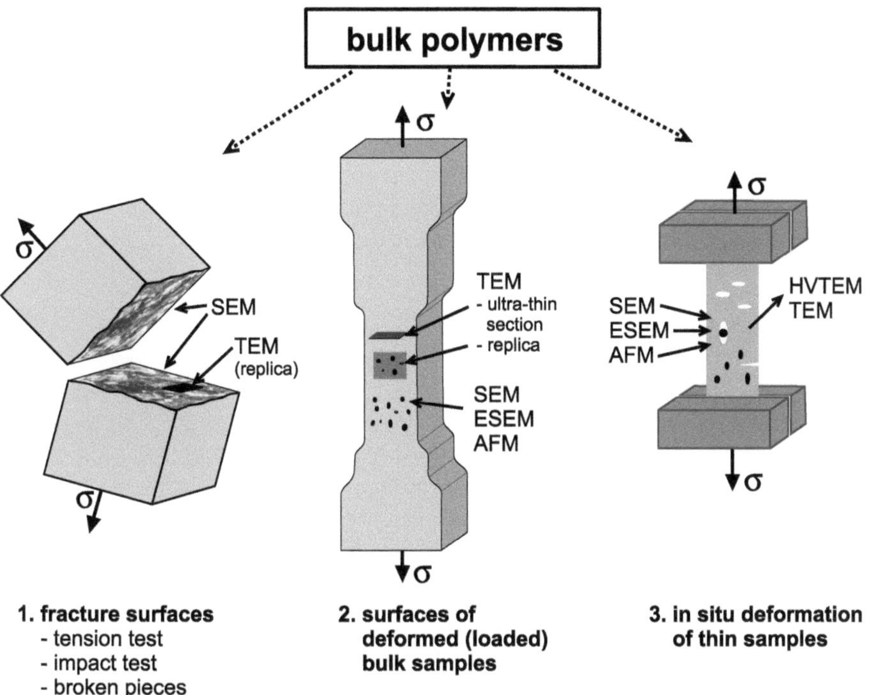

Fig. 12.2. Survey of electron microscopic methods for investigating micromechanical processes in polymers

Microfractography usually yields clear results if the fracture is brittle on a macroscopic scale. In general, the structures on the fracture surface can be influenced by the sample morphology, the deformation and fracture mechanism, or by secondary effects, such as secondary cracks or oscillations during crack propagation. The reasons that several surface structures arise are not very easy to differentiate.

Ductile materials show large amounts of plastic deformation before fracture occurs, making the identification of the whole process on the final fracture surface difficult or impossible. Such ductile mechanisms and, in general, processes that occur before fracture, such as crack initiation, microvoid formation, local fibrillation and crazing, can be seen with method 2.

Method 3 enables the morphology and deformation processes to be investigated with high resolution. As the sample size and thickness decreases, the resolution usually increases. One particularly useful method is the investigation of semi-thin sections in a HVTEM [1]. Several tensile stages that are applicable to SEM, HVTEM, TEM and AFM are shown in Fig. 6.2. The use of several tensile stages and sample sizes enables a sample with a "representative thickness", containing the characteristic morphological features of the material of interest, to be deformed. A general problem with this method, however, is radiation damage (see Sect. 8.2), since (micro)mechanical properties are drastically changed by electron irradiation. There are two solutions that can be used to overcome this obstacle:

- Perform the deformation outside the microscope or within the microscope in overview mode (at low magnification with a drastically reduced electron beam intensity), and then inspect at higher magnification.
- Perform the deformation and inspection in situ at one sample area and take photos of another nonirradiated area (i.e. use the low-dose technique; see Sect. 8.2).

12.2 Specimen Preparation

There are no special requirements for methods 1 and 2 in Fig. 12.2. Fracture surfaces or tensile bars are covered with a thin conductive layer (C or C/Au) for SEM inspection or they can be directly studied by ESEM or occasionally by AFM. For TEM inspection, ultrathin sections must be prepared using ultramicrotomy without destroying any of the deformation structures or adding any cutting artefacts (see Chap. 10).

The tensile stages used for method 3 for SEM, ESEM, AFM or HVTEM (see Fig. 6.2) can deform relatively thick specimens or semi-thin sections, whereas for TEM only ultrathin samples can be used. Semi-thin sections represent "bulk"-like properties better than ultrathin sections if they contain a "representative volume" of the material. Figure 12.3 schematically shows a particle-filled polymer matrix where ultrathin sections only contain small parts of the particles, whereas semi-thin sections contain undamaged particles in their typical surroundings. The preparation of semi-thin sections is generally performed via ultramicrotomy; see Chap. 10.

However, special cutting conditions and requirements are involved when cutting thicker sections. Sections with thicknesses from about 0.5 μm up to several μm should be relatively large (about $0.5 \times 4 \, \text{mm}^2$). An advantageous shape for the trimmed specimen is shown in Fig. 12.4.

After cutting, the section is set between two adhesive tapes; each tape has a void about 1 mm in diameter. This "sandwich" is easily mounted in the corresponding tensile stage of the microscope, and the narrow edges of the central void are cut off with

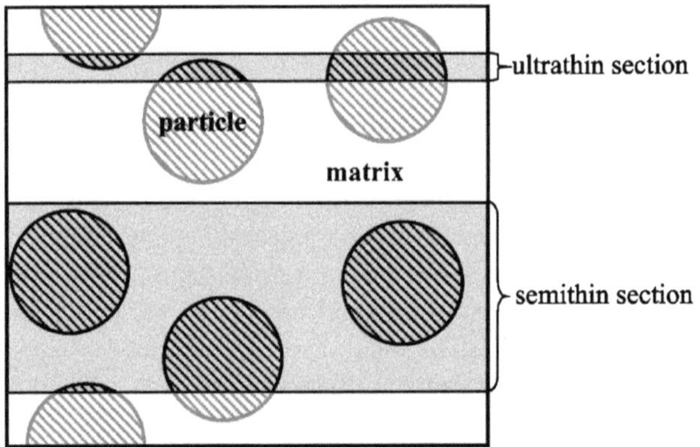

Fig. 12.3. Illustration of the effect that ultrathin sections include only small parts of particles in a matrix, whereas semi-thin sections contain a "representative volume"

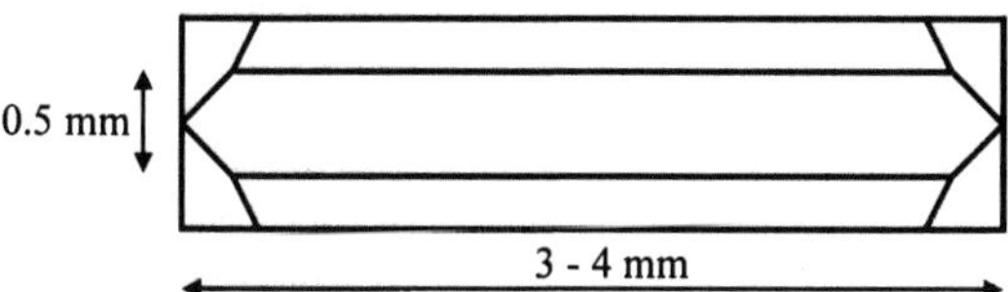

Fig. 12.4. Trimmed specimen shape for cutting semi-thin, large sections

very sharp scissors. The free-standing sample is then prepared and is ready to be deformed in the tensile stage [1].

The semi-thin sections can be cut by wet sectioning, using a knife with water filled truf (tray) – see Fig. 12.5, or by dry sectioning – see Fig. 12.6. Sectioning can be performed at room temperature if the material is stiff enough. Softer materials or polymers containing softer components cannot be fixed by chemical treatment, since the chemical reaction changes the (micro)mechanical properties of the material. Such materials must be cut at lower temperatures. Chemical staining is not possible for the same reason. However, in some cases it can aid the differentiation of structural details to perform staining after the deformation test [2]. Using an eyelash attached to a wooden needle, the sections can be directed away from the knife edge. In particular, the semi-thin sections exhibit a strong tendency to roll during dry sectioning (see Fig. 12.6). After cutting, the sections can be transferred using an eyelash to the

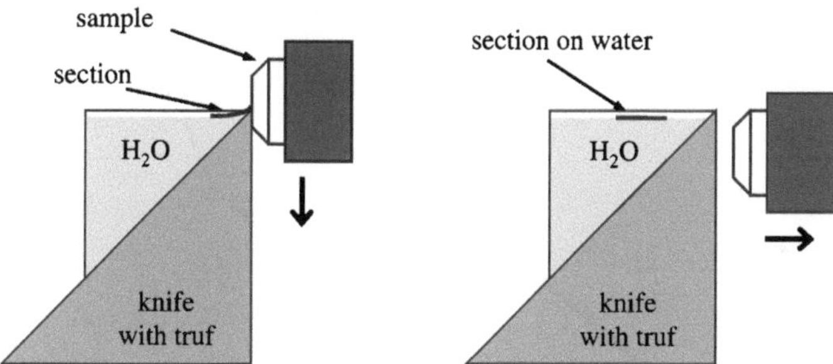

Fig. 12.5. Wet cutting of semi-thin sections using a knife with a water-filled truf: *left*, during cutting; *right*, after cutting

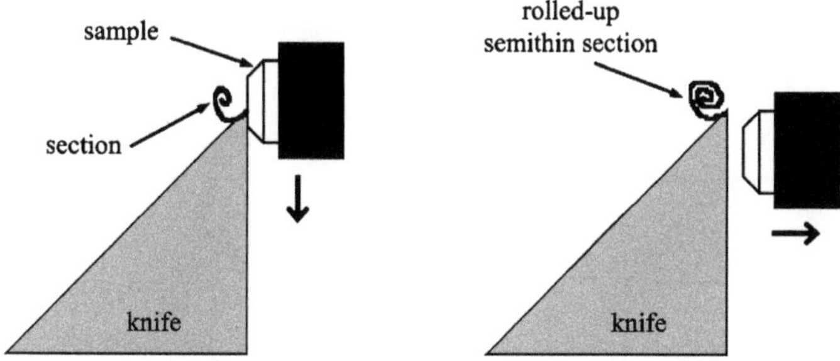

Fig. 12.6. Dry cutting of semi-thin sections: *left*, during cutting; *right*, after cutting

surface of a petri dish containing water (Fig. 12.7a). An adhesive tape with a small central void is placed on the section on the surface of the water to adhere it to the tape in a central position (Fig. 12.7b). Thicker sections can be also transferred from the knife directly onto the adhesive tape with the help of the eyelash (see Fig. 12.8) or using needle-like tweezers. Finally, the section is covered with a second adhesive tape containing a void (Fig. 12.8, right) to stabilise the sandwich structure. The result is shown in Fig. 12.9 in an optical micrograph. This procedure is described in more detail in [3].

The cutting of semi-thin sections is associated with some additional problems. The sections can be charged and thus spring away from the knife edge. Thickness determinations based on interference colours fail with semi-thin sections; thickness control can only be achieved by equipping the ultramicrotome with a sample feed

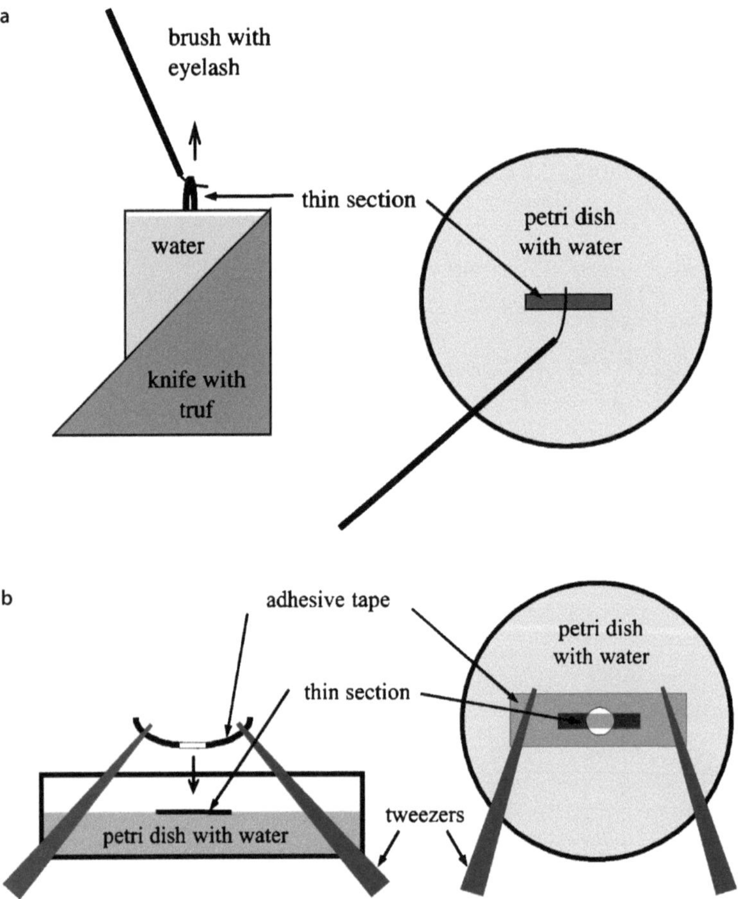

Fig. 12.7. a Transfer of a semi-thin section with the help of an eyelash from the knife to a water-filled petri dish; **b** fixation of a semi-thin section with an adhesive tape (central void diameter: 1 mm)

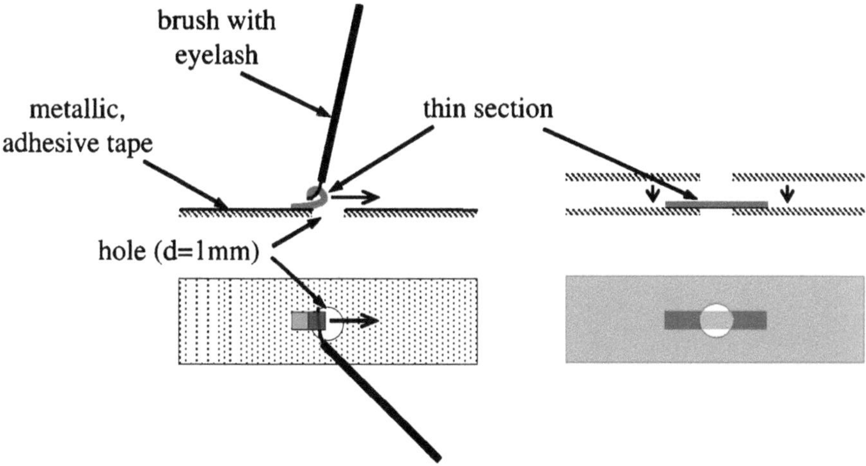

Fig. 12.8. A semi-thin section sandwiched between two adhesive tapes

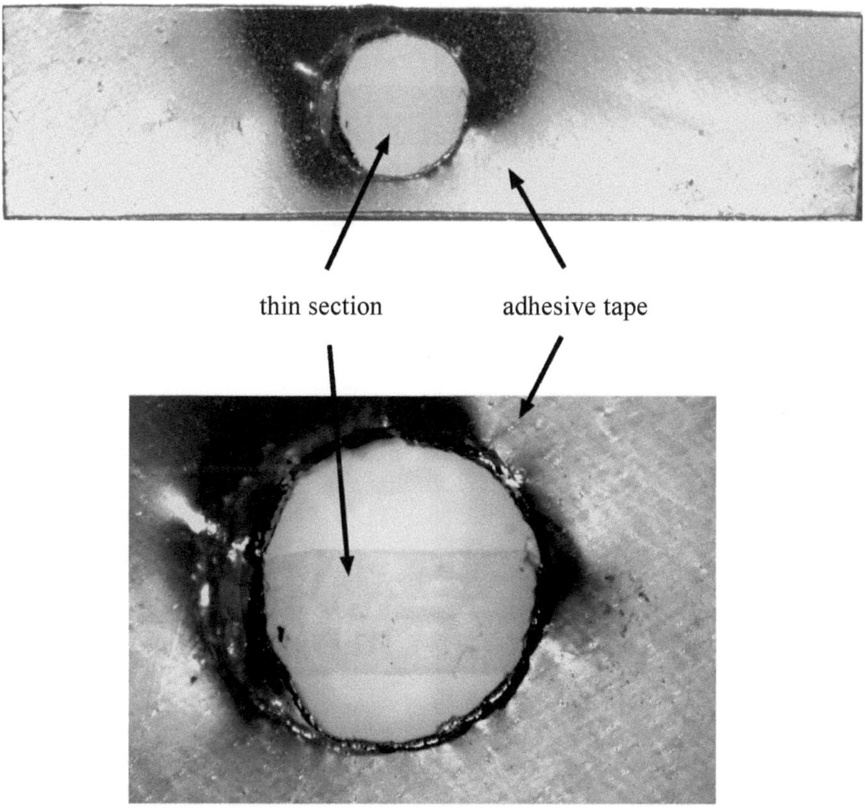

Fig. 12.9. Optical images of a specimen prepared for an in situ deformation test

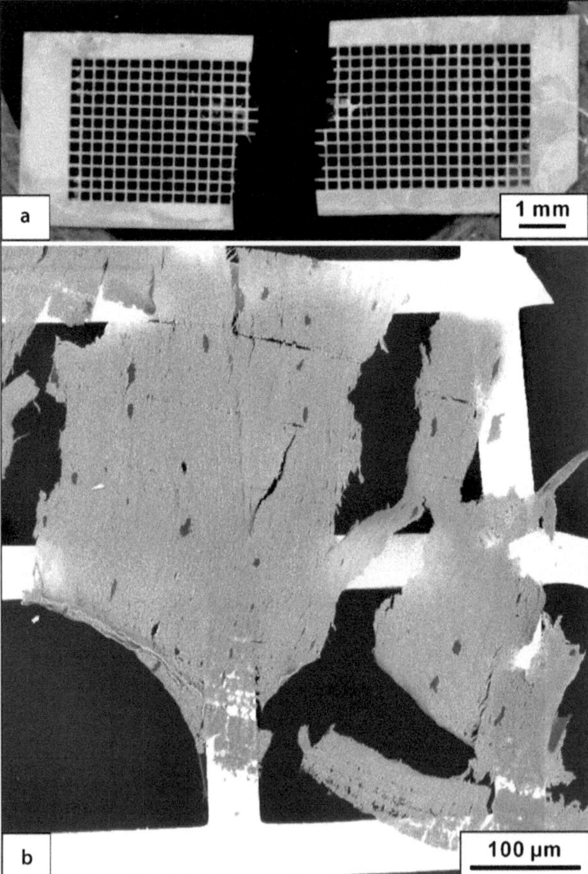

Fig. 12.10a,b. Examples of Cu meshes with attached films: **a** ultrathin PS film prepared by dip-coating on a custom-made copper mesh; **b** semi-thin section of cortical bone on ATHENE® TEM mesh after tensile deformation

controller. A second possibility for sample preparation is to produce ultrathin or semi-thin polymer solution-cast films that are between 60 nm and a few µm thick (see Chap. 11) and transfer them onto soft copper grids, which can be deformed together with the specimen in the tensile stage [4, 5]. An important advantage of this approach is that only a small amount of the polymer sample (1 g or less) is necessary to produce a specimen shape that is ready for micromechanical analysis. There are other advantages of using solution-cast films too. Larger areas without flaws and artefacts (which can be introduced during ultramicrotomy) can be easily realised. Also, these films lay smoothly on the supporting grids due to better adhesion, which can be improved by dipping the grids into a dilute solution of the corresponding polymer so that mechanical stresses can be transferred to the film. Film pieces of an appropriate size are transferred according the techniques described in Chap. 11. A variety of ductile copper meshes are available on the market. An appropriate product and supplier must be chosen considering the desired size and mesh size. For some products annealing is necessary to obtain the ductile modification of copper. Examples with different specimens are shown in Fig. 12.10; here (a) shows a grid in overview with an ultrathin PS film after rupture, and (b) depicts a semi-thin section of bone after deformation. The grids with the adherent films can be clamped to miniature tensile testing devices, as depicted in Fig. 12.11. During the tensile experiment the film can be observed through a light optical microscope. After stopping the experiment, the state of deformation of the polymer film is maintained by the supporting copper mesh. Subsequent staining is possible. Finally, the sample can be transferred to the TEM. If necessary, single sections of interest that will fit onto the TEM specimen holder can be selected and cut carefully from the mesh using a scalpel.

From a technical point of view, the following issues must be considered when preparing the samples for in situ electron microscopy:

- The approximate thickness of the sections should be known
- The thin sections must be reproducible
- The samples should not be damaged or mechanically stressed during microtoming or when mounting them on the tensile stage
- In multiphase heterogeneous polymer systems, the modifier particles must not be damaged before the deformation processes are initiated
- The thickness of the thin sections of samples to be studied should be larger than the diameter of the modifier particles (i.e., larger than a "representative thickness" of the material).

Miniaturised samples for deformation in ESEM and AFM can be produced using microinjection moulding (see Sect. 24.4). A special sample size has been developed

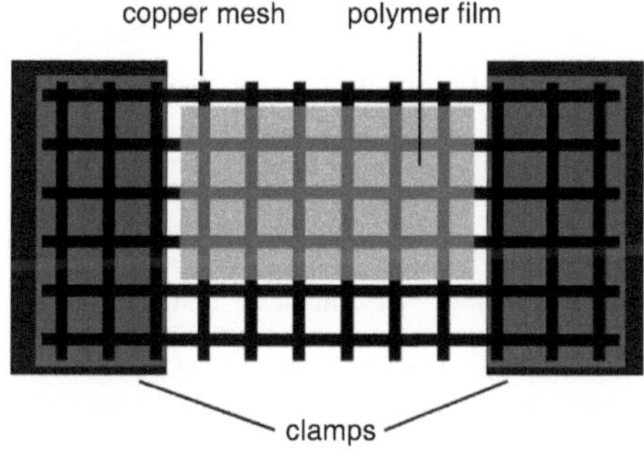

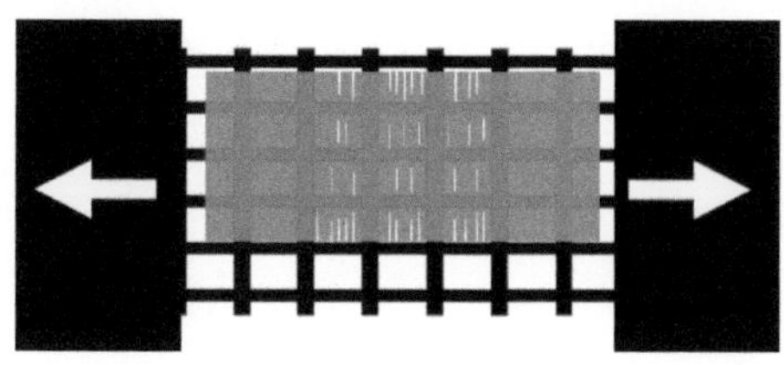

Fig. 12.11a,b. Clamping of copper grids: situation before (**a**) and after (**b**) tensile deformation

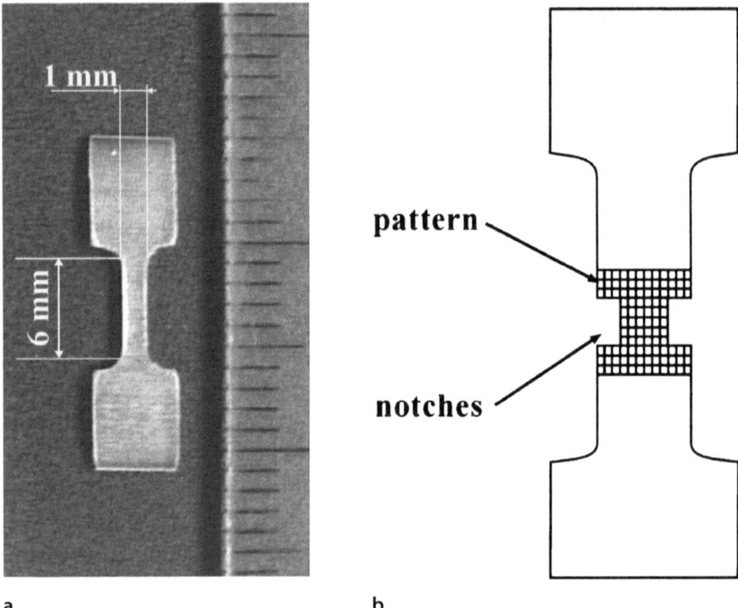

a   b

Fig. 12.12a,b. Special sample used for ESEM and AFM in situ tensile tests: **a** view of the whole sample; **b** notched sample with evaporated silver pattern

for in situ tensile tests. The deformation area of the sample measures 1×6 mm^2, as Fig. 12.12 shows. A silver pattern is evaporated in the middle of this area and the sample is then notched. The silver dots also help to identify the same area during deformation at higher magnifications, for instance in ESEM or AFM. An example of such an in situ deformation test is shown in Figs. 6.12 and 6.13.

References

1. Michler GH (1992) Kunststoff-Mikromechanik: Morphologie, Deformations- und Bruchmechanismen. Carl Hanser Verlag, München
2. Laatsch J, Kim GM, Michler GH, Arndt T, Süfke T (1998) Polym Adv Technol 9:716
3. Michler GH, Lebek W (eds) (2004) Ultramikrotomie in der Materialforschung. Carl Hanser Verlag, München, Chap 7.2, p 149
4. Kramer EJ, Berger LL (1990) Adv Polym Sci 51/52:1
5. Plummer CJG (1995) Macromolecules 28:7157

13 Contrast Enhancement

The contrast of different phases of polymer samples is often poor because polymers are mainly composed of the same light elements. In this chapter the procedures and methods required to prepare samples for cutting are described, starting with physical and chemical hardening/fixation. In particular, the potential to enhance contrast through chemical staining is discussed. Representative examples show the application of the staining agents most frequently used for polymers (osmium tetroxide and ruthenium tetroxide). Alternative contrast enhancements are described based on physical effects, using γ- or electron irradiation or an unconventional method called straining-induced contrast enhancement. In the final section, typical problems and artefacts associated with fixation and staining processes are discussed in detail.

13.1 Overview

Image contrast in transmission electron microscopy (TEM) results from differences in electron density between individual phases or structures. The contrast of different phases is often poor because polymers are mainly composed of the same light elements, such as C, H, N and O, and the differences in electron density between them are not large enough to cause sufficient contrast difference in heterogeneous polymers. Historically, contrast enhancement has been synonymous with performing chemical staining using osmium tetroxide (OsO_4) or ruthenium tetroxide (RuO_4), which is still widely practiced in contemporary polymer research. The hardening and staining of rubber phases with osmium tetroxide was introduced by Andrews and Stubbs [1, 2] for synthetic rubbers, and then further developed by Kato for rubber modified plastics and unsaturated latex particles [3–5]. Although this treatment yields satisfactory results in many cases, additional chemical staining procedures have been developed. Since chemical staining usually enhance polymer stiffness and hardness, fixation effects are also discussed in this chapter. Moreover, some physical effects also enhance the contrast.

13.2 Hardening (Fixation)

Soft polymeric materials, polymers containing soft components or organic substances (such as biological or biomedical materials) need to be fixed or hardened before sectioning by ultramicrotomy. This hardening/fixation can be done in the following two ways.

13.2.1 Physical Hardening (Fixation)

Physical hardening or fixation uses two different physical processes. In the first process, the sample is cooled below its glass transition temperature (T_g) prior to sectioning. Generally, the sectioning temperature should lie a few tens of K below the polymer's T_g in order to account for the local temperature rise caused by mechanical friction during sectioning. Alternatively, the sample can be bombarded with energetic radiation such as γ- or electron irradiation, with dosages of up to 10 MGy. This method is suited to polymers that undergo crosslinking rather than chemical degradation upon irradiation. Common examples are heterogeneous polymers containing polybutadiene (PB) as a rubber component or polyethylene (PE). The effect of crosslinking is illustrated in Fig. 13.1 by an ultrathin section of a blend of a PS/PB block copolymer after partial electron irradiation in TEM. The left hand side of the section is irradiated with electrons, inducing crosslinking. Upon treating this section

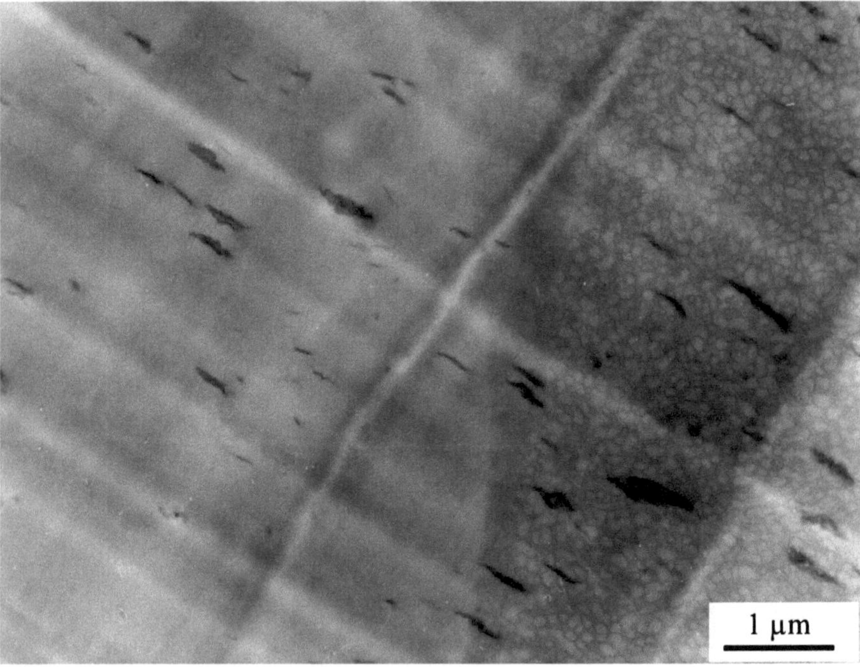

Fig. 13.1. TEM micrograph of a PS/PB star block copolymer after partial electron irradiation in the TEM (*left half*); treatment with OsO_4 vapour of the ultrathin section only yields a staining effect in the nonirradiated *right half*, while the PB phase is crosslinked on the left side due to the electron irradiation, and so cannot not be stained

with OsO_4 vapour, only the PB phase in the neat locations (nonirradiated areas on the right hand side) are effectively stained.

13.2.2 Chemical Fixation

Chemical treatment is the method most commonly employed for the fixation or hardening of polymeric materials. Following this route, the specimen is treated with one or several chemicals that preferentially react with or diffuse into specific polymer phases. The component that reacts with the chemical becomes hard or even brittle and thus contributes to the staining effect. This treatment is discussed in more detail in the following section.

13.3 Chemical Staining Procedures

As already stated in the preceding section, a typical problem encountered during electron microscopic examinations of polymers and biological specimens is the lack of intrinsic contrast between structural details. The reason for this is the presence of only very small differences in the electron densities of these structures. The main reason for using a chemical staining procedure is to artificially induce contrast by depositing high-density metal compounds into one or more of the components of the samples under investigation. The reactions of several chemical agents toward polymeric materials are often highly selective, i.e. different structural details react in different ways. If these reactions are associated with the absorption of elements with atomic numbers that are higher than those usually found in polymeric materials, a selective staining effect is possible. This effect may be enhanced by performing a special chemical pretreatment (e.g. "activation" of the material before staining). The pretreatment selectively changes the properties of constituents of the materials, such as amorphous regions of crystalline polymers. Different techniques can be employed for the selective staining procedure, either on the bulk material before sectioning or on the thin sections after cutting [6–8].

13.3.1 Media Used to Perform Chemical Staining of Polymers

Reactants often used for pretreatment include chlorosulfonic acid, tungsten-phosphoric acid, formalin, hydrazine, methanolic sodium hydroxide solution, etc. According to the procedure suggested by Kanig for PE [9], the specimen is first treated with a vapour or a solution of chlorosulfonic acid for a period lasting from several hours to a few days. The acid selectively attacks the amorphous phase, which, upon further treatment with uranyl acetate, makes the structural details of the polymer visible in TEM. Alternatively, osmium tetroxide can be used as the staining agent instead of uranyl acetate [10], as illustrated in Fig. 13.2. A major advantage of this staining procedure is that the specimen is sufficiently hardened to allow the sectioning of the PE sample to be performed at room temperature.

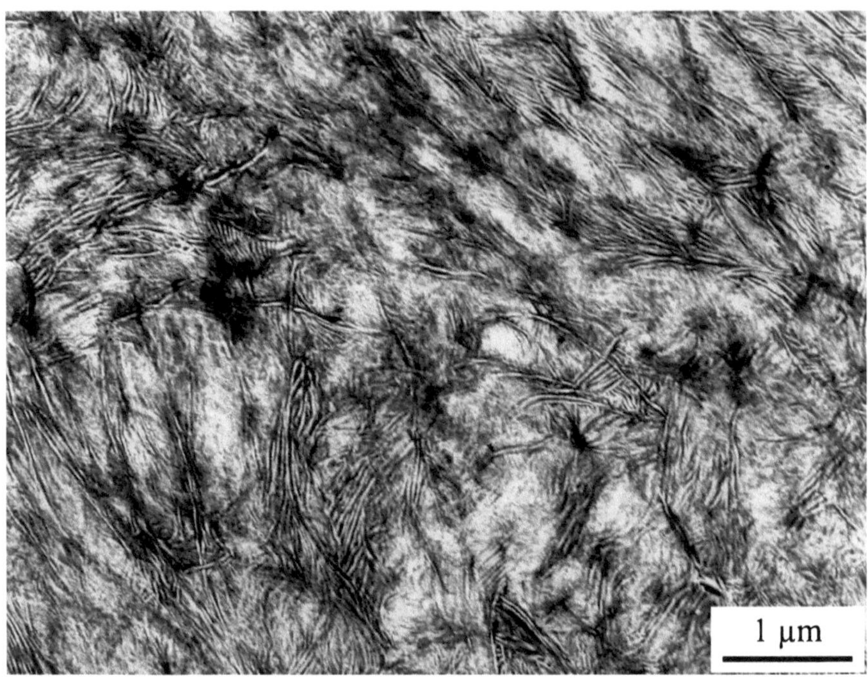

Fig. 13.2. TEM micrograph of HDPE, revealing lamellae (*white*) and interlamellar, amorphous regions (*dark*); OsO$_4$-stained ultrathin section after pretreatment with chlorosulfonic acid

Table 13.1. Chemicals successfully employed as staining agents in polymer research

Polymers	Staining agents
Polyolefins (e.g. PE, PP)	Chlorosulfonic acid/osmium tetroxide
	Chlorosulfonic acid/uranyl acetate
	Ruthenium tetroxide
Polyamides (nylons)	Formalin/osmium tetroxide
	Tungsten phosphoric acid/osmium tetroxide
	Ruthenium tetroxide
Polyacrylates	Hydrazine/osmium tetroxide
	Chlorosulfonic acid/osmium tetroxide
	Ruthenium tetroxide
Polystyrene, styrene copolymers	Ruthenium tetroxide
Polyurethanes	Chlorosulfonic acid/osmium tetroxide
	Ruthenium tetroxide
Polyvinyl chloride	Chlorosulfonic acid/osmium tetroxide
Polymers with double bonds (such as	Osmium tetroxide
PB, PI, HIPS, ABS)	Bromine solution
Polymers with OH groups	Osmium tetroxide

Today, the staining agents most frequently used for polymers are osmium tetroxide [9–16] and ruthenium tetroxide [17–26]. An overview of the staining agents that have been successfully employed in polymer microscopy is provided in Table 13.1.

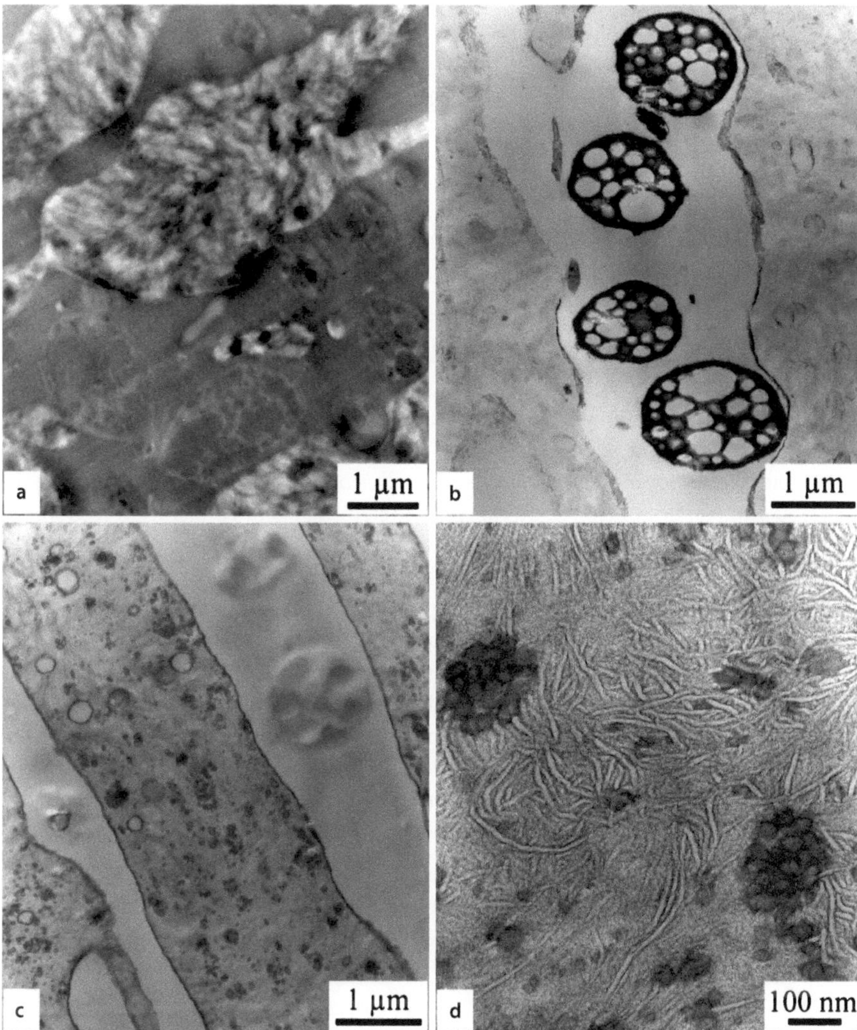

Fig. 13.3a–d. TEM micrographs of PE/PPE/PS/SEPS blend stained with different agents: **a** unstained; **b** OsO$_4$ staining mainly shows the HIPS particles (*dark*) embedded in the PS/PPE phase; **c** RuO$_4$ staining mainly shows the lamellae of PE and SEPS particles; **d** higher magnification of **c**

OsO$_4$ reacts with double bonds like those present in polyisoprene (PI) and polybutadiene (PB), but it does not react with conjugated double bonds [14]. The mechanism of staining with RuO$_4$ is different from that of OsO$_4$. The former does not react directly with chemical species; it forms clusters instead. It stains most of the polymers, but to different extents, depending on its diffusion rate into the polymer [20, 21]. For example, in semicrystalline polymers RuO$_4$ diffuses preferentially into the amorphous regions and stains them, whereas the crystalline regions remain unaltered.

The use of different agents to stain multicomponent polymers consisting of more than two phases can provide information about the structures of these phases (see Fig. 13.3). The effectiveness of the staining procedure can be significantly enhanced by performing the treatment at higher temperatures (below the glass-transition or melting temperatures of the polymers).

13.3.2 Chemical Staining of Compact Specimens Before Sectioning

Depending on the properties of the materials, the compact specimens are first treated with a chemical medium (such as formalin, sodium hydroxide, chlorosulfonic acid, etc.), as required. This step can be carried out using the vapours of the chemicals or by directly dipping the specimen into a solution with a suitable concentration. After pretreatment, the staining is performed using a solution or the vapour of osmium tetroxide (OsO_4), ruthenium tetroxide (RuO_4), uranyl acetate or some other similar heavy metal compound (see Fig. 13.4). The efficiency of the staining depends on the material and can be controlled by varying the staining agent concentration as well as the duration and the temperature of treatment.

The staining of the trimmed polymer block is performed at room temperature or at elevated temperature (e.g. 60 °C). The specimen block should be thoroughly rinsed with distilled water after staining. This can be done by immersing the specimen in distilled water for about an hour. Good results are obtained if multicomponent polymers are stained via this technique (see Fig. 13.5).

RuO_4 staining is similar to the staining of polymer blocks by OsO_4 vapour. It is important that the staining solution is always freshly prepared. The time required for staining may range from three to four hours. Typical examples of the TEM imaging of a semicrystalline polymer using this method are presented in Fig. 13.6.

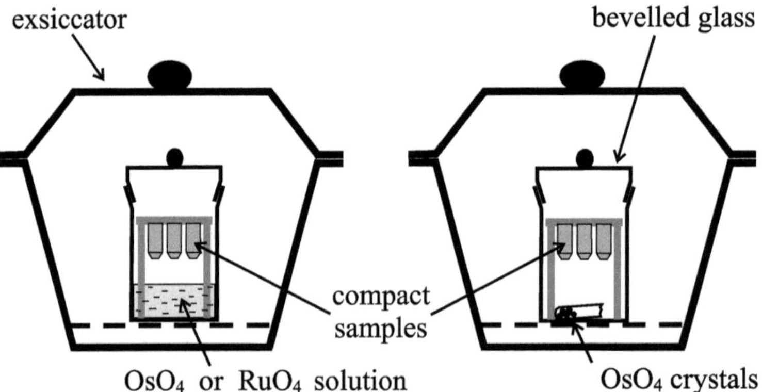

Fig. 13.4. Staining compact specimens using heavy metal vapour

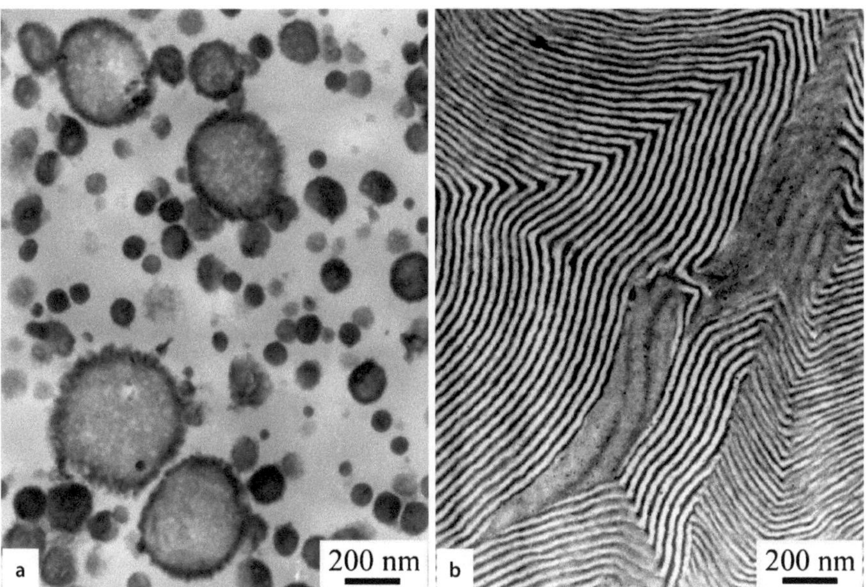

Fig. 13.5a,b. TEM micrographs of ultrathin sections; OsO$_4$ staining of bulk material and cutting at room temperature of: **a** ABS; **b** SBS triblock copolymer

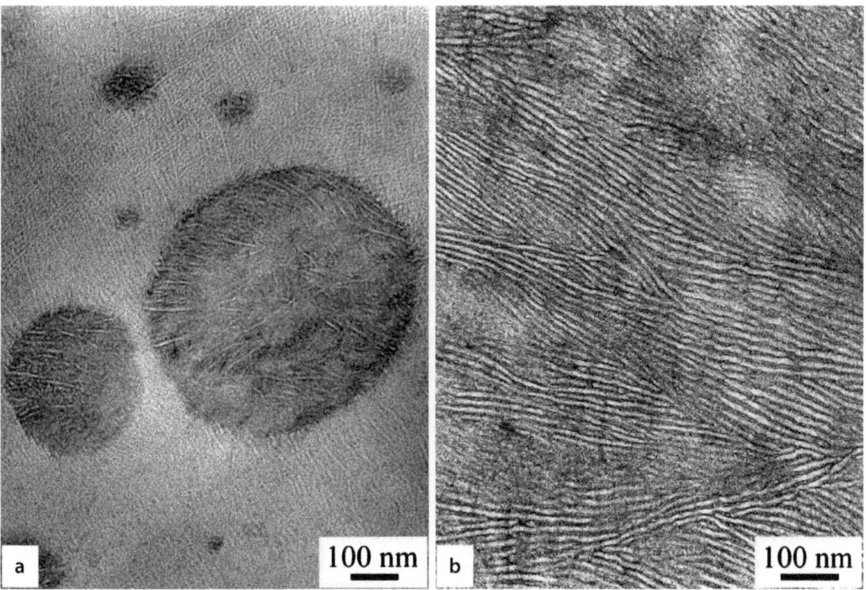

Fig. 13.6a,b. TEM micrographs of ultrathin sections; RuO$_4$ staining of bulk material and cutting at room temperature of: **a** PP/EPR blend; lamellae of PP matrix and PE lamellae in the dark EPR particles; **b** lamellae structure of beta PP

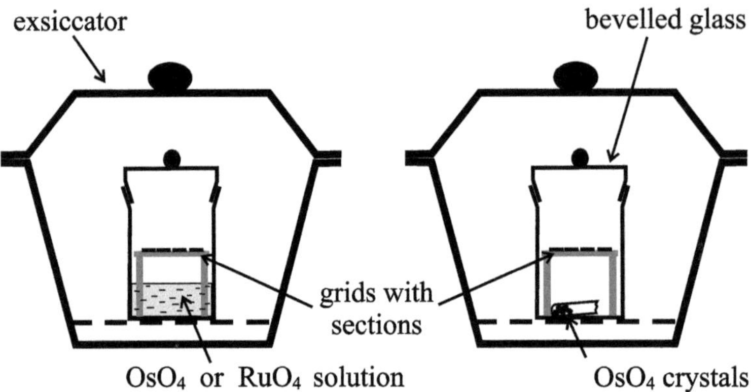

Fig. 13.7. Staining of thin sections using heavy metal vapour

13.3.3 Chemical Staining of Thin Sections

As an alternative to the staining of compact or bulk specimens, one can also directly stain sections of the samples produced by (cryo)ultramicrotomy or even by film-casting procedures. Ultrathin sections or thin films are usually stained using vapour (Fig. 13.7). In most cases, osmium tetroxide (OsO_4) and ruthenium tetroxide (RuO_4) are used. The time required for OsO_4 staining ranges from a few minutes to several hours or even days.

The staining of thin polymer sections with RuO_4 vapour takes place quite rapidly (a few seconds to a few minutes). Staining at temperatures that are slightly higher than room temperature (for example at 60 °C) provides good results. Examples of the staining of ultrathin sections of a flame-retardant PP with OsO_4 and of block copolymers with RuO_4 are presented in Figs. 13.8 and 13.9, respectively.

13.4 Enhancement of Contrast Through Physical Effects

Besides chemical staining, contrast enhancement of polymeric materials can also be achieved through physical effects. This is of particular interest if the polymers do not possess the double bonds or reactive groups needed for chemical staining to work (e.g. in the case of weather-resistant polymers for outdoor applications). Two techniques are generally used: irradiation-induced effects (γ- or electron irradiation) and straining-induced effects. An additional effect is particularly useful for elastomer blends. Here, one of the components can be thermally decomposed if the degradation temperatures of the components are very different (see Fig. 20.1).

13.4.1 Contrast Enhancement by γ- or Electron Irradiation

The gamma or electron beam irradiation of polymers causes different primary and secondary effects, which were briefly reviewed in Sect. 8.2. Secondary effects such

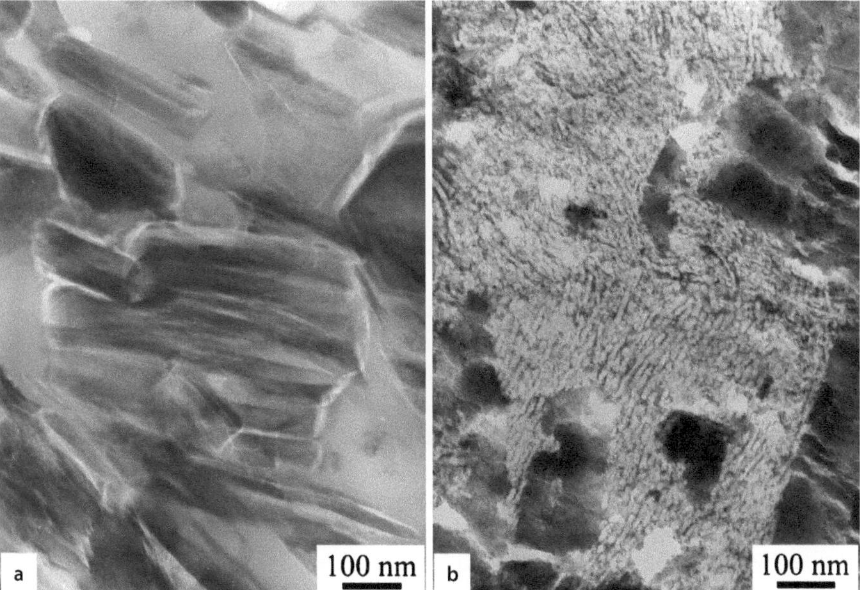

Fig. 13.8a,b. TEM micrographs of flame-retardant polypropylene: **a** cryoultrathin section without staining; **b** ultrathin section stained in a second step with OsO_4 vapour after pretreatment with chloro-sulfonic acid and OsO_4 solution

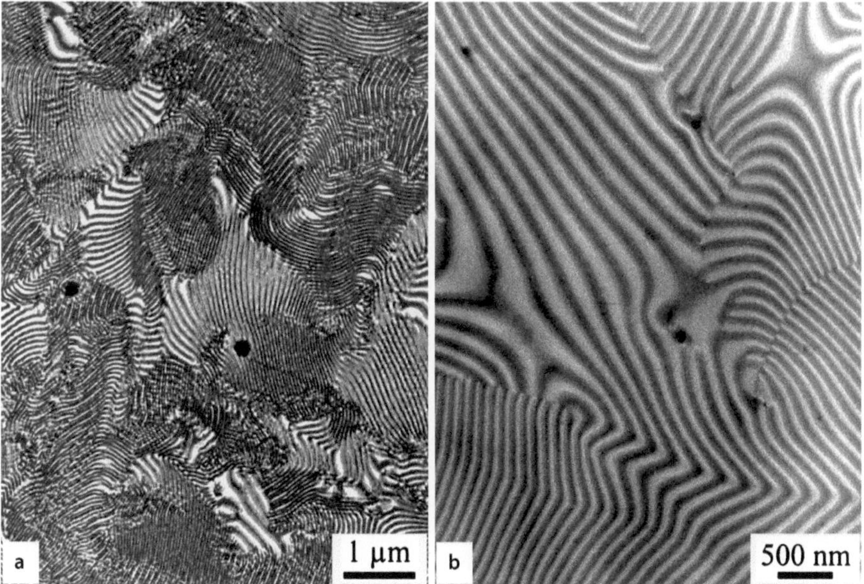

Fig. 13.9a,b. TEM micrographs of cryoultrathin sections; RuO_4 staining of sections after cutting: **a** PS/PBMA diblock copolymer, dark-stained PS lamellae; **b** PMMA/PBMA diblock copolymer, dark-grey-stained PBMA lamellae and brighter PMMA lamellae

as chain scission or crosslinking, mass loss, fading of crystallinity or heat generation can all be used to enhance contrast.

In the case of polymer blends, the different sensitivities of polymeric components can yield stronger mass loss in one component compared to the other, yielding the development of contrast at the start of irradiation in the electron microscope [27,28]. An example is shown in Fig. 17.13 for a blend of PVC (strong mass loss by dechlorination) and SAN (which is relatively stable during electron irradiation).

During γ- or electron irradiation, PE is known to show a stronger tendency to crosslink than to undergo chain scission, yielding a fixation effect [29] that was developed as a routine preparation technique [30]. Micrographs showing the results of applying this technique are shown in Fig. 13.10. Figure 13.10a shows spherulites with typical concentric rings of LDPE. Here, irradiation doses of up to 10 MGy are sufficient. Note that after irradiation doses of 20 MGy or more it is possible to attain a resolution higher than 10 nm and to identify individual lamellae; see Fig. 13.10b. The lamellae appear as bright lines in a surrounding darker amorphous area. The thicknesses of the lamellae are about 10–15 nm, which corresponds to the values for the lamellae measured in chemically stained sections.

The irradiation-induced crosslinking predominantly occurs in the amorphous interlamellar regions, yielding a reduction in the thickness of the amorphous layers between the lamellae and a dilatation in the lengths of the amorphous parts due to

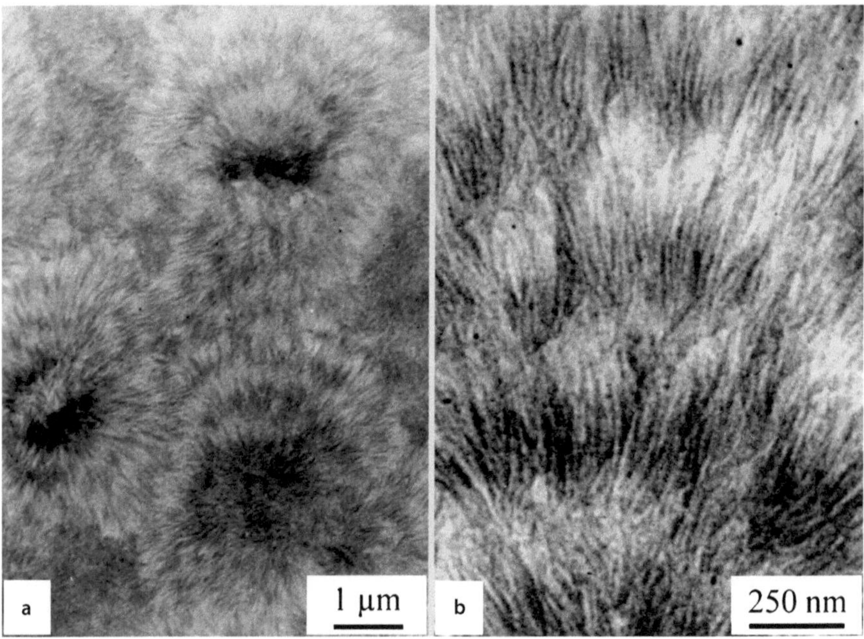

Fig. 13.10a,b. Contrast enhancement by γ-irradiation in semicrystalline LDPE (γ-irradiation with a dosage of 20 MGy): **a** banded spherulites; **b** lamellae inside the concentric bands (after [27], reproduced with the permission of Hanser)

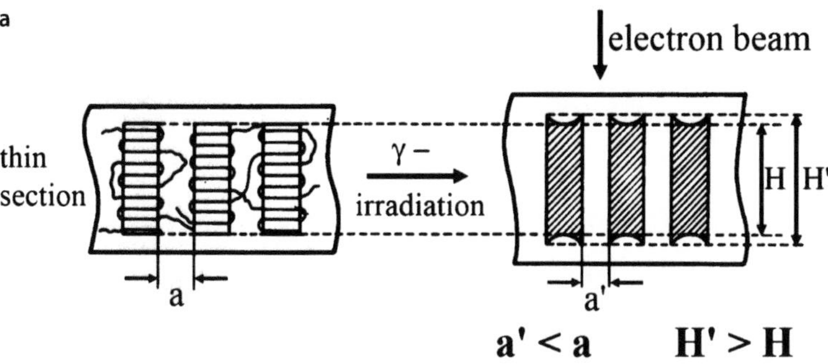

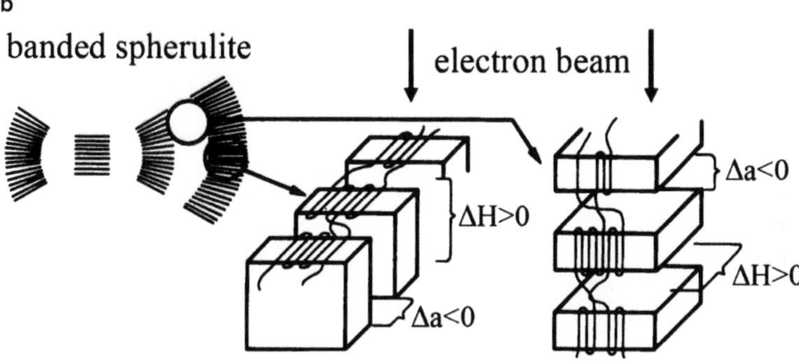

Fig. 13.11a,b. Scheme showing the effect of irradiation-induced contrast enhancement in semicrystalline polymers due to γ- or electron irradiation: **a** contrast development between amorphous zones and lamellae; **b** contrast development between concentric rings in banded spherulites

the requirement for constant volume or density; see Fig. 13.11a. In places where the boundary layers of the lamellae are oriented parallel to the normal direction of the ultrathin section and to the electron beam direction, the thickness increases in the amorphous layers, yielding darker amorphous zones and brighter lamellae. This interpretation of the predominant crosslinking in the amorphous parts is supported by the appearance of clear diffraction contrast in the electron microscope, even after irradiation doses of up to 10 MGy [30, 31].

The crosslinking of amorphous zones also induces a second contrast effect: the appearance of concentric rings inside spherulites, such as the typical banded spherulites of LDPE (Figs. 13.10a and 13.11b). In rings with the lamellae oriented parallel to the electron beam (the "edge-on" position), the enlarged amorphous zones cause a darker contrast of the rings. Due to the reduced thicknesses of the amorphous zones the circumferences of the rings decrease, whereas the circumferences of the intermediate rings (of lamellae in the "flat-on" position) remain constant. Thus, the thin section is likely to form a rippled surface where the rings are. This "ripple" effect was first discussed by Grubb and Keller [32] for the direct influence of an electron beam, but on the wrong assumption that crosslinking preferentially appears inside the lamellae.

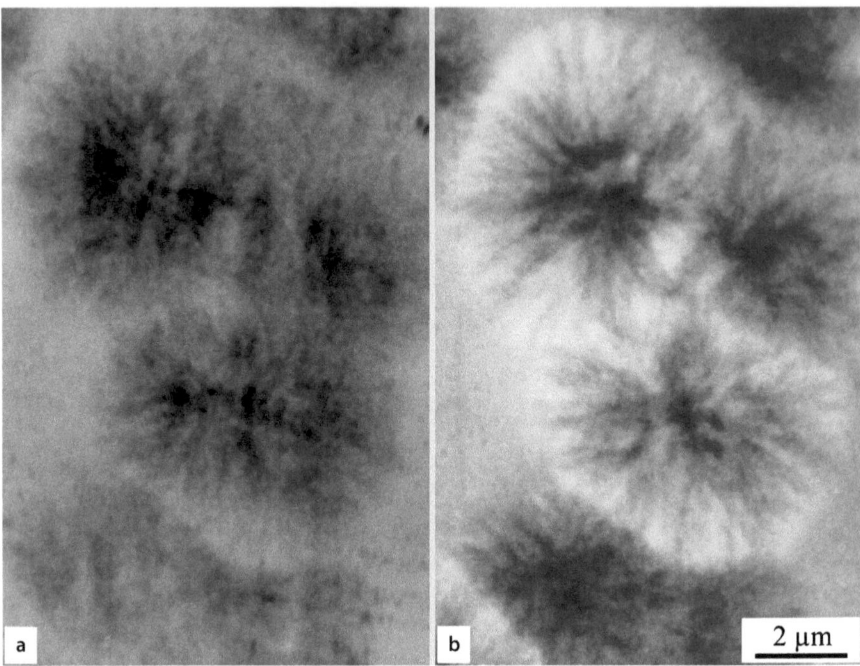

Fig. 13.12a,b. Contrast enhancement by electron irradiation in semicrystalline polymers during HVTEM studies: **a** PU spherulites at the beginning of the inspection; **b** the same PU spherulites after electron beam irradiation (after [27], reproduced with the permission of Hanser)

This effect, irradiation-induced contrast enhancement, also occurs in other polymers, and can be used during direct inspection in TEM. Figure 13.12 shows a thin cryo section of a segmented polyurethane [33]. At the start of electron beam irradiation the spherulites can only be seen with a weak contrast. After a few seconds of electron irradiation, the spherulites present rich contrast, with the radial elements inside the spherulites definitely appearing sharper. Other examples where this effect occurs include polymers containing polybutadiene rubber.

In general, this method is suitable for polymers that undergo crosslinking rather than chemical degradation due to irradiation. In Fig. 13.13, the rubber in the salami particles of high-impact PS (HIPS) appears brighter after in situ electron irradiation in TEM (micrograph a) and exhibits better contrast and resolution after γ-irradiation (micrograph b).

Combining electron irradiation and chemical staining can improve, for instance, the visibility of impurities or defect layers inside the lamellae in PE [11, 27]. Figure 13.14 shows lamellae with many of these defect layers, with individual crystalline blocks in between. There are actually two possibilities for achieving additional contrast enhancement [28]:

– The usual chemical staining and then specific in situ irradiation of the thin section in the TEM. It is assumed that during irradiation the mobility of the staining

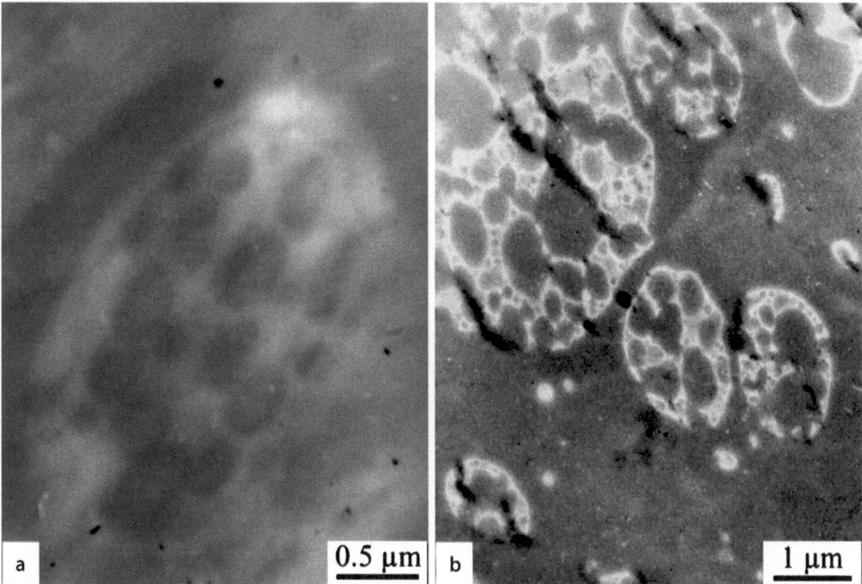

Fig. 13.13a,b. Fixation and contrast enhancement of rubber particles in unstained high-impact polystyrene (HIPS) due to: **a** irradiation with electrons in the TEM; **b** γ-irradiation (with a dosage of 10 MGy)

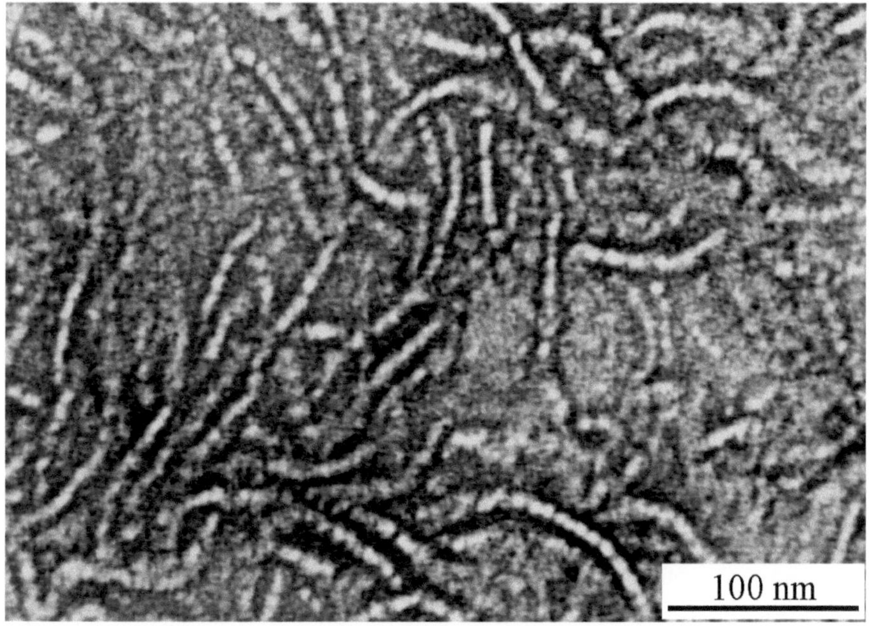

Fig. 13.14. Contrast enhancement of the amorphous layers between crystalline blocks inside the lamellae of branched PE through the combined attack of chemical staining and electron irradiation (after [27], reproduced with the permission of Hanser)

material is enhanced and reactions occur with radicals caused by the irradiation in the amorphous zones and in the defect layers.
- Irradiation with γ-rays or with electrons and subsequent chemical staining of the material. Irradiation activates the material, enabling it to react with the staining agent better.

13.4.2 Straining-Induced Contrast Enhancement

An unconventional method of contrast enhancement results from the micromechanical in situ analysis of polymers (see also Chap. 6). The stretching of thin samples (semi-thin sections) of heterogeneous polymers often leads to the observation that the phase with the lower modulus of elasticity (or higher deformability) appears brighter in the TEM images than the component with higher modulus [27, 28, 34].

Figure 13.15 is a schematic drawing of some structures represented by particles dispersed in a matrix. Figure 13.15a is a rather simple example of large particles with

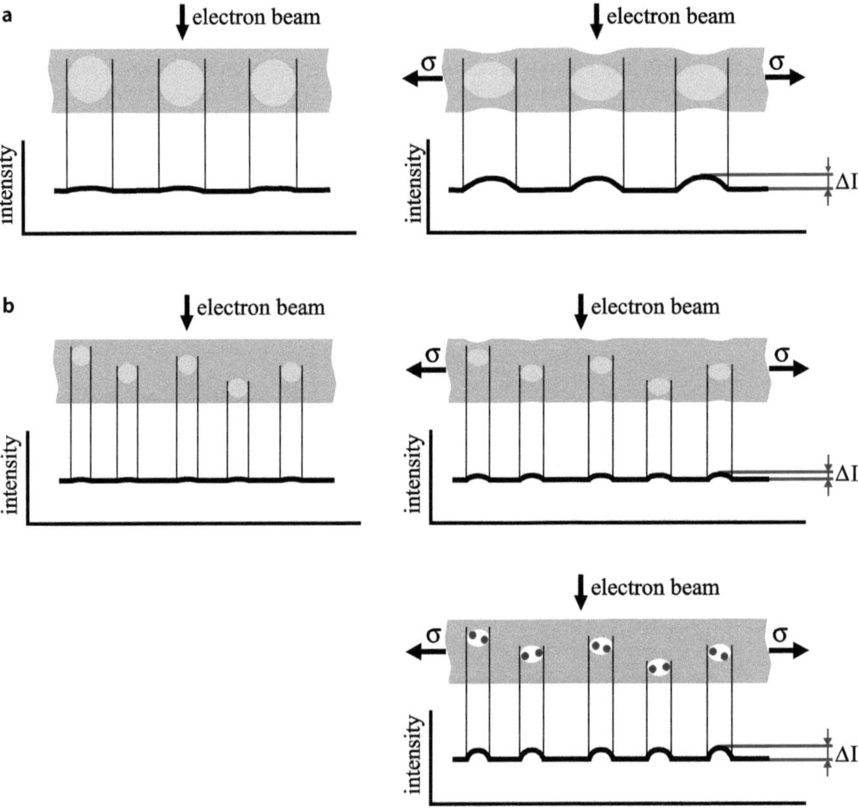

Fig. 13.15a,b. Scheme showing the effect of straining-induced contrast enhancement: **a** for larger particles; **b** for smaller particles

diameters that are only slightly smaller than the thickness of the film. The difference in density between the particles and the matrix are very small. Hence, the differences in the intensity of the electron beam are too small to yield visible contrast. The Young's modulus of the particles is assumed to be slightly smaller than the modulus of the matrix (or the extensibility of the particles is assumed to be somewhat higher); accordingly, the particles are deformed preferentially during the elongation of the film. In the direction of the electron beam, the thickness of the film decreases at the particles, thus causing a difference ΔI in the intensity of the electron beam, and a contrast between the particles and the matrix becomes detectable. Figure 13.15b, which shows smaller particles, is more interesting. The particles are also deformed to a higher degree than the matrix on average. Experience has shown that this effect even occurs with particles with diameters of less than a tenth of the film thickness. A modified case is when the particles do not have a smaller Young's modulus but do have a lower strength. The particles can cavitate and the microvoids provide a clear contrast with the matrix.

Figure 13.16 shows two more examples: TEM micrographs of PB rubber-toughened polystyrene (HIPS) and PBA–PMMA core–shell particles in a SAN matrix. Because the rubber has a lower modulus of elasticity than the matrix polymer, the rubber particles are stretched to greater extent. As a result, along the electron beam direction, the film thickness decreases at the locations occupied by the particles. Thus, the resulting electron intensity difference reinforces the contrast between the particles and the matrix.

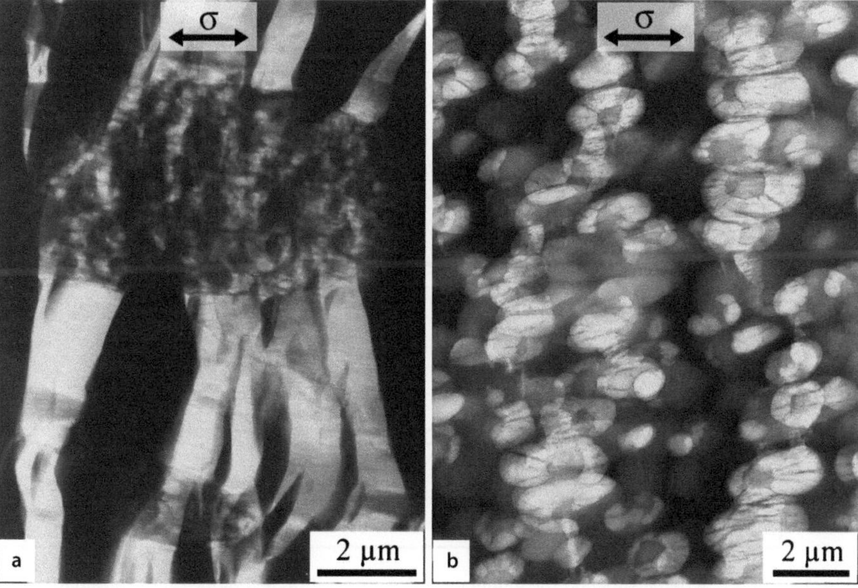

Fig. 13.16a,b. Straining-induced contrast enhancement in unstained semi-thin sections of rubber-particle-modified polymers; strain direction is shown by *arrows*: **a** strained rubber phase in "salami" particles (*dark grey*) and craze formation (*bright*) in HIPS; unstained semi-thin section; **b** strained PBA–PMMA core–shell particles (*dark grey*) in SAN matrix, and hole formation (*white*)

13.5 Problems and Artefacts

Because of the wide variety in structures observed for polymers, it is not possible to reveal all of structural details using just one of the many different preparation and investigation techniques discussed in the preceding chapters. This idea also holds for the different techniques of contrast enhancement described in this chapter. For instance, an overview of large spherulites can easily be obtained at lower magnification using thin sections and irradiation-induced contrast enhancement (see Figs. 13.10 and 13.12). Smaller details such as lamellae require chemical staining (see Figs. 13.2 and 13.6) or a combined method, as shown in Fig. 13.14. The various staining techniques available also enable different resolutions. The scheme of Fig. 13.17 demonstrates the average resolution ranges and the resolution limits possible when semicrystalline lamellar polyethylene is investigated via several preparation techniques and contrast enhancements. Thin sections prepared with cryosectioning yield an optimum resolution in TEM or SEM of about 0.1 µm; i.e. internal structures of spherulites or sticks of lamellae are detectable (Fig. 13.17a). Physical fixation due to γ-irradiation with dosages of up to 20 MGy enable resolutions of up to 10 nm, allowing individual lamellae to be detected (see Figs. 13.17b and Fig. 13.10). Ultramicrotomed thin sections that have undergone chemical fixation and staining provide details down to about 1 nm in size in TEM (see Figs. 13.17c and Fig. 13.2).

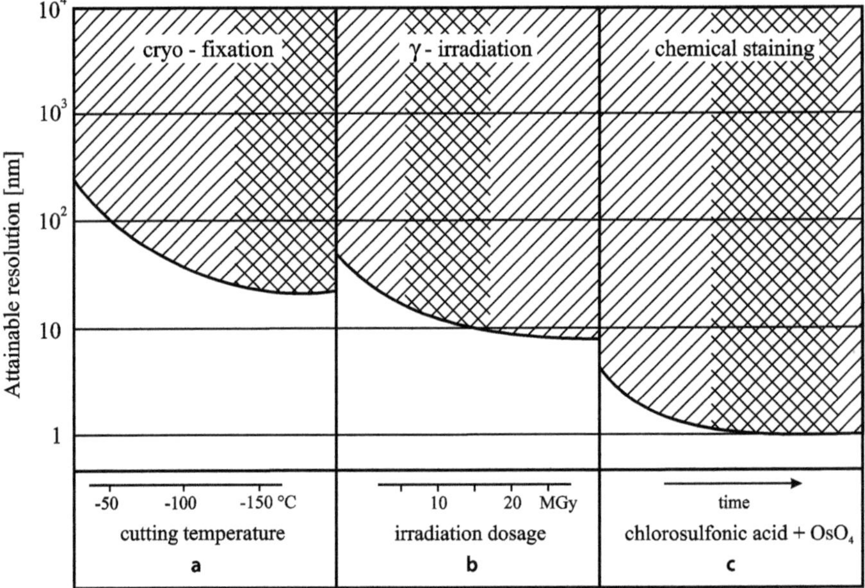

Fig. 13.17a–c. Scheme showing the influence of the preparation technique on the attainable resolution in semicrystalline polymers (PE): **a** cryoultramicrotomed thin sections in SEM or TEM; **b** thin sections after fixation by γ-irradiation, TEM; **c** ultrathin sections after chemical fixation and staining with chlorosulfonic acid and OsO₄, TEM

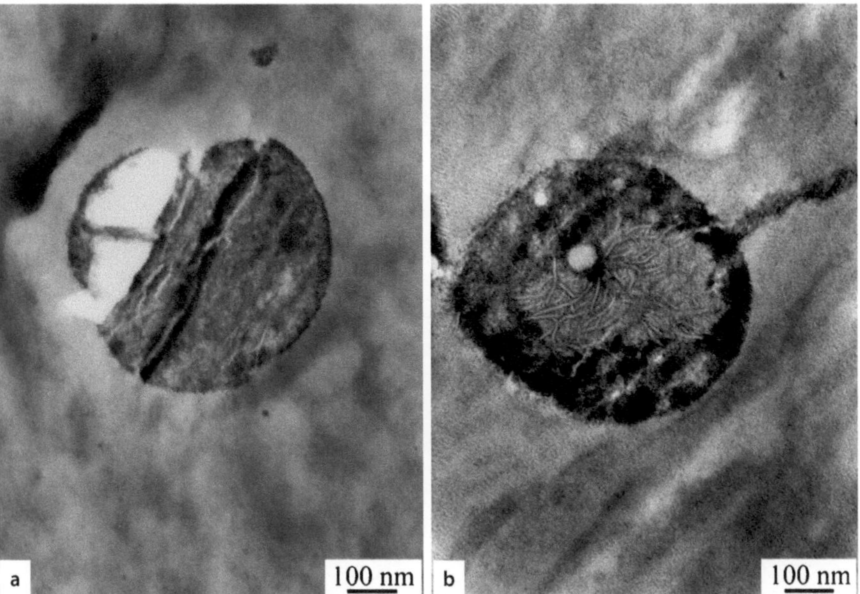

Fig. 13.18a,b. Effect of different durations of staining with RuO_4 on the sectioning of a PP/EPR sample: **a** duration of staining time is too short, resulting in the deformation of EPR particles, **b** same sample with optimised staining parameters

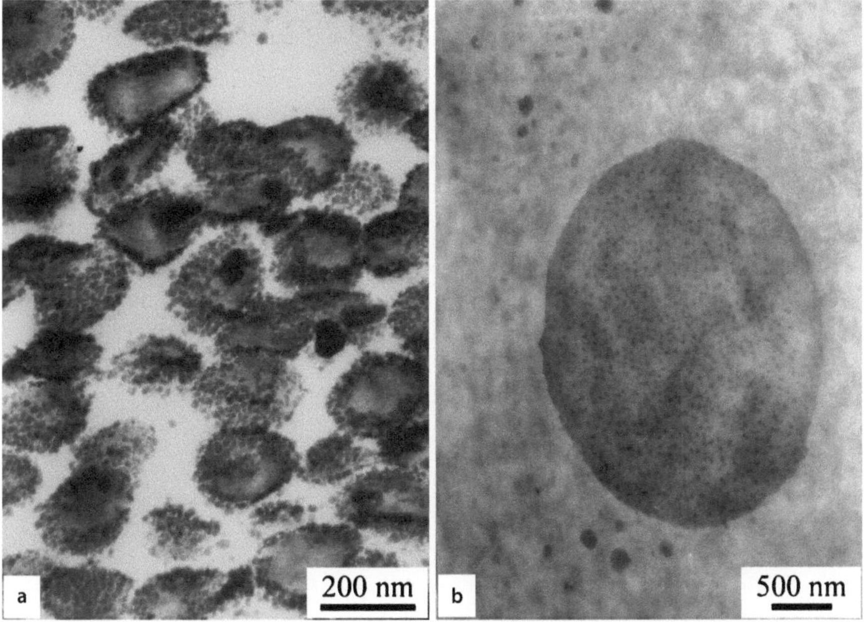

Fig. 13.19a,b. Deposition of staining agents on the sections after treatment of the bulk material and sectioning at room temperature: **a** osmium crystals at the interface between PB particles and the PMMA matrix; **b** ruthenium crystals in PA particles (PET matrix)

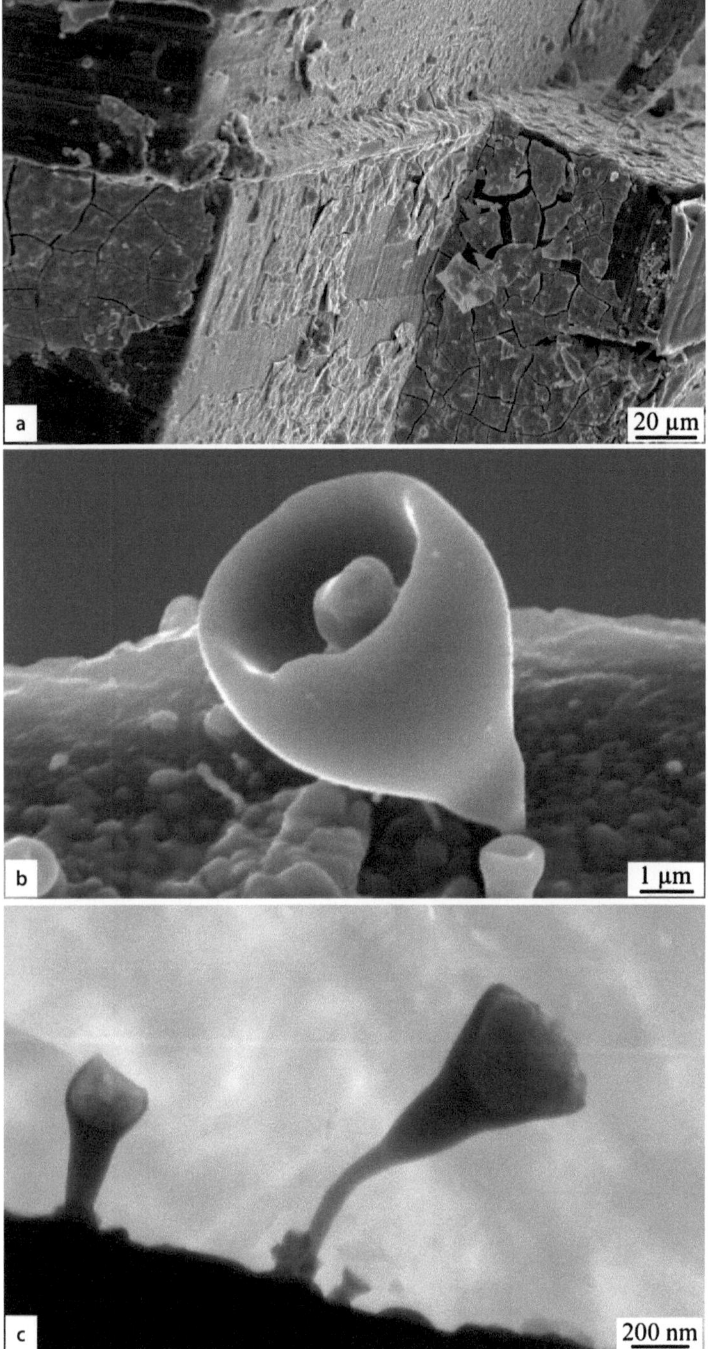

Fig. 13.20. Ruthenium deposition on a block surface (**a**) and "ruthenium trumpets" after an excessively long staining time (**b,c**); SEM micrographs (**a,b**) and TEM micrograph (**c**)

The electron microscopic investigations of polymeric materials should always provide error-free results. Therefore, sample preparation with sectioning and staining should not introduce undesirable artefacts. It is particularly important to choose suitable parameters, such as proper choices of the staining agent, the concentration of the solution as well as the duration and temperature of the chemical treatment.

The efficiency of a particular staining agent changes from material to material. The selection of wrong parameters leads to problems during sectioning and also introduces imaging artefacts [8, 35].

A staining process that is too short may be insufficient to harden the polymer, resulting in the formation of holes and deformation zones during sectioning (see Fig. 13.18). It is often particularly difficult to find optimum conditions for multicomponent polymers. On the other hand, a staining time that is too long results in embrittlement of the polymer. The staining agent may also be deposited onto the sample film in the form of small crystals. For instance, Fig. 13.19 shows the deposition of osmium and ruthenium particles onto the sections of different polymers.

In particular, an excessive staining time with RuO_4 results in the deposition of ruthenium on the surface and so-called "ruthenium trumpets", as shown in Fig. 13.20. This effect can appear upon the treatment of both bulk materials and their sections.

The optimum staining conditions for different materials should be tested by varying parameters experimentally. Finally, much depends upon the character and patience of the experimentalist. It is necessary to develop a "feel" for sample preparation in electron microscopy; the more you test conditions with enthusiasm and patience, the more you achieve.

References

1. Andrews EH, Stubbs JM (1964) J R Microsc Soc 82:221
2. Andrews EH (1964) Proc R Soc London 562
3. Kato K (1965) J Electron Microsc 14:219
4. Kato K (1965) J Electron Microsc 14:220
5. Kato K (1966) J Polym Sci Polym Lett 4:35
6. Sawyer LC, Grubb DT (1996) Polymer microscopy, 2nd edn. Chapman and Hall, London
7. Vastenhout JS (2002) Microsc Microanal 8:1238
8. Michler GH, Lebek W (2004) Ultramikrotomie in der Materialforschung. Hanser Verlag, München
9. Kanig G (1973) Colloid Polym Sci 251:782
10. Michler GH, Naumann I (1982) Acta Polym 33:399
11. Michler GH, Gruber K (1980) Acta Polym 31:771
12. Sue HJ, Garcia-Meitin EI, Burton BL, Garrison CC (1991) J Polym Sci Polym Phys 29:1623
13. Bozzola JJ, Russel LD (eds) (1992) Electron microscopy: principles and techniques for biologists. John and Bartlett, Boston, MA
14. Parker MA, Veseley (1993) Microsc Res Techniq 24:333
15. Huong DM, Drechsler M, Cantow H-J, Möller M (1993) Macromolecules 26:864
16. Ribbe AE, Bodycomb J, Hashimoto T (1999) Macromolecules 32:3154–3156
17. Vitali R, Montani E (1980) Polymer 21:1220
18. Trent JS, Scheinbeim JI, Couchman PR (1983) J Polym Sci Polym Lett 19:315
19. Trent JS, Scheinbeim JI, Couchman PR (1983) Macromolecules 16:589
20. Trent JS (1984) Macromolecules 17:2930
21. Morel DE, Grubb DT (1984) Polym Commun 25:68

22. Montezinos D, Wells BG, Burns JL (1985) J Polym Sci Polym Lett 23:421
23. Fischer H (1994) Macromol Rapid Commun 15:949
24. Li JX, Ness JN, Cheung WL (1996) J Appl Polym Sci 59:1733
25. Li JX, Cheung WL (1999) J Appl Polym Sci 72:1529
26. Chou TM, Prayoonthong P, Aitouchen A, Libera M (2002) Polymer 43:2085
27. Michler GH (1992) Kunststoff-Mikromechanik: Morphoplogie, Deformations- und Bruchmecha-
    nismen von Polymerwerkstoffen. Hanser Verlag, München
28. Michler GH (1993) Appl Spectrosc Rev 28:327
29. Hendus H (1970) Angew Macromol Chem 12:1
30. Michler GH, Gruber K, Steinbach H (1982) Acta Polym 33:550
31. Michler GH (1996) J Macromol Sci Phys 35:329
32. Grubb DT, Keller A (1972) J Mater Sci 7:822
33. Foks J, Michler GH (1986) J Appl Polym Sci 31:1281
34. Michler GH (2005) Micromechanical mechanisms of toughness enhancement in nanostructured
    amorphous and semicrystalline polymers. In: Michler GH, Baltá-Calleja FJ (eds) Mechanical prop-
    erties of polymers based on nanostructure and morphology. Tayler & Francis, Boca Raton, FL, p 379
35. Sitte H (1981) Ultramikrotomie - Häufige Fehler und Probleme. Reichert-Jung Optische Werke AG,
    Wien

Part III

Main Groups of Polymers

14 Structural Hierarchy of Polymers

Polymeric materials show a large variety of molecular and supramolecular structures, ranging in size from about 0.1 nm to 1 mm. In this chapter, an overview is initially provided of macromolecular structural parameters, including constitution, configuration, conformation, the coil diameter and the macromolecule size. This is followed by a discussion of the two basic macromolecular arrangements – amorphous and semicrystalline structures – and the morphologies of the various polymer classes. Some basic relationships between macromolecular parameters and properties are mentioned, along with the conclusion that mechanical properties are mainly determined by the supramolecular structures and morphologies of polymers. This insight is discussed in detail for various polymer classes in subsequent chapters in Part III.

14.1 Overview

Polymeric materials show a wide variety of molecular and supramolecular structures, depending on the types and arrangements of the monomers, the method of polymerisation used, the processing employed, and other factors. These structures range in size from 0.1 nm to 1 mm (or 10^{-10} m to 10^{-3} m), i.e. over orders of magnitude. A rough classification includes:

- Molecular level (monomers, macromolecules) $\approx$0.1 nm ... 10 nm scale
- Microscopic level (supramolecular structures; morphology) $\approx$10 nm ... 1 μm scale
- Mesoscopic level (larger morphological details) $\approx$1 ... 100 μm scale
- Macroscopic level >0.1 mm.

A rough overview of the important parameters of macromolecules and polymer morphology as well as their influence on (micro)mechanical properties is provided in the following sections.

14.2 Macromolecular Structures

Polymers consist of a large number of the same monomer molecules or various different monomer molecules, which are connected by covalent bonds. The number N of

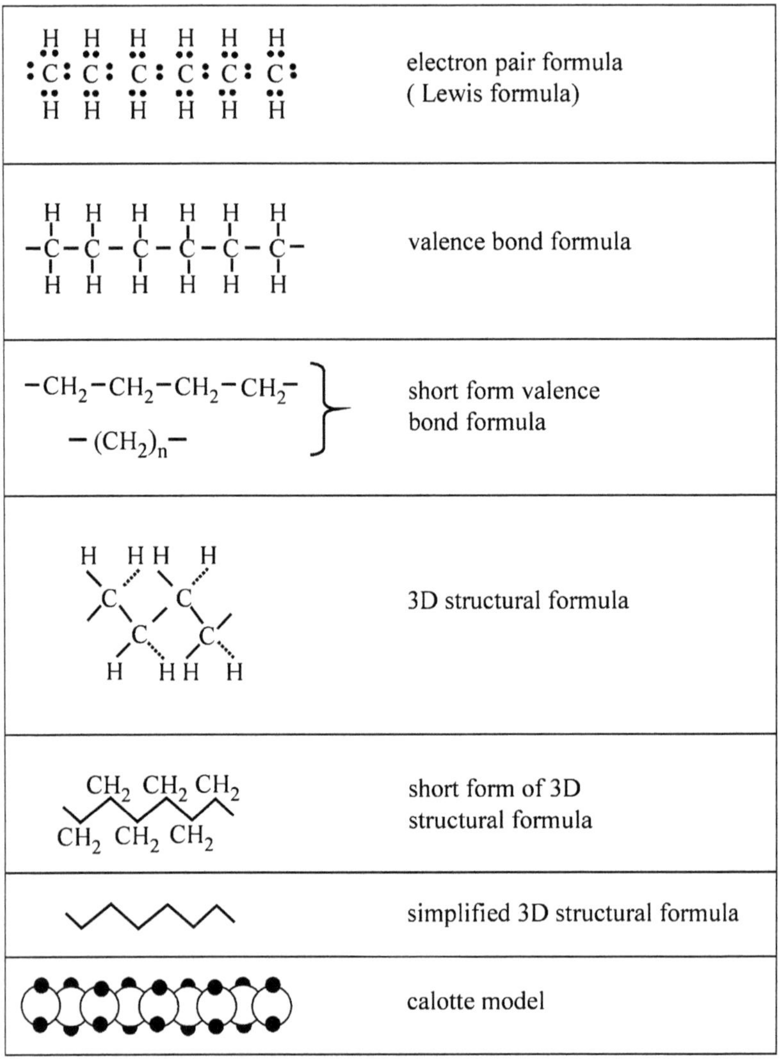
electron pair formula
(Lewis formula)
valence bond formula
$-CH_2-CH_2-CH_2-CH_2-$
$-(CH_2)_n-$
short form valence bond formula
3D structural formula
short form of 3D structural formula
simplified 3D structural formula
calotte model

Fig. 14.1. Macromolecular constitution: different descriptions of the chemical structures of polymers (example used is polyethylene)

monomers usually varies between 10^3 and 10^5, yielding molecular weights of macromolecules (the product of the number of and the molecular weights of monomers) of between 10^4 and more than 10^6. The chemical structure, the arrangement of the monomers and the shape of the macromolecules can be described by three parameters, the so-called "3C": constitution, configuration and conformation.

14.2.1 Constitution

The constitution is determined by the chemical nature and the compositions of the monomers. The simplest case is a polyethylene macromolecule consisting of identical ethylene monomers C_2H_4. There are various different model-like descriptions of the chemical structure; see Fig. 14.1. Homopolymers contain only one type of monomer in the chain, whereas copolymers, block copolymers, graft polymers or terpolymers consist of two or more different kinds of monomers.

14.2.2 Configuration

Asymmetric monomers or monomers with side groups can be arranged in different ways along the macromolecule, as described by the tacticity (see Fig. 14.2):

- The monomers in *atactic* polymers are distributed statistically
- The monomers in *syndiotactic* polymers exhibit an alternating arrangement (of the side groups)
- The monomers in *isotactic* polymers are all oriented identically.

Side groups or *branchings* at the main chain define another type of configuration; see Fig. 14.2. Short chains (short branches) or long chains (long branches) can be combined in different ways, with variations in the lengths and the positions of the branches along the main chain. In particular, short-chain branches can be distributed statistically, comb-like or concentrated locally. Macromolecules, which are crosslinked by side chains, form three-dimensional networks. Often, the branches are not distributed homogeneously along the macromolecules; in particular, there may be concentrations of branches at the ends of macromolecules or on shorter macromolecules (with lower molecular weight).

The configurations of copolymers vary greatly; see Fig. 14.3. In statistical or random copolymers, the different monomers are statistically distributed; in alternating copolymers the position of the monomers changes regularly; while longer sequences of monomers appear in block copolymers. In graft polymers, the monomer of the side chains is different from the main-chain monomer. Combinations of a main-chain copolymer with grafted side chains of a third monomer are called terpolymers. The configurations of block copolymers exhibit a great deal of variety, including two- and three-block copolymers, linear and star block copolymers, and others (compare Fig. 19.1 and Sect. 19.2).

14.2.3 Conformation

The conformation describes the shape or the form of the whole macromolecule. Because of easy rotation around the C–C bond, several shapes are conceivable for macromolecules and are illustrated as models in Fig. 14.4. Limiting cases are the *random* or *statistical coil* (model a) and *highly regular fold lamellae* (model d). A random coil with a statistical arrangement of the successive monomer segments (after

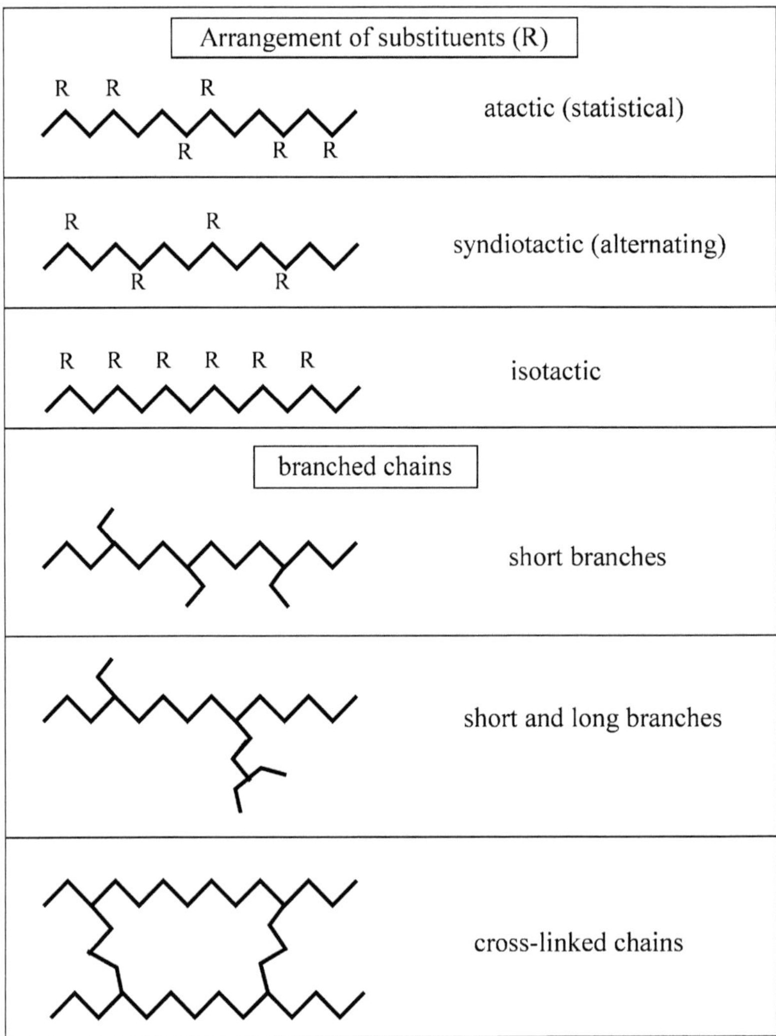

Fig. 14.2. Configuration of homopolymers: tacticity and branching

Flory [1]) corresponds to the coil conformations of macromolecules in a highly diluted θ solvent and is the basic element of amorphous polymers. The diameter of the coil is proportional to $\sqrt{M_w}$ and usually ranges from 10 to 30 nm; for instance, a PS macromolecule with a molecular weight of $M_w = 2 \times 10^5$ reaches a coil diameter of about 30 nm [2]. A tendency towards ordered structure is visible in the *bundle* or *micelle model*, which has a parallel arrangement of segments of neighbouring macromolecules, as shown in Fig. 14.4c [3, 4]. The so-called *meander model* in Fig. 14.4b with a meander-like folding of single macromolecules or bundles of macromolecules was proposed by Pechold [5]. The highly regular folded macromolecule (model d) is

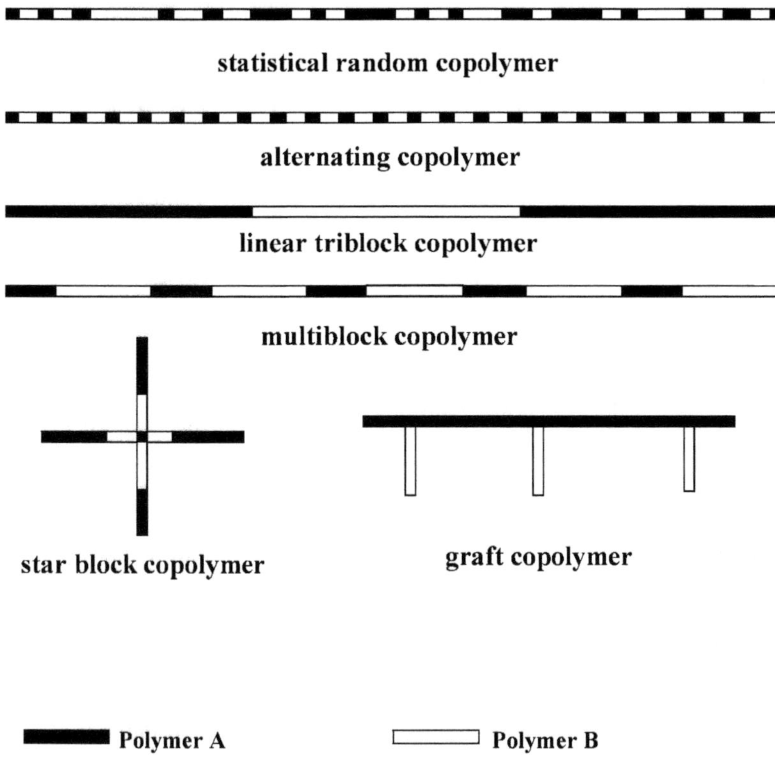

Fig. 14.3. Configuration of copolymers

the basic element of semicrystalline polymers [6–8]. However, this regular folding of a long macromolecule is not realistic (with the exception of crystal growth from solution), and so elements from models a, c, and d are combined into the *modified folding model*, which is most realistic for semicrystalline polymers [9,10]. *Highly oriented chains* (model f) appear in extended chain crystals due to high-pressure crystallisation [11], or in highly oriented fibres [12].

14.2.4 Macromolecule Size

The size or molecular weight of the macromolecules is determined by the number N of the monomers and the degree of polymerisation. The different macromolecules in a polymeric material usually vary in molecular weight somewhat. Because of this molecular weight distribution, two average values are generally used: the number average $\overline{M}_n$ and the weight average $\overline{M}_w$. The ratio

$$U = \frac{\overline{M}_w}{\overline{M}_n} - 1$$

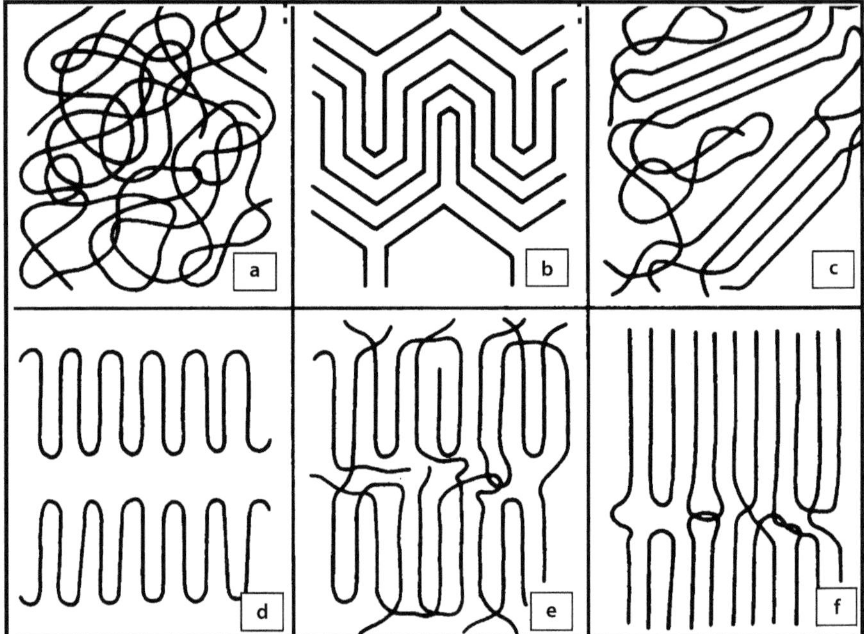

Fig. 14.4a–f. Different conformation models for macromolecules: **a** random or statistical coil; **b** meander model; **c** bundle or micelle model; **d** regular folding; **e** modified folding model; **f** highly oriented chains

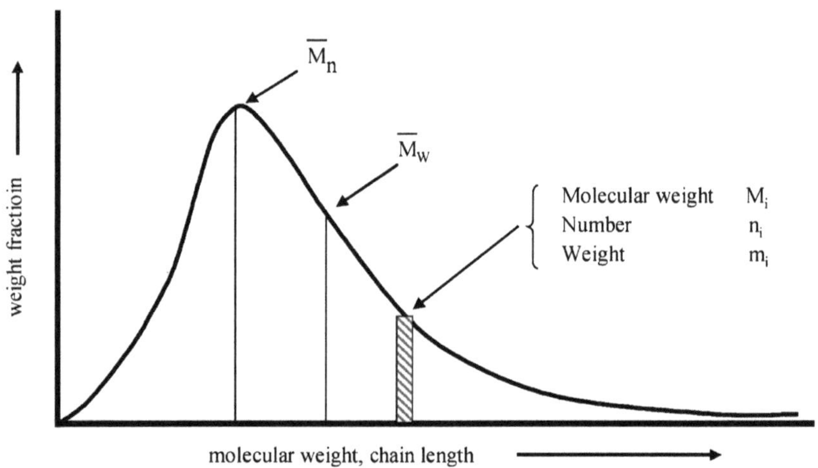

Fig. 14.5. Scheme showing molecular weight distribution with number average $\overline{M}_n$ and weight average $\overline{M}_w$

(the polydispersity) is a measure of the broadness of the molecular weight distribution; see Fig. 14.5. Polymers with narrow molecular weight distributions show small values of polydispersity (close to zero), while practical polymers with broad distributions have U values of >5.

The length of the macromolecules can easily be estimated as the product of the number of monomers and their length. As an example, for PE the distance between the C_2H_4 monomers is 0.252 nm, and with a degree of polymerisation of $N = 10^4$ the stretched length of such a macromolecule reaches 2.5 µm (for comparison, this would be a fibre 2.5 m long and 0.5 mm thick at a magnification of 10^6). The molecular weight of such a molecule is 2.8×10^5, a typical value for PE.

14.3 Supramolecular Structures

There are two basic macromolecular arrangements, which correspond to the limiting cases in Fig. 14.4, the *random coil model* (a) and the *modified fold lamellae* (e):

- Amorphous structures
- Semicrystalline structures.

In *amorphous structures*, the density of an individual random coil is in the range of 0.01 g/cm^3. Since the density of an amorphous polymer is about 1 g/cm^3, this means that roughly 10^2 macromolecular segments must exist in the same place in the polymer. This results in the formation of lots of topological links – *entanglements* – which largely hold the macromolecules together and provide the well-known strength and toughness of polymers. In comparison to these entanglements, the van der Waals forces between neighbouring segments of macromolecules are of only secondary importance for the mechanical strength. The entanglement points between the macromolecules can be connected to form a network with meshes. There are indications that the meshes inside the entanglement network contain unentangled macromolecular segments, chain ends or shorter macromolecules, which makes them a little bit weaker than the entanglement network [10,13]. Parameters of amorphous structures include the diameter of the macromolecular coil, the entanglement distance d and the molecular weight M_e of the macromolecular segments between the entanglements, as well as the mesh size (mesh diameter D); see Fig. 14.6. The packing density of coiled macromolecules is lower than that of conformations with parallel arrangements of molecular segments. This unoccupied volume is termed the "free volume", and has a great influence on the mechanical properties of the polymer [14,15].

Crystalline structures (lamellae) contain folded molecules and parallel arrangements of neighbouring molecular segments; see Fig. 14.4e. Sections of the macromolecules are included in crystallites/lamellae. This reduces the mobilities of neighbouring sections, thus hindering the formation of new crystals/lamellae and yielding a semicrystalline structure with amorphous regions between the crystalline lamellae. The characteristic elements of the semicrystalline structure are the crystalline

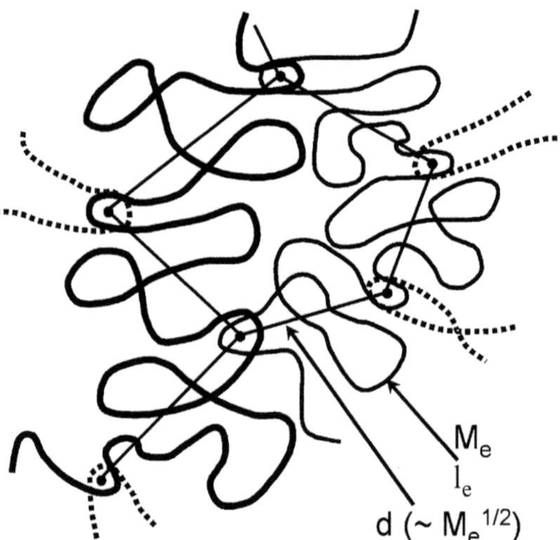

Fig. 14.6. Entanglements (connected to an entanglement network) in amorphous polymers; M_e, l_e are the molecular weight and the length of the macromolecular segment between entanglements, respectively; d is the distance between entanglements

lamellar phase, the lamella interface, and the interlamellar amorphous region; see Fig. 14.7 [10]. Crystalline defects are often incorporated inside the lamellae, such as chain ends or short branches, forming defect layers between the crystalline blocks. Because of the difference in density between the crystalline part (higher density) and the amorphous part (lower density), the macromolecular segments inside the crystalline blocks must be tilted (this is also confirmed via electron micrographs). In the interlamellar amorphous parts, the macromolecules are entangled (as in amorphous polymers) or form connections between the lamellae known as tie molecules. Density differences can be enhanced by means of chemical staining, which then yields the contrast distribution of a lamellar structure in an electron micrograph; see Fig. 14.7 on the left hand side.

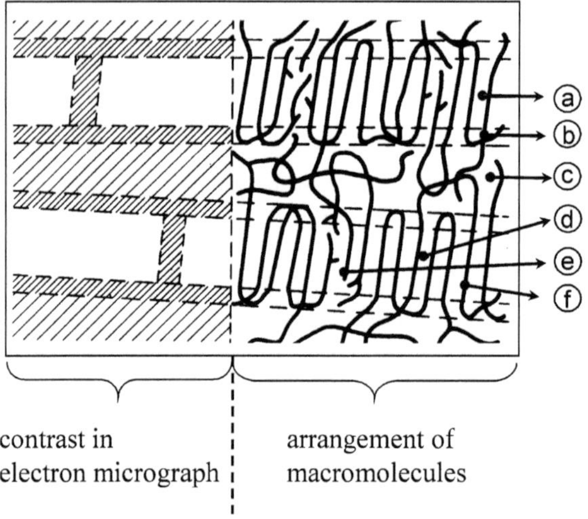

Fig. 14.7. Lamellar structure of semicrystalline polymers, as visible in electron micrographs (*left*), and model of the arrangement of macromolecules (*right*): *a*, crystalline; *b*, interface (amorphous); *c*, interlamellar (amorphous); *d*, microblock; *e*, defect layers between microblocks; *f*, inclined chains inside lamellae

Fig. 14.8. Schematic model of the shish kebab structure (after Pennings [16], reproduced with the permission of Wiley)

Alternative semicrystalline arrangements include the parallel arrangements of macromolecules seen in highly oriented fibres as well as shish-kebab structures; see Fig. 14.8 [16]. Such structures appear if highly oriented melts crystallise in the form of a bundle of oriented (longer) macromolecules (the "shish") and smaller perpendicular lamellae (the "kebabs").

14.4 Morphology

The basic features of *amorphous polymers* are the entanglements (Fig. 14.6); there is an absence of any crystalline, ordered structures (the word "amorphous" means "structureless" or "shapeless"). Typical examples of amorphous polymers include PS, PMMA, SAN, PC, COC, and others. However, small fluctuations in density and in macromolecular packaging are also assumed. Several models include small domains or granules of size <10 nm [17,18] or particles up to 100 nm in size [19,20]; see Figs. 15.1 and 15.2.

PVC is, however, a significant exception, since it exhibits a morphological hierarchy including microdomains of about 10–20 nm, domains around 100 nm in size, and micron-sized primary particles [21, 22]; see Fig. 15.4.

In *semicrystalline polymers*, the crystalline lamellae can be arranged in different ways, from the parallel arrangement of a few shorter lamellae into sticks or bundles and longer lamellae into dendrites or sheaf-like structures, up to the radial arrangement observed in so-called "spherulites", with diameters of up to several tens of microns; see Fig. 14.9. If the radial lamellae (or the radial segments) show periodic changes in their orientation, banded spherulites will arise [10]. At the bottom of Fig. 14.9, some preparation and microscopic techniques suitable for studying these different structures are mentioned [10].

In *block copolymers*, the blocks of different monomers with varying or defined lengths are usually incompatible and thus show microphase separation. Depending on the amounts (volume fractions) of the different components, typical morphological types of spheres, cylinders, lamellae, and others will appear. Moreover, the wide variety of possible configurations available produces an additional variation in the morphology (see Chap. 19).

In *polymer blends*, again, the different polymer components are generally incompatible and thus show phase separation. The minor component usually forms the dispersive phase and the major component the matrix. Interpenetrating structures or networks are often formed at close to a composition of 50:50. The sizes of the

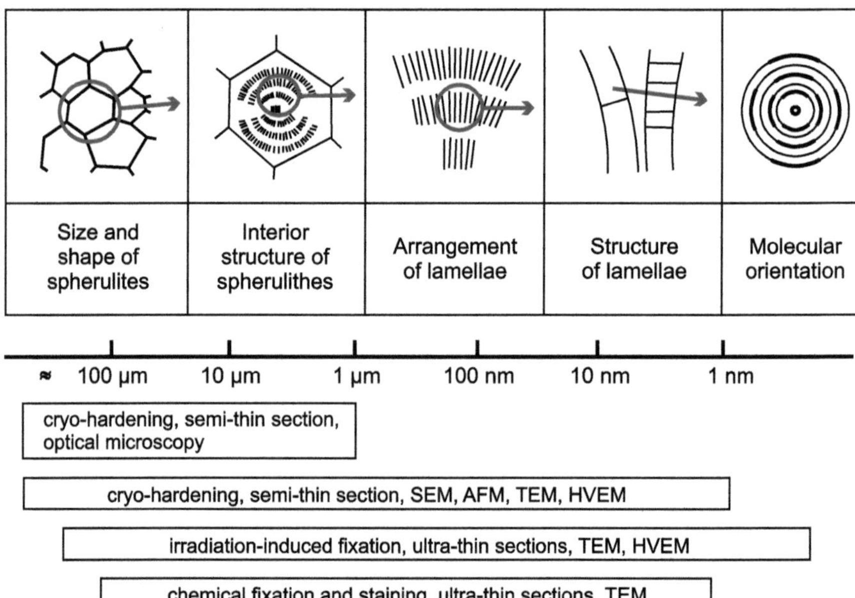

Fig. 14.9. Overview of the structural and morphological elements of semicrystalline polymers, together with preparation and investigation techniques that are suitable for investigating them

demixing structures vary considerably depending on the degree of incompatibility and the processing method used, with the interfaces a few nm in size often found between the components. Typical examples of the morphologies of polymer blends are shown in Chap. 17.

In *composites*, a polymeric matrix contains inorganic particles (*particle-filled polymers*) or fibres (*glass fibre* or *C-fibre reinforced polymers*). Here, the main morphological feature is the arrangement and distribution of the inorganic components. The polymer matrix adjacent to the inorganic parts can be modified structurally to some degree, resulting in modified interfaces. For instance, epitaxial growth of the lamellae can be initiated at particles in a semicrystalline matrix. The amount of interface increases drastically with decreasing particle diameter (in the nanocomposites), meaning that in such nanocomposites the usual morphology of the matrix polymer is transformed into an "interface material", with corresponding changes in properties (see Chap. 22).

14.5 Basic Relationships Between Macromolecular Parameters and (Micro)mechanical Properties

In general, the (micro)mechanical properties of polymers are determined to a very large degree by their supramolecular structures and morphologies. However, the types, sizes and shapes of the macromolecules exert some basic influence over the properties of polymers. For instance:

- Constitution influences packing density, chain cross-section, chain stiffness and mobility
- Configuration influences crystallisation, melting temperature, stiffness and mobility
- Conformation influences crystallisation, glass transition or melting temperature, and entanglement density
- Free volume influences density, mobility, yield stress and ageing.

The molecular weight has a basic influence on the mechanical properties of the polymer; see Fig. 14.10. With increasing molecular weight, important mechanical properties such as strength, elongation at break or toughness increase strongly [23]. At large M_w values the properties plateau somewhat. The strongest increase in the properties appears at the critical molecular weight M_c, which is about twice the molecular weight needed to create load-bearing entanglements between neighbouring macromolecules (entanglement molecular weight M_e; $M_c \approx 2\,M_e$). Short macromolecules (below M_c) are unable to form entanglements and yield brittle materials (wax-like behaviour). At around M_c, the molecular segments start to entangle but can also easily disentangle, and the segments only form strong topological connections above M_c. The critical molecular weight M_c is on the order of 10^4 for amorphous polymers [24, 25]; for instance 4×10^4 for PS. It is shifted to higher values for semicrystalline polymers (LDPE, HDPE), since some of the macromolecular segments (roughly 50%) are located in the crystalline lamellae and so only

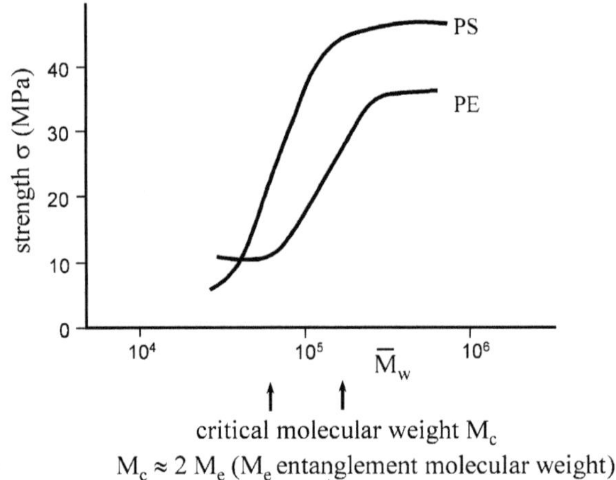

a

M_w	$\ll M_c$	$< M_c$	$> M_c$
σ_B	very low	low	high

b

Fig. 14.10a,b. Influence of molecular weight on mechanical properties: **a** increase in strength with critical molecular weight for an amorphous polymer (PS) and a semicrystalline polymer (PE); **b** sketch of the formation of entanglements between macromolecules (M_c is the critical molecular weight)

the rest can form entanglements in the amorphous interlamellar regions (which are thus an important influence on the strength and elongation of semicrystalline polymers).

The molecular mobility is strongly dependent on the temperature and loading time and so mechanical properties such as stiffness or yield stress show analogous dependences. The chain strength of a macromolecule is determined by its very strong C–C bonds, and a theoretical strength in the range of 10–20 GPa has been calculated. This is a factor of 100–1000 higher than the practical, experimental strength of 10–100 MPa. The reason for this lies – as it does for all other materials – in the morphology, which includes lots of defects, microvoids, and structural heterogeneities; see Fig. 14.11. Here, some types of molecular defects are sketched on the right hand side, including chain ends, weak entanglements (i.e. only short molecular segments are entangled), weak molecular connections or defects inside the lamellae. Some larger morphological defects are illustrated on the left hand side, including inter-

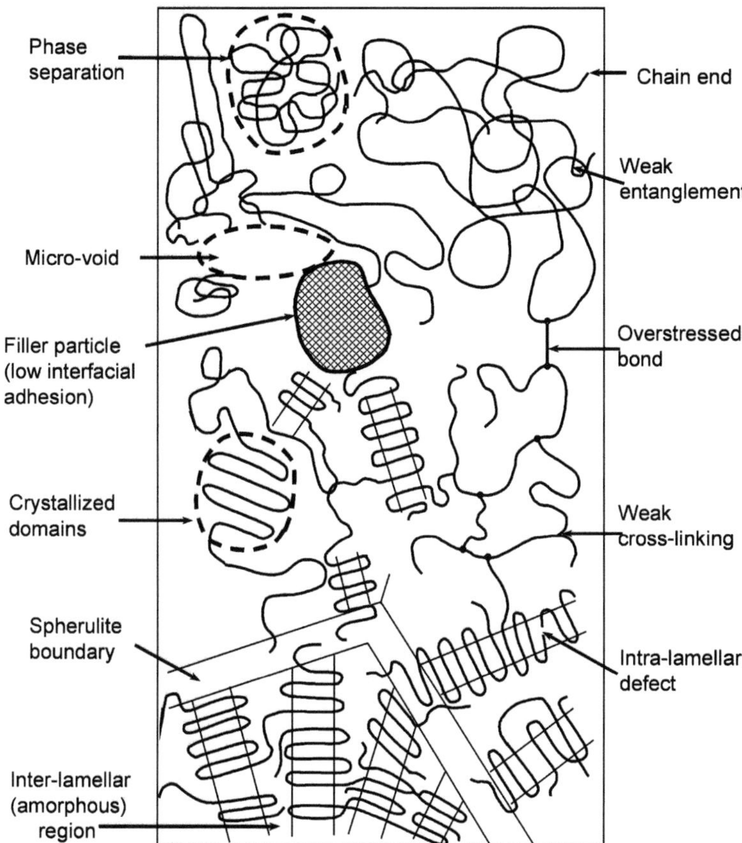

Fig. 14.11. Typical defects in polymers at the molecular level (*right*) and at the supramolecular level (*left*)

lamellar and interspherulitic defects, separately crystallised parts, inorganic filler particles, phase separations due to incompatibility, or microvoids. Under the application of mechanical loading, some molecular segments oriented parallel to the loading direction are overstressed and can therefore rupture, creating microvoids. All structural defects cause stress concentrations, which drastically reduce the strength of the material.

A general conclusion is that the molecular parameters provide only a broad indicator of the behaviour of a polymer; the exact mechanical properties are mainly determined by the supramolecular structures and the morphology, which can be resolved in detail by electron and atomic force microscopy.

References

1. Flory PJ (1969) Statistical mechanics of chain molecules. Wiley, New York (reprinted edition from 1989 was published by Hanser, Munich)
2. Fischer EW, Dettenmaier M (1978) J Non-Cryst Solids 31:181
3. Kargin VA (1958) J Polym Sci 30:247
4. Arschakov SA, Bakeev NF, Kabanov VA (1973) Vysokomol Soed A 15:1154
5. Pechold W (1971) J Polym Sci C 32:123
6. Rees DV, Bassett DC (1971) J Mater Sci 6:1021
7. Yeh GSY (1973) Polym Repr 14:718
8. Bassett DC (1981) Principles of polymer morphology. Cambridge University Press, Cambridge
9. Flory PJ (1962) Am Chem Soc 84:2857
10. Michler GH (1992) Kunststoff-Mikromechanik: Morphologie, Deformations- und Bruchmechanismen. Hanser Verlag, München
11. Wunderlich B (1973) Macromolecular physics, vol I. Academic, New York
12. Fischer EW, Goddar H, Schmidt GF (1968) Macromol Chem 119:170
13. Michler GH (1991) Plaste u Kautschuk 38: 268
14. Struik LCE (1978) Physical aging in amorphous polymers and other materials. Elsevier, Amsterdam
15. Donth E (1981) Glasübergang Akademie-Verlag, Berlin
16. Pennings AJ (1977) J Polym Sci Symp 59:55
17. Geil PH (1975) Ind Eng Chem Prod Res Dev 14:59
18. Kämpf G (1975) Prog Colloid Polym Sci 57:249
19. Großkurth KP (1972) Gummi-Asbest-Kunststoffe 25:1159
20. Lednicky F, Pelzbauer Z (1982) J Macromol Sci Phys B 21:19
21. Hattori T, Tanaka K, Matsuo M (1972) Polym Eng Sci 12:199
22. Menges G, Berndtsen N (1976) Kunststoffe 66:735
23. Robertsen RE (1976) In: Toughness and brittleness of plastics (ACS No 154). ACS, Washington, DC, p 89
24. Aharoni SM (1983) Macromolecules 16:1722
25. Aharoni SM (1986) Macromolecules 19:426

15 Amorphous Polymers

Amorphous polymers can be defined as polymers that do not exhibit any crystalline structures in X-ray or electron scattering experiments. They form a broad group of materials, including glassy, brittle and ductile polymers. In contrast to the word "amorphous", weak domain-like or globular structures can exist, which are often only visible after pretreatment of the material, e.g. using straining-induced contrast enhancement in TEM. The micromechanical behaviour of amorphous polymers is linked to the formation of localised deformation zones, such as crazes, deformation bands, or shear bands, which are characterised by representative HVTEM micrographs. The strong correlation between crazing behaviour and the existence of entanglements and an entanglement network is shown. After discussing PS and PVC, some additional examples (PMMA, SAN, COC, PC) are presented.

15.1 Overview

Amorphous polymers form a large group of materials, including glassy, brittle polymers (such as PS, PMMA, SAN, COC) and ductile polymers (such as PVC and PC). The characteristic that such polymers have in common is their "amorphous" structure, which means that they do not exhibit any crystalline structures in X-ray and electron scattering experiments. In accordance with this definition, as well as the thermoplastics mentioned above, resins (such as polyesters and epoxies) also belong to the family of amorphous polymers. However, because of the highly crosslinked molecules of resins (see Fig. 14.2), they form a separate group and are usually not used as homopolymers but as matrix material in fibre-reinforced polymers. Amorphous block copolymers and polymer blends, which are discussed in separate chapters, are also amorphous polymers.

15.2 Morphology

As mentioned in Sect. 14.3, it is often assumed that – in contrast to the word amorphous – these polymers show only weak structure, ranging from small nodules below 10 nm in size to 20–50 nm domains right up to particles in the range of 100–300 nm (overview in [1]). The problem is that such structural details produce very poor contrast, making them impossible to investigate using scattering techniques and very

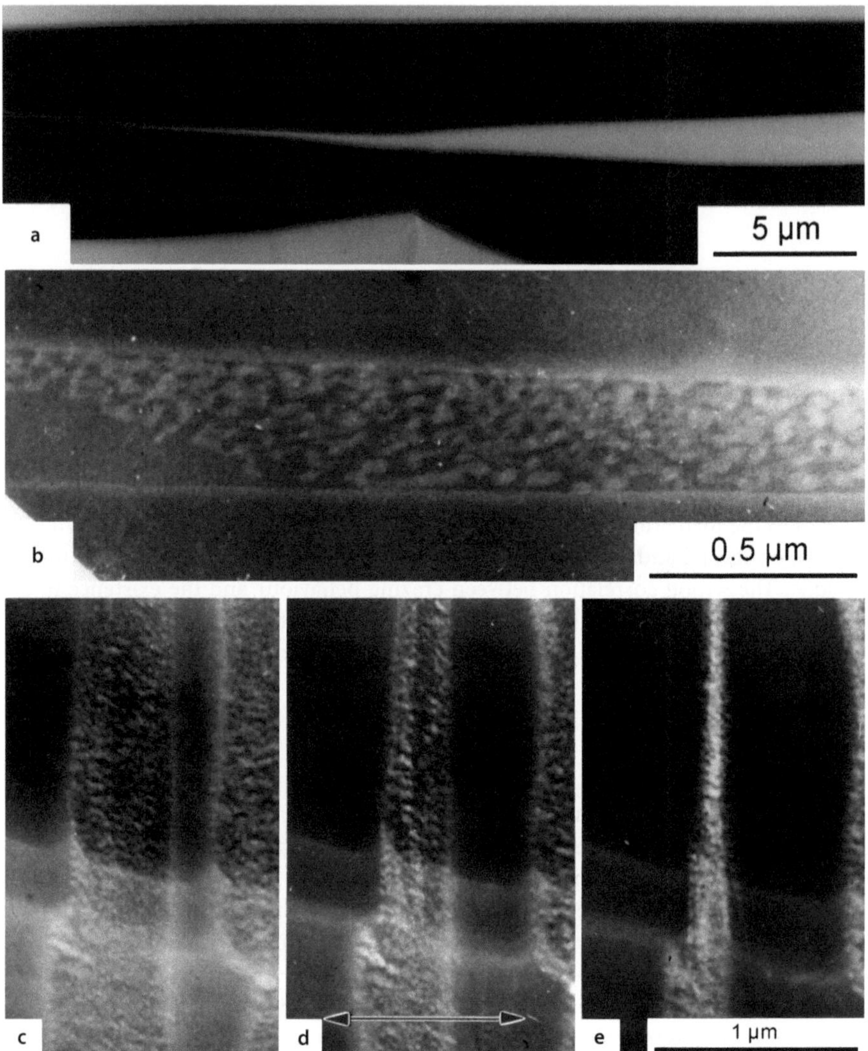

Fig. 15.1a–e. Craze with pre-craze in PS: **a** overview, the pre-craze is to the left in front of the craze tip; **b** larger magnification of the pre-craze, consisting of small deformed domains (deformation direction vertical); **c–e** pre-craze in a thin section before and after tilting the section in the microscope by 5° and 10° (deformation direction horizontal). Semi-thin sections of PS, deformed in a 1000 kV HVTEM (after [1], reproduced with permission of Hanser)

difficult to identify under the electron microscope. Often, these structures are only visible after pretreatment of the material, e.g. hot deformation, surface etching, chemical attack, or other methods. One successful technique of this kind is "straining-induced contrast enhancement" (see Sect. 13.3.2). Typical results obtained when this technique is employed are shown in the first two figures of this chapter. Figure 15.1

shows micrographs from a deformed PS film in a 1000 kV HVTEM, in which a craze (i.e. a typical deformation zone, very long and tiny, containing highly stretched material; see Sect. 15.3) is presented in overview (a) and the zone ahead of the craze is displayed at a larger magnification (b). This pre-craze is a narrow band of small domains with diameters of 20–50 nm. These domains are somewhat more plastically deformed than their surroundings and so they appear brighter (for details see [2]). This effect (that softer domains are plastically deformed to a higher degree than their surroundings) was also checked using another amorphous polymer, styrene-acrylonitrile copolymer (SAN). If the composition doesn't follow the azeotropic composition (76 mol% S: 24 mol% AN) during the polymerisation, molecular segments with higher S contents and other segments with lower S contents are formed. If the difference is larger than 4% [3], these different segments are not fully miscible and exhibit microphase separation with the formation of domains of softer AN-rich material in S-rich surroundings [4]. The domains are not visible in the usual transparent SAN copolymers; however, under mechanical loading they are plastically deformed to a higher degree than the matrix. Strongly plastically stretched domains (stretched by as much as 100% or even more) with diameters of about 100 nm are visible in the craze-like deformation zone ahead of the crack-tip in Fig. 15.2 [4,5].

The pre-craze shown in micrograph (b) of Fig. 15.1 appears to be a broad band several hundreds of nanometres wide in front of the craze tip. However, the pre-craze is thinner than 100 nm, which can be demonstrated by tilting the specimen in the microscope: In Fig. 15.1, a pre-craze is shown in micrograph (c) without tilting and in micrographs (d) and (e) after tilting by 5° and 10°. Therefore, the pre-craze is only a little thicker than the deformed domains (which are 20–50 nm thick), and the whole band is tilted in the sample by about 10°.

Using these results for strongly plastically stretched domains in PS, a "network model" of amorphous polymers was deduced [6]. The entanglements between macromolecules separated by an average distance d_e (proportional to $\sqrt{M_e}$; see Fig. 14.6) can be assumed to form a network containing meshes of average diameter $D = d_e \cdot \sqrt{2} \approx 1.4\,d_e$. The meshes inside the entanglement network correspond to the plastically deformed domains, indicating that they consist of mechanically softer material, such as chain ends, short molecules or localised free volume. Such a network with softer domains that are stretched in front of a localised stress concentration is sketched in Fig. 15.3a. A comparison with electron micrograph (b) reveals the similarity of the model to the structure inside a pre-craze. It must be assumed that there is a mesh size distribution. Larger meshes are deformed more easily than smaller ones, and so the domain diameters of 15–50 nm measured on the electron micrographs of the pre-crazes correspond to larger meshes [6].

The experimentally determined main distance between the domains of about 50 nm in PS gives the average distance of larger, mechanically active (i.e. convertible into pre-craze domains) meshes in the entanglement network. This network model was developed for PS with its relatively large entanglement distance (mean value of 9.6 nm). However, it can also be used for other amorphous polymers with smaller

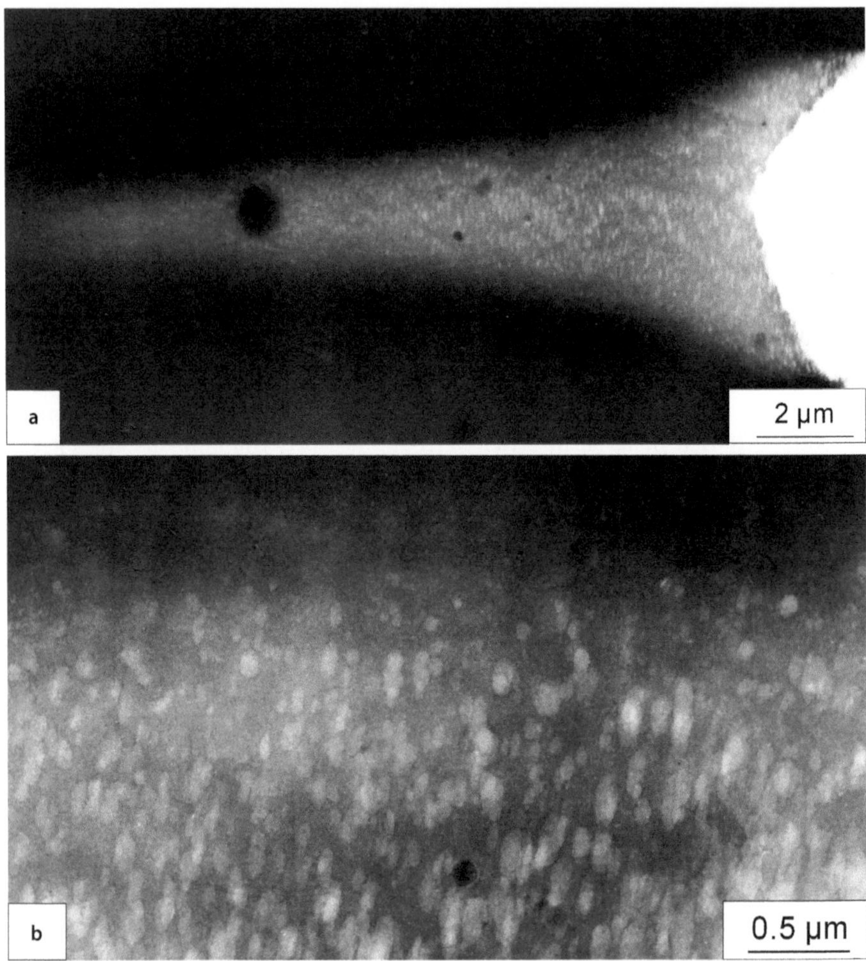

Fig. 15.2. Craze-like deformation zone ahead of a crack tip in a SAN copolymer in overview (**a**) and at larger magnification (**b**) with strong plastically deformed domains. Semi-thin section of SAN copolymer, deformed in a 1000 kV HVTEM, deformation direction vertical (after [1], reproduced with permission from Hanser)

entanglement distances (e.g. PMMA, 7.3 nm; PC, 4.4 nm [7]). PVC is an exception since it has a morphology of microdomains (10–20 nm), domains (about 100 nm) and primary particles (about 1 µm) that is clearly visible in electron micrographs [8]. The interfaces between these structural details are preferential sites for chemical staining and so they highlight the morphology of PVC; see Fig. 15.4.

The morphologies of amorphous polymers with different phases and thus phase separations, such as block copolymers, graft polymers and polymer blends, are clearly visible in the electron microscope (see Chaps. 17, 18, and 19).

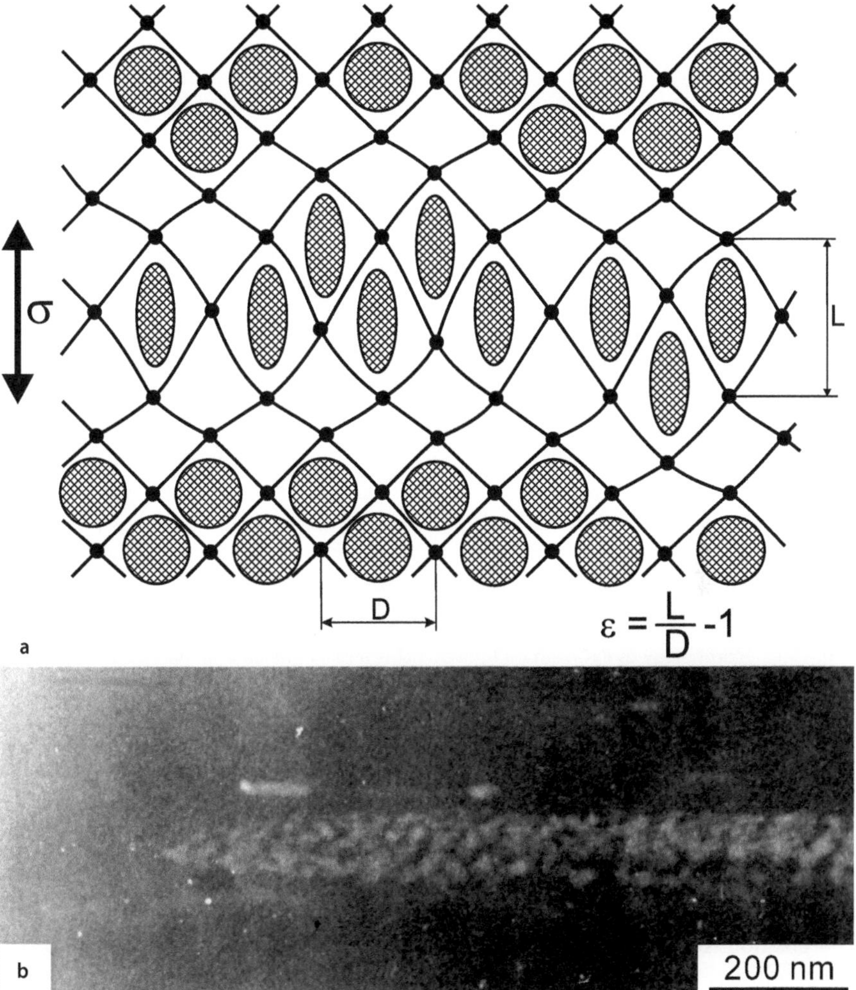

Fig. 15.3. **a** Scheme of a network of entangled macromolecules with softer meshes, which are more strongly plastically stretched in an area of stress concentration, resulting in the formation of a pre-craze structure; **b** electron micrograph of a pre-craze in PS (thin section in a 1000 kV HVTEM, deformation direction vertical)

15.3 Micromechanical Behaviour

Under load, amorphous polymers exhibit localised deformation zones, such as crazes, deformation bands or shear bands. The typical type of deformation seen in amorphous brittle ("glassy") polymers is the "craze". Crazes are often visible with the naked eye in reflected light, and the word "craze" recalls a crack-like appearance ("craze" is an old English word for hairline crack; synonyms include "craquele", "micro cracks" and "silver cracks"). The true nature of crazes and how they are different from cracks

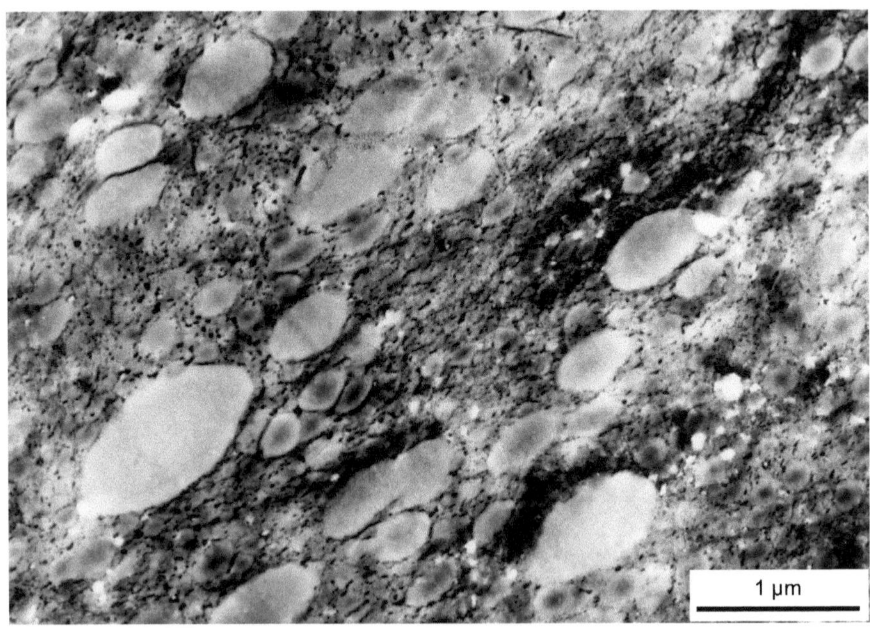

Fig. 15.4. Morphology of PVC with ~0.1 μm domains and ~1 μm large primary particles; ultrathin section, chemically stained, TEM (from [1], reproduced with permission from Hanser)

was revealed by electron microscopy (for summaries see [1, 9]). Several preparation and investigation techniques have been applied to study the structures of crazes in detail:

- Staining or filling the microvoids inside crazes with chemical agents in the bulk, preparation of ultrathin sections and inspection in a TEM [9,10]. The craze compositions of microvoids and stretched fibrils are thus made visible; see Fig. 15.5.
- Studies of stretched ultrathin sections or solution-cast films on mylar films or copper grids in TEM [11, 12], or the deformation of semithin sections directly in HVTEM [1,13]. The fibrillar structure inside the crazes then becomes clearly visible; see Fig. 15.6.
- Studies of surfaces of deformed bulk materials or of fracture surfaces in SEM. Since crazes are known to have a fibrillar structure, domain-like structures on fracture surfaces can be identified as broken and relaxed fibrils; see Fig. 15.7.

The crazes in PS can show wide variations in terms of structure. Some examples are shown in Fig. 15.8. These variations arise due to differences in the loading rate, stress state or sample geometry. Long-chain branched types of PS show small differences in terms of properties from the usual linear PS (such as a higher glass transition temperature T_g, which increases from 104 °C up to 111 °C, and yield stress σ_y, which increases from 57 MPa to 62 MPa). Small differences between the structures of crazes in these two types of PS were found using in situ TEM. Long-

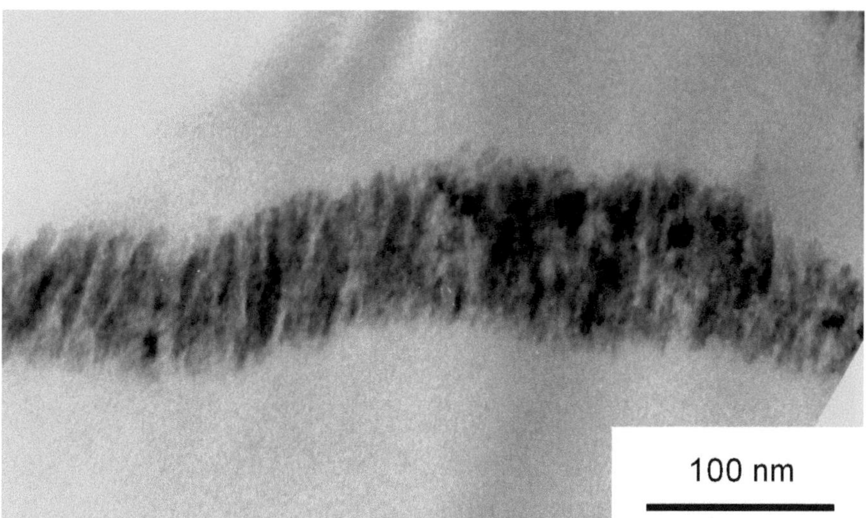

Fig. 15.5. Stained craze in deformed PS. Deformation direction vertical; deformed and chemically stained in the bulk; ultrathin section, TEM

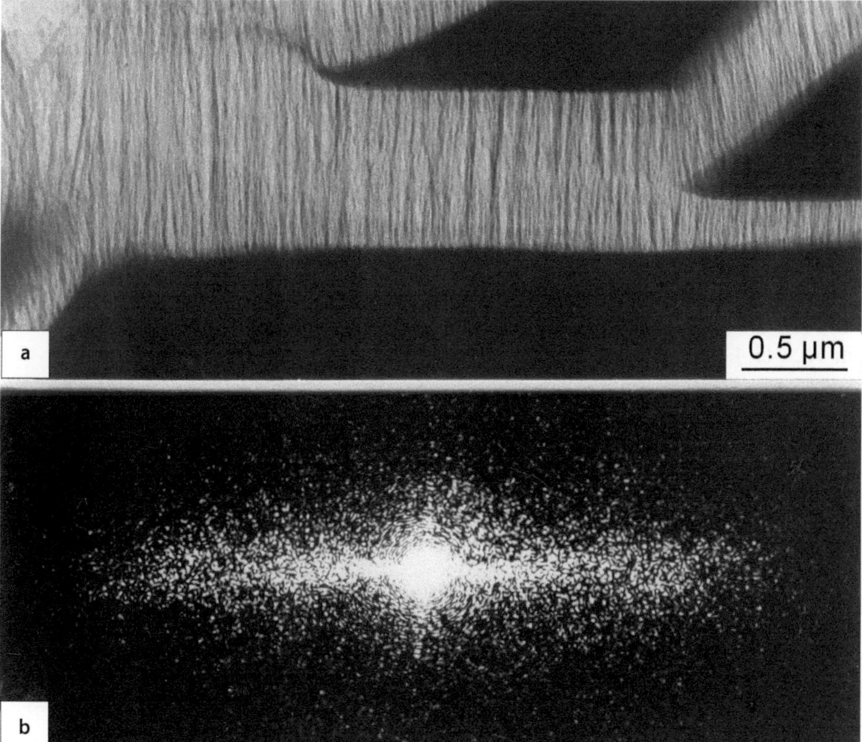

Fig. 15.6a,b. Interior of a craze in PS with clear fibrillation (stretching a semi-thin section in HVTEM, deformation direction vertical): **a** HVTEM micrograph; **b** laser diffraction pattern (from [1], reproduced with permission from Hanser)

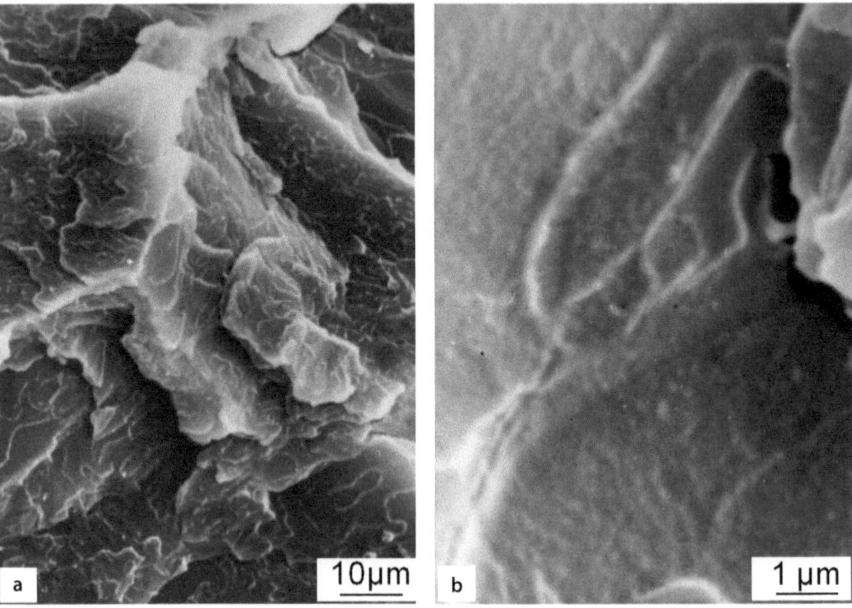

Fig. 15.7a,b. Fracture surface of PS in SEM: **a** overview with brittle fracture edges; **b** larger magnification of ruptured crazes with domain-like hills, revealing broken and relaxed craze fibrils (from [1], reproduced with permission from Hanser)

chain branched PS presents more finely fibrillated crazes, up to homogeneous crazes at room temperature [14]. If the temperature at which the deformation is performed in the TEM (in a heated straining holder) is increased below the glass transition temperature, the crazes thicken (i.e. coarser fibrils and greater distances between the fibrils are obtained) in both linear and long-chain branched PS [15]. This transition is associated with thermally induced disentanglement and is discussed in more detail in Sect. 15.4.

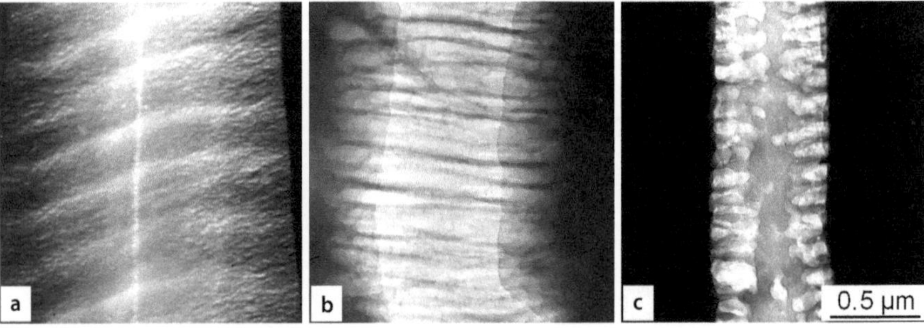

Fig. 15.8a–c. Crazes with different inside structures in PS: **a** fine network fibrils; **b** coarse fibrils; **c** microvoid concentration at the craze boundaries. (Deformed thin sections in HVTEM)

As mentioned above, a craze will first appear in the form of a pre-craze, a zone with stretched domains; see Fig. 15.1. With increasing load, the pre-craze zone is transformed into the fibrillar structure of the craze. The fibrillated craze grows through the continuous stretching of new material at the craze boundaries (surface drawing or pull-out mechanism); see Fig. 15.9. In the thick craze of micrograph (c), the first (thin) crazes are visible as relatively bright zones. Increasing the load further initiates craze rupture; see Fig. 15.10. After the craze is ruptured, the broken and relaxed fibrils are visible on fracture surfaces using SEM; see Fig. 15.7.

In addition to the fibrillated crazes, the more ductile amorphous polymers (PC, PVC, etc.) show homogeneous deformation bands and shear bands. The coexistence of homogeneous deformation bands (i.e. craze-like zones with homogeneously stretched material inside) with fibrillated crazes is shown in a SAN copolymer in Fig. 15.11. The coexistence of all three deformation bands is shown in the deformed PVC material in Fig. 15.12. The reason for the occurrence of these different deformation zones is connected to the entanglement density.

PS has a low entanglement density, i.e. a large entanglement molecular weight and a large entanglement distance, and so it only exhibits fibrillated crazes. With increasing entanglement density (decreasing entanglement molecular weight and distance), fibrillated and homogeneous crazes (as seen in SAN, COC) start to coexist and a transition to homogeneous crazes and shear bands (as seen in PVC, PC) occurs; see Table 15.1. This transition is sketched in Fig. 15.13 [16]]. Here, a fine entanglement network with small meshes deforms in a homogeneous manner (a); a coarser network

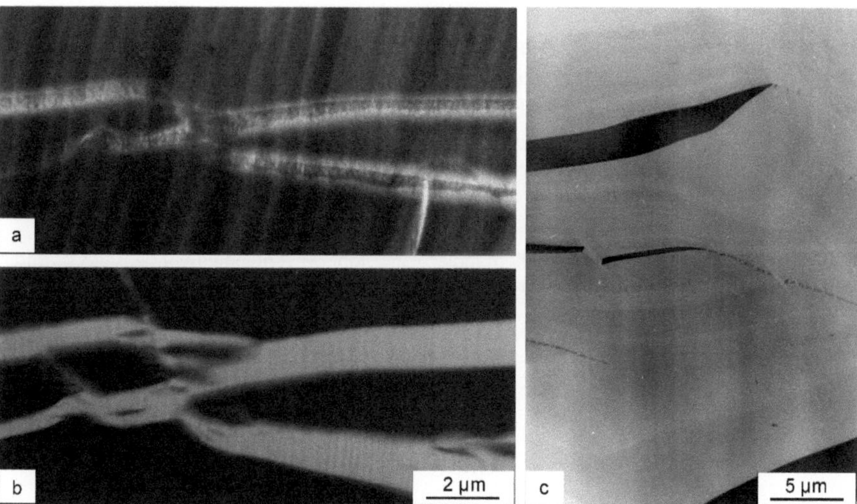

Fig. 15.9a–c. Growth of crazes in PS: **a** pre-craze domains; **b** pre-crazes in micrograph (**a**) transformed into fibrillar crazes; **c** growth in thickness (visible thickening of the first, thinner crazes). In situ deformation in HVTEM, vertical deformation direction (from [1], reproduced with permission from Hanser)

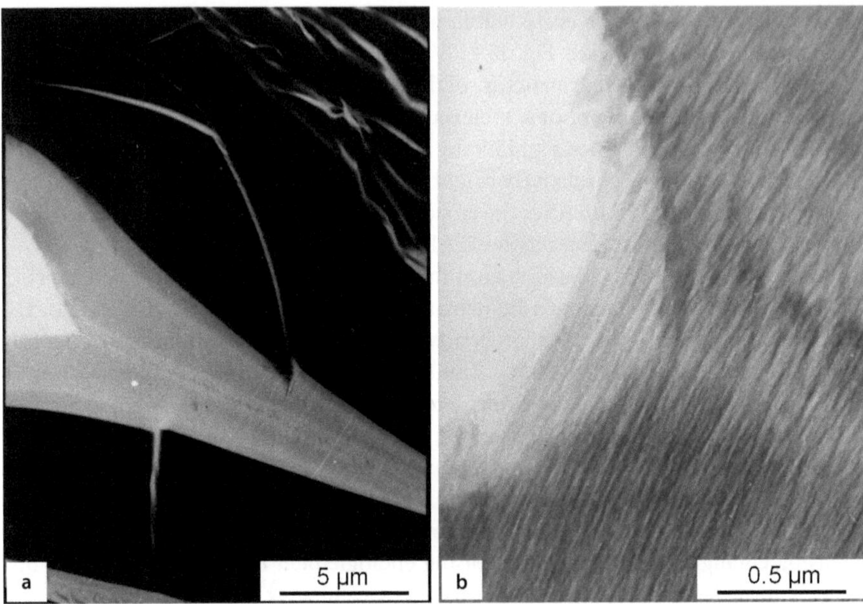

Fig. 15.10a,b. Crack propagation inside a craze in PS: **a** overview, crack propagates from left to right; **b** larger magnification of the crack tip with successive stretching, overstressing and rupturing of the craze fibrils shown. In situ deformation in HVTEM, approximately vertical deformation (after [1], reproduced with permission from Hanser)

with larger meshes deforms with stretching and rupturing of the largest meshes and fibrillation of the material between these microvoids (b).

Crazes (fibrillated as well as homogeneous) play an important role in improving the mechanical properties of polymers. One well-known example of this is multiple crazing in rubber-toughened polymers (see Chap. 18).

Table 15.1. Comparison of entanglement molecular weights M_e, entanglement densities v_e, entanglement distances d and typical deformation structures for different polymers (after [7, 16])

Polymer	M_e	$v_e[\mu m^3]$	$d[nm]$	Deformation structure
PS	19 100	3×10^7	9.6	Fibrillar
SAN	11 600	6×10^7	8.2	Coexistence
PMMA	9150	8×10^7	7.3	Coexistence
PPO	4300	15×10^7	5.5	Homogeneous
PC	3490	29×10^7	4.4	Homogeneous

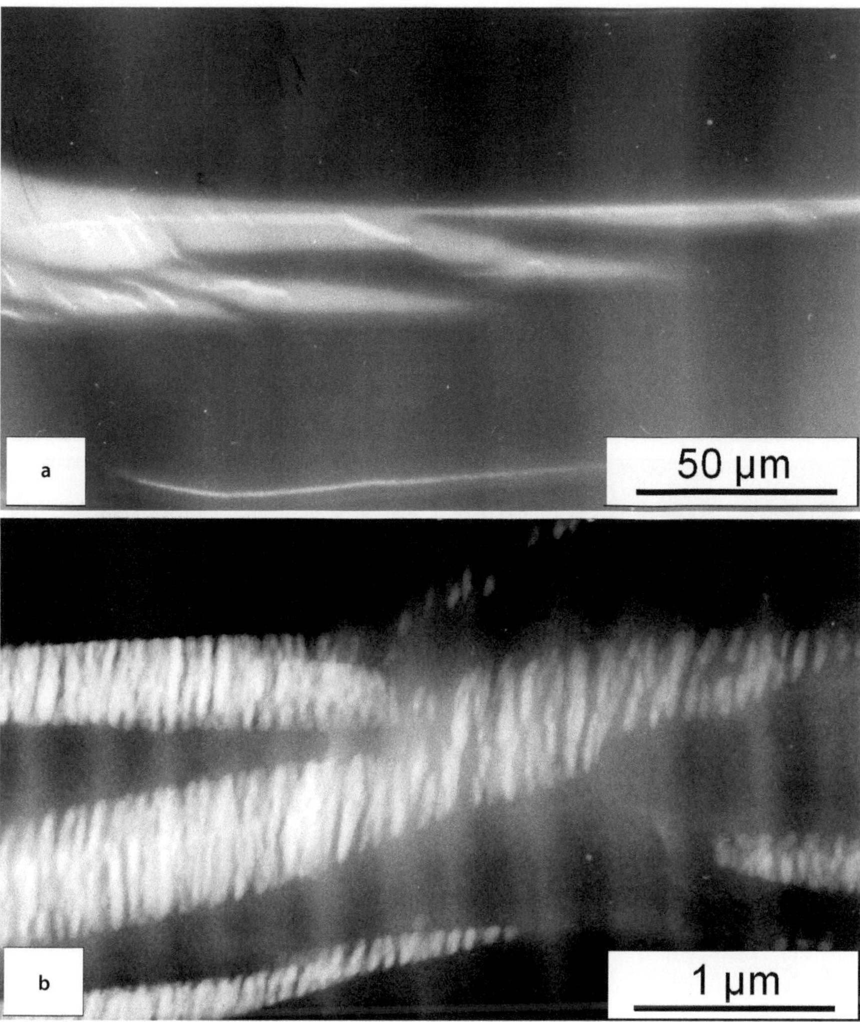

Fig. 15.11a,b. Coexistence of fibrillated crazes and homogeneous deformation zones in SAN: **a** overview; **b** larger magnification. Deformed semi-thin section in HVTEM

15.4 Additional Examples of Amorphous Polymers

a) PMMA

Like PS, PMMA is a typical example of an amorphous glassy polymer that exhibits crazes under loading. However, the fine structure of these crazes is only visible at high magnifications in the electron microscope, but PMMA is also very sensitive to electron irradiation. Therefore, very few results exist for PMMA in this context. The appearance of the beam damage has been studied by HVTEM [17]; see Fig. 15.14. At

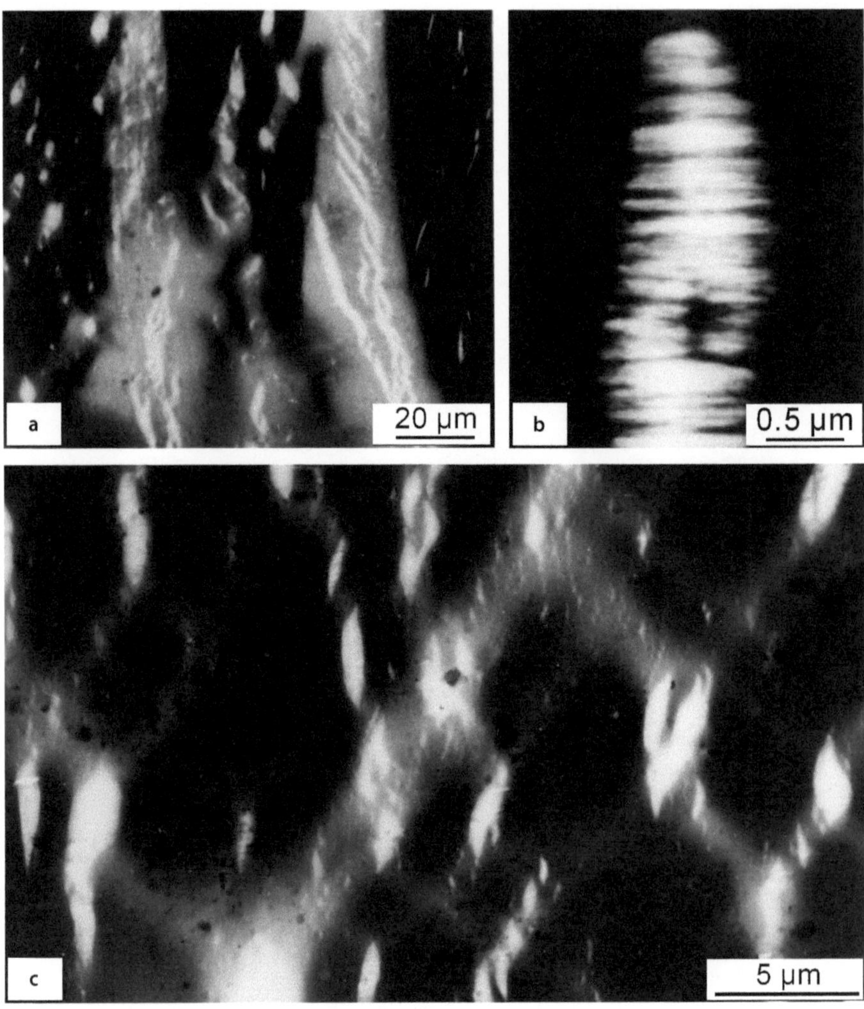

Fig. 15.12a–c. Coexistence of fibrillated crazes, homogeneous crazes and shear bands in PVC: **a** overview; **b** fibrillated craze inside a homogeneous craze; **c** crossed shear bands and crazes. Deformed semi-thin section, deformation direction horizontal, HVTEM (from [1], reproduced with permission from Hanser)

low irradiation dosages (before the damage is done), a typical craze pattern is visible in front of a crack (left micrograph). Damage appears inside the stretched parts of the sample, i.e. in the crazes, along with cavitation (micrographs in the middle and on the right). Higher beam intensities create larger voids and can destroy the whole sample.

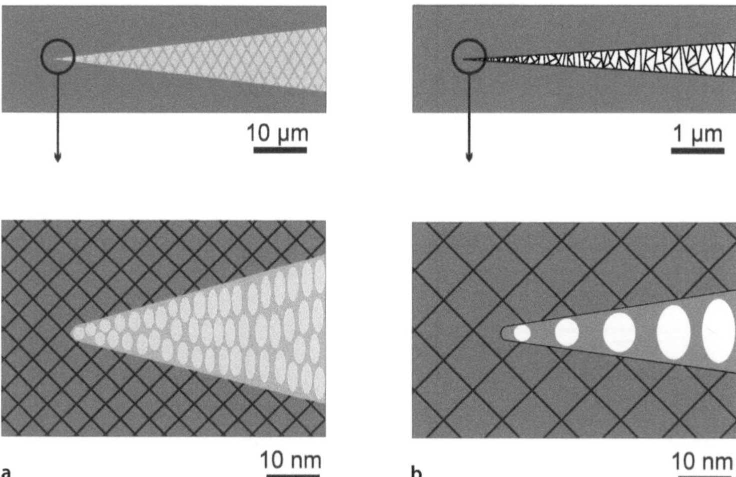

Fig. 15.13a,b. Scheme showing differences between homogeneous (**a**) and fibrillated crazes (**b**) in relation to the entanglement network: *top*, view as in electron micrographs; *bottom*, entanglement network with stretched meshes. **a** Dense pattern of small stretched meshes in homogeneous crazes; **b** coarse pattern with voiding and fibrillation in large meshes in fibrillated crazes

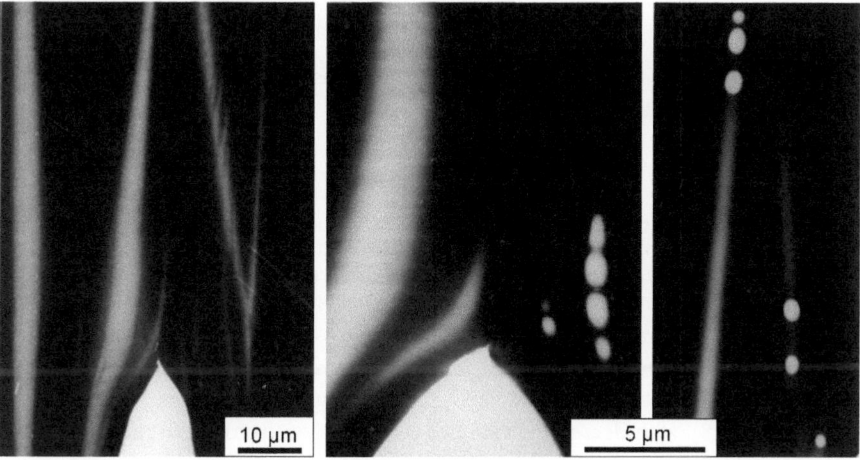

Fig. 15.14. Appearance of damage due to electron irradiation in PMMA. Before (*left*) and after (*middle* and *right*) damage is incurred; small voids are shown inside the deformation bands. Semi-thin section, deformation direction horizontal, HVTEM (from [1], reproduced with permission from Hanser)

b) Cyclic Olefin Copolymers (COC)

Cyclic olefin copolymers of ethylene and different types of cyclic monomers (e.g. norbornene) form a relatively new class of amorphous brittle polymers with excellent optical, thermal and permeation properties. Depending on the type of cyclic monomer used, COCs show a decrease in toughness and elongation at breakage with increasing

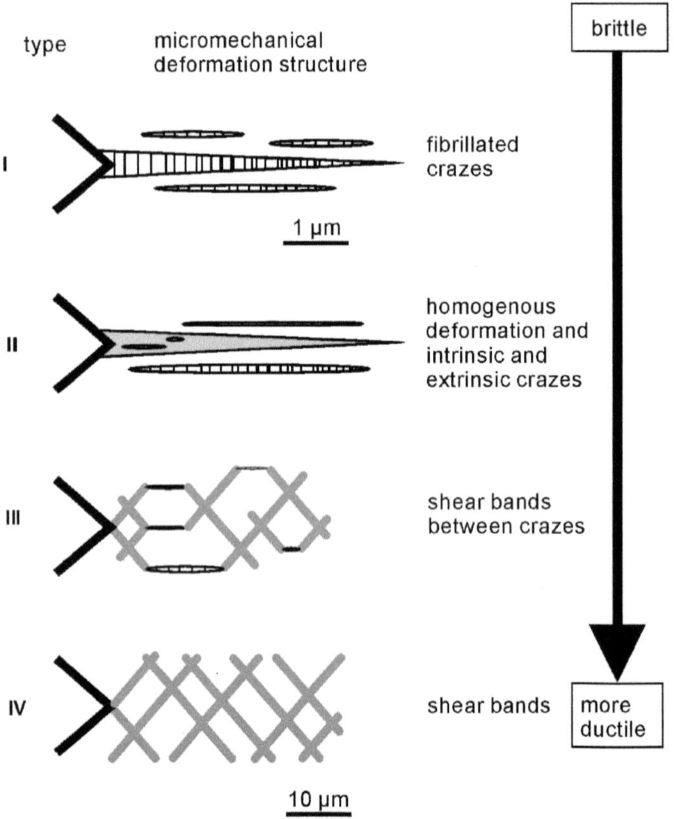

Fig. 15.15. Changes in the micromechanical deformation structures in different cyclic olefin copolymers (COC) in relation to cyclic monomer type and content with transition from brittle to ductile behaviour, as revealed by TEM. Deformation direction perpendicular and propagation of the deformation zones from *left* (from a notch tip) to *right* (from [18], reproduced with permission from Elsevier)

cyclic monomer content. This behaviour correlates very well with the micromechanical deformation mechanisms that occur in such polymers, as revealed using electron microscopy [18]: with decreasing toughness a transition from shear bands to homogeneous deformation zones and to fibrillated crazes (and occasionally a combination of them) appears; see Fig. 15.15. Some types of COCs behave like SAN copolymers in terms of deformation processes and mechanical properties (i.e. the coexistence of different deformation zones is observed).

Fig. 15.16a–f. Change from homogeneous crazes into fibrillated ones and reduction in size and amount of the crazes in SAN after annealing at 80 °C for different times. Annealing time *t*: **a** starting material; **b** 10 h; **c** 146 h; **d** 0.5 h; **e** 93 h; **f** change and decrease in length *L* of the crazes with annealing time *t*. All results for deformed semi-thin sections in HVTEM; **a–c** overviews; **d–e** larger magnifications (after [1], reproduced with permission from Hanser)

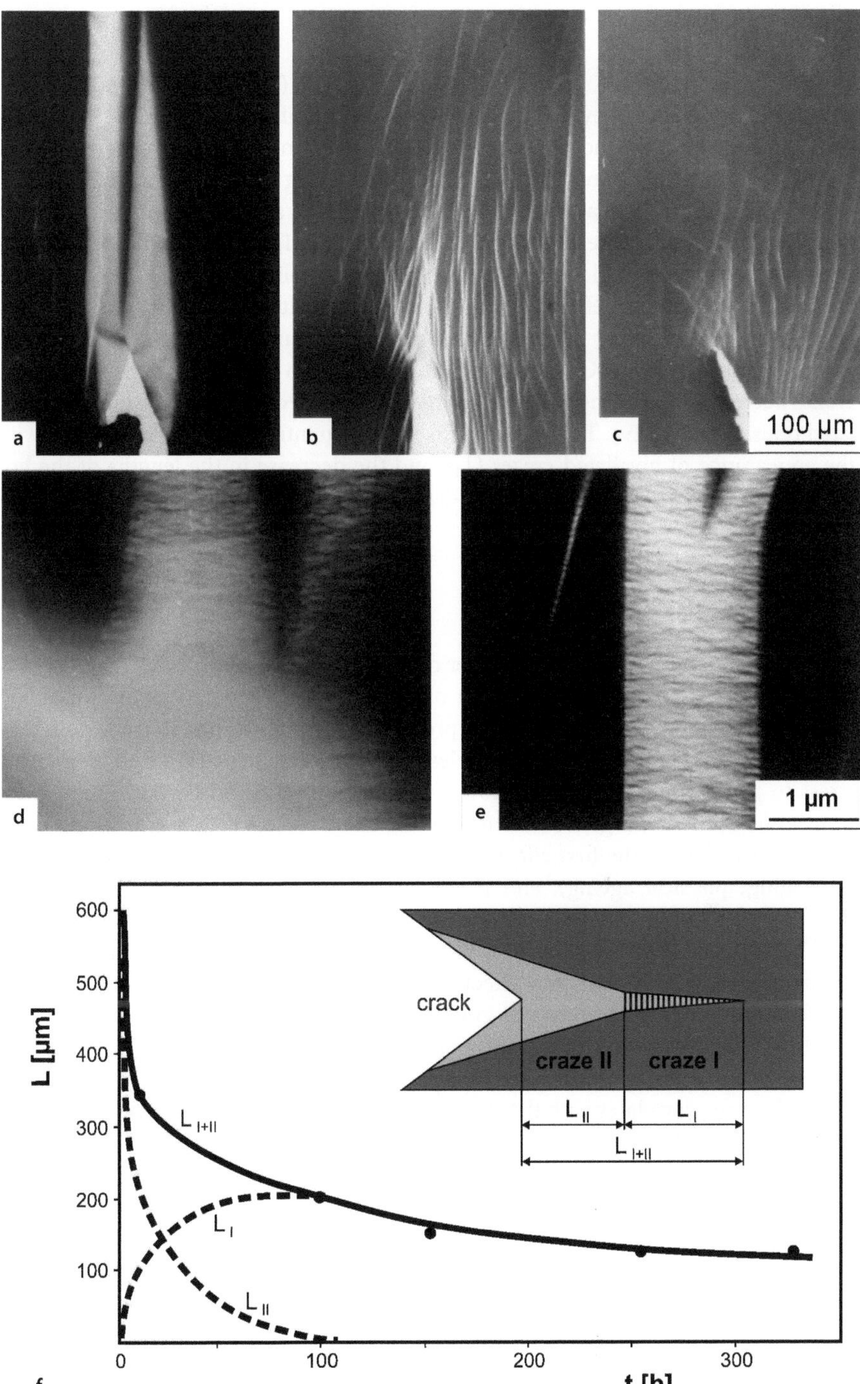

a
b
c
100 μm
d
e
1 μm
600
500
400
300
200
100
L [μm]
crack
craze II
craze I
L II
L I
L I+II
L I+II
L I
L II
0
100
200
300
t [h]
f

c) SAN Copolymers

Statistical copolymers of styrene (S) and acrylonitrile (AN) are amorphous, transparent polymers and they are usually a little bit tougher than PS. This correlates with the appearance of homogeneous deformation zones in coexistence with fibril crazes; see Fig. 15.11. Annealing SAN (i.e. maintaining the sample at an increased temperature below the glass transition temperature T_g; ageing) increases the brittleness. This is usually correlated with a decrease in the free volume. TEM studies of deformed annealed samples demonstrate not only a qualitative but also a quantitative correlation between the effects of annealing and embrittlement. Figure 15.16 shows, in overview and at larger magnifications, unannealed material (micrograph a) and material annealed at 80 °C. With increasing time (0.5 h up to 146 h), the sizes of the crazes decrease and a transition from (large) homogeneous crazes to (smaller) fibrillated ones occurs [5,19]. This transition is shown quantitatively in diagram (f), with the appearance of fibrillated crazes (L_I) and the decrease in the lengths of the homogeneous crazes (L_{II}) and the whole deformation zone (L_{I+II}) in front of the crack over time.

d) PC

At room temperature, PC shows higher ductility than the amorphous polymers PS, SAN and PMMA, with the formation of extended plastic deformation zones and shear bands. However, PC exhibits a pronounced tough-to-brittle transition with increasing sample thickness or in samples with sharp notches (transition from plane stress to plane strain condition) [20]. Two other tough-to-brittle transitions (with transitions from homogeneous deformation bands to fibrillated crazes) occur as temperature effects. The first effect appears as result of thermal treatment below T_g (annealing, physical ageing), and is similar to the effect shown in Fig. 15.16 for SAN [19]. The other effect appears due to deformation at higher temperatures below T_g. A decrease in toughness or an embrittlement with increasing deformation temperature is a fairly abnormal effect for common polymers, but it is known to occur for high-temperature-resistant grades of polymers [21]. If the deformation temperature T is near room temperature or far below T_g ($\Delta T = T_g - T > 40\text{–}80$ K), homogeneous yielding (shear bands and deformation zones) appears; see Fig. 15.17a. This behaviour correlates with the rule that polymers with high entanglement densities deform with homogeneous deformation zones; see Table 15.1. If the deformation temperature is closer to T_g, a transition from homogeneous yielding to fibrillated crazing and a coarsening of the craze structure with increasing fibril thickness and spacing occurs; see Fig. 15.17b,c.

This effect is based on an increased molecular mobility with increasing deformation temperature and thus a thermally induced disentanglement. The less entangled material deforms via fibrillated crazes (see Table 15.1), which are thinner than homogeneous deformation bands. Therefore, the total amount of plastically deformed

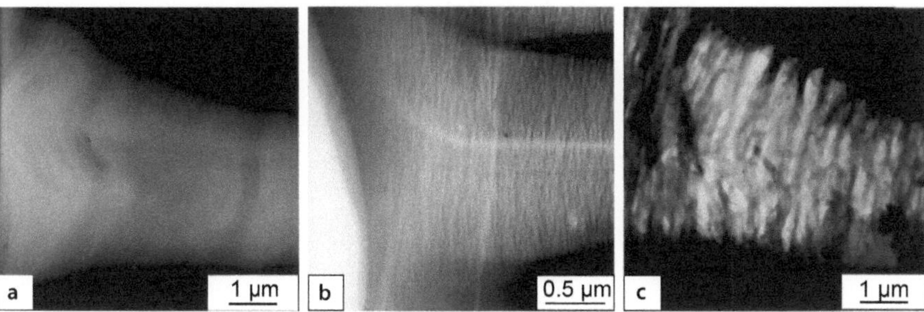

Fig. 15.17a–c. Relationship between internal structure of deformation zones and crazes in PC and the deformation temperature T ($\Delta T = T_g - T$): **a** $\Delta T = 45$ K, homogeneous deformation band; **b** $\Delta T = 37$ K, finely fibrillated craze; **c** $\Delta T = 16$ K, craze with coarse craze fibrils. Deformed thin specimens, HVTEM, deformation direction vertical

polymeric material is smaller for fibrillated crazes. The result of this is an observed macroscopic effect where the toughness decreases with increasing deformation temperature.

References

1. Michler GH (1992) Kunststoff-Mikromechanik: Morphologie, Deformations- und Bruchmechanismen, Hanser-Verlag, München
2. Michler GH (1992) Kunststoff-Mikromechanik: Morphologie, Deformations- und Bruchmechanismen, Hanser-Verlag, München, Sect 7.2.1, pp 142
3. Locatelli JL, Riess G (1973) J Polym Sci B 11:257
4. Michler GH (1977) PhD Thesis, Martin-Luther-Universität Halle-Wittenberg
5. Michler GH (1992) Kunststoff-Mikromechanik: Morphologie, Deformations- und Bruchmechanismen, Hanser-Verlag, München, Sect 7.4, pp 161
6. Michler GH (1991) Plaste Kautschuk 38:268
7. Kramer EJ (1984) Polym Eng Sci 24:761
8. Menges G, Berndtsen N (1976) Kunststoffe 66:735
9. Kausch HH (ed) (1983, 1990) Crazing in polymers, vols 1, 2. Springer, Berlin
10. Kambour RP (1973) J Polym Sci Macromol Rev 7:1
11. Beahan P, Bevis M, Hull D (1973) J Mater Sci 8:162
12. Kramer EJ (1983) In: Kausch HH (ed) Crazing in polymers, vol 1. Springer, Berlin, p 1
13. Michler GH (1979) Kristall Technik 14:1357
14. Garcia Gutierrez MC, Michler GH, Henning S, Schade C (2001) J Macromol Sci Phys B 40:797
15. Garcia Gutierrez MC, Henning S, Michler GH (2003) J Macromol Sci Phys B 42:95
16. Michler GH (1992) Kunststoff-Mikromechanik: Morphologie, Deformations- und Bruchmechanismen, Hanser-Verlag, München, Chap 6, pp 88
17. Michler GH (1992) Kunststoff-Mikromechanik: Morphologie, Deformations- und Bruchmechanismen, Hanser-Verlag, München, Sect 7.3, pp 157
18. Seydewitz V, Krumova M, Michler GH, Park JY, Kim SC (2005) Polymer 46:5608
19. Michler GH, Grellmann W (1989) Plaste Kautschuk 36:120
20. Hyakutake H, Nisitani H (1987) Jap Soc Mech Eng Int J 30:29
21. Michler GH (2001) In: Grellmann W, Seidler S (eds) Deformation and fracture behaviour of polymers. Springer, Berlin, Sect B.1.4, p 193

16 Semicrystalline Polymers

Understanding the correlation between morphology and micromechanical processes of deformation and fracture of semicrystalline polymers is essential for material development, modification, and failure analysis. Since semicrystalline polymers exhibit a hierarchical morphology with structures ranging from the nanometre to the millimetre scale, various electron microscopic techniques are applied to image typical structural units such as crystalline blocks, lamellae, spherulites, and fibrils. In this chapter, a general survey of methods of morphological and micromechanical analysis is first provided, with examples given for the most significant semicrystalline polymers, polyethylene (PE) and polypropylene (PP). The chapter presents methods for enhancing contrast by chemical staining and etching, the application of electron diffraction contrast, electron beam irradiation effects, studies of crystallisation and melting phenomena, as well as typical micromechanical phenomena that can be related to brittle, ductile, or high-strength behaviour, such as crazing, chevron formation, fibrillation, and others. Additional examples of semicrystalline polymers, such as polyamides, fluoropolymers, polyurethanes, biomedical polyesters, etc., are then given.

16.1 Overview

The importance of the morphologies, micromechanical mechanisms and mechanical properties of semicrystalline polymers can be exemplified by a discussion of polyethylene and polypropylene, two of the most prominent commodity plastics. These representatives of the family of polyolefins make up nearly 50% of the world's plastics market [1]. The rapid development of new catalysts has allowed the tailored design of macromolecules with defined semicrystalline morphologies and thus defined properties, such as toughness, stiffness, wear resistance, transparency, etc. Besides these commodity polymers, there are a number of technical and functional polymers (e.g. PEEK, PVDF, PTFE) that have typical semicrystalline structures. Very often, this semicrystalline morphology provides and controls the physical properties of technical relevance in polymers. Therefore, the analysis and control of the morphologies of semicrystalline polymers as well as an understanding of the micromechanical processes of deformation and fracture are crucial to material development, modification, and failure analysis.

16.2 Morphology

16.2.1 Structural Hierarchy in Semicrystalline Polymers

One common feature of semicrystalline polymers is a hierarchical morphology. The scales of structural details within them range from nanometres (or even less, i.e. Ångstroms) to millimetres. For this reason, it is important to select microscopic techniques and preparation methods that are suited to the actual problem. Figure 14.9 summarises a possible classification scheme for the structural details of semicrystalline polymers and the instrumentation that can be applied for structure determination.

Under certain conditions macromolecules are able to form periodic structures that involve adjacent chains or chain segments. Such preconditions include sufficient flexibility of the chains, stereochemical regularity (tacticity, see Fig. 14.2), and others. A parallel alignment of segments can be achieved, for instance, by repeated folding of a flexible polymer chain. Such a chain folding mechanism results in densely packed and highly ordered domains. Nevertheless, there are also polymer-specific factors that inhibit the formation of perfect crystals. Such limiting factors include:

- Restricted chain mobility due to entanglements of the coiled macromolecules (see Fig. 14.6),
- Variations in the molecular weights of single macromolecules (i.e. differences in the chain length, see Fig. 14.5)
- Chain-end defects resulting from the limited chain lengths of macromolecules (see Fig. 14.11).

Consequently, any polymer fulfilling the abovementioned conditions will only be partially crystalline; highly ordered domains will coexist with regions containing randomly coiled chains. In other words, a semicrystalline material consists of an amorphous phase and a crystalline fraction with identical chemical compositions but divergent physical properties. This phase separation results in a lamellar structure that is typical of semicrystalline polymers. The organisation of such a lamella is demonstrated in Fig. 14.7, where a three-phase model is proposed. Based on TEM investigations, one can distinguish a crystalline part (the lamella) that is not influenced by the staining agent and therefore appears in light shades of grey. The less ordered amorphous portion, on the other hand, is easily attacked by the chosen staining agent, resulting in dark shades of grey. Between these phases there is an interface that is usually stained even more strongly. It is worth noting that the appearance of a lamella in a TEM micrograph depends on the angle of tilt with respect to the image plane. The correlation between the tilt angle and the apparent lamellar thickness is depicted in Fig. 16.1a. The different measures used to characterise the thickness of lamellae are illustrated in Fig. 16.1b. The change in the visibility of lamellae with the tilt angle of the thin section is shown in Fig. 16.2. Only lamellae that are oriented with their boundary layers parallel to the electron beam ("edge-on" position) are visibly sharp

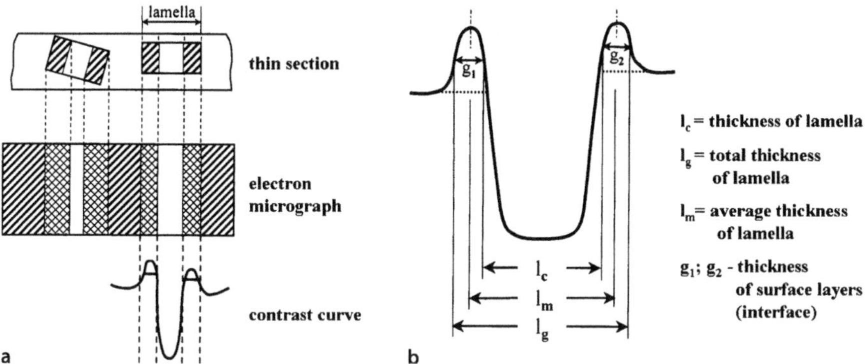

Fig. 16.1a,b. Determination of lamellar thickness by TEM: **a** lamellae with different tilt angles with respect to the imaging plane, corresponding TEM image, and contrast curve; **b** various ways to measure lamellar thickness

(show maximum contrast), whereas lamellae in the perpendicular "flat-on" position disappear in the micrograph. Only the use of a special preparation technique will permit flat-on lamellae to be distinguished (see Fig. 16.10).

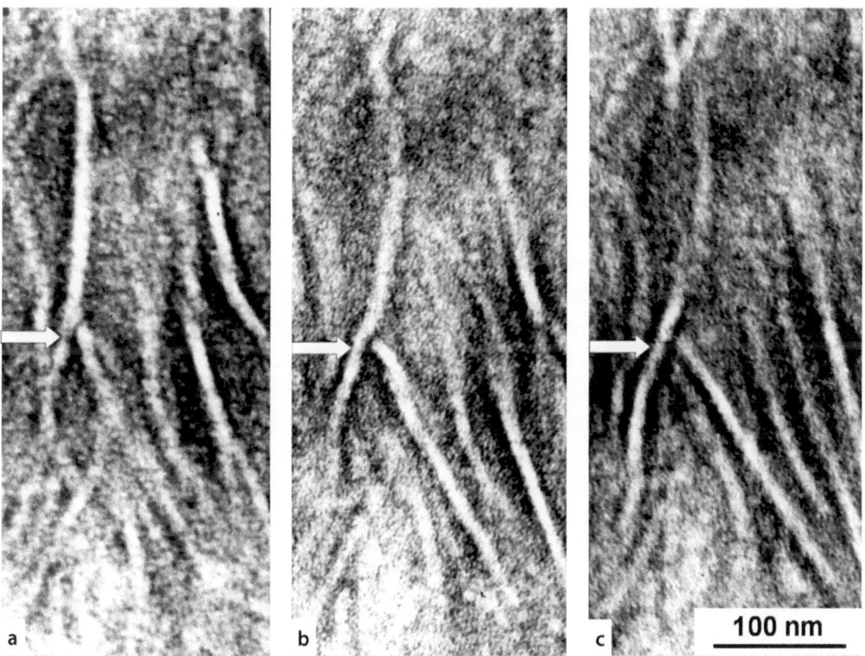

Fig. 16.2. Series of TEM images from lamellae in LDPE taken at different tilt angles **a–c**. (TEM images of ultrathin sections after staining are shown)

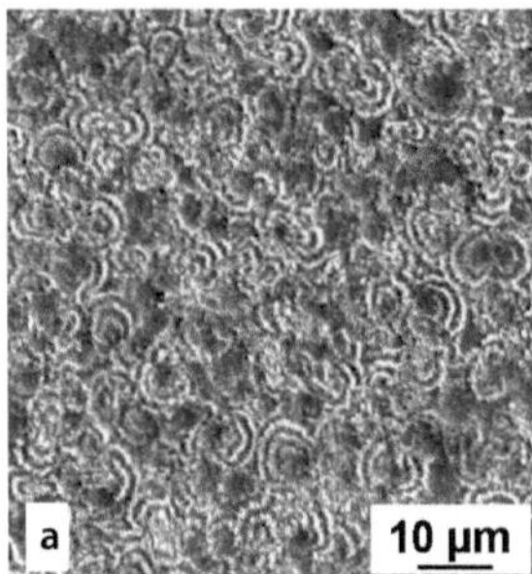
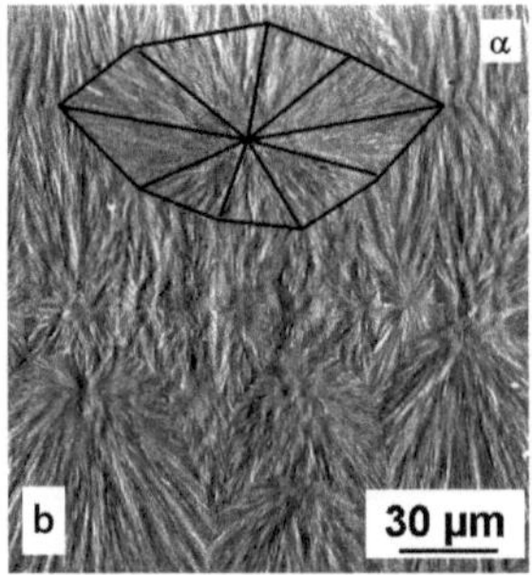
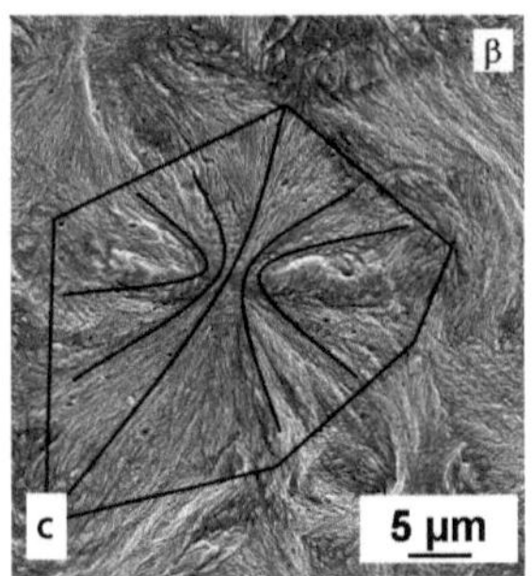

Fig. 16.3a–c. Spherulitic texture of LDPE and the two main types of isotactic polypropylene (iPP): **a** banded spherulites of LDPE; **b** α-iPP with lamellae radiating from the centre of the spherulite; **c** β-iPP showing spherulites with sheaf-like lamellar arrangements. (SEM images after permanganic etching)

The main morphological parameters, i.e. the lamellar thickness l_c (measured as shown in Fig. 16.1b) and the long period d_l, are typically in the range of 5–45 nm in size. Hence, these polymers, which typically exhibit the coexistence of two different phases within one chemically homogeneous polymer, can be considered to be nanostructured materials. In most cases and under normal conditions (atmospheric pressure, 21 °C, moderate deformation rate), the coexisting phases are a combination of a soft (amorphous) component and a relatively stiff (crystalline) component connected by covalent bonds (tie molecules and entanglements within the amorphous regions) that give rise to a good balance of stiffness and toughness for the polymeric material. Due to the polymer-specific factors that inhibit the formation of perfect crystals, the lamella itself also contains defects. One example is presented in Fig. 13.14. The lamellae in a highly branched PE comprise short crystalline blocks that are interrupted by imperfections.

When a polymer is cooled down from the melt, the (primary) crystallisation starts from initial points that are randomly distributed in the volume. Such starting points are either homogeneous or heterogeneous nuclei (e.g. nucleating agents, impurities, or filler particles). This primary crystallisation proceeds into the liquid melt until the melt freezes at significant undercooling ($T < T_g$) or until the whole volume is filled with solidified material. This radial growth results in a characteristic arrangement of lamellae. The most prominent superstructures are spherulites and sheaf-like bundles of lamellae that show rectilinear boundaries formed by colliding growth fronts (Fig. 16.3). These superstructures come in a variety of forms depending on the polymer and its crystalline nature. The sizes of these superstructures are typically in the range of 1–100 μm.

These superstructures generally form a texture consisting of one or more spherulite types with a characteristic spherulite size distribution. The texture may differ over the sample cross-section; in some cases these heterogeneities are visible to the naked eye.

It is important to note that the morphologies of semicrystalline polymers, at all hierarchical levels, are controlled by the architectures of their macromolecules

(chemical structure, configuration, conformation) and their processing history [2].
Some of the most influential factors are:

- The molecular weight and the molecular weight distribution
- The chain architecture (flexibility, side chain length, number and type of side chains/side groups)
- The tacticity
- The processing route (melt processing, solution casting, powder compaction, etc.)
- The cooling rate and the crystallisation temperature
- The nucleating agents, fillers, pigments, etc.
- The pressure
- The shear force.

Electron microscopic images are used to determine the dimensions of all of the structural items. Qualitative descriptions and the measurements of the following morphological parameters are widely used to characterise semicrystalline polymers:

- Type and size of the crystalline unit cell
- Degree of crystallinity (crystalline fraction in percent)
- Thickness and length of lamellae, lamellar thickness distribution, thickness of the amorphous layers, long period
- Fine structure of lamellae, lamellar arrangement
- Spherulite type, spherulite size distribution.

There is a huge number of structural and morphological studies of semicrystalline polymers, the results of which are summarised in many reviews (e.g. in [2–6]).

16.2.2 Methods of Morphological Analysis

As was mentioned earlier, the wide range and variety of structural details that can occur in semicrystalline polymers require that appropriate electron microscopic methods and accordant preparation procedures are applied to investigate them. On the other hand, equivalent results can be achieved in different ways. This fact means that it is possible to select the most effective (i.e. least expensive and least time-consuming) method to solve a given problem, which also takes into account the instrumentation that is available.

The electron microscopic imaging of any structural detail usually requires the application of contrast enhancement steps, since although the distinct phases differ in their physical properties, they offer no detectable differences in electron density which could then be used to produce a mass thickness contrast in the TEM (see Chap. 3, Fig. 3.1). Therefore, in most cases, chemical staining procedures (described in Chap. 13) are inevitable. In Fig. 16.4a, for instance, a staining procedure based on ruthenium tetroxide is applied to the two types of isotactic polypropylene (iPP). The less ordered amorphous portion is strongly stained and appears dark, while the

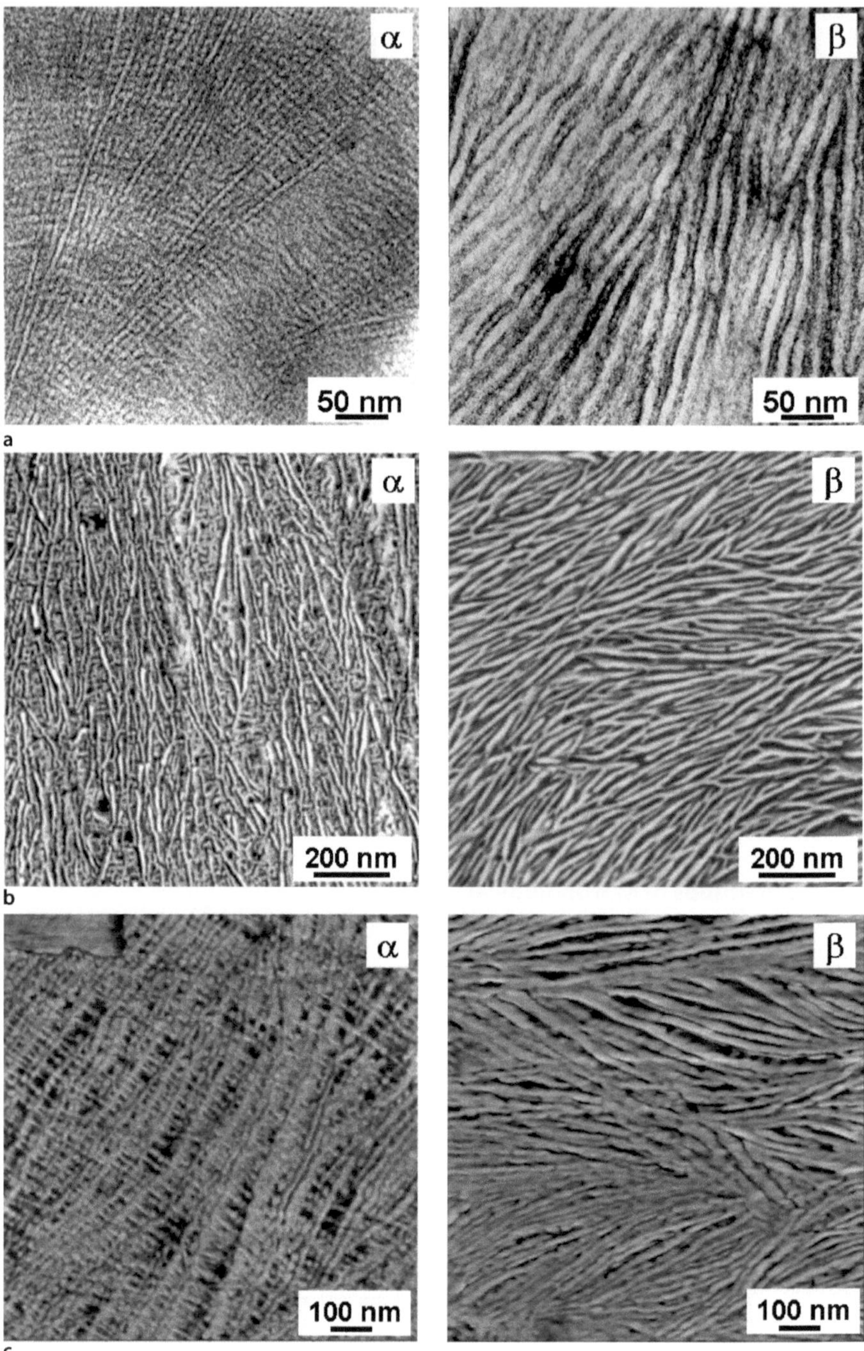
α
β
50 nm
50 nm
α
β
200 nm
200 nm
α
β
100 nm
100 nm
a
b
c

 Comparison of the results from different electron microscopic techniques applied to the two main types of isotactic polypropylene: **a** TEM images of ultrathin sections (RuO_4 staining); **b** SEM images after permanganic etching; **c** AFM tapping mode images. *Left hand images:* α-iPP, *right hand images:* β-iPP

densely packed crystalline part (the crystalline lamella) is not affected by the staining agent. It is worth mentioning that in the case of α-iPP the lamellae form a so-called "crosshatched" arrangement consisting of primary lamellae radiating from the spherulitic centre and secondary (or "daughter") lamellae formed by epitaxial growth. In contrast, the lamellae of β-iPP are aligned parallel to each other.

In order to elucidate details of the semicrystalline morphology by SEM investigations, the latent internal structures must be transformed into a topography that can be detected as a modulation of the secondary electron emission intensity. The corresponding techniques used to achieve this (e.g. chemical and physical etching procedures) are specified in Chap. 9. The iPP sample surfaces presented in Fig. 16.4b are prepared by a permanganic etching procedure. The lamellar arrangement inside the spherulitic superstructure only becomes observable after this treatment. Typically, the lamellar structures that are imaged using the SEM secondary electron signal appear coarser than they do in the corresponding TEM image. Clear lamellar structures are also visible in AFM images in Fig. 16.4c (see below).

There are only few techniques based on contrast formation mechanisms that work without the aforementioned pretreatments. These are described below.

Electron Diffraction Contrast in the TEM

In contrast to amorphous polymers, the structural details of semicrystalline polymers can be imaged using diffraction phenomena. In this case, unstained ultrathin sections or solution-cast films are used. Electrons of the impinging electron beam undergo defined diffraction as they interact with crystalline units (see Fig. 16.5). If only the undiffracted electrons are allowed to pass through a diaphragm a bright-field image is produced. In this case, crystalline fractions of the material appear dark in the TEM image (Fig. 16.5a). On the other hand, a dark-field image is formed if only diffracted electrons are allowed to contribute to image formation. In the latter case, the crystalline regions appear as bright objects on a dark background (Fig. 16.5b). Dark-field imaging can be used to suppress cutting artefacts (e.g. scratches, shatter marks) because mass thickness effects do not contribute to the image contrast; see Fig. 16.6.

Electron diffraction patterns allow structural analysis at the level of the crystalline unit cell, and they can be used to detect the crystal orientations of a given semicrystalline polymer (Fig. 16.5c, see Chap. 2).

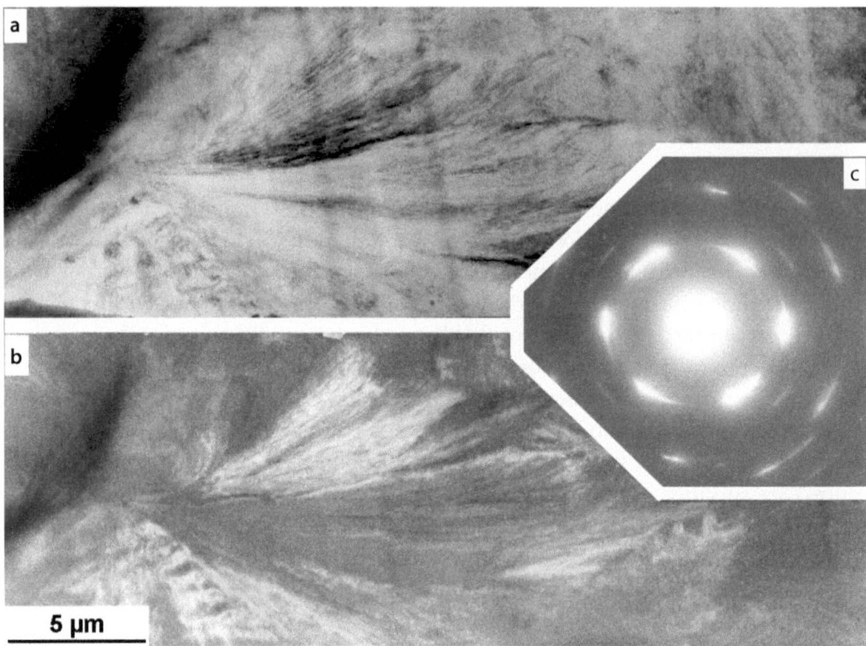

Fig. 16.5a–c. Images of a sheaf-like lamellar structure in LDPE obtained by different operating modes of the TEM: **a** bright-field image; **b** dark-field image; **c** electron diffraction diagram. (Cryosection, HVTEM)

Contrast Enhancement by Electron Beam Irradiation

Many polymers are quite sensitive to electron beam damage. The interactions of the impinging electrons with the macromolecules may lead to depolymerisation and crosslinking reactions. After chain scission, the more mobile chains of the less ordered amorphous phase are able to recombine, forming crosslinks, whereas the chains that are frozen in the crystalline environment do not change their conformations. If we consider an ultrathin polymer film prepared for TEM analysis, these effects will result in differences in thickness: there is certain degree of shrinkage of the amorphous domains while the crystalline portion is unaffected. A mass thickness contrast is therefore generated. Figure 13.12 in Chap. 13 shows how spherulitic structures of a polyurethane sample become visible after electron beam exposure in situ. An analogous effect can be achieved via γ-irradiation (see Figs. 13.10 and 13.11).

Very similar to the effects that were described for thin sections under electron bombardment in the TEM, a surface topography can be produced by electron beam irradiation in the SEM or ESEM. It should be mentioned that such a contrast enhancement is only achieved for a limited number of semicrystalline polymers. The most prominent example is polyethylene: as illustrated in a series of ESEM images of an untreated UHMWPE surface (Fig. 16.7), lamellar structures become visible

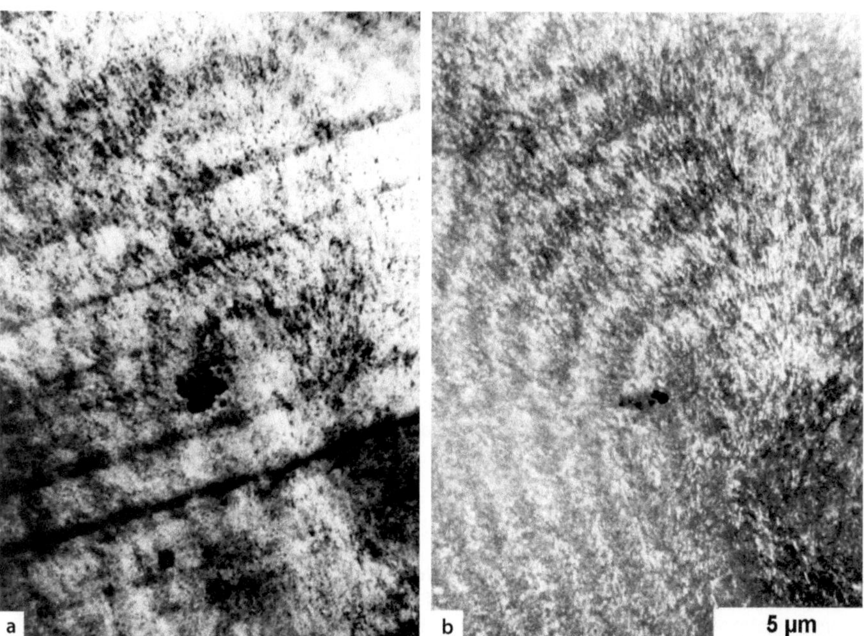

Fig. 16.6a,b. Comparison of two images of HDPE showing banded spherulites that were recorded using diffraction contrast: **a** bright-field image with visible cutting artefacts; **b** dark-field image. (Cryosection, HVTEM; reproduced from [2] with the permission of Hanser)

after only a few seconds of exposure. Unfortunately, the opposite effect is observed for many polymers (e.g. polypropylene): a surface topography that was produced by chemical or physical etching is levelled out during observations due to electron beam-induced degradation.

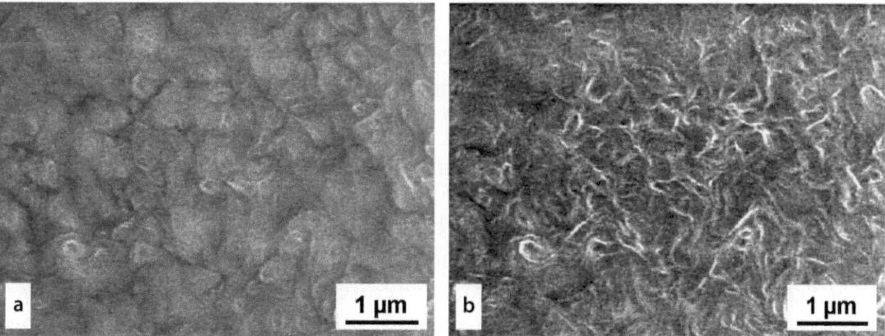

Fig. 16.7a,b. Series of ESEM images from a UHMWPE surface taken after different electron beam irradiation times: **a** at the start of irradiation; **b** after 15 s of observation

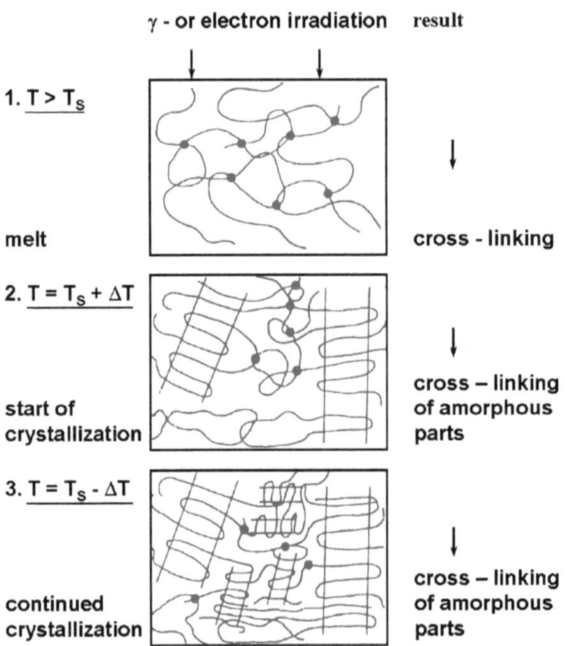

Fig. 16.8. Scheme for investigating structure formation in semicrystalline polymers through the use of irradiation-induced crosslinking and fixation at temperatures in the melting range. After fixation of the amorphous phase in each case, the structure formed can be studied by TEM. (Reproduced from [9] with permission from Marcel Dekker)

"Stiffness Contrast" in the AFM

The difference between the mechanical properties of the crystalline and amorphous fractions can be exploited to generate a contrast from untreated (but preferably smooth) surfaces using dynamic AFM modes (e.g. the *tapping mode*, see Chap. 5). The phase shift in the oscillation of the cantilever, which depends on the local stiffness of the substrate, can be transformed into an AFM image that provides structural information. An example is presented in Fig. 16.4c. The advantage of this technique is that the morphology of a semicrystalline polymer can be imaged without having to put the material through chemical or physical treatment. That allows the in situ imaging of morphological changes caused by deformation experiments or thermal treatments (e.g. melting and crystallisation kinetics [7]).

Study of the Kinetics of Crystallisation and Melting

An unusual preparation technique can be used to study the kinetics of crystallisation and melting [8, 9]. This technique is based on experiences that show that γ- or electron irradiation initiates pronounced crosslinking in the amorphous parts of semicrystalline polymers, whereas the structure inside the lamellae (the crystallinity) is not destroyed, so long as critical doses are not administered (for bulk PE this is about 10^{-3} to 10^{-2} C/cm^3 [10], or more than 20 MGy [11]); see also Sect. 13.3.1. The principles of this technique are illustrated in Fig. 16.8 [9]:

1. For irradiation above the melting range (temperature T sufficiently above the melting temperature T_s), the whole sample exists in the molten (i.e. in the amorphous) state. Crosslinking of macromolecules by irradiation yields a stabilisation of this phase. This is also the case after cooling the irradiated sample down to room temperature (i.e. any crystallisation is prevented), enabling the investigation of this state in the TEM; e.g. Fig. 16.9 shows that no lamellar structures are present in HDPE irradiated at 130 °C.

2. Irradiation within the melting range (or crystallisation range, temperature T somewhat above T_s) results in the initiation of sample crystallisation, with the formation of the first lamellae. The surrounding material, which is still in the molten (amorphous) state, is crosslinked, preventing any additional crystallisation during the subsequent cooling of the sample down to room temperature. Therefore, lamellae only form at the higher temperature and these can be made visible in the TEM after the usual chemical staining of the sample, as shown in Fig. 16.9 for some thick lamellae at 120 °C.

3. Irradiation at the lower part of the crystallisation range (temperature T somewhat below T_s) results in continued crystallisation, such that nearly all lamellae are formed and the residual amorphous zones are crosslinked by irradiation. This structure can be investigated in the TEM after cooling and staining the sample; Fig. 16.9 shows results obtained at 110 °C.

The example of the application of this technique shown in Fig. 16.9 depicts the stepwise melting of a sample of HDPE. At room temperature, the sample shows a typical arrangement of longer, thick lamellae and shorter, thin ones (upper part of Fig. 16.9) with a relatively broad distribution of lamella thickness (lower part of Fig. 16.9). As the temperature is increased to the melting range (to 110 °C and 120 °C, respectively), more and more lamellae are molten, leaving only relatively thick lamellae. The preferred melting of the thinnest lamellae can also be seen in the diagram of Fig. 16.9, with the shifting of the curve of the lamella thickness distribution to larger values. At 130 °C, all of the lamellae have disappeared and only small bright domains are visible, which could indicate the preorientation of the macromolecules in the melt. The drop in the number of lamellae results in an increase in the spacing (the long period) of the lamellae; Table 16.1 compares values of long periods determined directly from electron micrographs with those obtained from small-angle X-ray scattering (SAXS).

Table 16.1. Comparison of the long periods of lamellae in HDPE measured at room temperature and in the melting range, as determined directly on electron micrographs and by SAXS

Sample temperature	TEM		SAXS
	Range (nm)	Average (nm)	Average (nm)
RT	18.0–41.7	27.3	33
110 °C	30.1–47.5	35.5	36
120 °C	45.2–160	83.0	Diluted system
130 °C	–	–	Highly diluted system

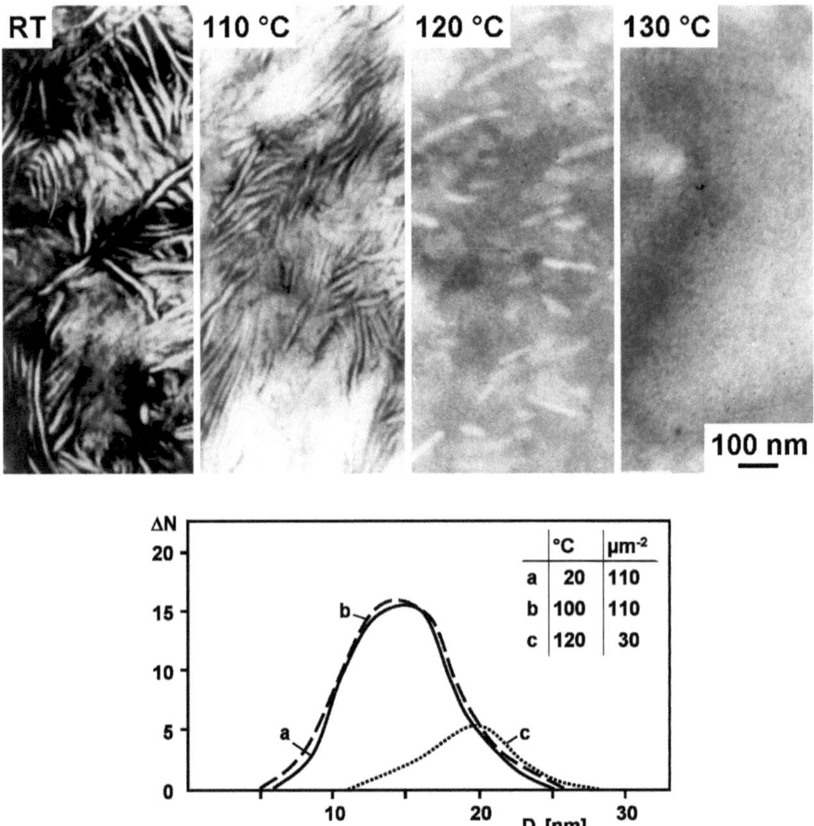

Fig. 16.9. Melting of lamellae in HDPE with increasing temperature (see Fig. 16.8). *Top*: sequence of electron micrographs showing the morphologies at the temperatures indicated; *bottom*: corresponding thickness distributions of lamellae and estimated numbers of lamellae per μm^2 at the temperatures indicated. (Partly from [9], reproduced with permission from Marcel Dekker)

The technique of irradiation-induced fixation in the melting range not only aids the study of the melting or crystallisation behaviour of semicrystalline polymers, but it also provides a very advantageous way of studying the structures of the lamellae themselves. Two effects can be employed (see Fig. 16.10 [2]):

1. Because of the reduced number of lamellae, the remaining lamellae are visible with better contrast. Therefore, lamellae are visible when they are in the usual "edge-on" position or in the "flat-on" position (these latter lamellae are usually not detectable).
2. Irradiation at elevated temperatures yields an additional staining effect inside the lamellae, resulting in an improved visibility of the intralamellar defect layers.

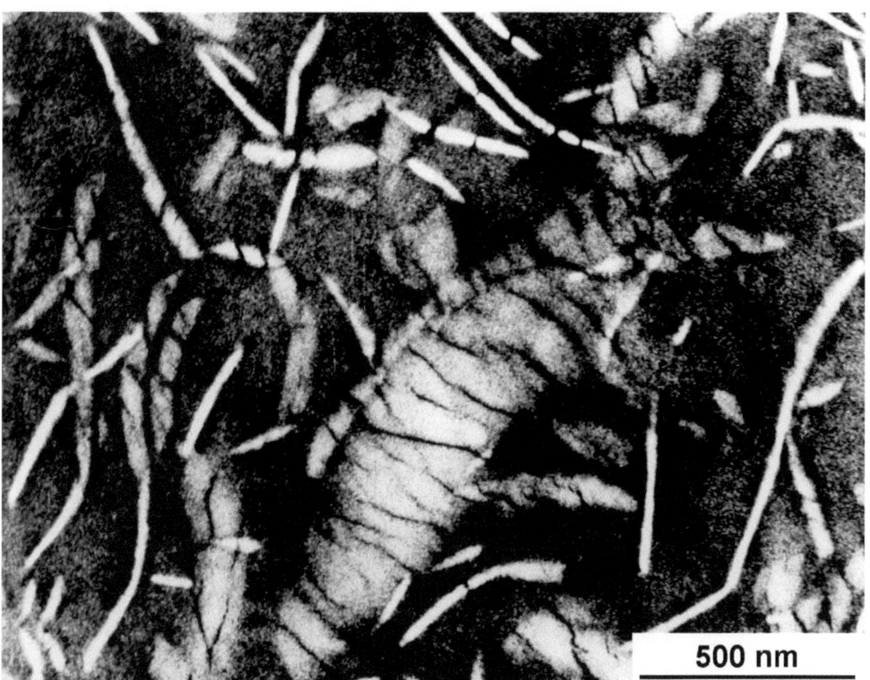

Fig. 16.10. Improved visibility of the shapes and internal structures (defect layers) of lamellae in HDPE in the "edge-on" and in the "flat-on" positions achieved by combining staining with irradiation. (Ultrathin section, TEM; reproduced from [2] with the permission of Hanser)

In polymer blends with crystallisable components, co-crystallisation of macro-molecules of both of the components or demixing (separation) as the melt cools are both possible processes. If the polymers are similar, such as different types of polyethylene, it is difficult to decide between these two processes. One approach that can be employed is to exactly determine the thickness distribution of the lamellae of each component and of the blend, and then to compare these quantitatively [12]; see Fig. 17.18, which refers to blends of LDPE/HDPE.

A more elegant approach is provided by a special application of the technique of melt fixation described above [9, 13]. The idea of this technique is illustrated in the scheme of Fig. 16.11 [9]. The method requires that the polymer blend has components that show a tendency to crosslink during irradiation and that differ in their melting temperatures.

1. If irradiation occurs above the melting temperatures T_s of both components, all parts of the blend are in an amorphous state and are crosslinked during irradiation. After cooling to room temperature and performing selective chemical staining, occasionally some bright domains are made visible (these are possibly pre-existing structures that were present in the melt).

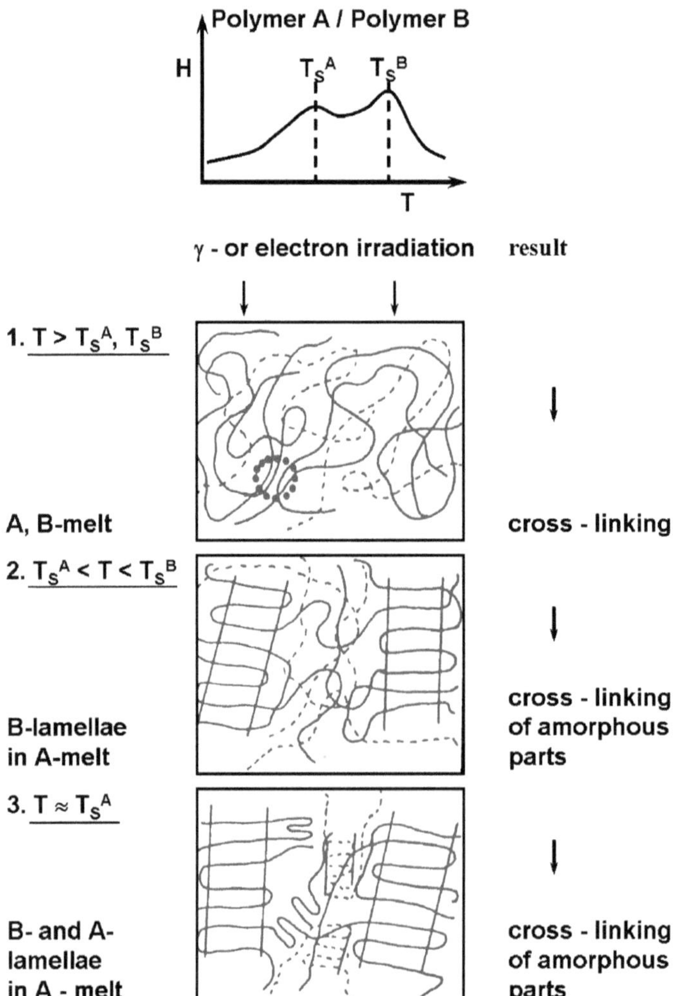

Fig. 16.11. Scheme for investigating structure formation in polymer blends using the effect of irradiation-induced crosslinking and fixation in the melt. (Reproduced from [9] with permission from Marcel Dekker)

2. At a temperature between the melting temperatures of the components, lamellae of the higher-melting component exist in a melt of residual material, which is crosslinked during irradiation. After cooling, only these lamellae are detectable in the TEM.

3. At a temperature near the melting temperature of the lower-melting component, the first few lamellae of this component form, whereas the higher-melting component is crystallised to a large extent.

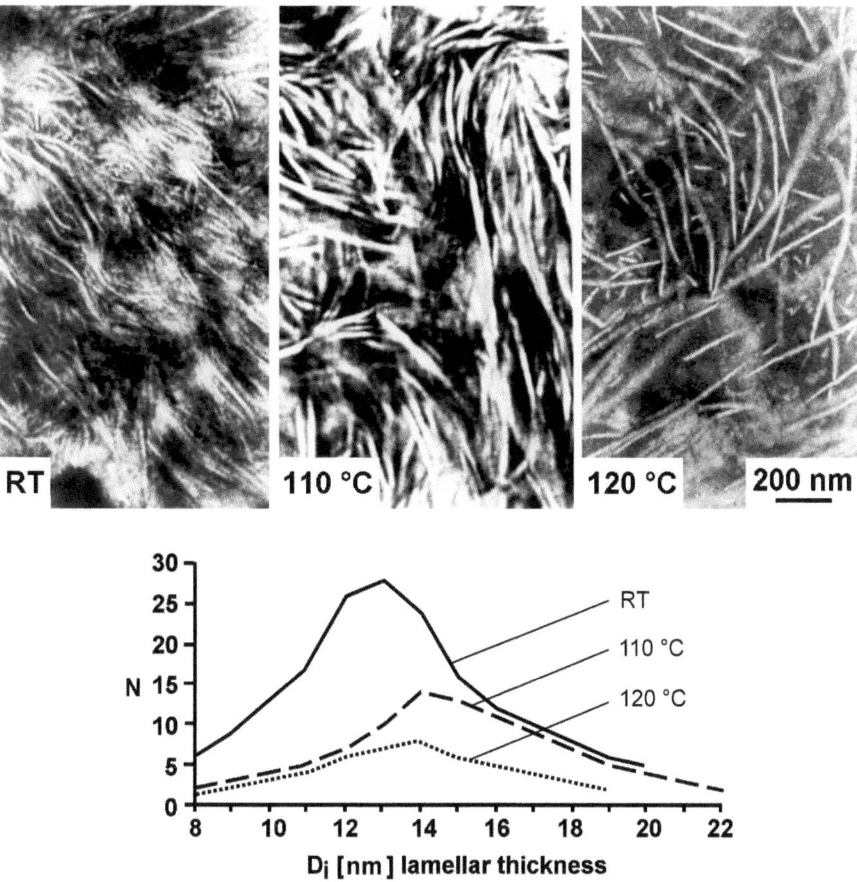

Fig. 16.12. Melting of lamellae in a LDPE/HDPE (80/20) blend with increasing temperature (see Fig. 16.11). *Top*: sequence of electron micrographs showing the lamellae that still exist at the temperatures indicated; *bottom*: corresponding thickness distributions of lamellae. (Reproduced from [9] with permission from Marcel Dekker)

As an example, Fig. 16.12 shows the lamellar morphology of a 80/20 blend of LDPE and HDPE at different temperatures. At a temperature of 120 °C (above the melting temperature of LDPE), only HDPE-type lamellae exist; at 110 °C LDPE-type lamellae appear between the thicker HDPE lamellae. This indicates a tendency towards the separate crystallisation of the unbranched (HDPE) macromolecules in the melt of the branched (LDPE) macromolecules [9]. Distribution curves of lamella thickness reveal this stepwise crystallisation of first the thicker (HDPE-like) lamellae and second (at lower temperatures) the thinner (LDPE-like) lamellae. Micrographs of the sample taken at room temperature only show many lamellae, which are hard to relate to the components.

16.3 Micromechanical Behaviour

All of the features of the semicrystalline morphology play a role as they ultimately determine the mechanical properties that are of technical interest. To understand how the morphology controls macroscopic mechanical properties, it is essential to recognise the micromechanical mechanisms of deformation and fracture. This means that the morphological changes that occur at microscopic and even nanoscopic levels when a load is applied to a sample must be followed. Typical micromechanical mechanisms, such as crazing, shear band formation and fibrillation, are structural transitions that can be explained by microscopic yielding, microcavitation and orientation effects, which are directly linked to the aforementioned morphological features.

In this context, it appears trivial to state that micromechanical mechanisms and the resulting mechanical properties depend strongly on the chemical structure of the macromolecules and the processing history of the polymer. They will also vary with the service (or test) conditions: sample geometry, loading rate, temperature and chemical environment will all influence the response of the polymeric system to an applied load. Therefore, the range of mechanical properties of semicrystalline polymers spans from brittle fracture to highly ductile behaviour, which includes cold drawing and strain hardening; see Fig. 16.13.

16.3.1 Brittle Behaviour

In semicrystalline polymers, brittle behaviour is characterised by fracture at low elongations, as shown in the stress–strain curve (b) in Figure 16.13. The micromechani-

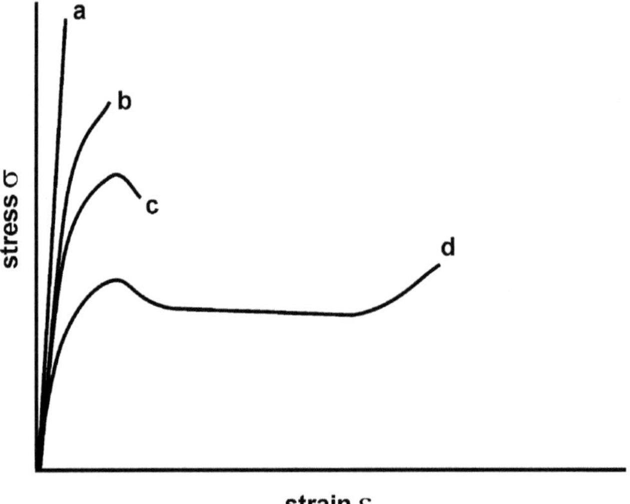

Fig. 16.13. Idealized stress–strain curves that can be recorded for semicrystalline polymers: *a*, oriented fibres of high strength and modulus; *b*, (semi)brittle fracture; *c*, necking, yielding; *d*, cold drawing. (Redrawn after [2])

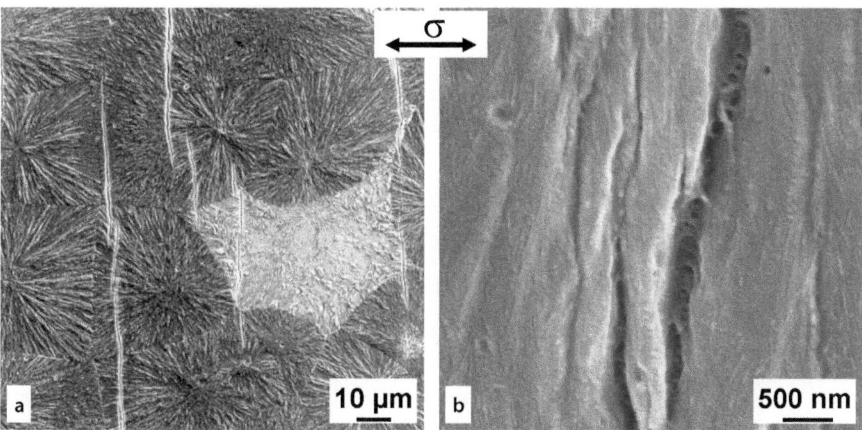

Fig. 16.14a,b. Crazing in α-iPP deformed at −40 °C: **a** overview image indicating that craze propagation is not influenced by spherulitic structures; **b** detail showing craze fibrils. (SEM images after permanganic etching)

cal mechanism that is typically associated with the brittle fracture of semicrystalline polymers is *crazing*. The craze structures that occur depend strongly on the type of brittle behaviour involved. There are different types of brittle fracture that occur under the following conditions.

Brittle Fracture at Low Temperatures

Brittle fracture occurs when a semicrystalline polymer is deformed at temperatures below the glass transition temperature of the amorphous fraction (T_g). The lack of chain mobility limits the ability of the polymer to undergo plastic deformation, which is largely governed by the amorphous phase. This typically leads to embrittlement when ductile semicrystalline polymers are cooled to low service temperatures.

Crazes propagate perpendicular to the direction of the applied load. The craze propagation is not influenced by morphological items like lamellar structures or spherulite boundaries (Fig. 16.14). The structure of a craze in a semicrystalline polymer is similar to that of a craze in an amorphous polymer, like polystyrene (see Figs. 15.5, 15.6).

Brittle Fracture of Materials with Low Molecular Weights

The craze propagation observed in low molecular weight iPP is comparable to the crazing observed at low temperatures (Fig. 16.15a). There is no influence of the lamellar orientations or spherulitic structures on the craze propagation.

Initiation of Brittle Fracture by Morphological Defects

Morphological defects in semicrystalline polymers can be categorised as follows (Fig. 16.16):

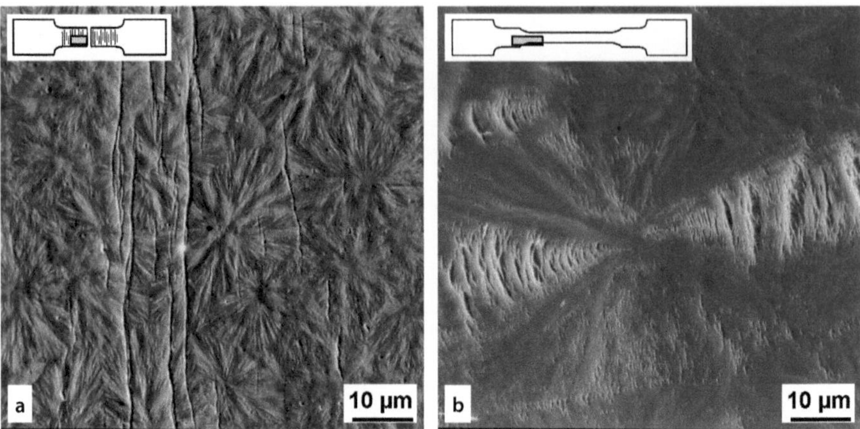

Fig. 16.15a,b. Deformation of α-iPP with different molecular weights: **a** brittle behaviour and craze formation in low molecular weight iPP (230 000 g/mol), showing no influence of morphology on craze propagation; **b** ductile behaviour in higher molecular weight iPP (482 000 g/mol). (SEM images after permanganic etching)

- *Interspherulitic defects*. A weak interface, especially one involving large spherulites, results in brittle behaviour of semicrystalline polymers (Fig. 16.17a).
- *Interlamellar defects*. A lack of tie molecules that link the crystalline lamellae and an absence of entanglements in the amorphous portion leads to brittle behaviour. This defect type occurs at very high degrees of crystallinity (e.g. in PHB) or at low molecular weights. The crack path is clearly determined by the lamellar orientation (Fig. 16.17b).

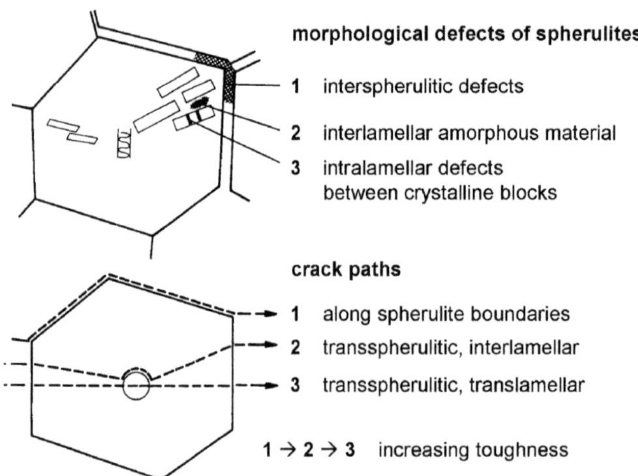

Fig. 16.16. Typical defects in the semicrystalline morphology that may result in brittle fracture; schematic drawing of possible intra- and interspherulitic defects and corresponding crack paths

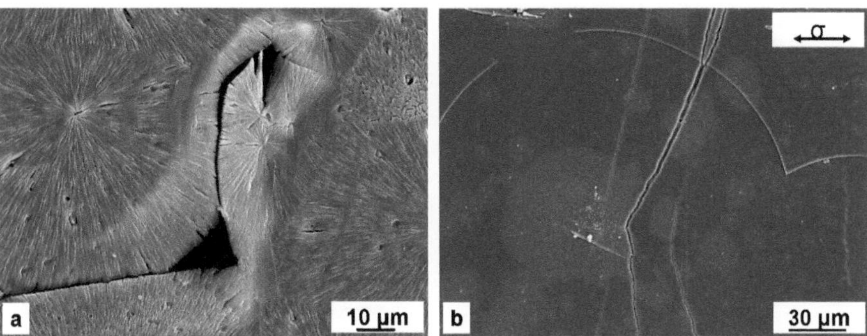

Fig. 16.17. a Interspherulitic defects (weak spherulite boundaries) due to a high degree of crystallinity, undeformed sample; **b** transspherulitic cracks generated during tensile deformation follow the orientation of the lamellae. (SEM images of PHB after chemical etching)

– *Intralamellar defects.* There are several morphological defects of the crystalline portion resulting from polymer-specific factors, such as low molecular weight fractions, chain ends, and chain branching, that can initiate fracture at temperatures above T_g.

Environmental Stress Cracking

This type of brittle fracture occurs when a semicrystalline polymer under external or internal mechanical stresses is exposed to aggressive media. Crazes that are observed in the course of environmental stress cracking are significantly different from crazes in amorphous polymers and from crazes that are generated at low temperatures. They have a very rough internal structure and thicker fibrils [14]. Again, there is no correlation between lamellar or spherulitic structures and craze propagation.

Physical Aging

In the simplest case, shelf storage of a polymer sample may result in physical changes linked to a decrease in the free volume or macromolecular mobility, and which therefore yields embrittlement. Another important damage mechanism is chain scission due to oxidation at higher temperatures (thermooxidation, e.g. during processing) or under the influence of ultraviolet radiation (e.g. outdoor applications). As these mechanisms decrease the molecular weight of the amorphous portion (i.e. decrease the entanglement and the tie molecule density) or crosslink the macromolecules (decreased the mobility of the amorphous fraction), the sample becomes brittle.

16.3.2 Ductile Behaviour

The ductile behaviour of semicrystalline polymers is usually linked to neck formation, cold drawing and strain hardening. In the stress–strain curves (c) and (d) in

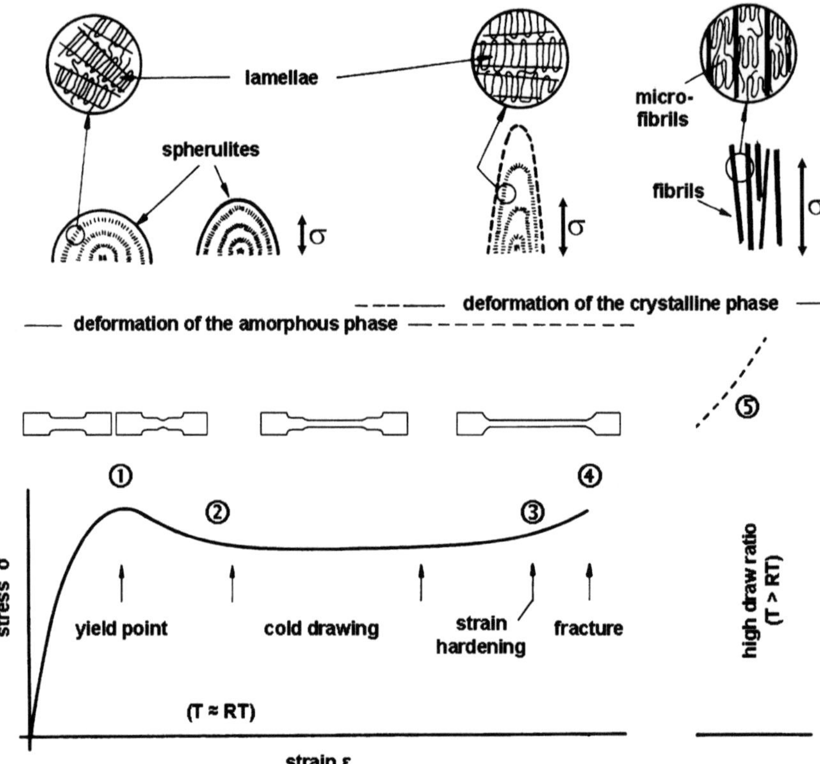

Fig. 16.18. Schematic drawing of the correlation between different stages of deformation and the distortion of spherulites

Fig. 16.13, these events are reflected by a distinct yield point that is followed by a plateau region where the sample is elongated by up to several times the original length. Under certain conditions the plastic deformation will proceed until strain hardening due to ongoing molecular orientation results in a significant increase in the mechanical stress.

All of these escalating phases of plastic deformation proceed along with drastic changes in the semicrystalline structures on all hierarchical levels. Micromechanical mechanisms include conformational changes of macromolecules in the interlamellar (i.e. amorphous) as well as in the lamellar (i.e. crystalline) regions, the destruction and/or reorganisation of the crystalline phase, and the orientation of macromolecules. Electron microscopic investigations enable us to distinguish the following micromechanical processes.

Plastic Deformation of Spherulites

The ductile behaviour of a semicrystalline polymer is associated with the plastic deformation of the spherulites, as shown in the upper part of Fig. 16.18. Starting with

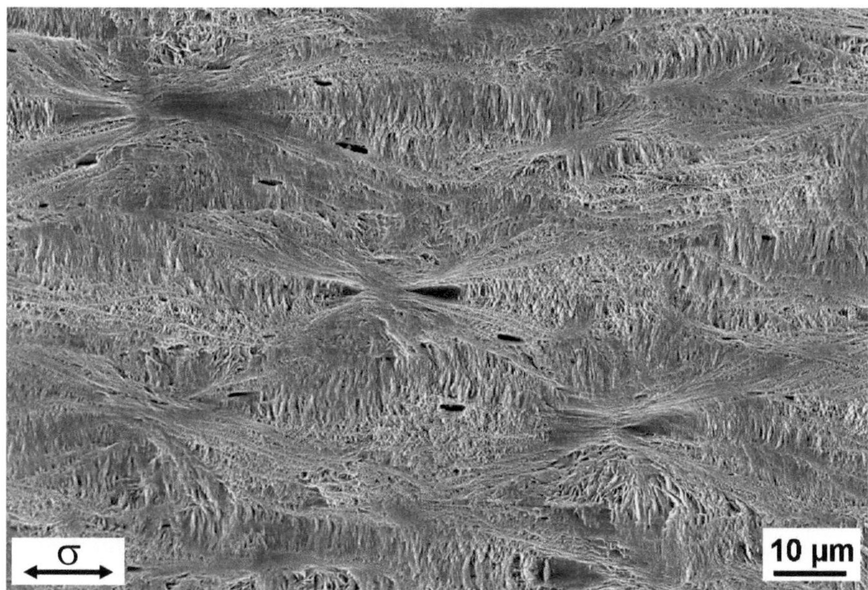

Fig. 16.19. Plastic deformation of spherulites: microvoid formation in spherulites of β-iPP. (SEM image from a deformed tensile bar; permanganic etching)

circular structures, one observes an increasing elongation of the spherulites in the direction of strain with increasing macroscopic deformation. The distortion of the spherulitic texture can be correlated with the deformation state, which is represented by the stress–strain curve in the lower part of Fig. 16.18. An example is given in Fig. 16.19: the deformation of spherulitic structures in β-iPP involves microvoid formation that is connected with pronounced stress whitening of the sample. Here, the intensity of microvoiding depends strongly on the position in the spherulite with respect to the direction of the applied strain. A similar result for α-iPP is presented in Fig. 16.15b.

Deformation at the Lamellar Level

The plastic deformation in the spherulites is not homogeneous. Diverse micromechanical mechanisms are observed in different sectors of the spherulites. In the first place, these processes depend strongly on the orientation of the lamellae with respect to the direction of applied stress. Figure 16.20 shows a simplified classification of the lamellar orientation, which depends on the position inside the spherulite. There are lamellae running parallel to the direction of strain (a), others are aligned perpendicular to the applied load (c), and some are tilted (b). Accordingly, the following types of lamellar deformation can be distinguished: interlamellar slip (pole regions, Fig. 16.20a), lamellar rotation (intermediate regions, Fig. 16.20b), and lamellar separation (equatorial regions, Fig. 16.20c). At the initial stages of deformation, these processes are mainly controlled by the amorphous phase.

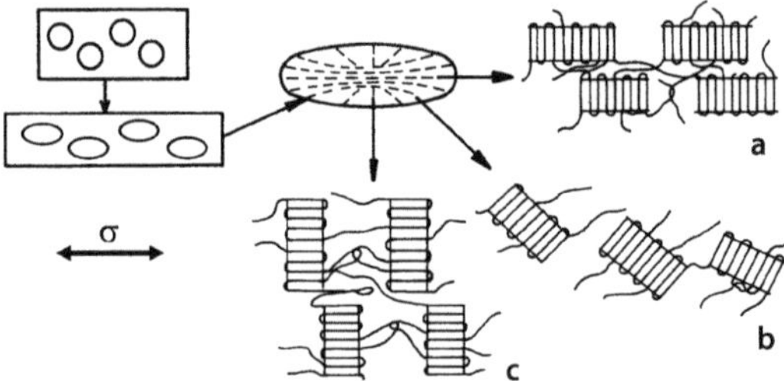

Fig. 16.20a–c. Possible lamellar orientations in the spherulite and resulting micromechanical mechanisms; see text. (After [2], reproduced with the permission of Hanser)

Electron microscopic studies of blown films of stretched HDPE reveal characteristic processes (Fig. 16.21). Starting from an isotropic distribution of lamellae (the electron diffraction pattern in Fig. 16.21a), the lamellae break and rotate in the deformation direction. After $\lambda=6$, more and more lamellae of the original shape have disappeared and are transformed into microfibrils (see Fig. 16.18) [2,15].

As the deformation proceeds, the crystalline regions also undergo dramatic changes. Different processes have been proposed to explain the transitions of the crystalline phase, based on electron microscopic investigations: the breaking of lamellae into mosaic blocks, tilting, unfolding (destruction of crystalline order), and reorganisation into a highly oriented fibrillar structure [16–18] (see summary in [2]).

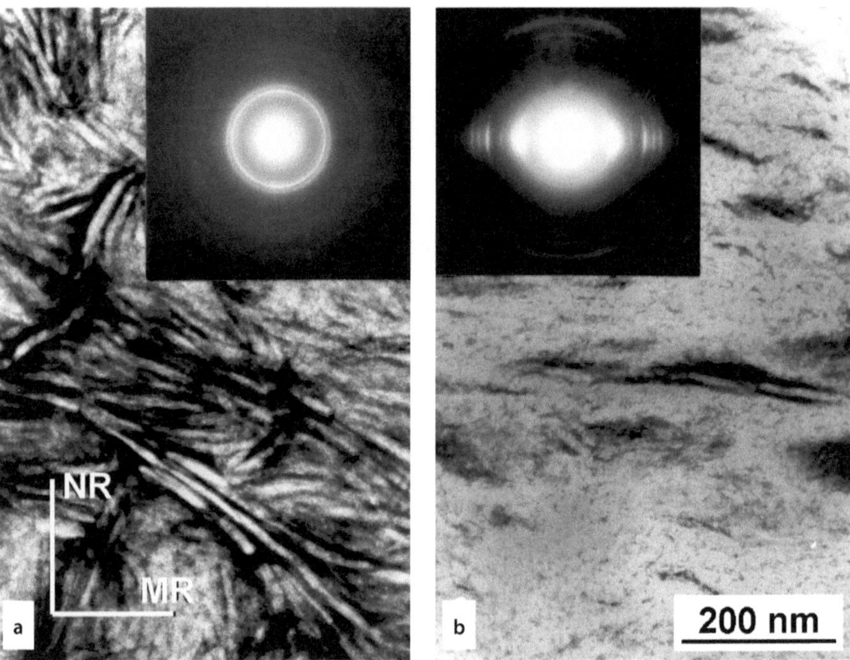

Fig. 16.21. TEM images of HDPE blown films at different stages of deformation: **a** isotropic distribution of lamellae in the undeformed sample; **b** at $\lambda=6$ the original lamellae have disappeared and have turned into microfibrils)

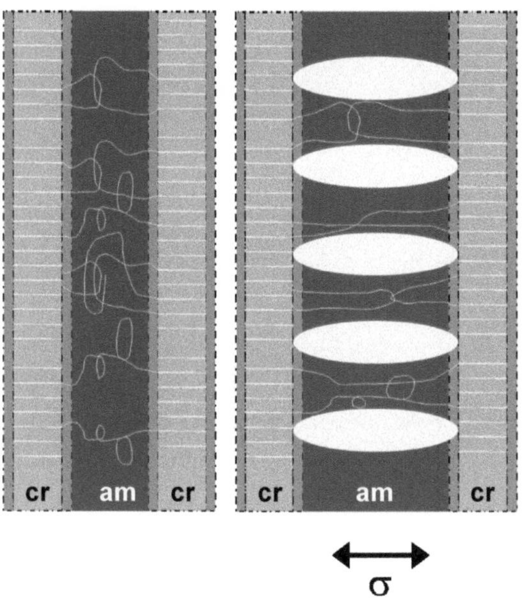

Fig. 16.22. Lamellar separation leading to craze-like structures; loading direction is perpendicular to lamellar orientation (*cr*, crystalline lamellae; *am*, amorphous interlamellar regions)

Craze-Like Phenomena at Plastic Deformation

In contrast to the crazes that occur with brittle fracture, there are craze-like phenomena that occur during plastic deformation that are triggered and controlled by semicrystalline micro- and nanostructures. These are connected to the lamellar separation processes that are observed at initial stages of deformation and are localised in the interlamellar amorphous regions. As in conventional crazes, there is a coexistence of microscopic voids and tiny fibrils that bridge the craze. However, in this case the craze formation and propagation are geometrically limited by the presence of the crystalline lamellae. In other words, the crazing phenomena are controlled by the actual nanostructure and are highly localised but homogeneously distributed over a large sample volume. The processes of lamellar separation, microvoid formation and fibrillation are illustrated in Fig. 16.22, and Fig. 16.23 shows examples for iPP.

Formation of Chevron Patterns

Chevron structures are formed in different polymers that have a lamellar morphology consisting of alternating soft and hard components, such as block copolymers (e.g. some SBS-triblock copolymers, see Chap. 19) and semicrystalline polymers [19]. At the initial stages of plastic deformation, lamellae that are aligned perpendicular to the straining direction undergo collective breaking and twisting processes; see Fig. 16.24. Chevron patterns are found in deformed polyethylenes [2, 20], in β-iPP [19, 21, 22], and in syndiotactic polystyrene deformed at higher temperature [19].

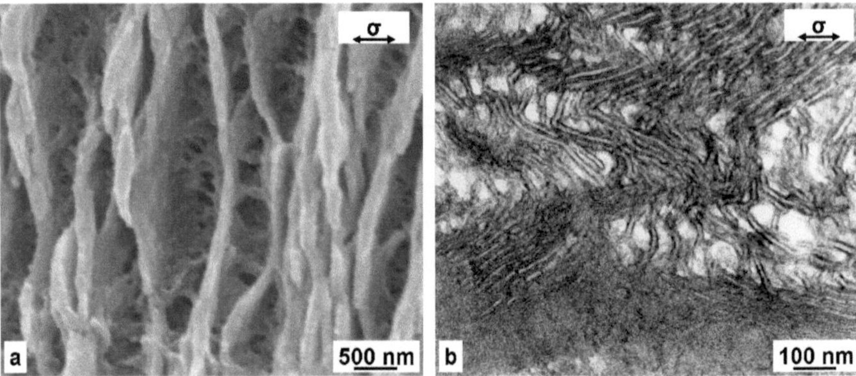

Fig. 16.23a,b. Lamellar separation process showing microvoid formation and fibrillation of the amorphous interlamellar region of iPP: **a** SEM image of deformed α-iPP after permanganic etching; **b** TEM image of deformed β-iPP; ultrathin section after RuO_4 staining

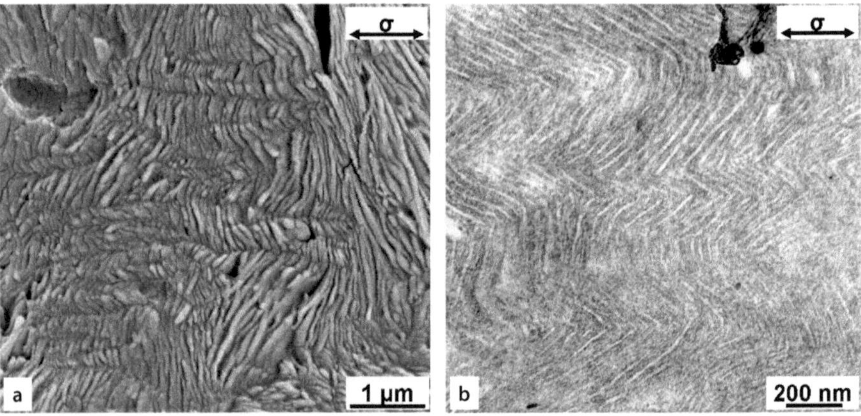

Fig. 16.24a,b. Chevron structures: **a** SEM image of β-iPP deformed at 21 °C (permanganic etching); **b** TEM image of the same situation. (Ultrathin section, RuO_4 staining)

Strain Hardening and Self-Reinforcement

The cold drawing process that is represented by the plateau in the stress–strain curve in Fig. 16.18 continues until point 4 is reached. The moment when the stress–strain curve starts to ascend is usually identical to the macroscopic completion of the necking process. That means that the lateral contraction that started at a certain point in the dog-bone-shaped tensile bar reaches both ends of the narrowed region between the shoulders. Further plastic deformation must lead to an amplified orientation of the macromolecules that completes the destruction of the remaining crystallites and creates a completely reorganised morphology that is characterised by microfibrils, as shown in Fig. 16.25. When a critical stress is reached, the sample will break (point 4 in Fig. 16.18). Provided the material has sufficient molecular weight (i.e. sufficient con-

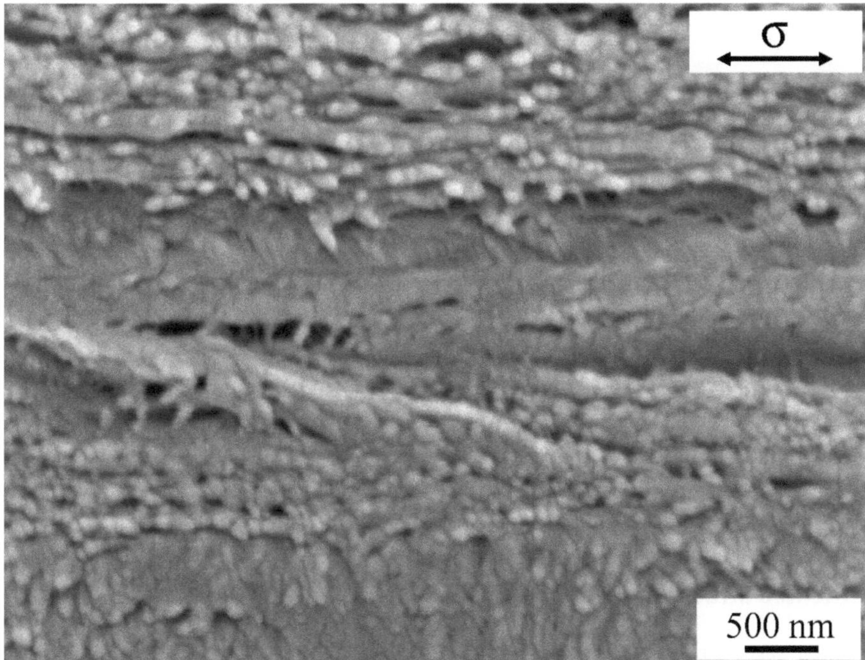

Fig. 16.25. Morphological changes in β-iPP at higher elongations: transition to a fibrillar structure at $\varepsilon \approx 600\%$. (SEM image from a deformed tensile bar after permanganic etching)

centrations of entanglements and tie molecules) and that the deformation takes place at low deformation rates or elevated deformation temperatures ($T_\mathrm{m} > T_\mathrm{defo} > \mathrm{RT}$), respectively, this process results in a fibrous structure consisting of macromolecules that are highly oriented in the drawing direction. The dense packing due to the parallel arrangement then resembles a crystalline morphology that, again, is interrupted by defects due to chain-end effects and other polymer-specific factors mentioned above. Since the load-bearing elements are the chains themselves, the mechanical load now largely acts on the covalent bonds of the backbone of the macromolecule. Thus, a high strength and a high Young's modulus in one direction (i.e. the straining direction, often also referred as the machine direction, MR) are achieved.

If the stretching of such a polymer is stopped before the ultimate elongation is reached (point 5 in Fig. 16.18), the resulting sample will maintain the integrity of its highly oriented structure and can be considered a self-reinforced material. Such fibres can be worked into sheets, semi-finished products, or even more sophisticated shapes by weaving and/or (subsequent) hot compaction (see Sect.24.2).

16.3.3 High-Strength Fibres and Parts

As discussed in the section above, the stiffness and strength of a semicrystalline polymer can be enhanced by creating highly oriented structures. The aim is to align most

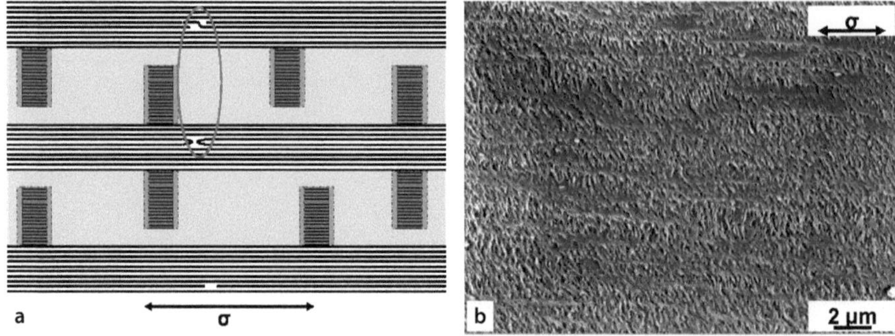

Fig. 16.26a,b. Morphology of semicrystalline polymers oriented by melt processing: **a** schematic drawing of a shish-kebab structure; **b** morphology of a deformed miniature polypropylene sample. (SEM image after permanganic etching)

of the macromolecules parallel to a given direction. In general, there are two more processes that lead to products with high macromolecular orientations.

Oriented Structures by Melt Processing

Under certain processing conditions resulting in high shear of the melt (e.g. a relatively low melt temperature, high extrusion pressure, high molecular weight of the polymer), a high degree of orientation of the macromolecules can be achieved. The orientation can be retained via rapid solidification (i.e. rapid cooling, an abrupt increase in pressure). The crystallisation of such an oriented melt leads to a specific morphology known as a shish kebab structure, as shown in Fig. 16.26 (see also Fig. 14.8). The "sticks" (the "shish") consist of chains that are elongated in the processing direction, forming a crystalline extended chain structure. These crystals that are induced by extensional flow of the melt act as nuclei for the epitaxial growth of lamellar crystals (the "kebabs"). When the melt is cooled down, lamellar crystals form that have their layer normals parallel to the processing direction [23].

There is usually a gradient of shear force over the cross-section of the melt, yielding a corresponding gradient in the orientations of the macromolecules. Very often one observes a thin skin layer of highly oriented material that does not influence the mechanical properties of the sample, since the oriented fraction is marginal compared to the overall material volume. The situation changes when the parts become smaller and smaller. In very small samples, such oriented layers dominate the mechanical properties (see Sect. 24.4).

Oriented Structures by Solution or Gel Spinning

Oriented polymer fibres can be produced by spinning a liquid polymeric system with very high viscosity, i.e. a concentrated polymer solution or gel. The most successful technical application is the gel spinning of ultrahigh molecular weight polyethylene [25–28]. The chain length and the entanglements of the polymer in the solution influence the properties of the gel, the process of fibre formation, and therefore the resulting fibre strength. The degree of orientation is restricted by entanglements. For UHMWPE fibres, the degree of parallel orientation can be up to 95% and the degree of crystallinity up to 85%. Such materials are marketed under different tradenames and are used, for instance, in bullet-proof vests, high-strength lines and ropes and other high-performance products.

One special case of solution spinning is the electrospinning process (described in Sect. 24.3). It has been shown that the macromolecules in the micro- and nanofibres produced by electrospinning are highly oriented. Moreover, their small size leads to new micromechanical mechanisms that can be used to enhance the toughness of polymers.

16.4 Additional Examples of Semicrystalline Polymers

Polyamides (PA)

The term "polyamides" encompasses a group of condensation products of amines and carboxylic acids containing amide groups. Their chemical structures can be deduced by the numbers that are added to the abbreviation PA in each case. The most common representatives, marketed under the tradenames Nylon (PA66) and Perlon (PA6), are fibre products that differ in the constitutions of their monomers. Besides synthetic fibres, there are a growing number of technical applications of the compact

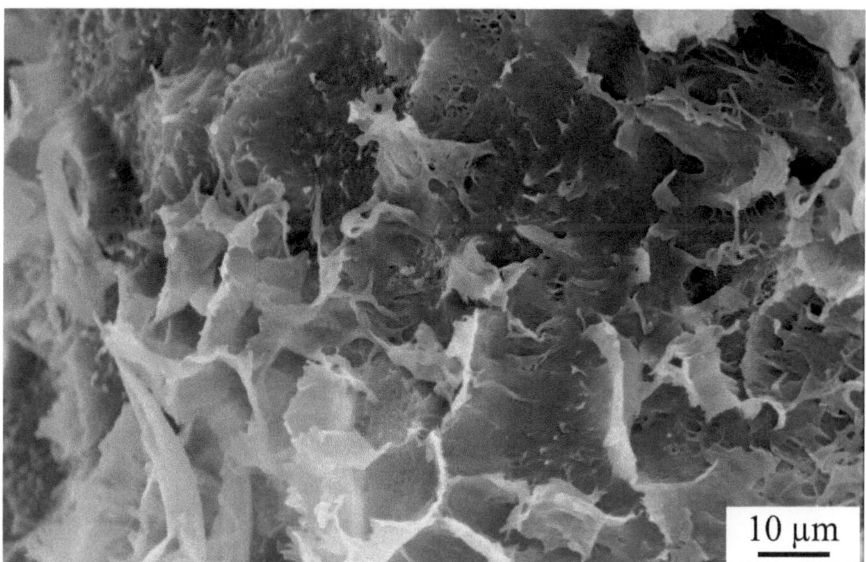

Fig. 16.27. Fracture surface of polyamide (PA6) after environmental stress cracking, exhibiting craze-like deformation zones with very rough fibrils

material. Most polyamides (except, for instance, PA6I) are semicrystalline possessing the features of morphological hierarchy discussed in Chap. 14 (Fig. 14.9). Due to their relatively high capacity to take up water, their mechanical properties additionally depend on the humidity: the uptake of water by the amorphous phase yields a softening effect, resulting in enhanced toughness. Figure 16.27 shows the fracture surface of a PA6 that was produced due to environmental stress cracking after exposure to water and different salts. In the lower part of the image, the material exhibits ductile behaviour.

Syndiotactic Polystyrene (sPS)

Syndiotactic polystyrene is a new type of PS that was first marketed by the DOW Chemical Company in the late 1980s. It is produced using a new generation of catalysts (i.e. single-site, metallocene). As a result, sPS is a semicrystalline polymer that exhibits both a lamellar morphology and a spherulitic superstructure (Fig. 16.28). It has excellent dimensional stability and good chemical resistance. Since the glass transition temperature of the amorphous phase is significantly higher than room temperature ($T_g{\approx}100\,°C$), the material is brittle under normal conditions. This situation is comparable to the mechanisms of brittle fracture at low temperatures that are explained in Sect. 16.3.1. Accordingly, a sample that is deformed at room temperature will show crazes that are not influenced by the lamellar orientations inside the spherulites (i.e. transspherulitic fracture, Fig. 16.29a). In some cases, the fracture surfaces show spherical calottes that are produced by fractures along the borders of spherulites, as demonstrated in Fig. 16.29b.

Following the discussions in Sects. 16.3.1 and 16.3.2, the material should become ductile when the deformation experiment is performed at elevated temper-

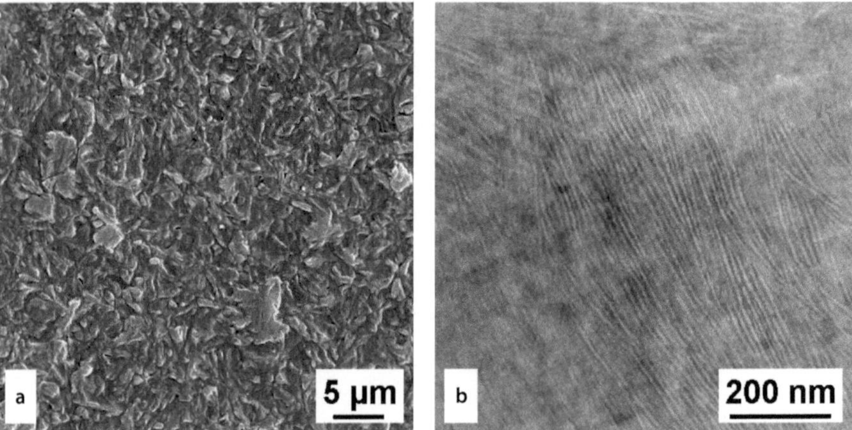

Fig. 16.28a,b. Morphology of syndiotactic polystyrene (sPS): **a** spherulites (SEM image after permanganic etching); **b** lamellar structure (TEM image of an ultrathin section after RuO$_4$ staining)

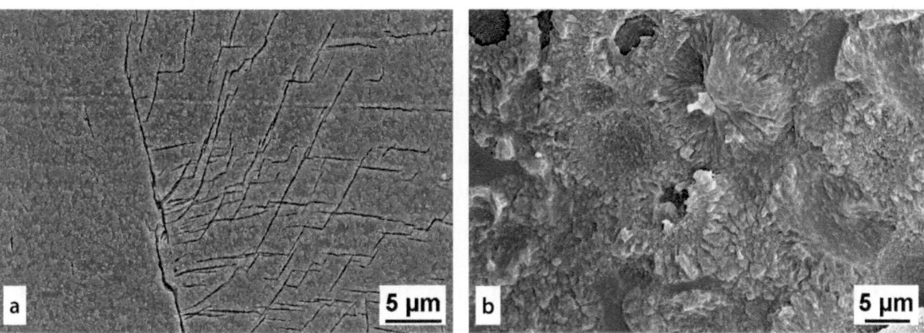

Fig. 16.29a,b. Brittle deformation of sPS deformed at RT: **a** crazing showing no influence of spherulitic superstructures on the craze propagation; **b** calottes on the fracture surface indicate interspherulitic crack propagation. (SEM images)

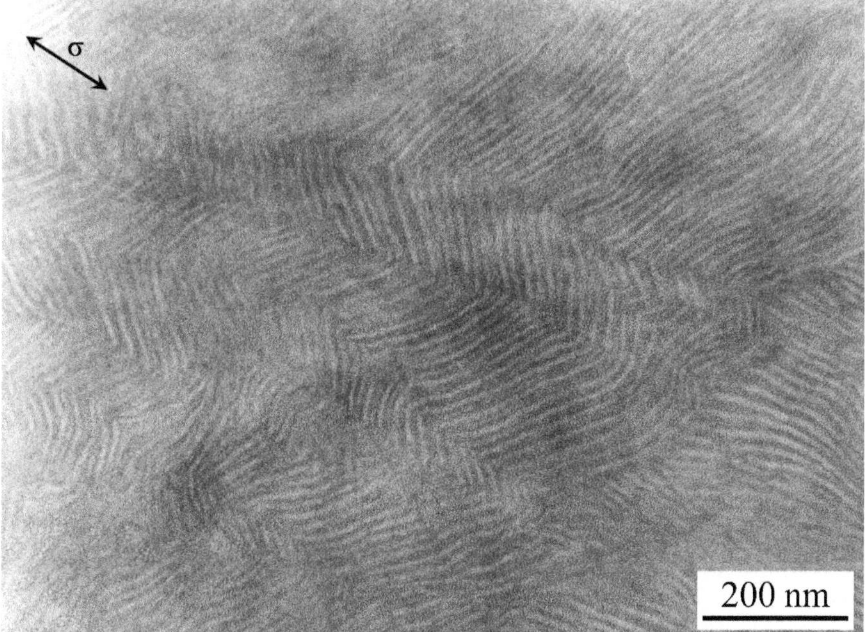

Fig. 16.30. TEM image of sPS deformed at 110 °C showing chevron morphology. (Ultrathin section, RuO_4 staining; from [19], reproduced with permission from Taylor & Francis)

atures, i.e. above the T_g of the amorphous phase. Indeed, during tensile tests at 110 °C sPS develops a chevron morphology that indicates a complete change in the micromechanical behaviour, connected to a macroscopic brittle-to-ductile transition (Fig. 16.30, [19]).

Polyesters for Biomedical Applications: PHA and PLA

The controlled degradation of polymers is sometimes desired for biomedical applications and environmental purposes. For instance, poly(hydroxy alcanoates) (PHAs) and polylactides (PLAs) are families of biodegradable and biocompatible polyesters. It was stressed in the sections above that the degree of crystallinity and the actual morphology control the important physical properties of semicrystalline polymeric materials. In the case of degradable biopolymers, the degradation rate is another important parameter that is also controlled by the semicrystalline structure [29, 30]. The crystallinity and the resulting morphology are usually controlled by using either different proportions of stereoisomers (enantiomers, e.g. L-lactide and D-lactide) of the same monomer or a defined ratio of comonomers of related polyesters. The latter can be illustrated using an example. Poly(hydroxy butyrate) (PHB) is the most elementary representative of the poly(hydroxy alkanoates) (PHAs). PHB is able to develop high degrees of crystallinity, and it forms a very coarse spherulitic structure. Additionally, the glass transition temperature of the amorphous phase is close to RT. Therefore, the pure PHB is extremely brittle (see Fig. 16.17). Moreover, a high degree of crystallinity yields low degradation rates since it has been shown that hydrolytic degradation (cleavage of ester bonds) preferentially occurs in the amorphous regions. The mechanical properties can be significantly enhanced through the incorporation of hydroxy valerate (HV) comonomers. The glass transition temperature gradually decreases with increasing HV content, yielding greater mobility of the amorphous phase. Additionally, the degree of crystallinity decreases with increasing HV content. There is a distinct brittle-to-ductile transition due to comonomer incorporation. Unfortunately, the addition of HV comonomers further decreases the rate of biodegradation. The semicrystalline morphology of P(HA-co-HV) with a comonomer ratio of 70:30 is shown in Fig. 16.31.

Fluoropolymers: PTFE, PVDF

The most important polymeric substances containing fluorine atoms are polytetrafluoroethylene (PTFE, trademark Teflon) and polyvinylidenefluoride (PVDF, trademark Kynar). The two of them show outstanding resistance to corrosive media. PTFE is known for its low coefficient of friction and its relatively high melting point, which makes this polymer a candidate for high-temperature applications. PVDF has a piezoelectric coefficient that is much higher than that of other polymers. Of course, the semicrystalline structure determines mechanical properties. Figure 16.32 demonstrates the spherulitic morphology of PVDF.

Polyurethanes (PU)

The collective term polyurethane encompasses polymers where the repeating units are coupled by urethane links. There are a multitude of variations depending on the starting substances and processing routes. Besides the most common applications, PU foams, microfoams and varnishes, there are also polyurethanes that are used

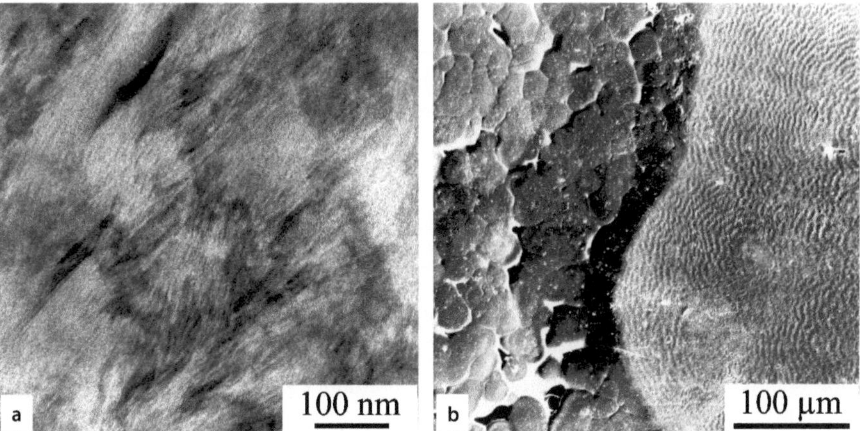

Fig. 16.31a,b. Morphology of P(BH-*co*-HV) (70:30): **a** lamellae of P(BH-*co*-HV) becoming visible after staining for 60 h in RuO_4 vapour (UDS, TEM); **b** spherulitic structures generated at different isothermal crystallisation temperatures (*left hand side*: smaller spherulites grown at 25 °C, *right hand side*: giant banded spherulites formed at 90 °C). (SEM images after methylamine etching)

in their compact form. They can exhibit semicrystalline morphologies with typical lamellar and spherulitic features. One beautiful example is given in [2], where the lamellae form a unique spiral pattern (see Fig. 20.7). Figure 13.12 (see Chap. 13) shows PU spherulites that have become visible after electron beam irradiation in the TEM.

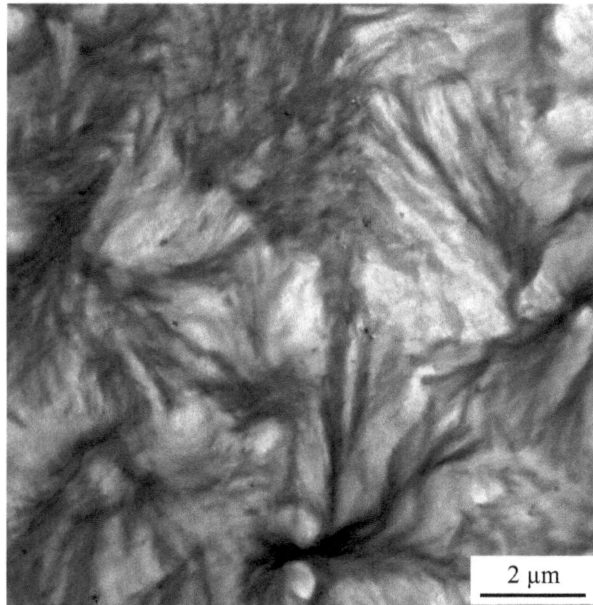

Fig. 16.32. Spherulites of PVDF. (TEM image, ultrathin section without staining)

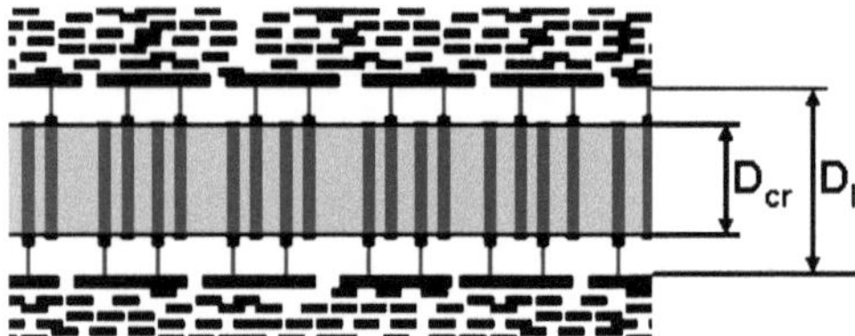

Fig. 16.33. Schematic drawing of the construction of crystalline lamellae from interdigitating side chains

Block Copolymers with Crystallisable Components

A special case of crystallisation occurs when a polymeric system is formed, for instance, from graft polymers consisting of an amorphous main chain and crystallisable side chains, or from macromonomers that contain a polymerisable backbone and crystallisable side chains [31–35]. The formation of the morphologies of such polymers can be described by three steps. Firstly, a microphase separation process orders the components so that the crystallisable parts are brought close to each other. Secondly, the crystallisable side chains start to order themselves within this confinement, forming crystalline regions. If the microphase separation yields a lamellar morphology, crystalline lamellae are formed, for instance by the interdigitation of crystallisable side chains, as shown in Fig. 16.33. This process resembles the primary crystallisation in conventional semicrystalline polymers: it will stop when the available volume is filled or if the molecular movements are frozen by undercooling or crosslinking. Thirdly, there is a secondary crystallisation, i.e. the incorporation of crystallisable material into the crystalline regions that were formed earlier; this results in lamellar thickening [31, 32].

To give just one example: a macromonomer constructed from a Bis-GMA backbone and oligomeric side chains of L-lactide or its copolymers will form lamellar structures by microphase separation [36]. The oligomeric side chains are concentrated in their own environment, which is confined by acrylic backbone components.

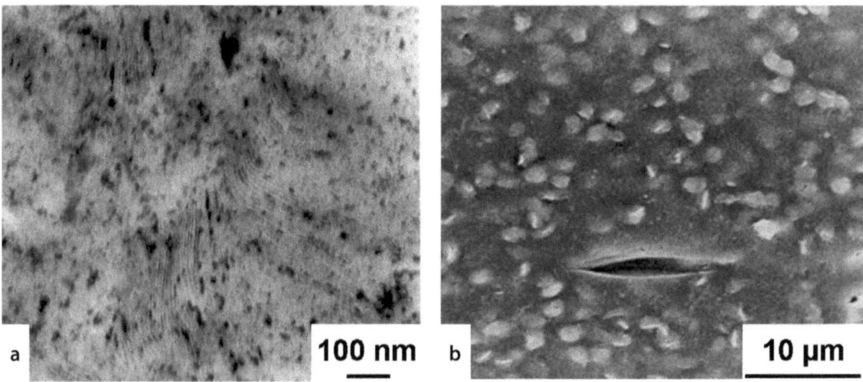

Fig. 16.34a,b. Bis-GMA with side chains of oligo(L-lactid-*co*-valerolactone): **a** lamellae, TEM image of a UDS after two-step staining procedure using formic acid (first) and OsO$_4$(second); **b** spherulitic superstructure. (SEM image, fracture surface, acetic acid etching)

The side chains are forced to interdigitate and to crystallise. The crosslinking reaction of Bis-GMA freezes these structures. Figure 16.34 shows the morphology of just such a system.

References

1. Editor (2005) Kunststoffe 95:35
2. Michler GH (1992) Kunststoff-Mikromechanik: Morphologie, Deformations- und Bruchmechanismen. Hanser Verlag, München
3. Bassett DC (2005) The morphology of crystalline polymers. In: Michler GH, Baltá Calleja FJ (eds) Mechanical properties of polymers based on nanostructure and morphology. CRC Press (Taylor and Francis), Boca Raton, FL, Chap 1, p 3
4. Bassett DC (1981) Principles of polymer morphology. Cambridge University Press, Cambridge
5. Bassett DC (1984) CRC Crit Rev Sol State Mater Sci 12:97
6. Hall IH (ed) (1984) Structure of crystalline polymers. Elsevier, London
7. Zhou JJ, Liu JG, Yan SK, Dong JY, Li L, Chan CM, Schultz JM (2005) Polymer 46:4077
8. Michler GH, Steinbach H, Bühler K (1988) Proc 12 Tagung Elektronenmikroskopie, Dresden, 18–20 Jan 1988, p 395
9. Michler GH (1996) J Macromol Sci Macromol Phys B35:329
10. Grubb DT (1974) J Mater Sci 9:1719
11. Michler GH, Gruber K, Steinbach H (1982) Acta Polym 33:550
12. Michler GH, Steinbach H, Hoffmann K (1986) Acta Polym 37:289
13. Michler GH, Naumann I, Steinbach H (1992) Prog Colloid Polym Sci 87:16
14. Michler GH (1992) Kunststoff-Mikromechanik: Morphologie, Deformations- und Bruchmechanismen. Hanser Verlag, München, Sect 8.3.2, p 205
15. Hofmann D, Geiß D, Janke A, Michler GH, Fiedler P (1990) J Appl Polym Sci 39:1595
16. Kajiyama T, Okada T, Takayanagi M (1974) J Macromol Sci Phys B9:35
17. Peterlin A (1981) J Macromol Sci Phys B19:401
18. Krug H, Karbach A, Petermann J (1984) Polymer 25:1687
19. Krumova M, Henning S, Michler GH (2006) Phil Mag 86:1689
20. Michler GH, Marinow SL, Naumann I (1989) Plaste Kautschuk 36:432
21. Henning S, Adhikari R, Michler GH (2004) Macromol Symp 214:157
22. Henning S, Michler GH (2005) In: Michler GH, Baltá Calleja FJ (eds) Mechanical properties of polymers based on nanostructure and morphology. CRC Press (Taylor and Francis), Boca Raton, FL, Chap 7, 245
23. Ania F, Baltá Calleja FJ, Bayer RK, Tshmel A, Naumann I, Michler GH (1996) J Mater Sci 31: 4199
24. Hine PJ, Ward IM, Jordan ND, Olley R, Bassett DC (2003) Polymer 44(4):1117
25. Barham PJ, Keller A (1985) J Mater Sci 20(7):2281
26. Smith P, Lemstra PJ (1980) J Mater Sci 15(2):505
27. Smook J, Pennings AJ (1984) J Mater Sci 19(1):31
28. Ivankova E, Vasilieva V, Myasnikowa L, Marikhin V, Henning S, Michler GH (2002) Epolymers 044:1
29. Pitt CG (1992) In: Vert M (ed) Biodegradable polymers and plastics; the proceedings of the Second International Scientific Workshop on Biodegradable Polymers and Plastics (Montpellier, 1991). Royal Society of Chemistry, Cambridge, p 7
30. Doi Y (1992) In: Vert M (ed) Biodegradable polymers and plastics; the proceedings of the Second International Scientific Workshop on Biodegradable Polymers and Plastics (Montpellier, 1991). Royal Society of Chemistry, Cambridge, p 139
31. Beiner M, Huth H (2003) Nature Mater 2:595
32. Hempel E, Budde H, Höring S, Beiner M (2006) J Non-Cryst Solids 352:5013
33. Al-Hussein M, de Jeu WH, Vranichar L, Pispas S, Hadjichristidis N, Itoh T, Watanabe J (2004) Macromolecules 37:6401
34. Chen W, Wunderlich B (1999) Macromol Chem Phys 200:283
35. Inomata K, Nakanishi E, Sakane Y, Koike M, Nose T (2005) J Polym Sci B Polym Phys 43:79
36. Sandner B, Steurich S, Gopp U (1997) Polymer 38(10):2515

17 Polymer Blends

In polymer blends, improved properties can be only realised if the blend exhibits optimum morphology. Therefore, the study and control of morphology by electron microscopy is an important task in polymer research as well as in the plastics industry. The wide variation in morphology exhibited by polymer blends is illustrated using various preparative and staining techniques and through inspection by SEM, TEM, and AFM. These techniques reveal not only the polymer phase distribution but also many structural details of the components and occasionally of the interface between the polymers. Micromechanical deformation processes are revealed by studying deformed specimens in TEM and in a very impressive way by AFM. Finally, some examples of blends of amorphous polymers, of amorphous and semicrystalline polymers, and of semicrystalline polymers are shown along with different preparation techniques.

17.1 Overview

Polymer blends consist of two or more different polymers. They are called "blends" because they are mainly created using a process called *blending*. However, other terms, such as "polymer mixtures" or "polymer combinations", are also used to describe them. The word "composite" should only be applied to polymers with inorganic components, e.g. particle fillers or glass- or carbon-fibres. The original reasons for preparing polymer blends were (i) to reduce costs by combining high-quality polymers with cheaper materials (although this approach is usually accompanied by a drastic worsening of the properties of the polymer) and (ii) to create a polymer that has a desired combination of the different properties of its components (such as the stiffness and ductility of rubber-modified polymer blends). This idea is illustrated by curve (a) in Fig. 17.1, which represents the simple combination of a property of polymer A with the corresponding property of polymer B for various compositions (this curve is termed the *additivity line*). Curve (c) shows an improvement in this property – i.e. it rises above the level of the additivity line (a) – within a special composition range (note that this is usually accompanied by a decrease in the property below the additivity line within another composition range). Now consider curve (d). This illustrates a synergistic effect that produces a property that is better than the properties of either of the polymer components; in other words, the blend with this composition has special new properties.

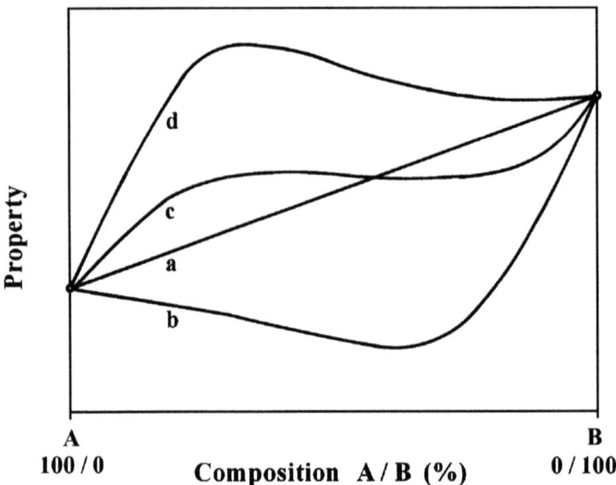

Fig. 17.1. Variation in mechanical properties with blend composition for blends of two polymers A and B (different cases *a, b, c, d*; see text)

The behaviour associated with curve (d) is an important aim of modern polymer research, and one that requires polymers with defined morphologies to lead to optimised or new microscopic property effects [1, 2].

Curve b in Fig. 17.1 illustrates the normal case of the simple mixing of different polymers. Usually, since different polymers are incompatible, large "demixing morphologies" appear, which are responsible for property deterioration (e.g. drastically reduced strength, ductility, toughness, or transparency). Therefore, improving compatibility between the different polymers and optimising the morphology are the main issues to address when producing polymer blends. Several processes are used to enhance the compatibility and connection between the polymeric phases:

–   Chemical connection by grafting (chemical coupling of macromolecules of one polymer onto those of the other polymer, e.g. grafting of SAN molecules onto rubber in ABS blends)
–   Mixing with compatibilisers, such as block copolymers
–   Chemical coupling during processing, e.g. by reactive blending in an extruder.

The results of these processes are reflected in the morphology. From a theoretical point of view, the compatibility between polymers is clearly thermodynamically defined [1–3]. However, from a practical point of view, a good indicator of the degree of compatibility is the heterogeneity of the polymer system (e.g. how the sizes of the dispersed domains depend on the processing conditions), as assessed via an electron microscopic inspection. It is well established that improved properties can be only realised if the blend system exhibits an optimised morphology. Therefore, studying and controlling the morphology are the main tasks to perform when creating polymer blends, not only in polymer research but also in the plastics industry. One special group of polymer blends, rubber-modified polymers, are particularly important and so they are discussed separately in Chap. 18.

17.2 Morphology

The morphology of the polymer blend is determined by the incompatibility of the polymer components (which causes phase separation) and the processing (e.g. intensity of blending, shear forces during mixing, viscosities, temperature). Usually, the minor component (which is typically present at contents lower than 30–40 vol%) is dispersed in the major component, which forms the matrix. At a composition of around 50:50 interpenetrating networks can appear.

Coarse structures such as larger particles in a matrix can easily be studied at low temperature (brittle) fracture surfaces in the SEM, since the fracture path follows the shape of the particle (see Fig. 17.6a,b). Here, the sizes, shapes and spatial distribution of the particles are easily detected, but smaller details in the matrix or inside the particles are not visible. One component may be chemically etched on a smooth surface, revealing the structure of the blend in SEM observations. Figure 17.2 shows the sizes, shapes and internal morphology of the rubber particles in the PP matrix in a PP/rubber blend. Chemical etching can reveal not only the phase distribution but also structural details of the components (matrix morphology, internal structures of particles, etc.). Smooth surfaces also reveal their morphologies in AFM tapping mode because the microhardnesses of the components often differ. In Fig. 17.3 a blend of SBS block copolymer (matrix) and PS

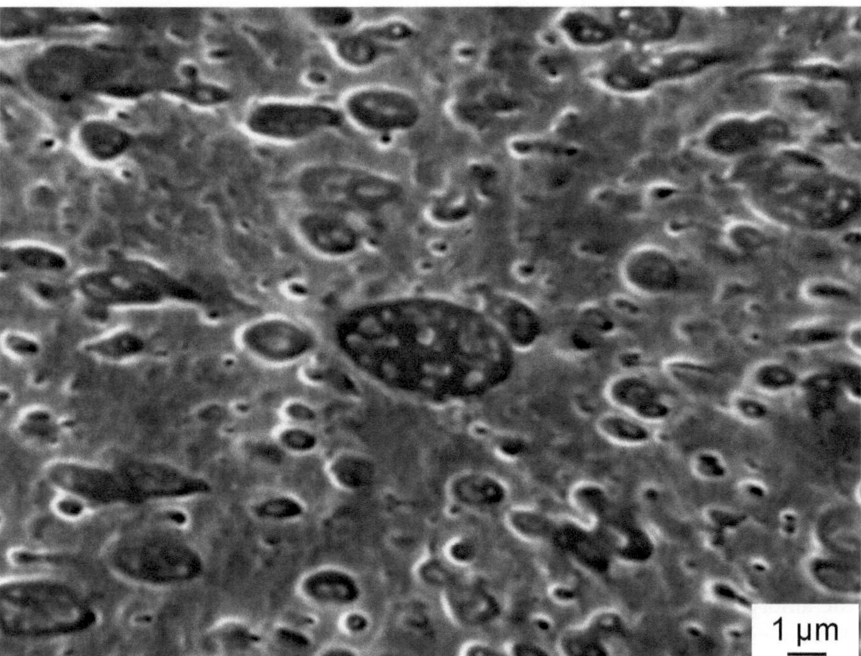

Fig. 17.2. Smooth surface of a PP/rubber blend after chemical etching of the rubber particles (permanganic etching; SEM micrograph)

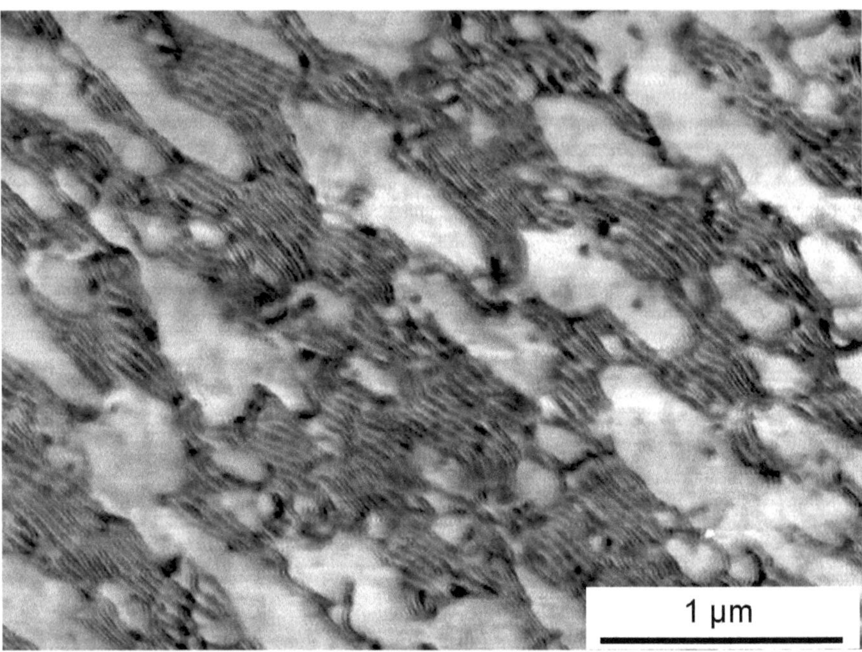

Fig. 17.3. Smooth surface of an SBS-block copolymer/PS blend observed in AFM tapping mode (phase image)

homopolymer (40 vol%) is shown as a matrix of parallel lamellae which contains PS particles.

Small details of the phase morphologies of blends are visible in chemically stained ultrathin sections in TEM. An example is shown in Fig. 17.4 for a blend of PA/ABS (volume ratio 70/30). The rubber particles inside the SAN phase of the ABS component are strongly chemically fixed and stained with OsO_4, and thus appear black. The PA matrix is stained with formalin/OsO_4, appearing grey. Besides the larger ABS particles (which would also be detectable in SEM), many smaller SAN particles with diameters of below a few hundreds of nanometres are visible.

Very similar polymeric materials – i.e. polymers with similar chemical constitutions and relatively good compatibility – form fine morphologies. In Fig. 17.5, a blend of 80% slightly branched HDPE (<0.4 mol% 1-hexene, ρ=0.930 g/cm^3) with 20% highly branched VLDPE (17.2 mol% 1-hexene, ρ=0.867 g/cm^3) is shown. The VLDPE appears as spherical particles about a few hundreds of nanometres thick that are well distributed in the HDPE matrix, which has a definite lamellar arrangement [4]. Some of the HDPE lamellae have grown from the matrix into the amorphous VLDPE particles, indicating a good interfacial connection. In the TEM micrograph (Fig. 17.5a), the amorphous, rubber-like VLDPE particles are darkly stained and the HDPE matrix shows a typical semicrystalline morphology, with bright crystalline lamellae and dark amorphous zones. The AFM tapping mode

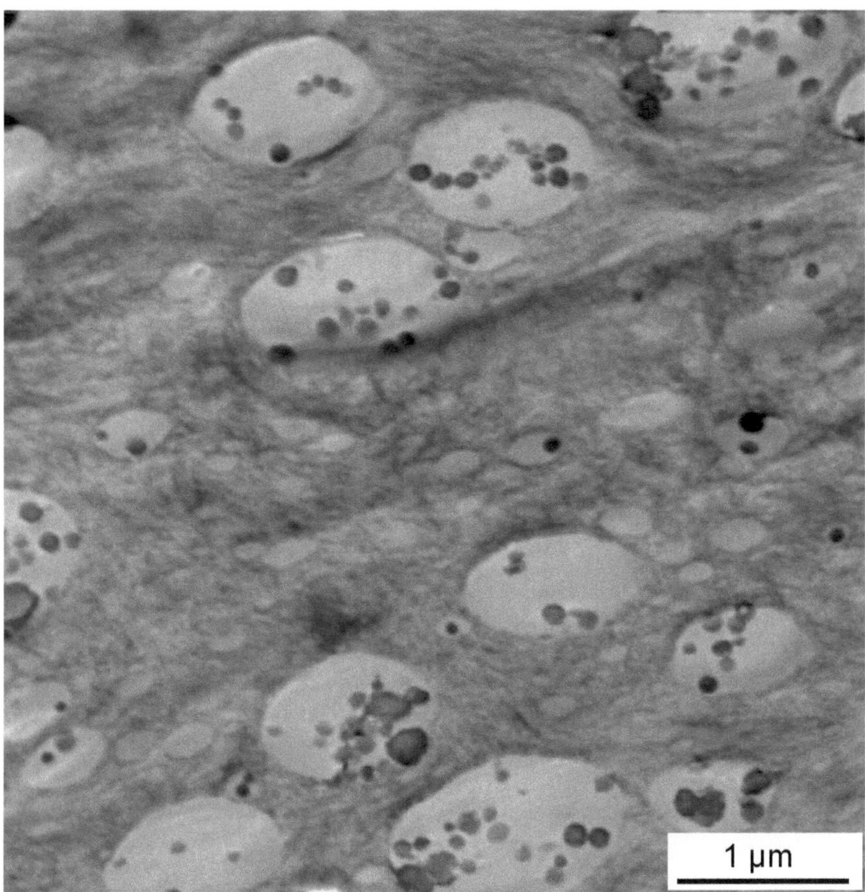

Fig. 17.4. PA/ABS (70/30) blend after chemical fixation and staining, revealing larger ABS particles and smaller SAN particles in the PA matrix. (Ultrathin section, staining of rubber with OsO_4 and of PA with formalin/OsO_4, TEM)

phase image (Fig. 17.5b) shows bright, soft, rubber-like particles, grey amorphous zones in the HDPE matrix and dark lamellae, revealing that these zones all have different local microhardnesses. Additionally, a pearl-like fine structure is clearly visible for the lamellae, which is difficult to detect in the TEM.

Incompatible polymers, such as PS with PE or PP, and PS with PBD, usually show strong phase separation with the formation of large particles. The contact between polymeric components can be improved by introducing compatibilisers. Figure 17.6 shows SEM micrographs of fracture surfaces of a PP/PS (85/15) blend without and with a PP/PS-block copolymer included as compatibiliser. In the pure PP/PS blend (Fig. 17.6a,b), large PS particles are visible in the PP matrix. The spherical shapes of the PS particles and debonding from the matrix are good indicators of low compatibility.

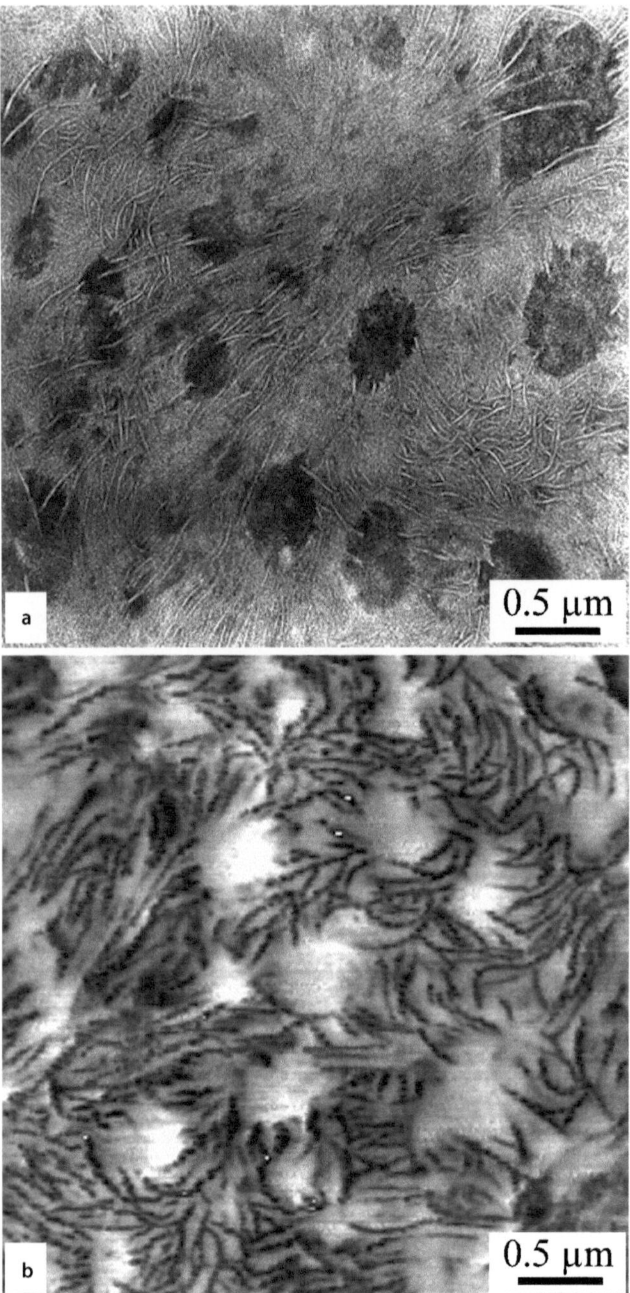

Fig. 17.5a,b. Blend of the components HDPE (80%) and VLDPE (20%): **a** TEM micrograph of an ul-trathin section after selective chemical staining; **b** AFM tapping mode phase image of a thick section with a smooth surface

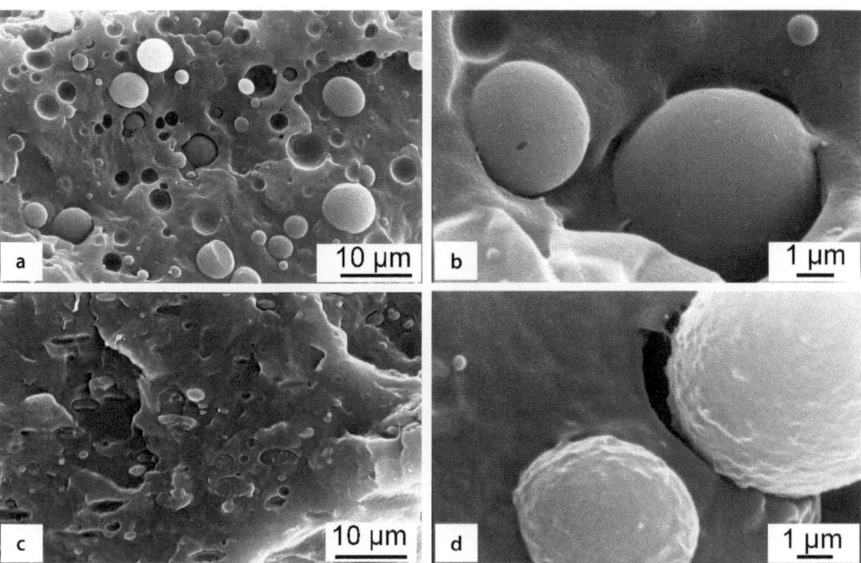

Fig. 17.6a–d. Contact at the interface in PP/PS (85/15) blends: **a,b** without compatibiliser; **c,d** with 3% PP/PS blockcopolymer used as compatibiliser (enhanced contact is apparent). (Fracture surfaces, SEM micrograph; from [5])

Under load, the crack follows the particle interface and the fracture path is around the PS particles. When 3% PP/PS block copolymer is added it covers the PS particles and improves the contact with the matrix (Fig. 17.6c,d). The weakest part is no longer the interface and so cracks usually propagate across the PS particles [5].

Compatibilisers is also cause a reduction in the particle size of the minor component in the matrix. Figure 17.7 shows blends of PA66 and syndiotactic PS without (micrograph a) and with (micrograph b) 10% sPS/PA modifier [6]. The large sPS particles in the blend without modifier (micrograph a) are drastically reduced in size when 10% modifier is added. In the blend with 40% sPS and 10% modifier (micrograph c), the sPS particles are coalesced and are therefore relatively large, but contain small PA particles inside them due to the improved compatibility.

The grafting of PS onto LDPE via γ-irradiation allows the formation of small PS domains in the LDPE matrix; see Fig. 17.8 [7]. The smallest PS domains (about 40 nm in size) in micrograph (a) correspond to PS macromolecular coils with molecular weights of about 10^6. As the degree of grafting increases the PS domains grow in size, as can be seen in micrograph (b), which presents a blend with 30% grafted PS.

The morphology of the blend is more complex, if each of the polymer components possesses its own morphology, as shown in the blends in Figs. 17.4 and 17.5. To differentiate between the various structural details, it is often helpful to apply a combination of different staining media (as shown in Fig. 17.4, where different staining procedures have been applied to the rubber component of ABS and the PA component).

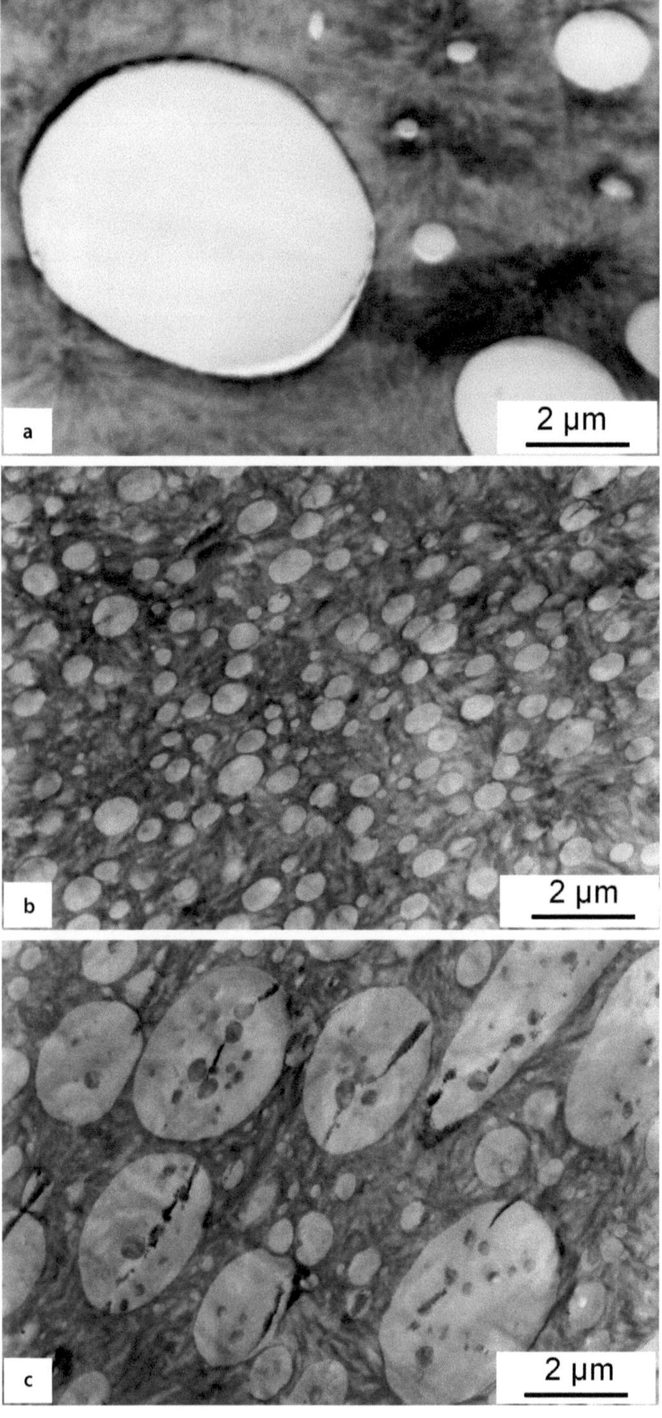

Fig. 17.7a–c. Influence of a compatibiliser on the morphologies of PA66/sPS blends: **a** PA66/sPS (80/20) blend without compatibiliser; **b** PA66/sPS (70/20) with 10% compatibiliser; **c** PA66/sPS (50/40) with 10% compatibiliser. (Selectively stained ultrathin sections, TEM; from [6], reproduced with the permission of Hanser)

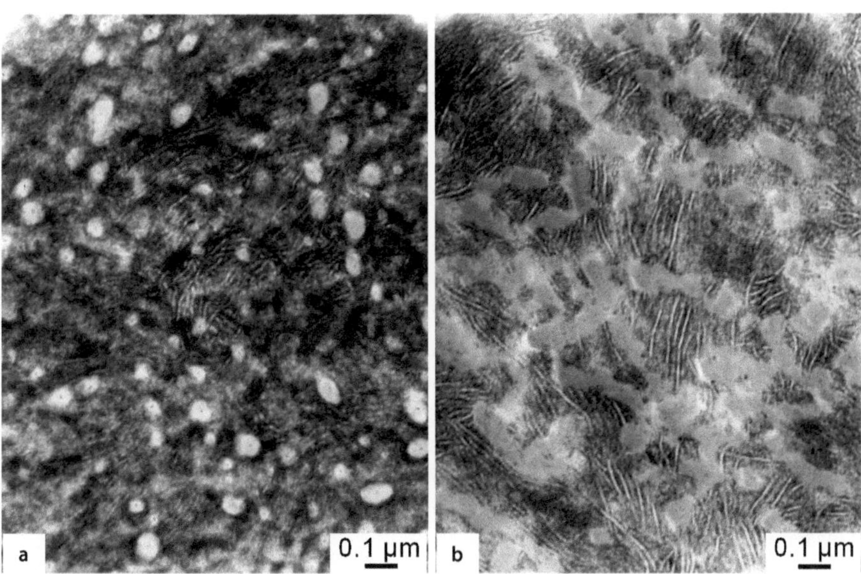

Fig. 17.8a,b. PS/LDPE blends after grafting styrene onto PE with small PS particles (*bright*) in the LDPE matrix. Grafting efficiency: **a** 10%; **b** 30%. (Selectively stained ultrathin sections, TEM)

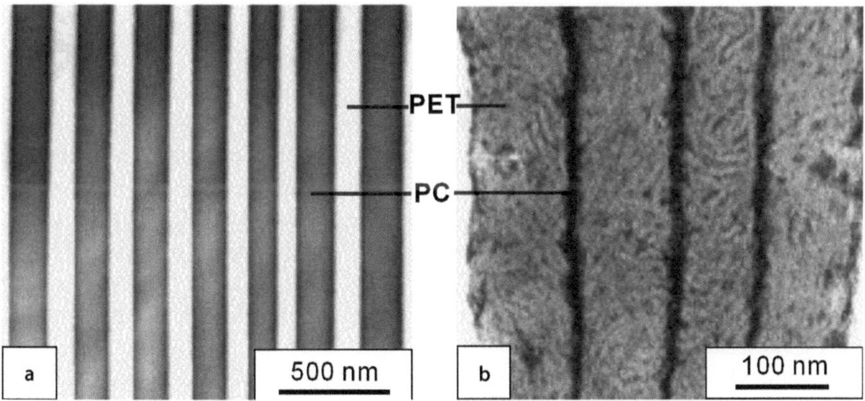

Fig. 17.9a,b. **a** Cross-section of a PET/PC nanolayered film with 4096 layers; the thicknesses of most of the layers are between 110 and 140 nm; the PC layers appear dark due to staining with RuO_4. **b** PET/PC nanolayered film showing the lamellar structure of the PET layers; the PC layers and the amorphous regions of PET appear dark due to staining with RuO_4. (TEM micrographs)

Using the coextrusion technique and a series of multiplying die elements, it is possible to produce polymer bands or films composed of hundreds or even thousands of layers with individual layer thicknesses at the micro- or nanoscale [8]. The synergistic combination of two or more polymers in a layered structure can enhance the overall properties of the material. An example is shown in Fig. 17.9 for a PET/PC nanolayered composition containing continuous layers with nearly uniform thicknesses of about 110–140 nm. The semicrystalline structure of the PET layers appears after annealing at 160 °C for 12 h and is made visible by a strong staining procedure with RuO_4; see Fig. 17.9 b [9,10]. The PET lamellae form in a region of confined crystallisation between adjacent PC layers.

17.3 Micromechanical Behaviour

In the case of incompatibility between the polymer components and thus strong phase separation, larger structural details (e.g. large particles in a matrix) usually show only a weak interfacial connection or no connection at all. This results in interfacial decohesion or debonding during mechanical loading, followed by the formation of larger cracks, crack propagation along morphological details and brittle fracture. An example is shown in Fig. 17.6a,b for a blend of PP and PS. Here, the fracture surfaces provide a clear picture of the phase morphology, similar to the low-temperature fracture surfaces.

Improved properties (higher strength, larger deformation at break, etc.) are only possible after improving compatibility, interfacial strength, and reducing the particle size below a critical limit. Under optimum conditions, the typical deformation behaviour of one component (e.g. the matrix) can be intensified by the second component (e.g. particles). This effect is illustrated in Fig. 17.10 for a HDPE/VLDPE (80/20) blend (the morphology is shown in detail in Fig. 17.5) [4]. An AFM in situ deformation experiment was performed in a Dimension 3000 microscope from Digital Instruments equipped with a deformation unit. Figure 17.10 shows the original image and images of three steps during the tensile test, for elongations of 50 µm, 100 µm, and 200 µm, respectively. Relative to the original length, this corresponds to mean local strains of 8, 16, and 56%. To illustrate the strain distribution within the imaged area, the changes in four segments are compared, as marked by arrows and the letters a, b, c, and d. While the length of segment a (in the matrix near the poles of the particles) does not change, the length of segment b (within an elastomeric particle) increases significantly in all of the micrographs (increases of ε = 16, 30, and 146%). Segments c and d (between the particles) show increased strains of ε = 8, 16, and 69%, respectively, which are also larger than the average strains. The AFM in situ tensile test reveals additional details for segments c and d (zones of locally increased stress due to stress concentrations at the rubbery particles). Figure 17.11 shows an area of the blend before (first row) and after deformation with a mean local strain of 35% (second row) [11]. The changes are more visible after processing the image using a high-pass filter, as seen in micrographs (b) and (d). The typical behaviour of the lamellae under stress is illustrated by

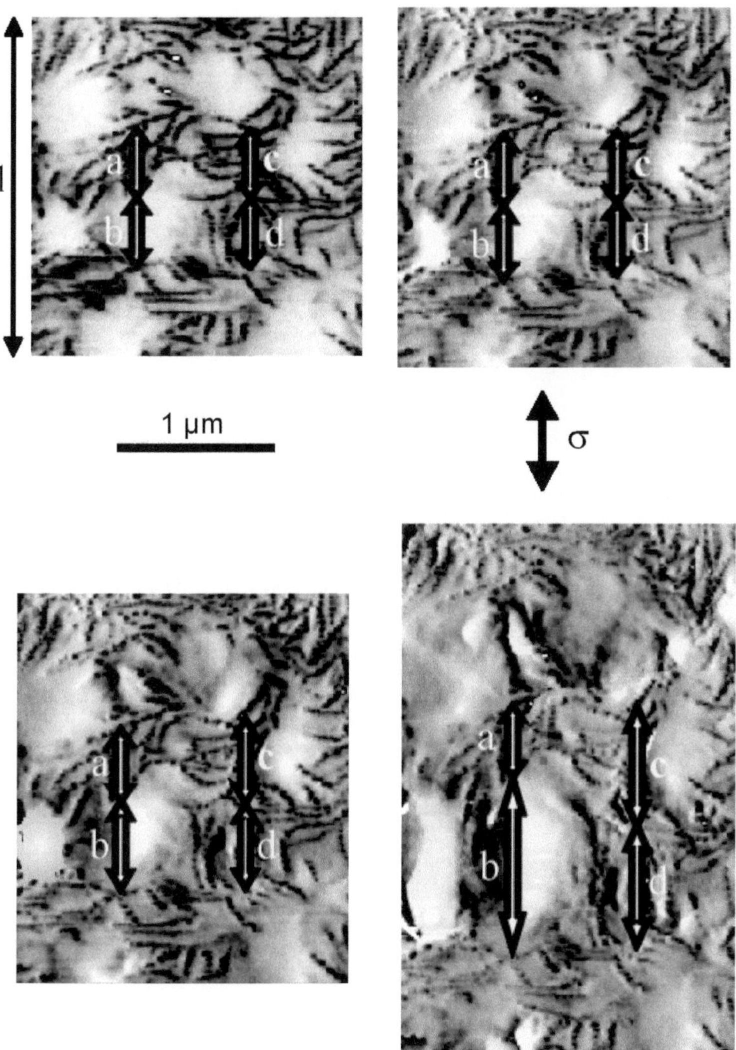

Fig. 17.10. AFM in situ deformation of a HDPE/VLDPE (80/20) blend (see morphology in Fig. 17.5); AFM tapping mode phase images show the same specimen area before deformation and at three stages of elongation. Elongation direction is vertical (from [4])

tagging lamellae segments with capital letters in the images. Originally, the lamellae segment AB is almost perpendicular to σ. During the deformation it rotates into the load direction by an angle of 14° without changing the segment length. The lamellae segment CD is already oriented in the load direction and so it enlarges due to microblock separation. A local enlargement of 55% occurs for GH, whereas a combined rotation and enlargement is observed for the lamellae segment EF.

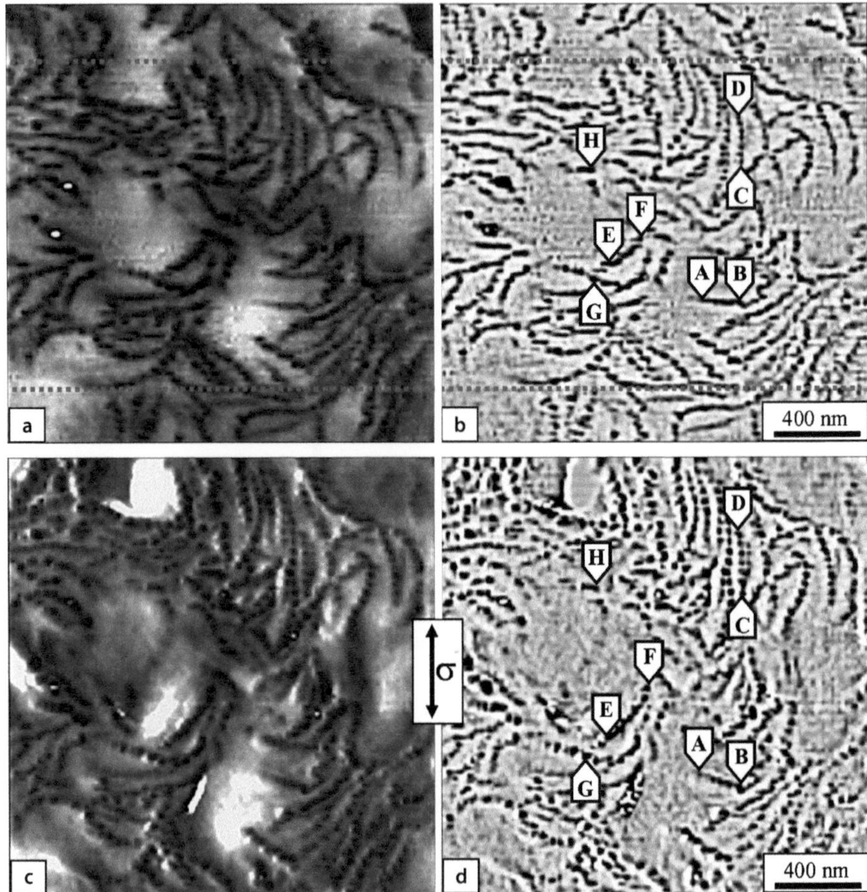

Fig. 17.11a–d. HDPE/VLDPE (80/20) blend (see Figs. 17.5 and 17.10); SFM tapping mode phase signal images before deformation (**a**), after deformation (**c**), and after additional image processing using a high-pass filter (**b,d**). Elongation direction is marked by σ

There are other micromechanical processes that can improve the mechanical properties of polymer blends. Examples include:

- The *bridging mechanism*. Weaker (rubber-like) polymer particles in a matrix can be plastically stretched in front of a crack tip and – after propagating the crack – connecting and holding together both of the crack surfaces (bridging).
- The *inclusion yielding mechanism*. Stiffer polymer particles are distributed in a matrix with a slightly lower yield stress, e.g. PSAN particles embedded in a PC matrix [12]. Under load, stress transferred to the rigid inclusions via the softer matrix can exceed the yield stress of the inclusions, and so the particles are forced to deform plastically and absorb energy.

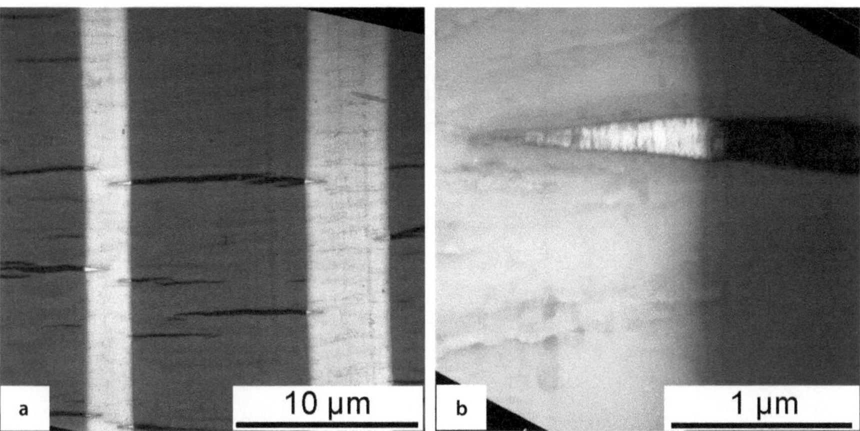

Fig. 17.12a,b. Deformation structures in a PS/PMMA nanolayered material showing the formation of numerous crazes in the PS layers. Due to selective staining by RuO_4 the PS layers appear *grey*, the crazes inside the PS are *black*, while the PMMA layers appear *bright*. (Deformed semi-thin section, chemically stained, HVEM)

– The *multiple yielding mechanism*. Similar to the effect shown in Figs. 17.10 and 17.11, rubber-like particles initiate local stress concentrations, which initiate local plastic yielding at and between the particles in the form of crazes ("multiple crazing") or shear deformation ("multiple shear yielding"). Since this is the typical effect associated with rubber-toughening, it will be demonstrated in the next chapter.

The micro- and nanolayered structures prepared by the coextrusion technique show enhanced mechanical properties with decreasing layer thicknesses. This is attributed to changes in micromechanical deformation mechanisms. If the layers are thick, the polymers deform in a similar way to the bulk materials. In Fig. 17.12, a PS/PMMA nanolayered material with 64 layers (each a few microns thick) is deformed. Crazes (with fibrillar interior structures) in the PS penetrate into neighbouring PMMA layers and then stop. Therefore, the propagation of crazes and their transformation into long cracks are prevented, with the result that many crazes can form. As the layer thickness decreases, an increased volume of the material is involved in the plastic deformation (i.e. more crazes) and enhanced toughness results [13].

17.4 Examples

17.4.1 Blends of Amorphous Polymers

One of the least compatible blends across the whole concentration range is PS/PPO [14]. With increasing PPO content, the micromechanical deformation character changes from the formation of fibrillated crazes (as in PS) to homogeneous deformation bands and shear deformation (as in PPO) [15]. This transition correlates with an increasing entanglement density in the blend (compare Table 15.1).

There are some other (partially) compatible blends; for instance SAN/PMMA blends if the AN content in SAN is 10–30% [16], PMMA/PVF blends [17] and PC/PAR blends [18].

Occasionally it can be difficult to study the morphology of an amorphous polymer blend in detail if there are no clear interface boundaries or the components show low staining abilities. In this case, irradiation effects in the electron microscope can play a useful role. For instance, blends with PVC are very easy to investigate, since PVC possesses (because of its Cl content) a higher density than other polymers and also shows a marked irradiation-induced mass loss and contrast change [19, 20]. Figure 17.13 shows a PVC/SAN blend where the PVC matrix appears dark at the beginning of irradiation and becomes brighter than the large SAN particle in the middle after intense electron irradiation. Mass loss also occurs in PMMA and so when it is in a blend PMMA becomes brighter with increasing irradiation [21].

Blends of PC with SAN or PMMA may show larger yield stresses and elongations at breakage compared with unmodified PC if the SAN or PMMA is dispersed

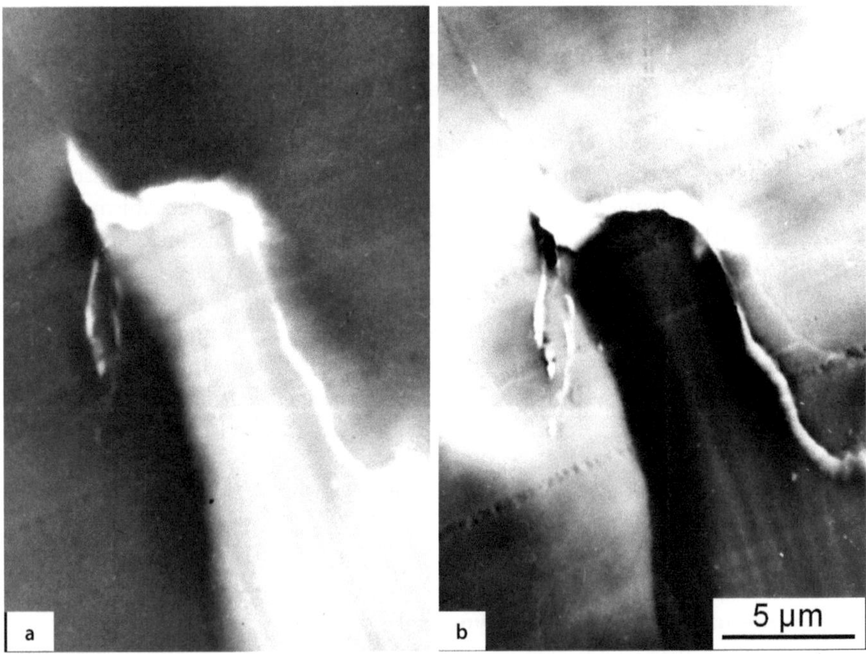

Fig. 17.13a,b. Differentiation of the polymer phases in a PVC/SAN blend due to different densities and mass loss after electron irradiation: **a** at the start of irradiation (PVC darker than SAN); **b** after intense irradiation (PVC brighter than SAN). (Semi-thin section, HVEM; from [20], reproduced with the permission of Hanser)

as small particles about 1 μm in size. In such systems, these particles can be plastically deformed and can absorb energy via the abovementioned "inclusion yielding" mechanism [22, 23]; see also Chap. 18. Practical application of this effect has led to blends of PC and ABS with improved processing, temperature stability, elongation at breakage and toughness [22].

17.4.2 Blends of Amorphous and Semicrystalline Polymers

Systems of PE and PS are almost the archetypal incompatible polymer blend. Blends of these polymers form clearly distinguishable phases and thus yield poor mechanical properties. Improving phase connection – e.g. by mixing with PS-g-PE graft-copolymers [24], other block copolymers or by irradiation-induced grafting of PS onto PE [6]; see Fig. 17.8 – can enhance the properties of such blends. A distribution of smaller PS particles can be also realised by polymerising styrene in PE [25]. A phase morphology with about 0.1–1 μm PS particles in a PE network is presented in Fig. 17.14. The higher magnification of micrograph (b) reveals small lamellae in the PE network. The boundaries between the PS particles and the PE network appear broadened, but this is only due to the curved surfaces of the particles. Under load, such a PE network with PS inclusions fails, yielding brittle fractured PS particles surrounded by a plastically deformed PE network; see Fig. 17.14c, d.

Both characteristics of incompatible polymer blends – large particles or irregularly shaped structural details and low interfacial strength – are also visible in blends of PMMA with segmented thermoplastic PU (TPU). Figure 17.15 presents SEM micrographs of fracture surfaces where the effect of debonding around PMMA particles and crack initiation (in a 5/95 and a 50/50 PMMA/TPU blend) are apparent [26]. If the sizes of the hard polymer particles can be reduced, the stress-initiated plastic deformation of the ductile matrix strands between the particles can be realised. An example is shown in Fig. 17.16 of a PBTP/PC blend with small PC particles a few tenths of a micron in size (a) and as a fracture surface (b). Under load, the spherical PC particles debond from the PBTP matrix and the microvoids initiate plastic yielding of the matrix strands up to fibrillation.

17.4.3 Blends of Semicrystalline Polymers

Typical semicrystalline polymer blends include PE and PP, as well as recycled polymers. Incompatibility between the components yields large particles and poor mechanical properties, but this situation can be improved by performing intense mixing in the melt [27]. The properties of such blends depend not only on the phase morphology but also on possible changes in the semicrystalline morphologies of the components themselves due to processing (e.g. crystallisation behaviour, depression of spherulites, size and orientation of lamellae). In particular, epitaxial effects can appear at the interfaces (this is also true for blends of amorphous and semicrystalline polymers described above if only one component is a semicrystalline polymer). Such an effect is illustrated in Fig. 17.17 for a HDPE/rubber blend [28]. The PE matrix shows

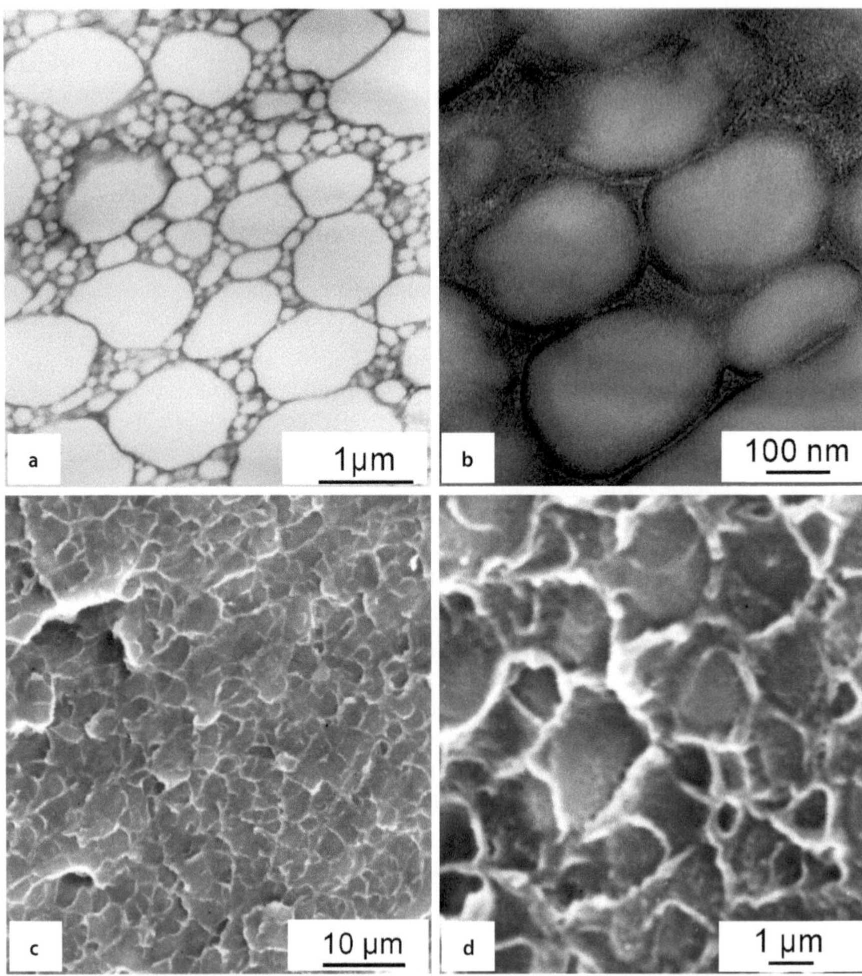

Fig. 17.14a–d. TEM micrographs of a PS/PE (75/25) blend with a PE network containing embedded PS particles. **a,b** Smaller and larger magnifications [25]; **c,d** fracture surfaces of a PS/PE (75/25) system with PS particles fractured in a brittle manner and surrounded by a plastically deformed PE network

oriented lamellae perpendicular to the boundaries of the rubber particles (micrograph a). Under loading in the direction perpendicular to the lamellae, the lamellae break, twist into the load direction and form chevron patterns (micrograph b, see also Figs. 16.24 and 16.30).

In blends of HDPE and LDPE, separate phases appear upon weak mixing. Strong mixing conditions only yield a separation at the molecular level between linear and branched macromolecules, with the formation of separate lamellae, or they yield molecular level mixing with the formation of "mixed lamellae". Such conclusions result from the detailed TEM analysis of selectively stained ultrathin sec-

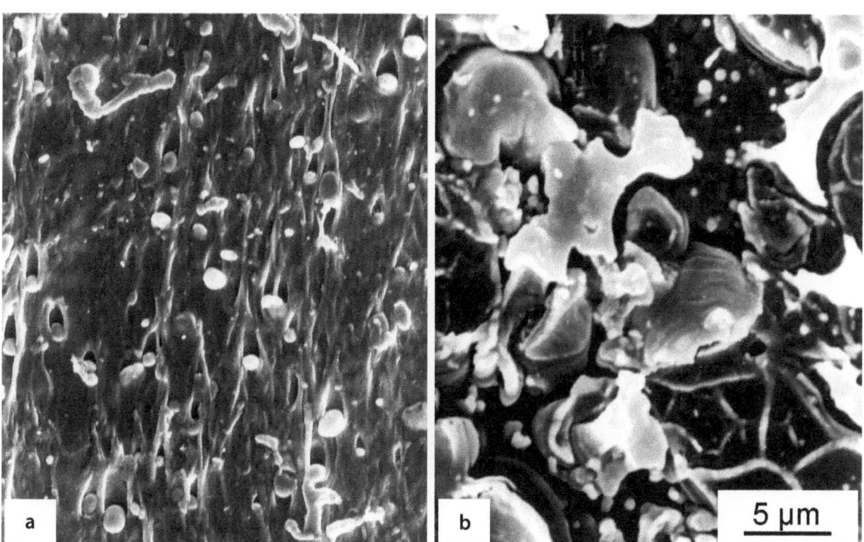

Fig. 17.15a,b. SEM micrographs of fracture surfaces of PMMA/TPU blends with PMMA particles debonding at the interfaces; particle diameter increases with increasing PMMA content: **a** 5%; **b** 50%

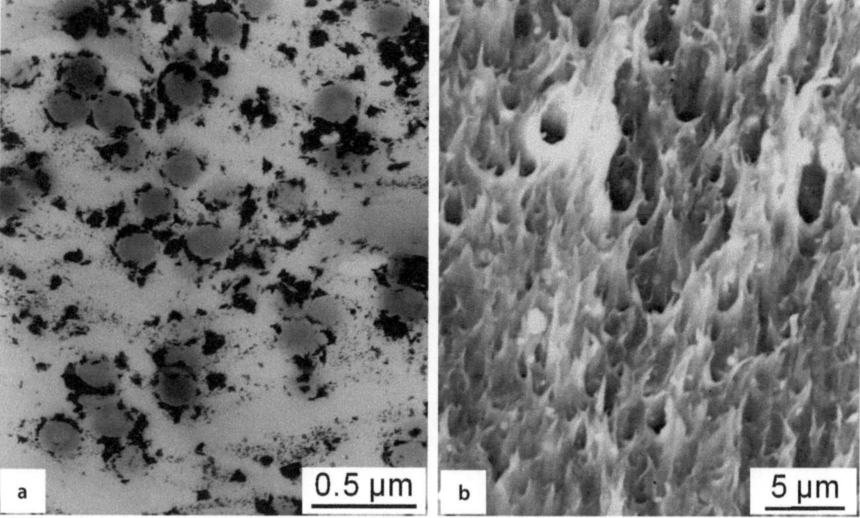

Fig. 17.16a,b. PBTP/PC blend with dispersed PC particles; PC particles are debonded from the PBTP matrix and plastically stretched PBTP fibrils are apparent. **a** TEM micrograph from selectively stained ultrathin section; **b** fracture surface in SEM

tions where the thickness distribution of the lamellae is measured exactly [29]. An example is shown in Fig. 17.18, which presents TEM micrographs of HDPE, HDPE/LDPE 70/30, 30/70 blends and LDPE along with the corresponding lamellae thickness distributions. A comparison shows that some of the linear macromolecules

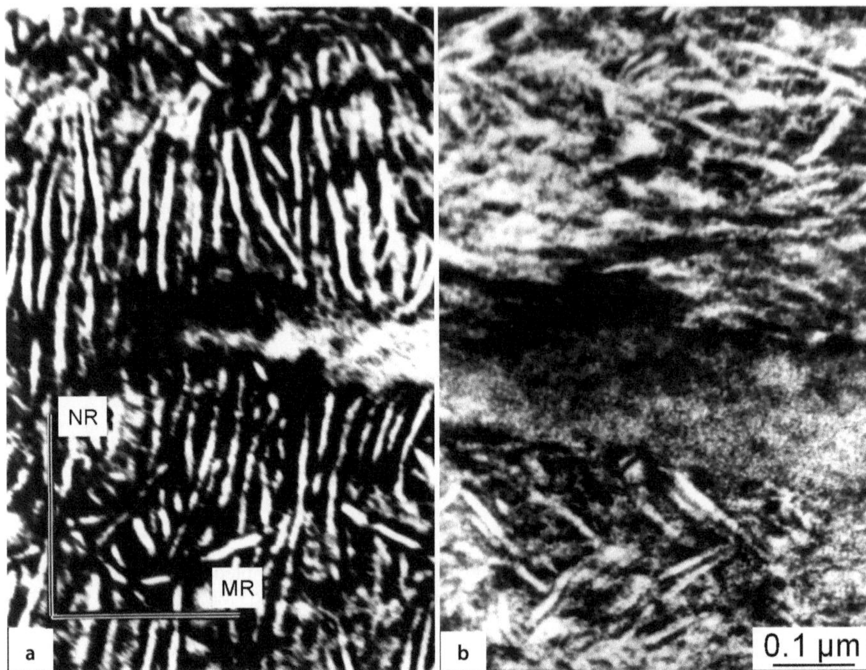

Fig. 17.17. HDPE/rubber blend showing a morphology with lamellae oriented in the direction perpendicular to the rubber particle interface (micrograph **a**); chevron patterns are formed during loading (micrograph **b**). *NR*, normal direction; *MR*, machine direction. (Stained ultrathin section, TEM; from [20], reproduced with the permission of Hanser)

Fig. 17.18a–g. TEM micrographs (**a–d**) of selectively stained ultrathin sections, revealing shapes, structures and size of lamellae in LDPE/HDPE blends of different compositions: **a** 100% LDPE; **b** LDPE/HDPE (70/30); **c** LDPE/HDPE (30/70); **d** 100% HDPE. Frequency distributions of thicknesses D_i of lamellae (**e**); schematic illustration of total thickness (D_g), internal crystalline thickness (D_i) and crystal block size B of lamellae (**f**); average size B and minimum size B^{min} of crystal blocks inside the lamellae (**g**)

are crystallised in lamellae of the same type as observed for linear PE (HDPE); the remaining linear macromolecules (probably the sequences that are slightly branched) co-crystallise with the branched molecules (from LDPE) at widely varying compositions, as indicated by the large variation in the thicknesses of the lamellae. The co-crystallisation can also be derived from the continuous increase in the average size B and the minimum size B^{min} of the crystalline blocks inside the lamellae. The higher the content of branched molecular segments contributing to the lamellae, the smaller the thickness D of the lamellae and the crystal block length B inside the lamellae. In the HDPE/VLDPE (30/70) blend of Fig. 17.19, phase separation while cooling from the melt is visible, involving the separation of linear (HDPE) macromolecules (resulting in long, thick lamellae at higher temperatures) and branched

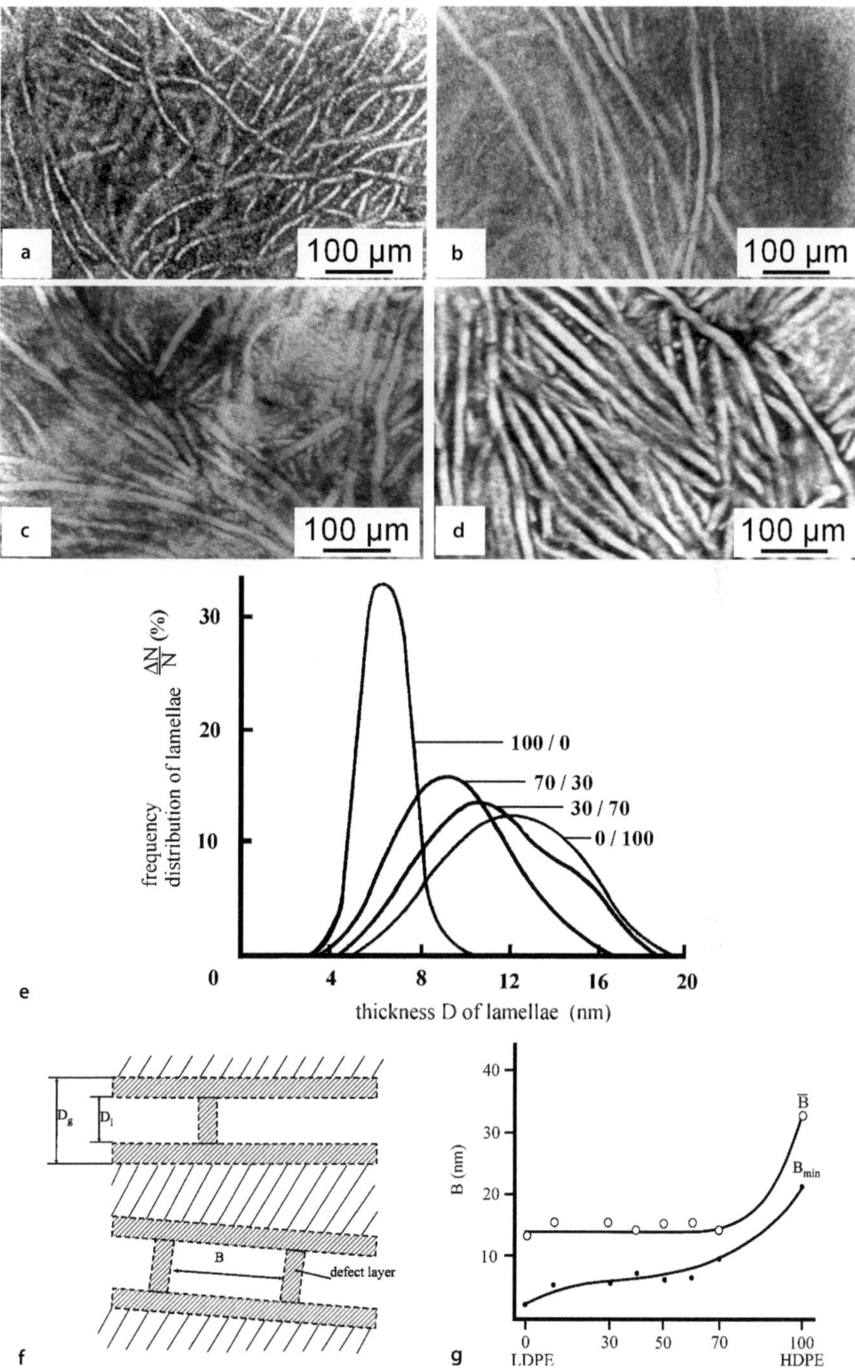

100 µm
a
100 µm
b
100 µm
c
100 µm
d
30
20
10
$\frac{\Delta N}{N}$ (%)
frequency distribution of lamellae
100 / 0
70 / 30
30 / 70
0 / 100
0
4
8
12
16
20
thickness D of lamellae (nm)
e
D_g
D_i
B
defect layer
f
40
30
20
10
B (nm)
$\overline{B}$
B_{min}
0
30
50
70
100
LDPE
HDPE
g

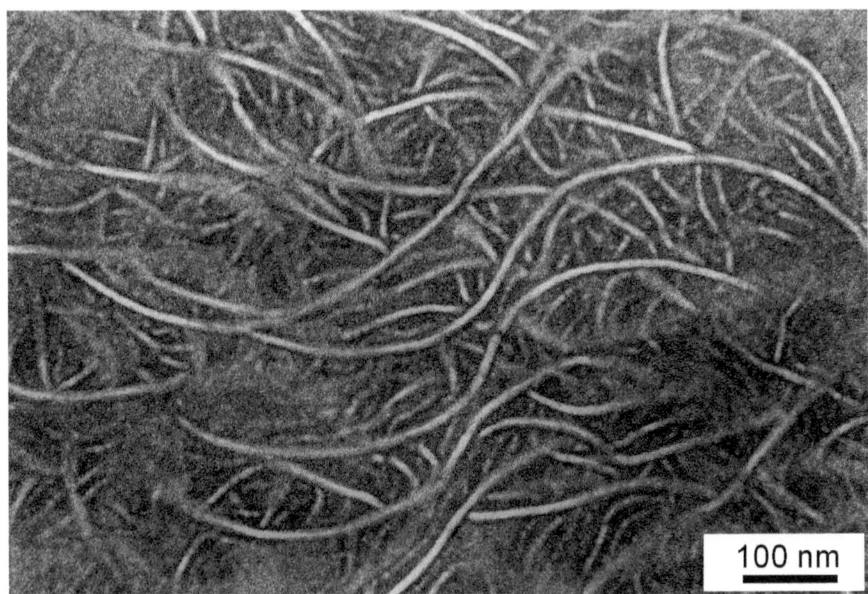

Fig. 17.19. HDPE/VLDPE blend (30/70) showing separation between thick, long lamellae and thin, short lamellae. (Stained ultrathin section, TEM)

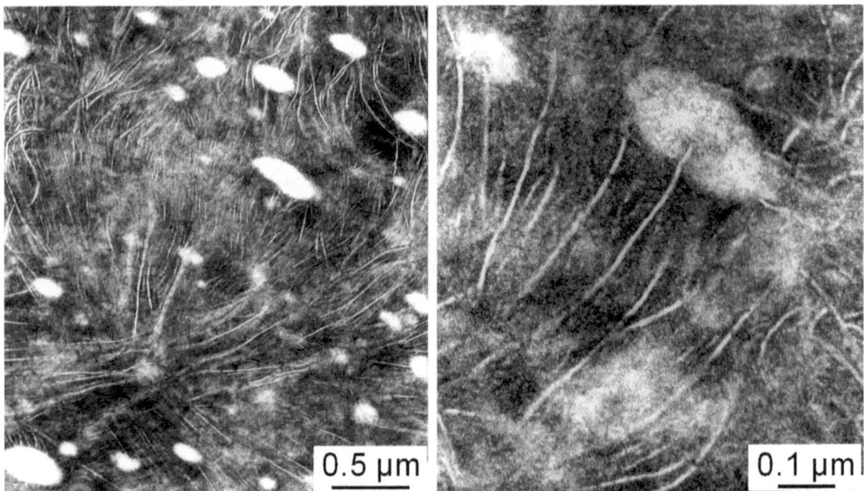

Fig. 17.20. LDPE/HDPE/PP composition, revealing the PP phase (*bright particles*), the HDPE phase (*thick, long lamellae*) and the LDPE phase (*thin lamellae*). (Stained ultrathin section, TEM)

macromolecules (which form short, thin lamellae consisting of small crystalline blocks at lower temperatures), which occur between the long lamellae. The broadening of some of the lamellae occurs due to the twisting of these lamellae in the thin section.

Information on the separation of the lamellae of different components can help to identify the morphologies of blends containing three or more components. Figure 17.20 shows the phase morphology of a LDPE/HDPE/PP (75/20/5) blend. The bright particles represent the PP phase, the long, thicker lamellae the HDPE, and the thinner lamellae with internal defect layers the LDPE component.

References

1. Paul DR, Bucknall CB (eds) (2000) Polymer blends, vols 1, 2. Wiley, New York
2. Pascault JP (ed) (2003) 7th European Symposium on Polymer Blends (Macromolecular Symp No 198). Wiley-VCH, Weinheim
3. Flory PJ (1953) Principles of polymer chemistry. Cornell University Press, Ithaca, NY
4. Godehardt R, Rudolph S, Lebek W, Goerlitz S, Adhikari R, Allert E, Giesemann J, Michler GH (1999) J Macromol Sci Phys B38:817
5. DFG-Innovationskollegs (1997–2000) Abschlußbericht des DFG-Innovationskollegs: Neue Polymermaterialien durch gezielte Modifizierung der Grenzschichtstrukturen/Grenzschichteigenschaften in heterogenen Systemen. Martin-Luther-Universität Halle-Wittenberg, 1997–1999, Merseburg 2000, Projekt B.3:119
6. Seydewitz V, Häussler L, Rapthel I, Michler GH (2004) Kunststoffe 94:98; Kunststoffe Plast Eur 1
7. Friese K, Michler GH, Steinbach H, Hamann B, Runge J (1986) Angew Makromol Chem 141:185
8. Bernal-Lara TE, Ranade A, Hiltner A, Baer E (2005) In: Michler GH, Baltá-Calleja FJ (eds) Mechanical properties of polymer based on nanostructure and morphology. Taylor & Francis, Boca Raton, FL, 15:629
9. Ivankova EM, Michler GH, Hiltner A, Baer E (2004) Macromol Mater Eng 289:787
10. Adhikari R, Lebek W, Godehardt R, Henning S, Michler GH, Baer E, Hiltner A (2005) Polym Adv Techn 16:95
11. Michler GH, Godehardt R (2000) Cryst Res Technol 35:863
12. Kolařik J, Lednicky F, Locati G, Fambri L (1997) Polym Eng Sci 37:128
13. Ivankova EM, Krumova M, Michler GH, Koets PP (2004) Colloid Polym Sci 282:203
14. Yee AF (1976) Polymer Prepr 17:145
15. Berger LL (1989) Macromolecules 22:3162, (1990) 23:2926
16. Suess M, Kressler J, Kammer HW (1987) Polymer 28:957
17. Roerdink E, Challa G (1978) Polymer 19:173
18. Katagawa M, Kitayama T (1990) Proc Benibana Int Symp Yamagata p 272
19. Michler GH (1986) In: Sedlaček B (ed) Morphology of polymers. Walter de Gruyter & Co, Berlin, p 749
20. Michler GH (1992) Kunststoff-Mikromechanik: Morphologie, Deformations- und Bruchmechanismen. Carl Hanser Verlag, München
21. Thomas EL, Talmon Y (1978) Polymer 19:225
22. Kurauchi T, Ohta T (1984) J Mater Sci 19:1699
23. Inoue T, Fujita Y, Angola JC, Sakai T (1986) Proceed 1. Dresdner Polymerdiskussion. Institut für Technologie der Polymeren und Technische Universität Dresden, Dresden p 134
24. Barentsen WM, Heikens D, Piet P (1974) Polymer 15:119
25. Borsig E, Fiedlerova A, Michler GH (1996) Polymer 37:3959
26. Michler GH, Grellmann W, Naumann I, Schierjott U, Haudel G (1989) Plaste Kautschuk 10:333
27. Greco R, Mucciariello G, Ragosta G, Martuscelli E (1980) J Mater Sci 15:845
28. Michler GH, Marinow Sl, Naumann I (1989) Plaste Kautschuk 36:432
29. Michler GH, Steinbach H, Hoffmann K (1986) Acta Polym 37:289

18 High-Impact Rubber-Modified Polymers

Rubber toughening generally means the modification of a hard, brittle polymer with soft, rubber particles. High toughness is a synergistic effect that arises due to a defined morphology with special toughening mechanisms. TEM micrographs taken from stained ultrathin sections contain a variety of information on the size, size distribution, arrangement and internal structure of the modifier particles and have therefore led to an understanding of the toughening effect. From the morphological perspective, two types of rubber-toughened polymers can be distinguished: "disperse systems" that contain homogeneous, heterogeneous or core–shell particles, and "inclusion systems" or "network systems" that present a network arrangement of the rubber phase. After these differences in morphology are illustrated, three main micromechanical deformation mechanisms are described: the multiple crazing mechanism, the multiple shear yielding mechanism and the network yielding mechanism. The chapter closes with a description of some additional toughening mechanisms.

18.1 Overview

As the utilisation of polymers for structural applications increases, a new class of polymeric materials with a unique combination of high strength, high modulus and high toughness is now being demanded. The original idea for producing polymers with enhanced toughness was to combine hard, brittle polymers with soft, rubbery ones, according to the equation "brittle + soft = tough". Toughness is, however, one of the most complex properties and one that is difficult to control, because it is greatly influenced by many morphological and micromechanical parameters. A special family of polymer blends known as *rubber-modified polymers*, exhibit very high toughness. The toughness of a rubber-modified polymer does not simply result from combining the properties of its components – see curve (a) in Fig. 17.1 in the previous chapter – but arises because the polymer has a pronounced morphology with special toughening mechanisms – i.e. synergistic effects corresponding to curve (d) in Fig. 17.1. The basic aim of any toughening enhancement is to increase the area below the stress–elongation curve in tensile or deformation tests (i.e. the energy absorption during deformation until fracture). In general, it is possible to increase σ by suppressing all critical defects and avoiding critical stress concentrations, although this causes a decrease in elongation at breakage (e.g. in highly oriented fibres or thin

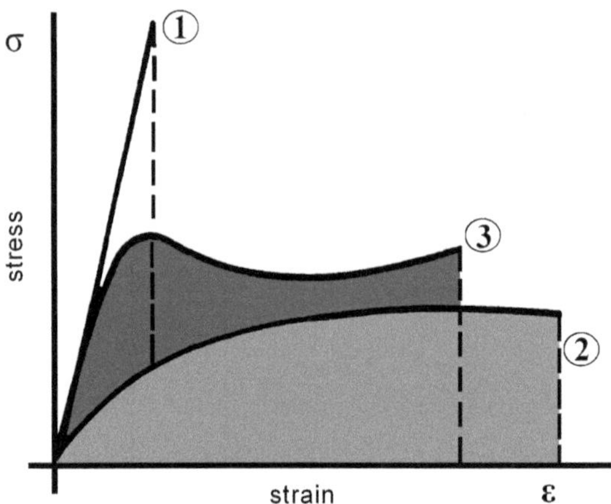

Fig. 18.1. Basic cases of enhancing toughness: 1, σ very high, ε very small; 2, σ small, ε very high; 3, σ high, ε high (true "toughening")

films); see curve (1) in Fig. 18.1. The opposite case, a large elongation ε, is usually connected to low strength ("softening"; see curve (2) in Fig. 18.1). Therefore, the optimum polymer to use for many applications is one with relatively high values of stiffness (modulus E), strength σ, and elongation ε, as shown in curve (3). This situation requires that large, critical defects and stress concentrations are avoided, but on the other hand a particular morphology is needed that permits lots of very small local yielding events (true "toughening"). Such criteria require a morphology that is strongly defined at the submicron level and therefore exact microscopic morphology control.

Rubber toughening generally refers to the modification of a hard, brittle, glassy polymer (e.g. PS, SAN, PMMA, PVC) with soft, rubber particles, yielding HIPS, ABS, ACS, etc. A comparable situation appears if a typically ductile polymer is used at lower temperatures in the brittle state (below the glass transition temperature, e.g. PP below 0 °C), yielding so-called "low temperature toughness". Both types of toughness can be realised by homogeneous rubber particles, core–shell particles, or heterogeneous modifier particles in so-called "disperse systems". An alternative is a network arrangement of the rubber in an "inclusion system" or "network system". These different rubber-modified polymers are the focus of several books and many reviews (e.g. [1–6]).

18.2 Morphology

The morphologies of different types of modifier particles greatly influence the local stress field, which plays a major role in activating plastic deformation in the matrix material. Therefore, the development of polymer systems with improved mechanical properties requires detailed knowledge of the inter-relationships between the morphology, micromechanical deformation processes and the macroscopic mechanical properties of modified polymer systems. As is well known for polymer blends, incompatibility between the rubber component and the hard polymer yields to phase separation and dispersed rubber particles ("disperse systems"). The rubber content typically varies between 5 and 30 vol%. TEM micrographs taken from stained ultrathin sections of ABS polymers containing PB particles in a SAN copolymer matrix contain a great deal of information on the size, size distribution, arrangement and also internal structure of the particles; see Fig. 18.2. In this case the rubber particles are prepared separately, grafted with SAN macromolecules and then mixed with SAN. The effect of grafting is visible on the dotted surface (surface grafting) and in small white SAN inclusions inside the particles (internal grafting). There are two types of rubber particles, smaller ones with diameters of around 100–200 nm and larger ones of about 400–600 nm (see the frequency distribution of particle diameters in Fig. 18.2c). When determining exact frequency distributions, it is important

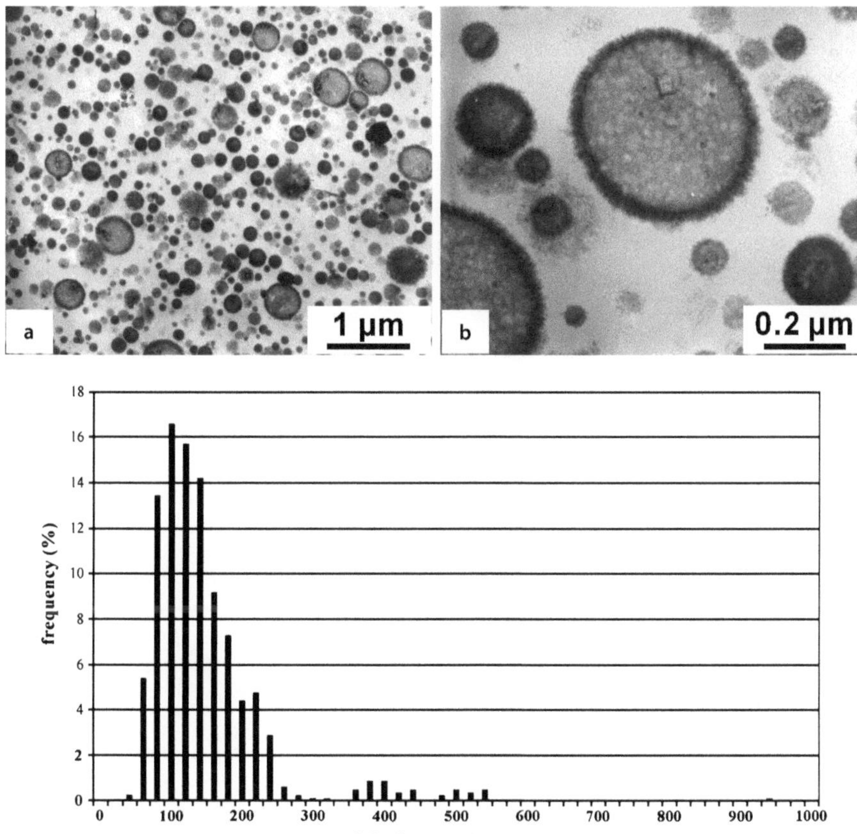

Fig. 18.2a–c. Size and arrangement of rubber particles in ABS polymers. **a,b** image of spherical particles between 50 and 600 nm (**b** is image at larger magnification, showing grafted surfaces of rubber particles and small SAN inclusions inside particles); **c** size distribution of the rubber particles. (Rubber phase selectively stained, ultrathin sections, TEM)

to note that larger particles are visible in the (about 100 nm thick) ultrathin sections only as cross-sections, which are usually smaller than the real diameter. Therefore, the measured frequency distribution is shifted to apparently smaller diameters (this effect is known as the "tomato salad problem"; see Fig. 12.3).

Figure 18.3 shows the phase structures of high-impact PS (HIPS). During the polymerisation of PS in the presence of the resolved rubber, phase inversion takes place, yielding rubber particles with PS inclusions, and various types of particles can be created depending on the stirring rate, ranging from core–shell particles to so-called "salami particles". The grafted surface layer of the rubber particles can be clearly identified in the larger magnification of Fig. 18.3b, which shows a lot of grafted dots. A quick overview of the size and distribution of the rubber particles can be obtained through the SEM inspection of chemically stained surfaces; see Fig. 18.4. Due to staining, the rubber particles appear bright in the SEM images. The thin white lines

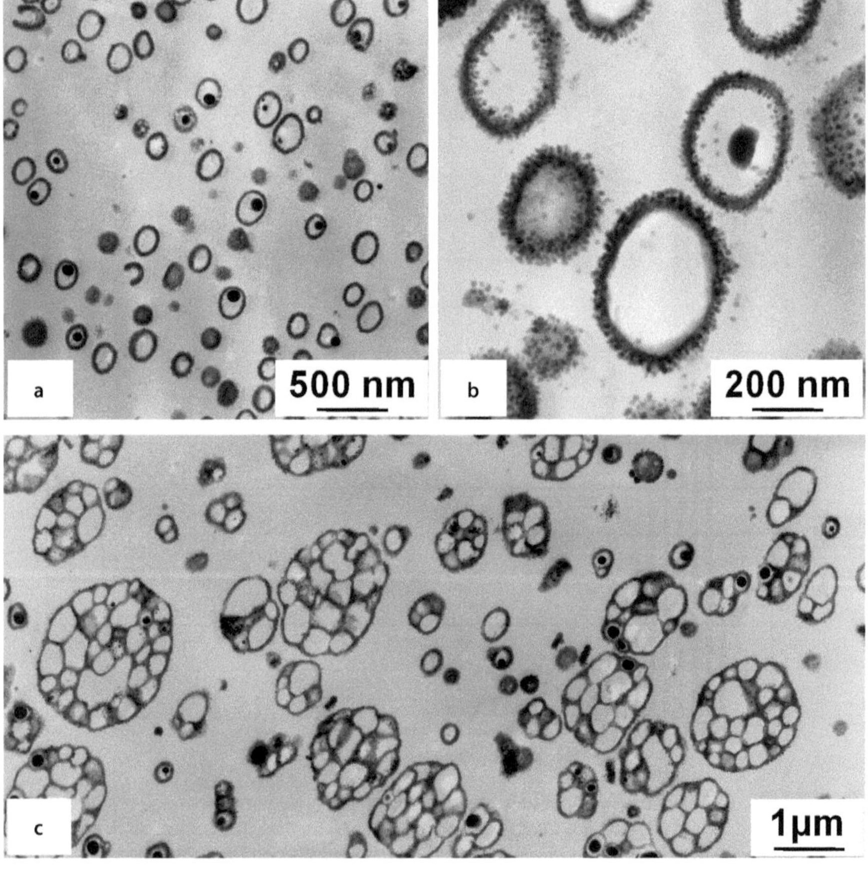

Fig. 18.3a–c. HIPS with rubber particles in form of: **a,b** rubber shells (core–shell) with SAN-grafted surfaces; **c** "salami particles". (Rubber selectively stained, ultrathin sections, TEM)

in Fig. 18.4a,b are crazes that appear when the samples are bent. Rubber particles with a heterogeneous internal structure exist in rubber-toughened PVC or in the form of core–shell particles in PMMA (see Fig. 18.13c).

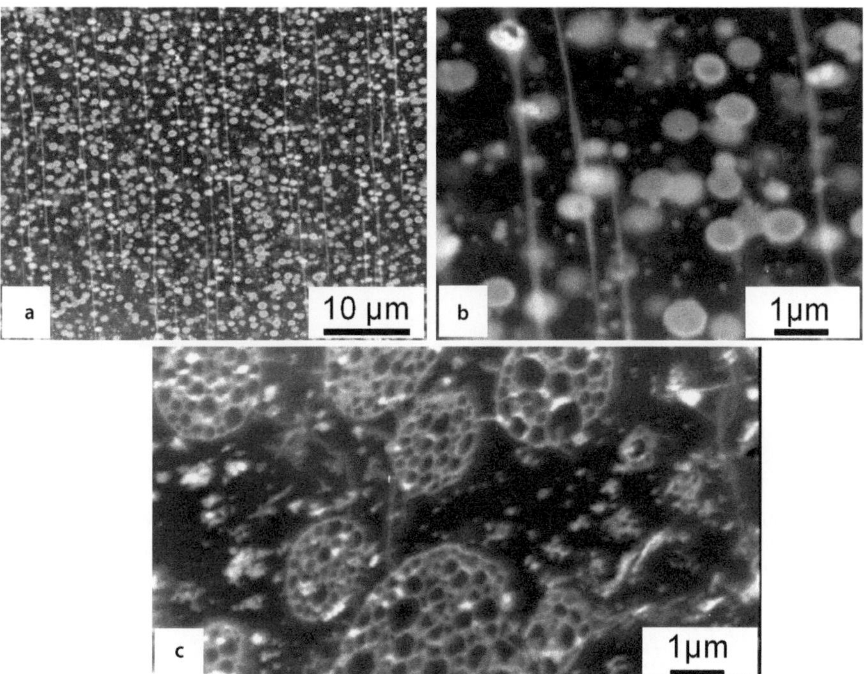

Fig. 18.4a–c. Quick inspection of high-impact polymers: **a,b** ABS, the white lines are crazes; **c** HIPS. (Surface layers stained with OsO_4, SEM)

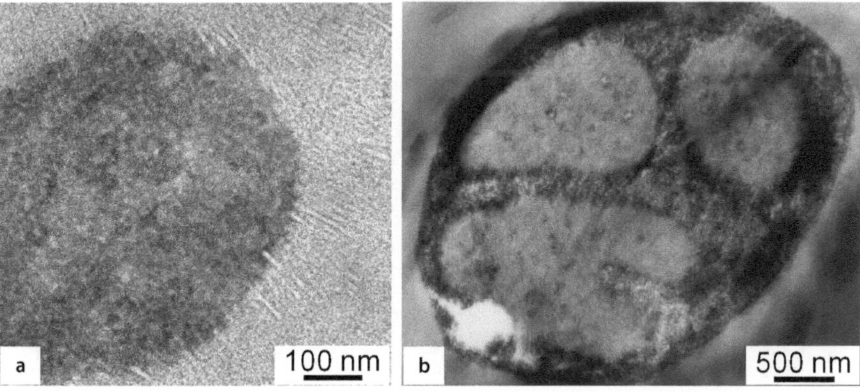

Fig. 18.5a,b. Rubber-toughened PP: **a** EPDM particle; **b** EPR core–shell particle. (Selectively stained ultrathin sections, TEM)

Different types of modifier particles are used in semicrystalline polymers such as PP. Figure 18.5a shows a typical phase structure for EDPM particles in the matrix of semicrystalline PP containing lamellae. Inside the EPDM particles there are a few lamellae because the EPDM particles possess non-negligible but low crystallinity.

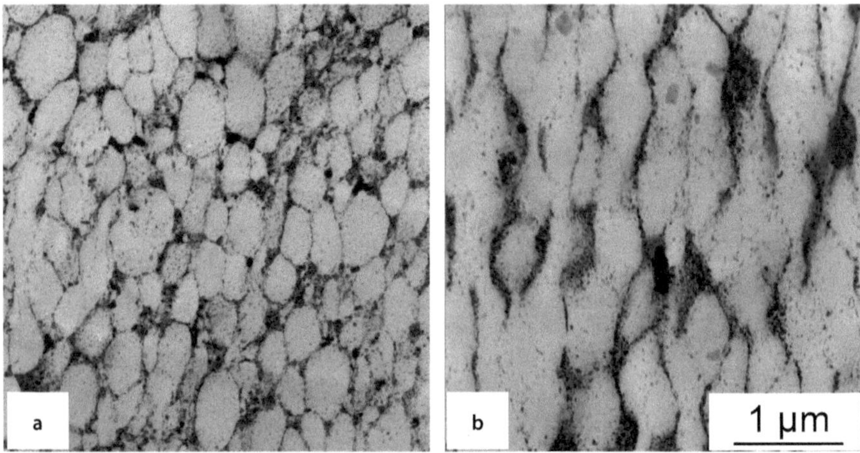

Fig. 18.6a,b. Rubber-toughened PVC with network structure of EVAc: **a** total network morphology; **b** partial network arrangement. (EVAc phase stained dark, ultrathin section, TEM; reproduced from [2] with the permission of Hanser)

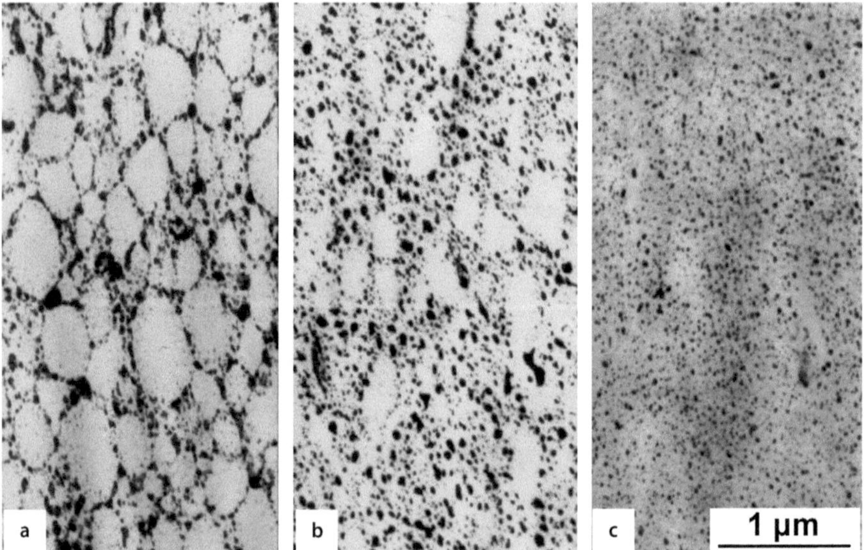

Fig. 18.7a–c. Destruction of the network structure in PVC/EVAc graft polymers due to processing. Mixing time: **a** 7.5 min; **b** 15 min; **c** 20 min. (EVAc stained dark, ultrathin sections, TEM; reproduced from [2] with the permission of Hanser)

Figure 18.5b is taken from EPR particles with a 21 mol% ethylene content. The phase structure corresponds to core–shell particles with shells made of a dark, structureless, amorphous ethylene–propylene rubbery phase and cores consisting of semicrystalline PE (identifiable through the difference in the thicknesses of the lamellae). Of particular interest is the interface between the particles and matrix and the question of whether it has an influence on the crystallisation behaviour of the matrix polymer (e.g. on the size and orientation of the lamellae).

A quite different type of morphology is presented by the network arrangement of toughened PVC: here very thin rubber layers wrap around the primary particles of PVC; see Fig. 18.6 (and compare with the morphology of PVC in Fig. 15.4). An advantage of these network systems is the resulting extraordinary enhancement in toughness, even for low rubber contents (about 6–8 vol%). However, they also have a strong disadvantage: the network morphology is highly sensitive to processing [2, 7, 8]. Larger shear forces or higher processing temperatures destroy the network arrangement and result in the usual distribution of small rubber particles in a PVC matrix, which in turn leads to a drastic decrease in toughness; see Fig. 18.7 [8]. Therefore, these types of network systems have no practical application today.

18.3 Micromechanical Processes

In the last few years, intensive studies of toughening mechanisms in rubber-modified polymers have been performed, mostly via electron microscopy. Based on the results from these studies, we can now distinguish three main mechanisms that are strongly related to the morphology (disperse or network structures) and the micromechanical deformation behaviour of the matrix [2]:

- Multiple crazing (e.g. in HIPS, ABS, ACS)
- Multiple shear yielding (e.g. in rubber-toughened PMMA, PA, PP)
- Network (particle) yielding (e.g. in toughened PVC).

Multiple Crazing Mechanism

The polymer crazing process can be distinguished by noting that (1) crazes are usually initiated in zones of chain segments that are weakly bonded, loosely packed, and contain structural defects under a dilatational stress field, and (2) that craze growth gives rise to strain-softening of the craze/bulk interface and strain-hardening of craze fibrils. It was observed that the fracture of HIPS is usually preceded by an opaque whitening of the stress area. This whitening is associated with the absorption of a large amount of energy. At low TEM magnification, these stress-whitened areas exhibit many whitening bands perpendicular to the loading direction (Fig. 18.8a). These small whitening bands, when observed at larger magnifications, are bridged by many tiny fibrils, revealing the nature of the crazes. In other words, the crazes are microcracks filled with voids and fibrils.

Crazes are usually initiated at zones of stress concentration at the rubber particles, i.e. in the equatorial zones of the matrix perpendicular to the loading direction; see Fig. 18.8b. When crazing occurs under well-controlled conditions, as in HIPS or ABS, it provides a mechanism for inelastic deformation that improves the material's

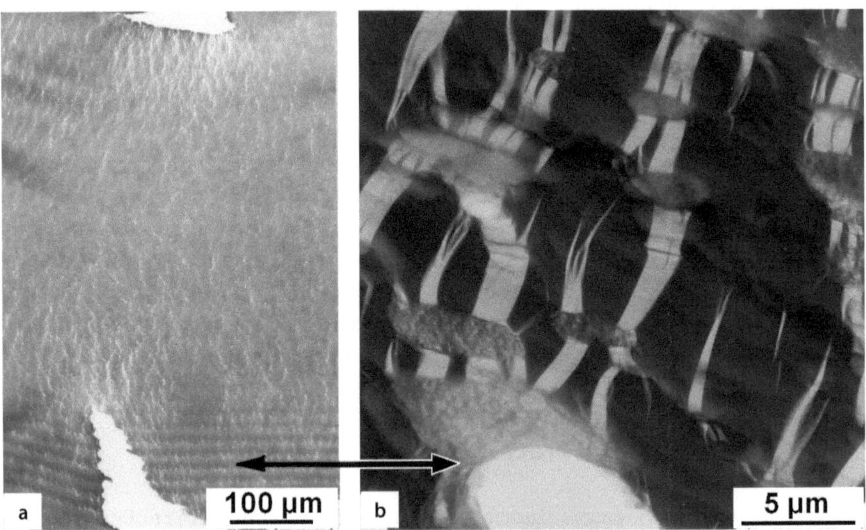

Fig. 18.8a,b. Deformation structures in HIPS: **a** overview of deformation area; **b** area in front of a crack tip with rubber particles (*grey*) in a matrix (*black*) with crazes (*bright*). (2 μm thick deformed section, for deformation direction see *arrow*, HVTEM; reproduced from [2] with the permission of Hanser)

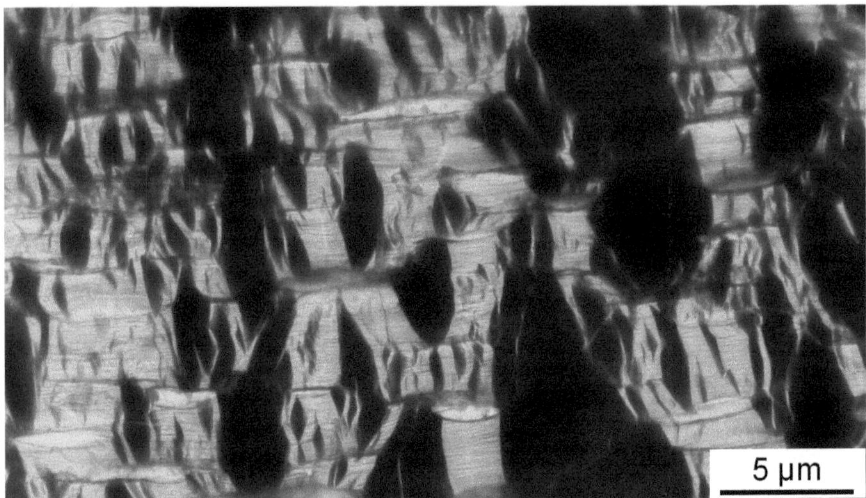

Fig. 18.9. Broad crazes and craze bands between rubber particles in HIPS. (Semi-thin section, deformation direction horizontal, HVTEM; reproduced from [2] with the permission of Hanser)

toughness. The energy dissipated by crazing can be divided up into the energy dissipated by yielding during fibril formation and the energy stored as surface energy in the matter in the craze. Figure 18.9 shows a high intensity of crazes with broad bands, propagating from one rubber particle to the next.

Crazes, however, are also the precursors to cracks and, ultimately, failure. Therefore, an important additional process is needed to stop the cracking by rubber particles; see Fig. 18.10. The rubber particles retard the rapid propagation of cracks and prevent the premature fracture of the sample. According to the number of rubber particles, many crazes can be created in larger volumes of the sample in order to achieve increased toughness. These individual processes can be summarised in a "three-stage mechanism of toughening", as shown in Fig. 18.11:

1. *Craze initiation.* Each rubber particle generates the stress concentration in the matrix surrounding it. In general, crazes start at points of highly concentrated stress, and then propagate perpendicular to the tensile direction. In many cases, the crazes are accompanied by cavitation inside the particles, but this is not considered to be a precondition for craze initiation.

2. *Superposition effect.* The stress fields around rubber particles overlap when the particle content is highly enough, i.e. more than 15 vol%. As a consequence, plastic strain-softening, which is characterised by a local yielding of matrix, also takes place, and is often followed by multiple crazing (fibrillated crazes, homogeneous crazes or combinations of them).

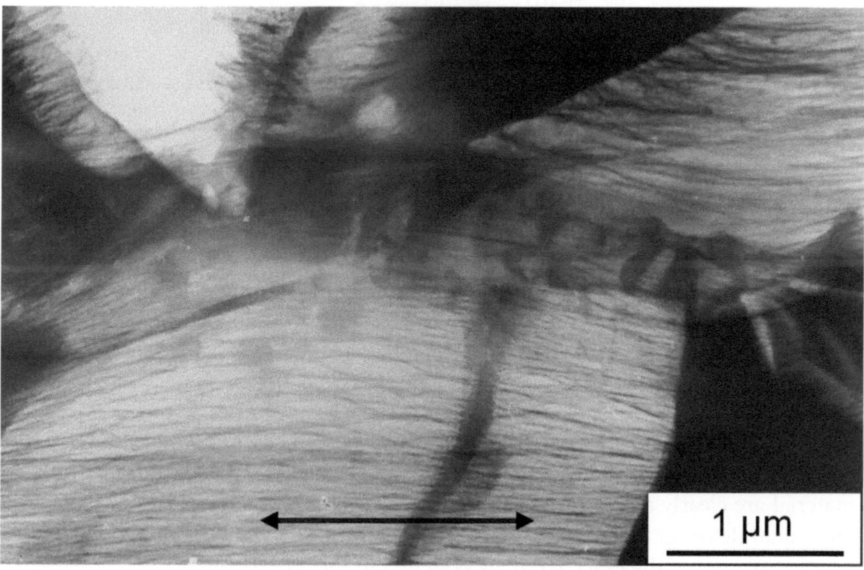

Fig. 18.10. Ruptured craze with crack stop in a rubber particle in HIPS. (Deformed semi-thin section, for deformation direction see *arrow*; HVTEM; reproduced from [2] with the permission of Hanser)

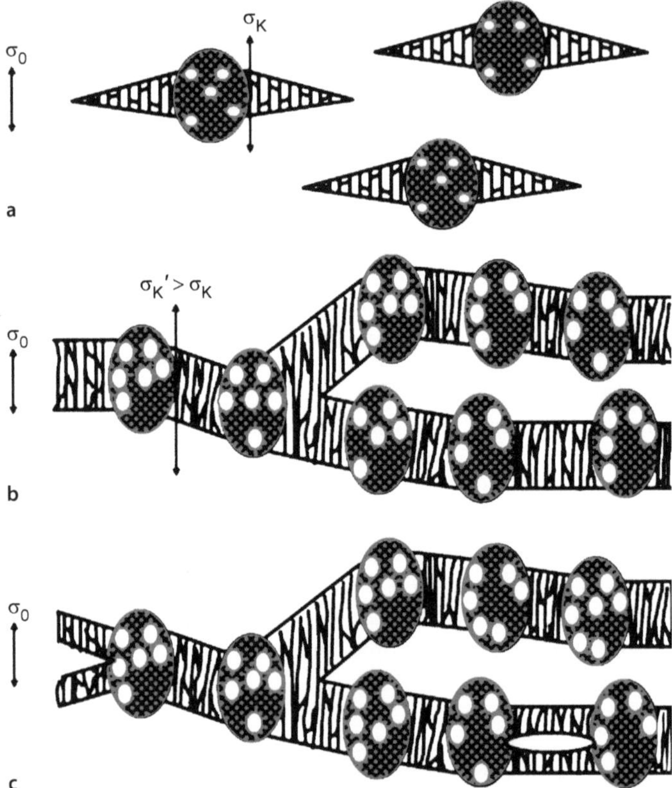

Fig. 18.11a–c. Three-stage mechanism of toughening (multiple crazing): **a** stress concentration σ_K at individual rubber particles; **b** the superposition of stress concentration fields occurs at a larger particle volume content (>15 vol%), and the resulting increased stress concentration σ_K' creates thicker crazes and craze bands; **c** crack stop at/in rubber particles

3. *Crack propagation.* Once the cracks have formed within the crazes, the crack propagation can be stopped and the crack tip blunted by neighbouring rubber particles. Consequently, the strain hardening of the yield zone, a process caused by stretching the rubber phase to very high strain, also contributes to the enhanced toughness.

After the failure of the material, its fracture surfaces can be examined by SEM. A typical example is shown in Fig. 18.12a, where the plastically stretched fibrils of matrix material are clearly apparent. It should be noted that these surfaces do not show the individual deformation steps (as illustrated in Fig. 18.11), only the final deformation process in front of the crack front, as sketched in Fig. 18.12b.

Maximum toughening – maximum craze formation due to the three-stage mechanism (Fig. 18.11) – can be attributed to several parameters, such as:

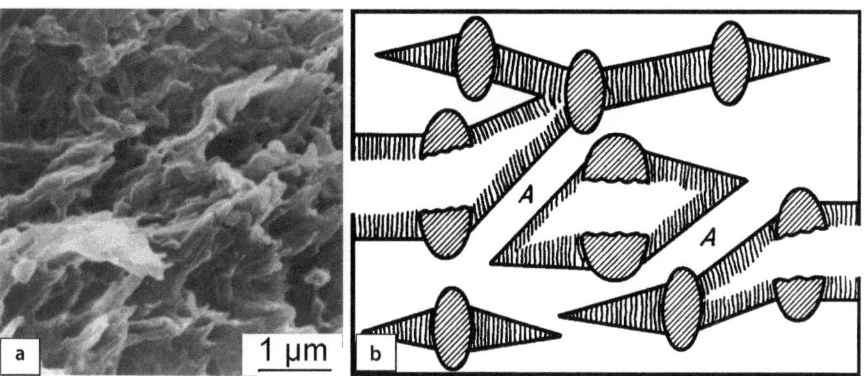

Fig. 18.12. **a** Fracture surface of ABS with highly plastically stretched matrix fibrils (SEM); **b** scheme of crack propagation in a highly crazed area; matrix strands (areas marked "A") are stretched to fibrils between ruptured crazes visible in micrograph (**a**)

–   Rubber particle content by volume
–   Rubber particle modulus
–   Particle size and size distribution
–   Shape and internal structure of particles
–   Degree of grafting at and in particles
–   The properties of the matrix itself.

From a morphological point of view, the particle size and the size distribution appear to be the main parameters responsible for enhanced toughness. The optimum rubber particle size is between 0.05 µm and 1 µm, especially between 0.1 and 0.5 µm, with a small size distribution. The influence of these particle parameters can be clarified in detail by performing ex situ or in situ deformation tests using electron microscopy [2]. Very small modifier particles below 200 nm in size (half the wavelength of visible light) are used to prepare transparent toughened polymers (e.g. for SAN or PMMA). One example of transparent rubber-toughened PMMA is shown in Fig. 18.13; this polymer contains 10 vol% PBA modifier of the core–shell type. The higher magnification image (c) clearly shows the deformation process in the particles: cavitation and fibrillation of the rubbery shell without deformation of the core. The matrix is found to be deformed in the highly stressed zones between the particles due to crazes or homogeneous yielding [15]. The electron microscopic investigation was performed in a 1000 kV HVEM, since PMMA is difficult to study due to its high irradiation sensitivity.

Several electron microscopic techniques have been used to study crazes in HIPS and ABS over the years:

–   Investigations of the surfaces of deformed samples have been realised in a TEM using replicas [9] or after selective etching directly in a SEM [10], which have yielded only limited results

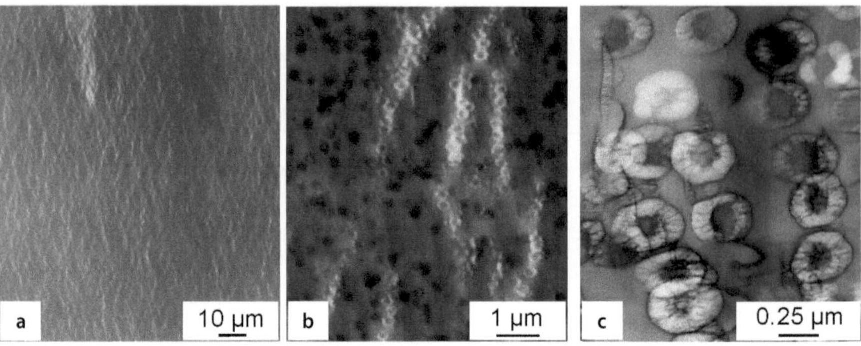

Fig. 18.13a–c. PMMA toughened with PBA core–shell particles: **a** overview of stress-whitened area; **b** craze-like deformation bands; **c** cavitation and fibrillation of the particle shell. (Deformed semithin section, stained with RuO_4, HVTEM)

- Filling crazes with chemical media (works best with OsO_4) and preparing thin sections for TEM provides a rough idea of the internal structure of crazes [11,12]; see Fig. 18.14
- Studying deformed semi-thin sections in SEM [13] has revealed the size and distribution of crazes

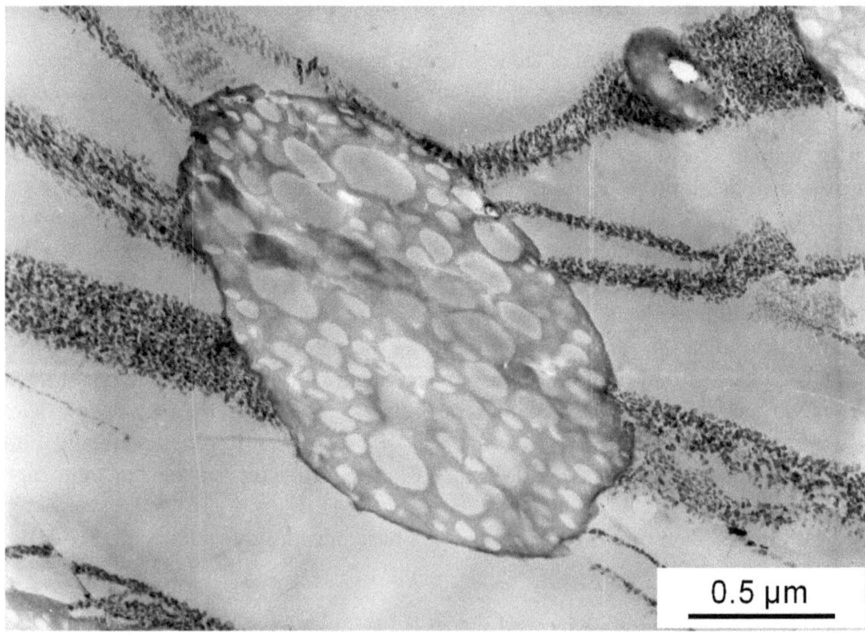

Fig. 18.14. Crazes around rubber particle in HIPS after deformation of a bulk sample and staining with OsO_4. (Loading direction vertical, TEM micrograph, reproduced from [2] with the permission of Hanser)

- Studying deformed solution-cast films by TEM [14] has given good results, but they are not representative of the bulk material.

To summarise, when studying rubber-toughened polymers, the best results are obtained by combining two approaches:

- Study (ultra)thin sections of deformed and selectively stained bulk material (see Fig. 18.14
- Deform thin sections directly in the microscope (as shown in the examples in Figs. 18.8–18.10).

Multiple Shear Yielding Mechanism

The micrographs in Fig. 18.15 show deformed and elongated rubber particles (the bright particles) in a PA matrix. The material between the particles is highly deformed and appears as bright, diffuse zones. All of the material between the particles is involved in plastic deformation in a form of homogeneous yielding without internal structure or cavitation (as seen in the crazes in HIPS). In this material, the plastically deformed areas are spread over a large volume of the sample, which indicates high toughness. For these systems with shear yielding, it was established that a sharp brittle-to-ductile transition occurs when surface-to-surface interparticle distances become lower than a critical value $ID_{crit.}$, which depends on the type of polymer matrix. $ID_{crit.}$ was found to be independent of particle size and rubber volume fraction [16]. The individual steps of stress concentration, initiation of multiple yielding and crack stopping are very similar to the three-stage mechanism in Fig. 18.11 [17]. A basic difference compared to the "multiple crazing" mechanism is the need for local cavitation in or at the rubber particles to enable yielding of the adjacent matrix strands. Extensive experimental studies have confirmed that microvoids form due to internal cavitation within the modifier particles or interfacial debonding at the interface between the matrix and the modifier particle.

In the following, characteristic deformation structures of different types of blends will be presented in more detail, as derived from in situ HVEM tensile tests.

Figure 18.16a,b shows typical deformation structures in PP/EPR blends with a low concentration of ethylene in the EPR particles. In these blends, the modifier particles are of a core–shell type with only one inclusion and are finely dispersed in the matrix. The shell phases consist of an amorphous ethylene–propylene block copolymer, which enhances the interfacial adhesion between the core and the matrix. At the early stage in the deformation, the modifier particles slightly deform in the tensile direction, together with the matrix. When the stress reaches a certain critical value, voids appear in the form of cavitation with or without fibrils in the interface between the modifier particles and the matrix. This process is strongly dependent upon the inherent properties of the rubbery shell. Along with successive void formation and the continuous growth of the voids themselves, weak shear bands form in the matrix ligaments between particles. When the polymer specimen is strained still further, shear yielding is induced in the whole specimen. Figure 18.16c,d

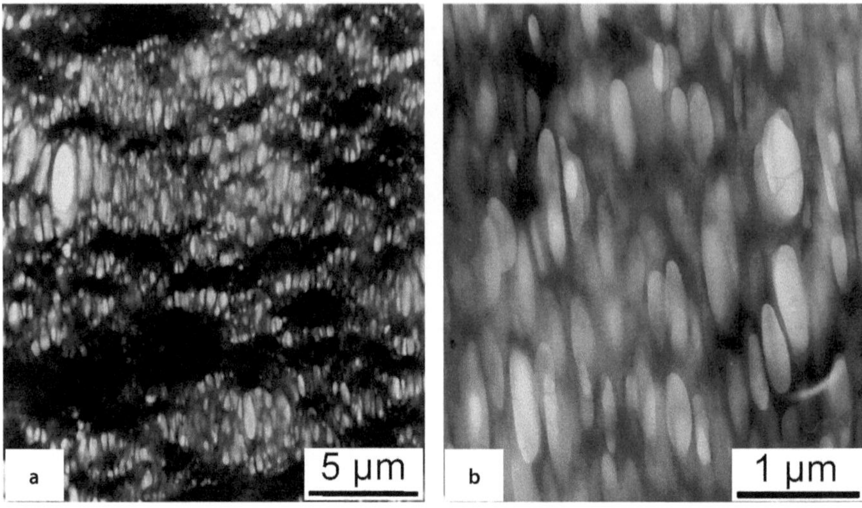

Fig. 18.15a,b. Deformation structures of rubber-modified PA (PA66, 22 vol% butyl acrylate): **a** plastic deformation of the matrix in band-like deformation zones with highly deformed particles; **b** microvoid formation inside plastically elongated particles

shows a HVEM micrograph taken during the deformation of a ethylene–propylene block copolymer containing a high ethylene content. In this blend, the modifier particles possess several inclusions in one rubbery shell. Void formation clearly occurs predominantly in the strongly plastically deformed EPR particles at the interface of the PE inclusions. In the next step, shear bands form in the matrix between the modifier particles. As the strain is increased, the sizes of the voids gradually increase, resulting in an acceleration of the shear flow in the matrix.

As demonstrated above, the main energy absorption mechanism that occurs under loading at room temperature is shear deformation of the matrix. When the loading occurs at lower temperatures, two changes must be considered:

1. If the temperature is below the glass transition temperature of the modifier particles (e.g. $T_g \approx -30\,°\text{C}$ for EPDM), the particles can no longer act as rubbery stress concentrators, the initiation of plastic deformation is lost and the materials break in a brittle manner. However, a low-temperature toughness is required for many practical applications.
2. Polypropylenes, which are commonly used as matrix materials, possess glass transition temperatures T_g of about $0\,°\text{C}$, and it is well known that below T_g the deformation mechanism changes from shear yielding to craze formation; see Fig. 18.17 [18].

Electron microscopic studies have revealed that PP with a low-temperature toughness can be realised using core–shell modifier particles [18]. Figure 18.18 shows a PP blend containing core–shell particles which was strained in situ on the cooling-

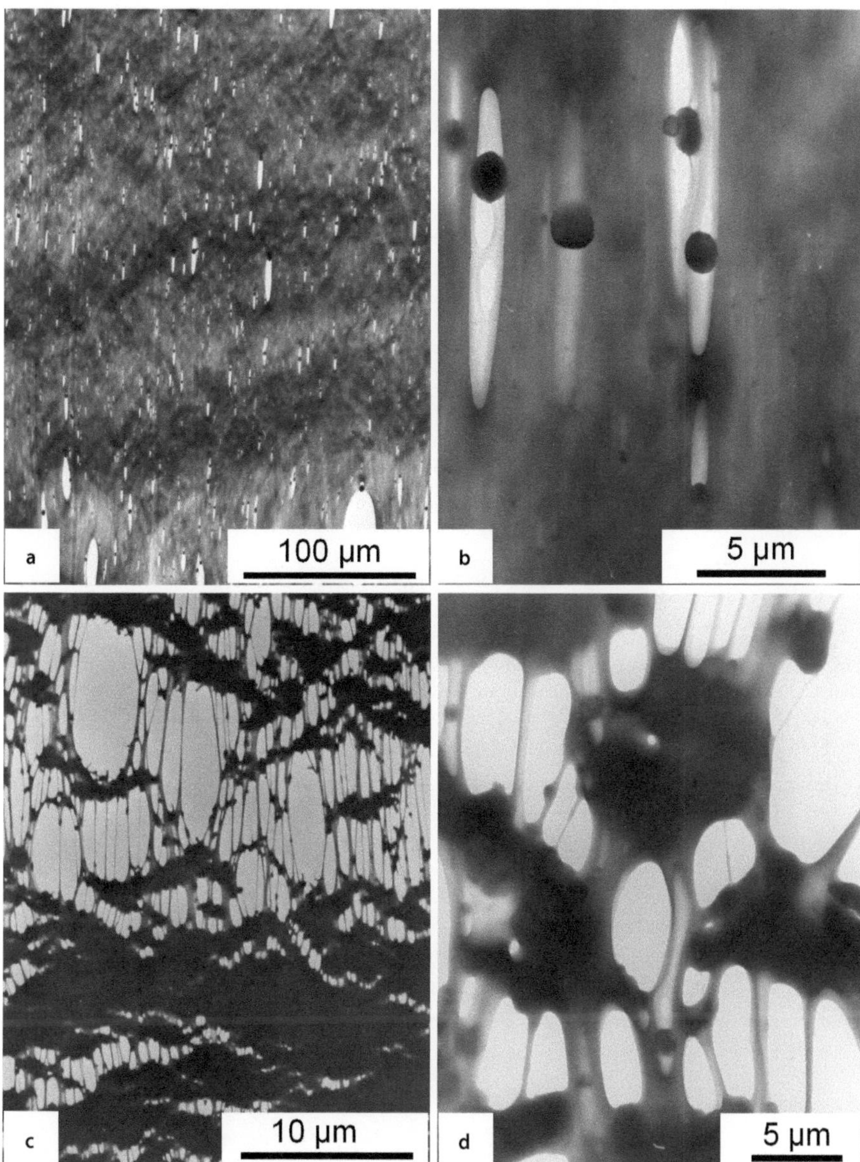

Fig. 18.16a–d. Micromechanical deformation processes of PP/EPR blends: **a,b** for EPR modifier particles with a low concentration of ethylene; void formation and fibrillation in the plastically deformed EP-rubbery shell; **c,d** for EPR modifier particles with a high concentration of ethylene; void formation in the EPR particles at the interface of PE inclusions. (Semi-thin sections, deformation direction vertical, HVTEM)

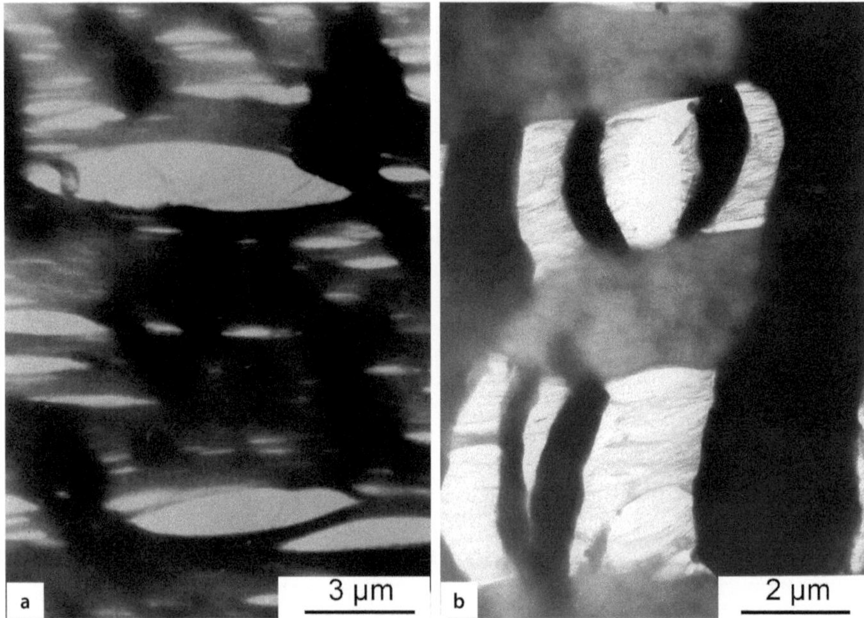

Fig. 18.17. a,b. PP/EPDM blend deformed at room temperature (*a*: cavitated EPDM particles with matrix shear yielding between particles) and at −40 °C (*b*: extension of EPDM particles and initiation of fibrillated crazes at the particles). (Ultrathin sections deformed in a cooling-tensile device, deformation direction horizontal, TEM)

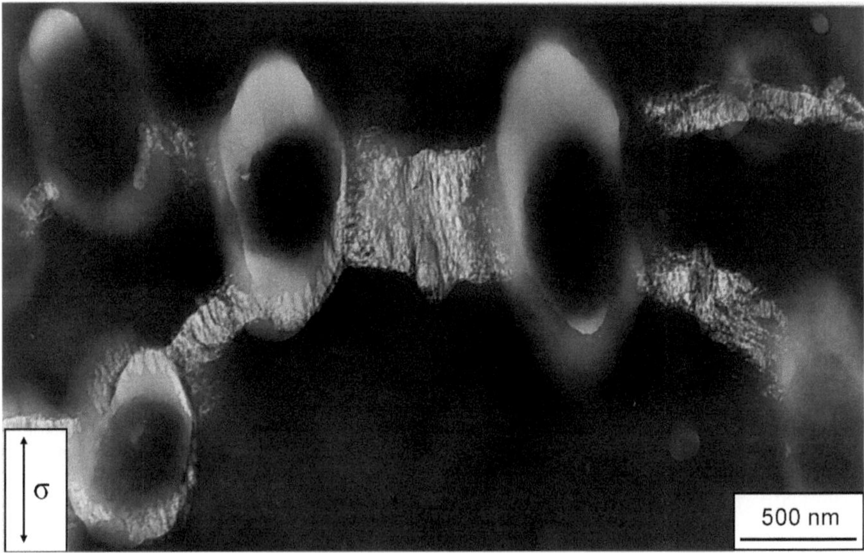

Fig. 18.18.   PP modified with EPR core–shell particles; thin section deformed at −100 °C in a 200 kV TEM, showing cavitated, fibrillated and stretched particle shells around PE cores and crazes in PP matrix, resulting in low-temperature toughness. (From [18], reproduced with the permission of the PRC, Institute of Materials)

tensile stage of a TEM at the very low temperature of −100 °C. The EP copolymer shell was stretched, cavitated and fibrillated, while crazes were formed in the adjacent matrix. This response exhibits some similarities to the toughening mechanism in HIPS (see Figs. 18.8 and 18.9). However, experiments carried out at a temperature well below the glass transition temperature T_g of the EP phase (ca. −50 °C) demonstrate that the toughening process is not limited by the T_g of the modifier particles.

One quick, cheap and accessible, in comparison to TEM, technique that can be used to study deformed bulk samples is SEM. A typical tensile bar is deformed until fracture and then embedded in epoxy. The middle part is then cut using a microtome and etched (permanganic etching); see Sect. 9.3.1. The results of this procedure for a PP/EPR blend are presented in Fig. 18.19, from the beginning of deformation with the formation of microvoids in the modifier particles (micrographs a, b) to a later stage with strong enlargement of the microvoids and intensive stretching with fibrillation of the PP matrix strands (micrographs c, d).If the details of the micromechanisms are known, this quick technique can be very useful, particularly when testing industrial materials.

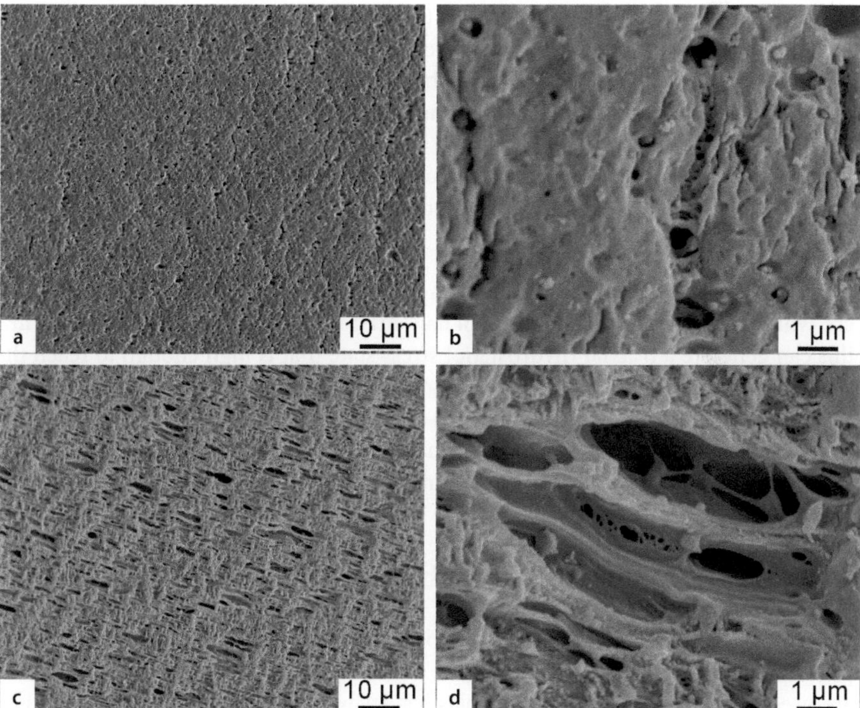

Fig. 18.19a–d. Deformed tensile bar of PP/EPR blend: **a,b** at the start of deformation with microcavitation in EPR particles; **c,d** enlargement of microvoids and stretching of PP matrix strands. (Deformation direction horizontal, internal surfaces of tensile bar microtomed, selectively etched, SEM)

Network Yielding Mechanism

Over several decades, it has been well established that there is an alternative and often very effective approach to the rubber toughening of amorphous polymers – the use of "rubber networks". This method involves embedding small particles of thermoplastic into a rubber network to form a honeycomb structure with thin layers of rubber separating the thermoplastic particles. As an example, Fig. 18.6 shows a network of a rubbery EVAc phase containing small particles of PVC. Since there are very thin network layers, the rubber content is usually kept below 10 vol% [7, 19]. When this rubbery network is tensile loaded, the PVC particles start to yield and absorb energy, which results in an enhancement in the overall toughness of the system. In the micrographs of Fig. 18.20, the two phases are clearly visible without any chemical staining, which is due to the effect of straining-induced contrast enhancement during the stretching of the sample. Due to the lower density and the predominant deformation of the rubber phase, the rubber network appears bright and the PVC particles appear dark. Under uniaxial tensile load, the following deformation processes occur [19, 21]:

- At the start of deformation, the weak rubber phase is stretched, resulting in the growth of a triaxial stress state in the (whole) network.
- The rubber phase transfers stresses from one PVC particle to another. When the stress transfer is high enough to reach the yield stress of PVC, the particles start to deform plastically; the yielding of numerous PVC particles mainly absorbs the total fracture energy.
- Through the partial rupture of the rubbery network, microvoids are generated in the specimen, yielding intense fibrillation of the network and an additional plastic yielding of the PVC particles.

One critical parameter of this mechanism is the thickness of the rubber network layers, which must be around a few tens of nanometres. Only a thin-walled network like this can generate a triaxial (hydrostatic) stress state that is high enough to reach the yield stress of PVC. However, the application of this polymer industrially is greatly hindered by its process sensitivity, which causes the destruction of the network and its transformation into an excessively fine rubber particle distribution. This phase separation results in a drastic reduction in toughness; see Fig. 18.7 [8].

18.4 Additional Toughening Mechanisms

Electron microscopic inspections of rubber-modified polymers have revealed some peculiar micromechanisms. One of these is connected to core–shell particles, which are used to toughen several thermoplastics, such as PS, SAN, PMMA. Previously, the hard cores in core–shell particles or the PS subinclusions embedded in the rubber in salami particles were not thought to actively participate in the energy absorption mechanism. They were generally considered to be passive participants in the large strain deformation of the blend, remaining within the elastic region up to

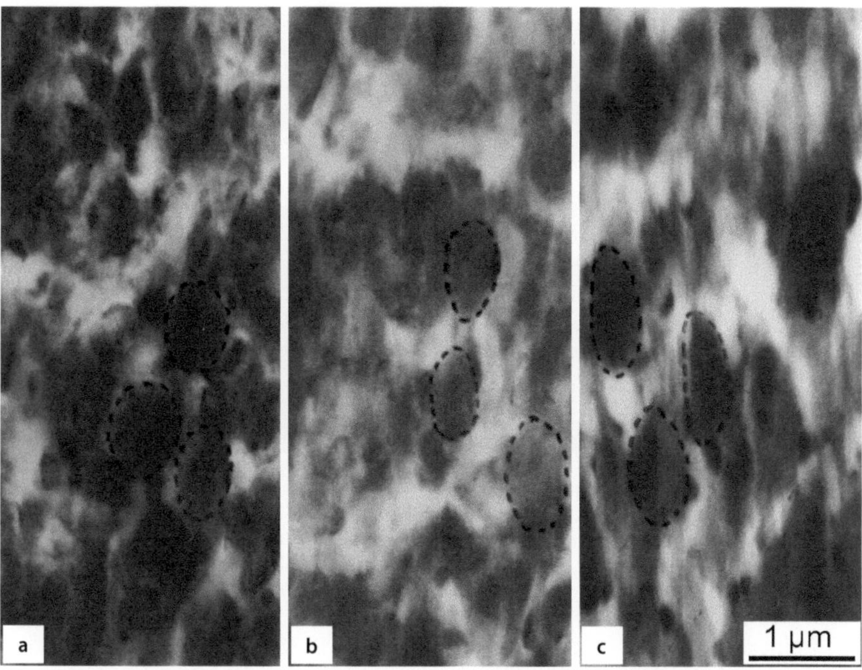

Fig. 18.20a–c. Rubber-toughened PVC with a network structure of rubber (EVAc, *bright appearance*) around PVC particles (*dark*); **a**, **b**, **c** show increasing elongation. (Deformation direction is vertical, HVTEM)

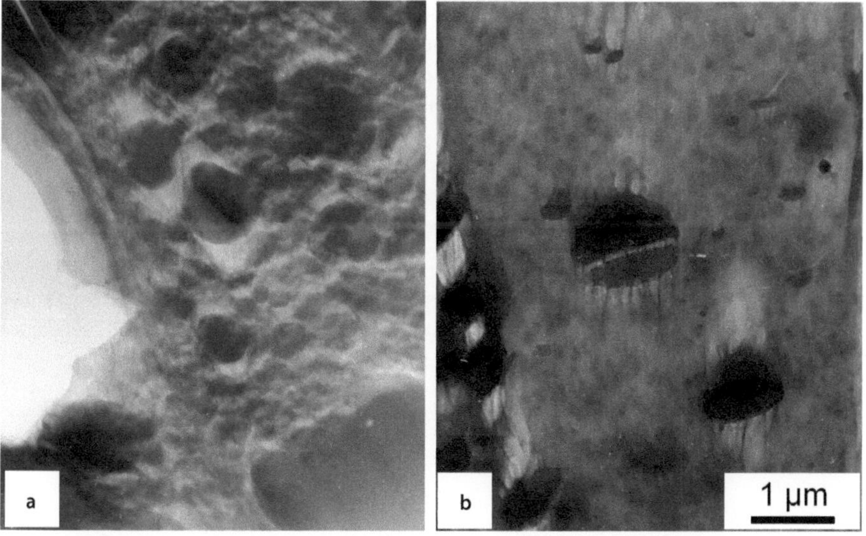

Fig. 18.21a,b. Stretched salami particles in front of a crack tip in HIPS: **a** only the rubber is fibrillated; **b** some large PS subinclusions are flattened. (Semi-thin section deformed, tensile direction vertical, HVTEM)

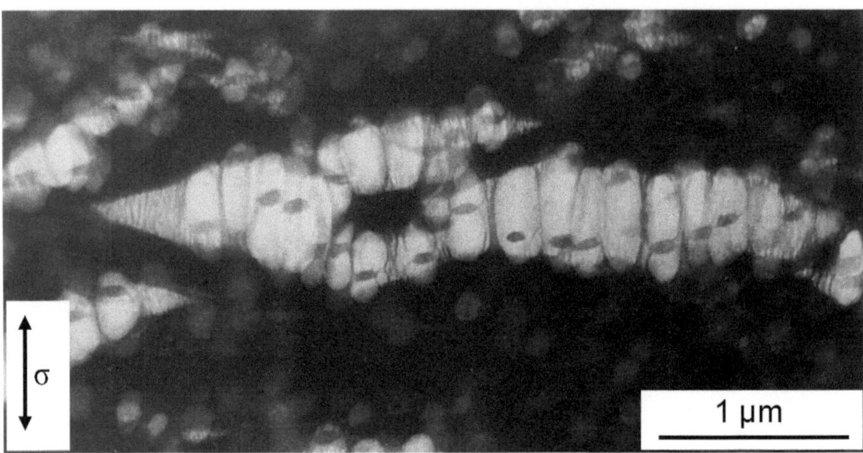

Fig. 18.22. Stretched ASA blend with crazes running between extended and fibrillated rubber particles and flattened PMMA cores. (Semi-thin section stretched to high strains in situ on HVTEM, tensile direction marked with an *arrow*; from [18], reproduced with the permission of PRC, Institute of Materials)

the point at which the specimen fractures. Figure 18.21 a shows a highly stretched salami particle in HIPS with fibrillated rubber, where the rigid PS subinclusions increase the volume fraction of the particles but do not deform plastically. In micrograph (b), some of the large PS subinclusions are flattened by pulling out fibrils from the polar regions of the subinclusions. Recently, a comparable effect has been observed in an ASA polymer containing acrylic core–shell particles. Figure 18.22 shows a pattern of light dilatation bands formed as a result of crazing in the PSAN matrix and cavitation in the modifier particles. Most of these particles feature dark, lens-shaped PMMA cores with highly stretched rubber fibrils connecting the matrix to the rigid cores [18]. This drawing process is almost identical to the drawing of fibrils from the walls of a craze (see Figs. 15.6, 15.8, 15.9). This process of large strain plastic deformation in the glassy thermoplastic cores of modifier particles has been called "core flattening", and is proposed to be a new toughening mechanism [18].

Another toughening mechanism revealed by electron microscopic deformation experiments is so-called "thin-layer yielding", which appears in layered PB–PS block copolymers; these will be discussed separately in Chap. 19.

References

1. Bucknall CB (1977) Toughened plastics. Applied Science Publ., London
2. Michler GH (1992) Kunststoff-Mikromechanik: Morphologie, Deformations- und Bruchmechanismen. Hanser-Verlag, München
3. Riew CK, Kinloch AJ (eds) (1996) Toughened plastics. American Chem. Soc., Washington, DC
4. Bucknall CB (1997) In: Haward RN, Young RJ (eds) The physics of glassy polymers, 2nd edn. Chapman & Hall, New York, p 363

5. Bucknall CB (2000) In: Paul DR, Bucknall CB (eds) Polymer blends, vol 2. Wiley, New York, p 83
6. Heckmann W, McKee GE, Ramsteiner F (2005) In: Michler GH, Baltá-Calleja FJ (eds) Mechanical properties of polymers based on nanostructure and morphology. Taylor & Francis, Boca Raton, FL, Chap 11, p 435
7. Menges G, Berndtsen N, Opfermann J (1979) Kunststoffe 69:562
8. Michler GH (1981) Plaste Kautschuk 28:191
9. Matsuo M (1966) Polymer 7:421
10. Bucknall CB, Clayton D, Keast WE (1972) J Mater Sci 7:1443
11. Matsuo M (1969) Polym Eng Sci 9:206
12. Lee D (1975) J Mater Sci 10:661
13. Michler GH, Gruber K (1976) Plaste Kautschuk 23:496
14. Donald AM, Kramer EJ (1982) J Mater Sci 17:2351
15. Laatsch J, Kim G-M, Michler GH, Arndt T, Süfke T (1998) Polym Adv Technol 9:716
16. Wu S (1985) Polymer 26:1855
17. Kim G-M, Michler GH (1998) Polymer 39:5689, 5699
18. Michler GH, Bucknall CB (2001) Plast Rubber Compos 30:110
19. Michler GH, Gruber K (1976) Plaste Kautschuk 23:346
20. Michler GH (1986) Polymer 27:323
21. Michler GH (2001) J Macromol Sci Phys B40:277

19 Block Copolymers

In block copolymers, there is enormous variety in terms of the type, size and arrangement of the different blocks due to microphase separation, and so electron microscopy and scanning force microscopy have been established as the most powerful analytical tools in block copolymer research. After an overview of the different configurations of block copolymers, some representative classes of diblock and triblock copolymers and block copolymer/homopolymer blends are discussed, with special emphasis placed on the application of electron microscopy to their structural characterisation at very small length scales. Three EM techniques are described that offer excellent opportunities to investigate deformation-induced local structural changes. As well as the usual deformation structures, such as shear bands and crazes, block copolymers with a lamellar block arrangements revealed the new mechanisms of thin layer yielding and chevron formation.

19.1 Overview

Block copolymers constitute a special class of heterogeneous polymers in which two or more homopolymer chains (blocks) exist in a single molecule (Fig. 19.1; see also Fig. 14.3). The incompatibility of the constituent blocks leads to intramolecular phase separation, but the chemical connectivity restricts the spatial dimension of the phase segregation to the nanoscale. As a result, at sufficiently high molecular weight, monodisperse block copolymers form an array of periodic nanostructures (periodicity 10–100 nm) commonly referred to as microphase-separated structures.

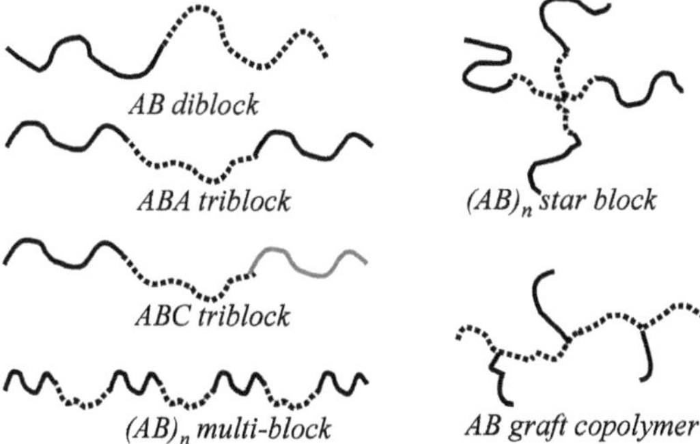

Fig. 19.1. Architectures (configurations) of frequently studied block copolymers

By changing the relative composition, the compatibility between different components and the architecture of the copolymer molecules, the size and the type of nanostructure can be precisely controlled.

Since the synthesis of the first block copolymers by living anionic polymerisation
in the mid-1950s, electron microscopy (EM) has provided an incomparable contribution to our understanding of the structure–property correlations of block copolymers, and so electron microscopy has been established as the most powerful analytical tool in block copolymer research. Today, various scattering techniques, scanning force microscopy (AFM) and simulation provide valuable supplement to the
structural characterisation achieved by electron microscopy.

Generally speaking, there are two kinds of block copolymers with respect to their
molecular construction: linear and branched; see Fig. 19.1. The linear ones can be AB
diblocks, ABA triblocks, ABC triblocks, $(AB)_n$ multiblocks, etc. (where the letters
A and B stand for individual polymer blocks). Graft copolymers and star-shaped (also
called radial) ones belong to the branched category. The tailored block copolymers
are generally prepared by ionic polymerisation [1], but recently controlled radical
polymerisation has been successfully used to synthesise well-defined block copolymer architectures [2].

Thus research interest in block copolymer materials continues to increase. These
materials act not only as model systems for the study of self-assembly phenomena
in soft matter, but also offer a number of new potential applications in nanoscience
and technology. An up-to-date review of block copolymer-based nanotechnology
has been collected in a recent monograph [3]. These polymers have long been used
as compatibilisers for incompatible polymer blends (Chap. 17) and as thermoplastic elastomers [4] (Chap. 20). As block copolymer properties are generally governed
by their solid state nanostructures, it is of practical importance to controllably vary
their morphologies. This can be done by several methods, such as by modifying the
molecular architecture at constant composition, blending with homopolymers and
other block copolymers, and even developing new polymerisation techniques and
changing the processing conditions.

Combinations of electron microscopy (especially transmission electron microscopy (TEM, see Chaps. 2 and 3) with scattering techniques have been used in the
characterisation of block copolymer nanostructures. In addition, the recently developed techniques of scanning or atomic force microscopy (AFM, Chap. 5) can be successfully employed to map the structure and properties of block copolymers. The
most significant aspect of electron microscopy is its ability to image the highly local structure of a bulk sample, enabling a straightforward elucidation of the specimen morphology. As in other polymeric systems, the selective staining of one (or
more) of the components by a heavy metal oxide is required for the transmission electron microscopic examination of block copolymer systems. In contrast,
the sample topography or differences in the viscoelastic properties are
employed to map the structure and properties of the polymers by means of AFM
[5–7]. Similar to other polymers, the specimens suitable for those studies are prepared by various standard techniques including ultramicrotomy, spin-coating,

dip-coating, shadowing, etc. (see Chaps. 10 and 11). In the following sections, some representative classes of the block copolymers are discussed, with special emphasis placed on the application of electron microscopy to their structural characterisation.

19.2 Morphology

19.2.1 Block Copolymer Nanostructures via Self-Assembly

The driving force for the microphase separation in an AB diblock copolymer that leads to periodic nanostructures is the delicate balance between the enthalpic (repulsive interaction between the monomers A and B, which increases as the value of χ increases) and entropic (configurational entropy of mixing, which decreases as N increases) components (where χ and N represent the Flory–Huggins interaction parameter and the total degree of polymerisation, respectively). Thus, the phase behaviour of a binary block copolymer is well illustrated by a phase diagram in which the product χN is plotted against the copolymer composition f_A (the effective volume fraction of component A) [3].

At sufficiently high values of N, on cooling from the melt (which increases the value of χ), body-centred cubic spheres (S) of A, hexagonal packed cylinders of A, co-continuous gyroid (G) structure and alternating A and B lamellae (L) are formed in a block copolymer with increasing volume fraction of A. On further increasing the volume fraction of A, the structures are formed in reverse order. The solid state morphology formed in the ABA triblock copolymers exhibit similar characteristics to their AB counterparts. Generally speaking, the change in the relative A/B volume fraction leads to a variation in the curvature of the interface (which is only a few nanometres thick) between the A and B homopolymer chains. It should be noted however that a change in molecular architecture can significantly alter the behaviour of the block copolymer phase [8].

Polystyrene-block-polybutadiene-block-polystyrene (SBS) triblock copolymers belong to an important class of polymers called thermoplastic elastomers (see Chap. 20). These are examples of polymers that have been modified to enhance the toughness through the use of a defined nanostructure and morphology. However, in contrast to classical toughened polystyrene (generally termed "high-impact polystyrene" or HIPS, see Chap. 18), the block copolymer-based products are optically transparent due to the nanoscale phase separation. At low polystyrene contents, the copolymers behave like crosslinked rubbers, but they can be processed like thermoplastics at elevated temperatures, as the physical crosslinks of the microphase-separated PS domains are destroyed. By changing the polystyrene/polybutadiene (PS/PB) volume ratio, a wide variety of morphologies can be produced. As a result, material properties ranging from those of an elastomer to those of thermoplastics or to those of impact-modified polystyrene can be selected.

Figure 19.2 illustrates TEM micrographs of some of linear SBS triblock copolymers cast from their toluene solutions. The periodicity of the nanostructures is in

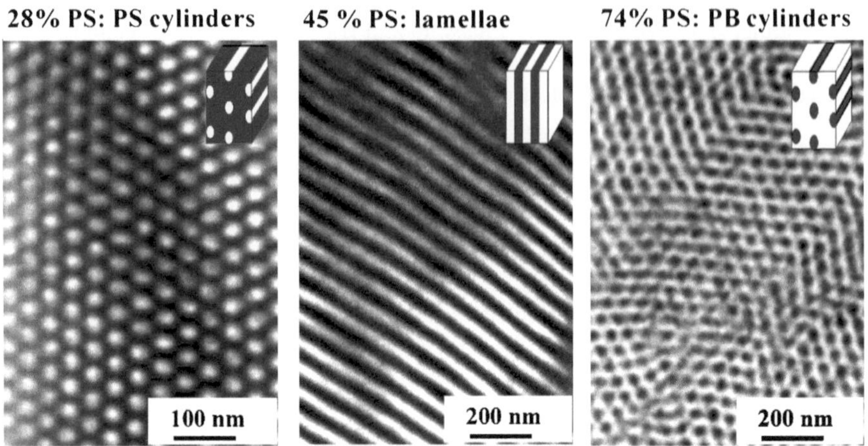

Fig. 19.2. TEM micrographs showing the nanostructures observed in SBS triblock copolymers with different compositions; the molecular weight of the materials is approximately 100 000 g/mol, the PS volume fraction (Φ_{PS}) is indicated. (Solution-cast films, PB phase is stained by OsO_4)

the range of 30–40 nm. The butadiene phase is selectively stained by osmium tetroxide (OsO_4) and hence appears dark in the TEM micrographs. One should note that the individual polymer chains are well phase-separated by a narrow interface a few nanometres in thickness. The appearance of the phase separation in the block copolymer is also visible in the existence of two separate glass transition temperatures corresponding to two homopolymers. The selection of staining agent used to increase the mass contrast during the TEM investigation of the block copolymer depends, as it does for other polymeric systems, on the nature of the monomers present. Sometimes treating the specimens with more than one staining agent can also lead to fruitful results.

The partially hydrogenated versions of styrene/diene block copolymers are well-known for their improved UV and oxidative stabilities. Thus, styrenic block copolymers with hydrogenated polybutadiene (PB) or polyisoprene (PI) mid-blocks have been commercialised by various companies. Due to the absence of double bonds in the molecules, OsO_4 cannot be used as staining agent. Therefore, in order to image nanostructures of a SEBS triblock copolymer (where EB stands for the ethylene/butylene copolymer formed by the hydrogenation of the polybutadiene block) by TEM, for instance, RuO_4 is used. The latter reacts preferentially with aromatic rings of the polystyrene (PS) blocks, making the PS domains appear dark in the micrographs (see Fig. 19.3, left). On the other hand, AFM may yield better results for this copolymer without any chemical treatment (see Fig. 19.3, right). Thus, this technique, which measures the differences in local mechanical properties of nanostructures, can be used for practically all block copolymers (see Chap. 5).

The AFM technique even works successfully for those fully hydrogenated SBS block copolymers (where the PS blocks and PB blocks are converted into polyvinyl cyclohexane or PCH and ethylene/butylene copolymer or PEB, respectively) for

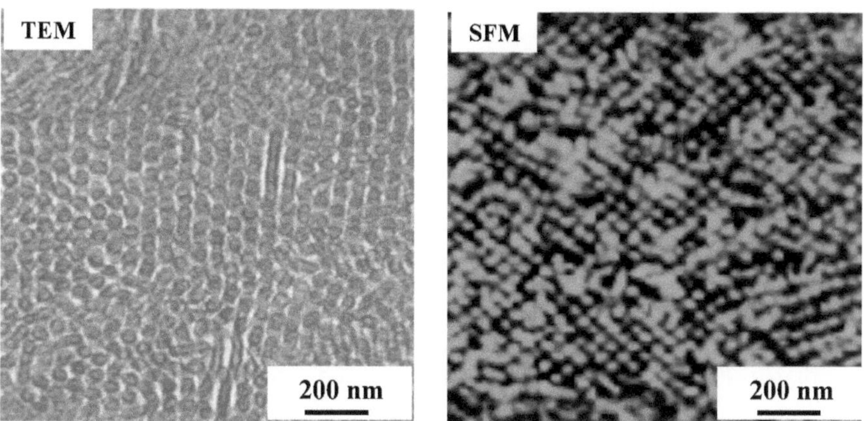

Fig. 19.3. *Left*: TEM image of a RuO$_4$-stained ultrathin section of a bulk Kraton SEBS triblock copolymer with about 30% PS showing cylindrical PS domains. *Right*: Tapping-mode AFM phase image of a dip-coated film of SEBS

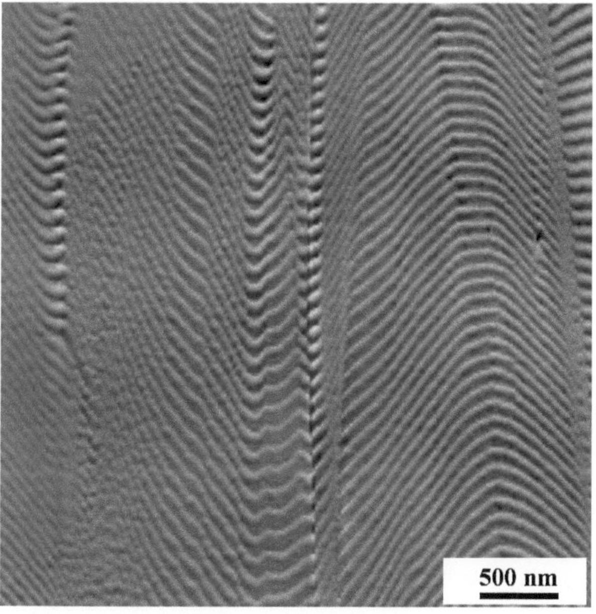

Fig. 19.4. Tapping-mode AFM phase image of dip-coated film of PCH–PEB block copolymer showing lamellae

which electron microscopy does not offer a straightforward solution to structural characterisation. These polymers have even been found to possess properties superior to those of partially hydrogenated block copolymers [4]. Figure 19.4 shows the AFM phase image of just such a block copolymer prepared by a dip-coating method. The excellent arrangement of alternating lamellar nanostructures of polyvinyl cyclohexane (bright regions) and ethylene/butylene copolymer (dark regions) can easily be identified.

The AFM contrast between the components of the block copolymer shown in Fig. 19.4 arises from differences in the local nanomechanical properties [6]. Using AFM tapping-mode phase imaging, other systems such as styrene/butadiene, styrene/isoprene, and methyl methacrylate/alkyl acrylate systems can be successfully imaged, where one of the components is much stiffer than the other at room temperature. It should be noted that the AFM can even be employed for block copolymers with components that have very similar properties. Figure 19.5 shows, for example, AFM height (left) and phase (right) images of a compositionally asymmetric polystyrene-block-poly(methyl methacrylate) diblock copolymer (PS/PMMA volume ratio about 70:30) with PMMA cylinders embedded in PS matrix. Even in the height images, the hexagonal arrangement of the cylinders is clearly visible.

Provided that one of the blocks contains metal atoms in it, the block copolymer nanophases can be imaged by the TEM even without chemical treatment. Due to the intrinsic mass contrast between the blocks, nanostructures of block copolymers based on poly(ferrocenyl dimethylsilane) and poly(methylmethacrylate) (PFS-block-PMMA) were successfully imaged by normal transmission electron microscopy using ultrathin sections of virgin samples [9]. In this case, the iron atoms in the PFS molecules act as intrinsic staining agent and cause the PFS phase to appear dark in the TEM micrographs. Using special TEM techniques such as Lorenz microscopy and electron holography (see Chap. 3), other unstained block copolymer samples have also been imaged [10, 11].

As mentioned earlier, the nanostructures of almost all of the varieties of block copolymers can be mapped successfully by AFM. This technique is especially suited to the study of thin films deposited onto a flat surface. However, skilled experimentalists can also prepare very flat surfaces by ultramicrotomy without damaging the

Fig. 19.5. Unfiltered tapping-mode AFM height (*left*) and corresponding phase (*right*) image of a compositionally asymmetric PS/PMMA diblock copolymer (70/30) showing hexagonally arranged PMMA cylinders in the PS matrix

nanostructures. Thus, the technique can also be used to study the bulk morphologies of block copolymers [12].

19.2.2 Examples of Tailored Block Copolymer Morphology

Influence of Chain Architecture

Based on many experimental [12–16] as well as theoretical [17, 18] studies on various kinds of block copolymer systems, it has been demonstrated that changing the block copolymer molecular architecture can have a significant impact on its phase behaviour and physical properties. Modifying the chain architecture has been found to shift the stability window for different block copolymer morphologies [13] and even the effective volume fractions of the components and the properties of individual domains [14–16]. As an example of this new route to morphological control, Fig. 19.6 collects TEM images of three asymmetric styrene/butadiene block copolymers (designated SB1, SB2 and SB3; PS volume fraction ~ 0.70). A detailed account of this method is discussed elsewhere [8]. Their respective molecular structures and schemes of the morphologies are presented at the side of each micrograph.

Each copolymer represented by the micrographs in Fig. 19.6 is highly asymmetric; i.e. the lengths of the outer PS blocks are very different. The centre PB block of each arm in SB1 has a tapered interface with the inner PS blocks. The other polymers lack a pure PB block and have a random copolymer of PS and PB (PS-co-PB) as the centre block instead.

For a symmetric SBS block copolymer with a styrene component comparable to that of these materials, one would expect a morphology comprising PB cylinders in PS matrix (similar to Fig. 19.2, right). However, the unusual architecture of these copolymers leads to morphologies that are quite different: special types of lamellae in SB1, well-defined lamellae in SB2 and even bicontinuous structures in SB3. In this way, this path to architectural modification gives rise to a wide range of morphologies at constant composition. For the block copolymers imaged in Fig. 19.6 (middle and right), the soft phase (i.e. the PS-co-PB copolymer) is stained by OsO_4. As the soft phase contains polystyrene as well, the TEM micrographs do not reflect the actual PS/PB composition quantitatively.

A very important consequence of modifying the block copolymer chain architecture is that a wide spectrum of mechanical properties can be tailored without changing the overall chemical composition. For instance, all of the high polystyrene content block copolymers presented in Fig. 19.6 showed much higher toughness than the usual symmetric SBS with the equivalent composition [8].

Block Copolymer/Homopolymer Blends

Another approach to fine-tuning the block copolymer nanostructure is also provided by blending a two-component block copolymer with other block copolymers or constituent homopolymers [19]. It was shown for both AB diblock and ABA triblock copolymers that different microphase-separated morphologies can be produced by simply mixing two block copolymers with highly asymmetric but complementary compositions [20,21]. Thus, this route makes it possible to design all possible block copolymer morphologies by simply mixing two block copolymers of varying compositions.

The relative volume fraction of the constituents of a two-component block copolymer can be altered by incorporating homopolymers identical to or miscible with a particular block. In this case, the molecular weight of the added homopolymer relative to that of the corresponding block plays a key role. In order that the homopolymer is accommodated into the corresponding domain of the block copolymer, the molecular weight of the former should be sufficiently smaller than that of the latter.

Figure 19.7 illustrates the effect of mixing low molecular weight PS (PS015) with a block copolymer. The block copolymer chosen was the asymmetric star block copolymer (SB1: see Fig. 19.6, left). The average molecular weight of PS015 is about 15 000 g/mol while that of the longest block of the block copolymer is approximately 70 000 g/mol. The star block copolymer has an essentially lamellae-like morphology, with each PB lamella embedding a thin discontinued PS layer (Fig. 19.6, left). The addition of 20 wt% of PS015 to the star block copolymer leads to the partial thickening of PS lamellae as the PS015 chains are accommodated by the PS block chains of SB1 (Fig. 19.7, left). At 40 wt% PS015, cylindrical and lamellar morphologies co-

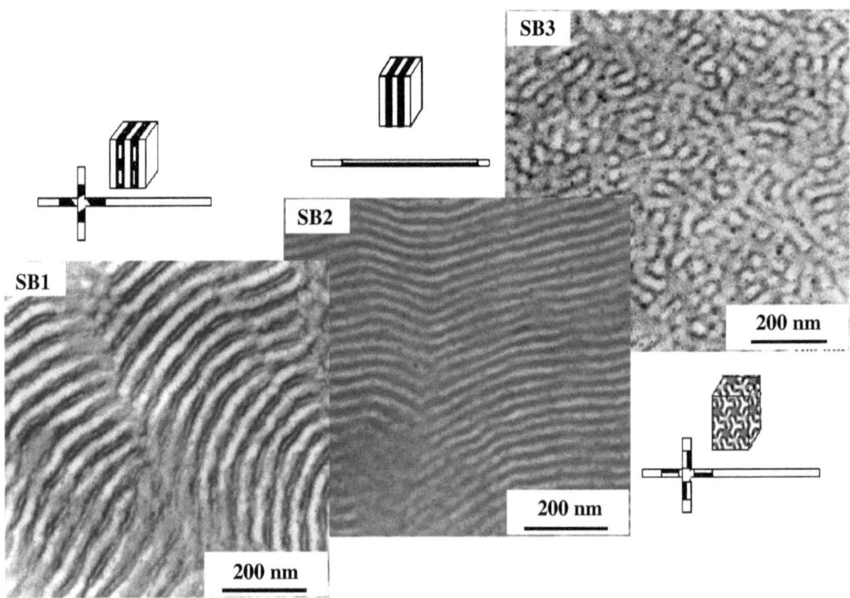

Fig. 19.6. TEM micrographs showing morphologies of asymmetric styrene/butadiene block copolymers; OsO_4 makes the butadiene-rich phase appear *dark*; PS volume fraction ~0.70

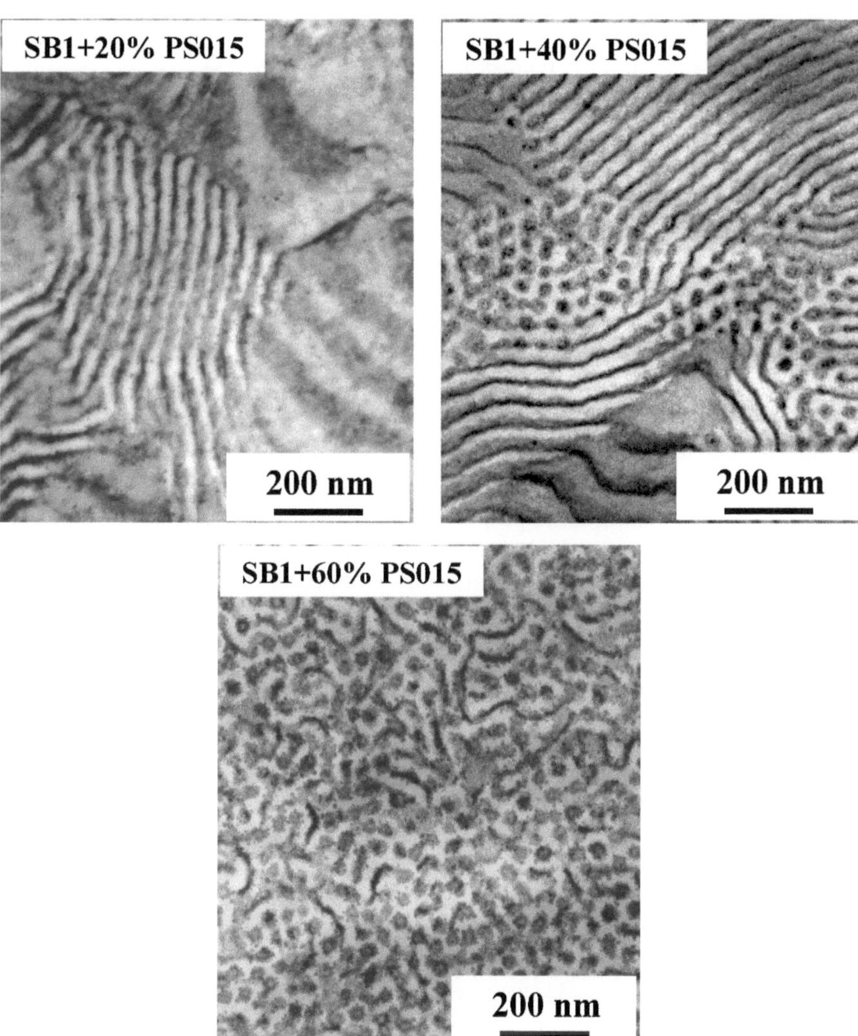

Fig. 19.7. TEM micrographs showing morphologies of solution-cast blends of a star block copolymer (SB1) and different amounts of low molecular weight homopolystyrene PS015 (M_w~15 000 g/mol); OsO_4 makes the butadiene-rich phase appear *dark*

exist (Fig. 19.7, middle) and the morphology transforms into a co-continuous one at 60 wt% PS015 (Fig. 19.7, right).

The morphological transition observed is associated with the change in interfacial curvature and packing density that results from the variation in chain configuration (discussed in [19]). A careful examination of the TEM micrographs reveals that the tiny PS domains embedded in the PB phase always exist independent of the PS015 content in the blends. It should be noted that the coexistence of cylindrical and lamellar morphologies in these blends is a special case, as this behaviour cannot be expected in monodisperse diblock copolymer systems. The observation discussed here is actually the result of the special chain architecture of the star block copolymer.

The possibility of designing different block copolymer morphologies by adding low molecular weight homopolymers is useful for applications where the nature of the nanostructure is of prime importance. However, viewed from a mechanical properties perspective, this may weaken the entanglements (i.e. the spontaneously formed molecular knots of polymer chains) and worsen the strength and ductility of the product. Therefore, for applications that demand a high level of toughness and stiffness, styrenic block copolymers are mixed with relatively high molecular weight PS. In this case, the corresponding PS block of the copolymer cannot accommodate the large homopolymer chain and so the added polystyrene forms a separate phase. Figure 19.8 shows a TEM micrograph of a blend of an SBS star block copolymer (SB1 in Fig. 19.6, left) and general-purpose polystyrene (GPPS) with a weight average molecular weight of about 190 000 g/mol. The blend contains 20 wt% of GPPS. At higher GPPS contents, the mechanical properties may again worsen due to stress concentrations and void formation at the periphery of the particles [22]. The morphologies of such blends should be optimised through the proper control of the processing parameters.

Block Copolymers via Controlled Radical Polymerisation

The practical application of styrenic block copolymers has two major drawbacks. First, the unsaturated polybutadiene block is very sensitive to photodegradation and can easily be oxidised by atmospheric oxygen unless the inherent double bonds have already been hydrogenated. The second disadvantage is the low service temperatures (60–70 °C) of these polymers, which are limited by the glass transition temperature of polystyrene (T_g–$_{PS}$ ~ 100 °C). Fully saturated versions of these block copolymers certainly have a wider service temperature range [3]. However, this requires an additional step during the polymerisation process. Further, the monomers must be absolutely pure in order to avoid the premature termination of the reaction. Thus one would be interested in producing block copolymers, via relatively convenient polymerisation techniques, that offer a broad service temperature range and that are not too susceptible to thermally and radiation-induced degradation.

With the development of controlled living radical polymerisation (CLRP), the number of molecularly engineered materials, including potential TPEs, has expanded exponentially [23]. In this respect, the use of PMMA-block-PnBA-block-PMMA triblock copolymers [PMMA: poly(methyl methacrylate) and PnBA: poly(n-butylacrylate)] can solve the drawbacks mentioned above. The higher T_g of PMMA (~120 °C) than that of PS in its SBS counterpart can clearly enhance the upper service temperature range of the copolymer. On the other hand, the rubbery nBA block (T_g ~ −50 °C) is insensitive to ultraviolet radiation and is more resistant to thermal oxidation.

Scanning force microscopy was successfully employed to probe the nanostructures of those block copolymers using thin films. In Fig. 19.9, the morphologies of two different block copolymers that contain 44 and 32 wt% PMMA are shown. In accordance with the phase diagram [3] established for monodisperse block copolymer systems, alternating lamellae (Fig. 19.9, left) and PMMA domains dispersed in a PnBA matrix (Fig. 19.9, right) are discernible.

This new class of block copolymers possesses sufficiently high application potential to allow it to compete with classical styrenic block copolymers in the market.

Processing-Induced Nonequilibrium Morphologies

So far we have discussed the morphologies of various block copolymer systems prepared by solvent casting followed by post-annealing procedures. Generally, a good solvent for the copolymer is used which is selective for none of the components. The films are prepared by allowing the solvent to evaporate over several days (see also

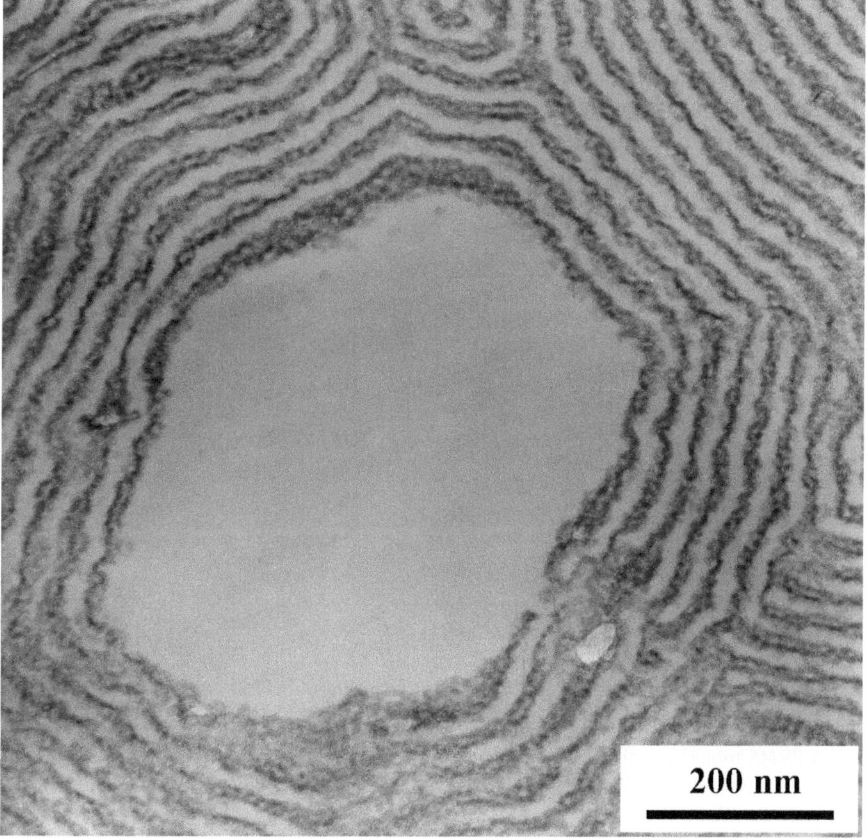

Fig. 19.8. TEM micrograph showing the morphology of a solution-cast blend of a star block copolymer (SB1) and GPPS (M_w~190 000 g/mol); OsO_4 makes the butadiene-rich phase appear *dark*; PS particles remain unstained

Chap. 11). Thus the morphologies are close to the equilibrium state. The equilibrium morphologies can also be obtained by prolonged annealing of the block copolymer melt, but this may lead to thermal degradation of the sample. Thus the more convenient and practical technique of solution casting is usually employed to study fundamental issues of block copolymer phase behaviour.

For practical purposes, common processing techniques like extrusion, injection, compression or blow-moulding are used, where nonequilibrium morphologies of the block copolymers may evolve. The morphology may alter during processing under the influence of external shear [24]. Even for solution processing, the morphologies formed may vary depending on the selectivity of the solvent towards a particular block. Under specific circumstances, one significant effect arising from the application of external forces such as shear [25], electric field, etc. [26] is the orientation of nanostructures along the direction of external force. This orientation leads to anisotropic mechanical properties [25]. Figure 19.10 shows the TEM micrograph of a lamellar block copolymer (SB2 in Fig. 19.6), the melt of which was subjected to steady shear during extrusion processing. In this way, macroscopic alignment of the lamellae was achieved. There are certainly several defects, but the morphology (consisting of continuous and uniform lamellae) extends over several microns and is therefore largely single-crystal-like in nature. Other block copolymer morphologies (such as cylinder and gyroid) can also be aligned using an external shear field. However, due to the parabolic profile of the shear field, the extruded sheet does not possess a uniform morphology across the whole cross-section.

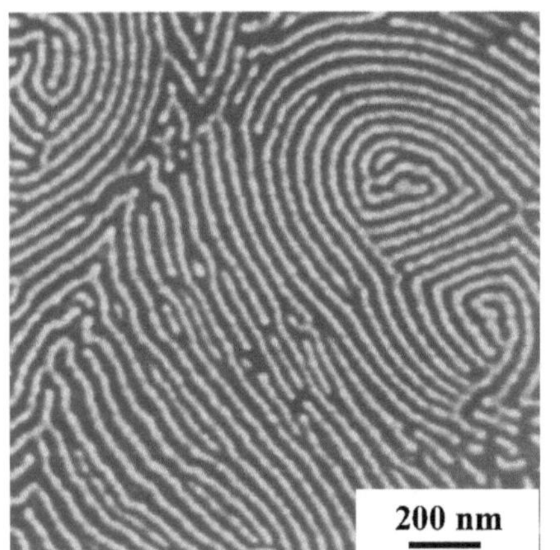

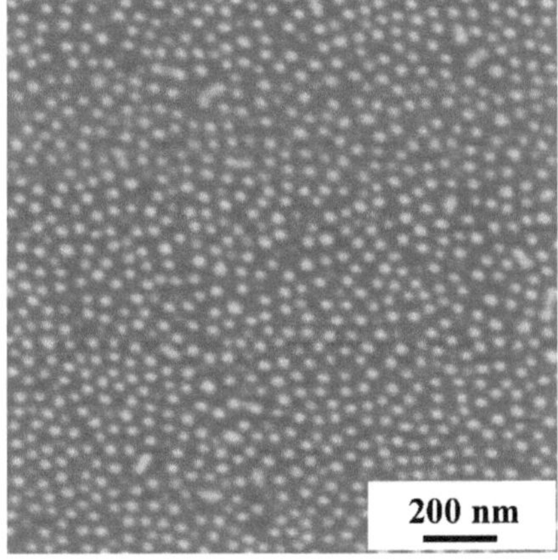

Fig. 19.9. Tapping-mode AFM phase micrographs showing morphologies of PMMA-block-PnBA-block-PMMA triblock copolymers, prepared by controlled radical polymerisation, that have different PMMA weight contents (*left*: 44 wt%, *right*: 32 wt%); thin films prepared by dip-coating

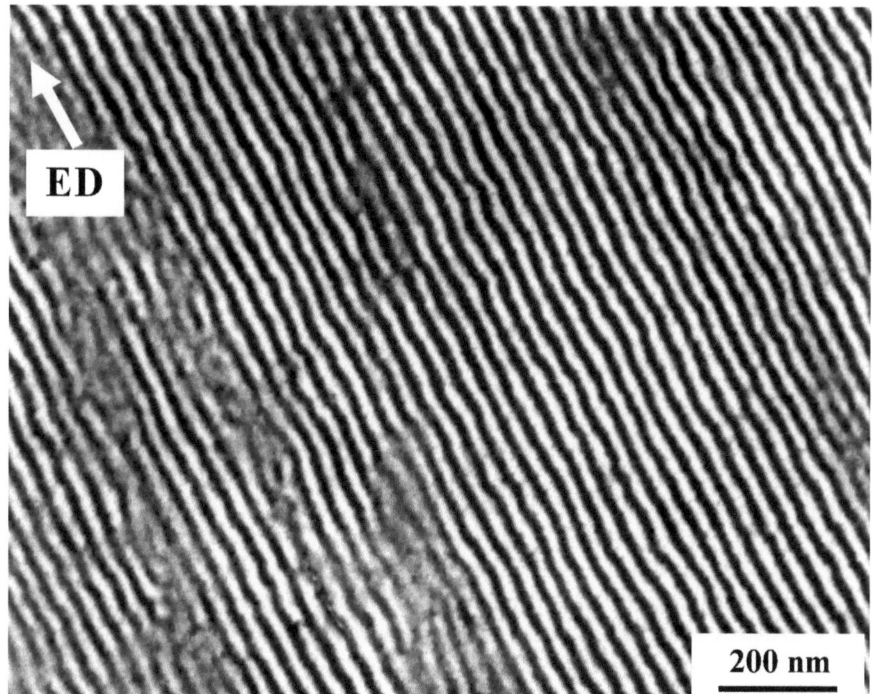

Fig. 19.10. TEM micrograph of a lamellar styrene/butadiene block copolymer oriented by extrusion; the butadiene phase is stained by OsO_4; the lamellae lie normal to the plane of the paper and are aligned along the extrusion direction (*ED*)

19.3 Deformation Mechanisms in Block Copolymers

The types of deformation structures found in polymer blends and composites such as shear bands or crazes are observed in block copolymers as well. However, due to the presence of structures on much smaller length scales, there are a number of unique characteristics of the deformation mechanisms of block copolymers. The most remarkable effect that can be noticed in block copolymers is a strong tendency towards local flow processes. Obviously, EM techniques (especially TEM) offer an excellent opportunity to investigate deformation-induced local structural changes of these polymers. Three main methods have been employed for the study of deformation mechanisms in block copolymers by TEM (compare with Chap. 6):

1. Stretching of semi-thin sections (a few hundred nanometres thick) followed by the staining and microscopic analysis of deformation structures.
2. Examination of the deformed sample after mechanical testing (post mortem study). For this purpose, thin sections are prepared from specimen locations close to the fracture surface.
3. In situ study of deformation processes of thin sections directly inside the microscope at larger magnifications.

To a first approximation, the deformation mode of a block copolymer depends on the size of the difference between the glass transition temperatures of the individual components and the test temperature. If the molecular weight is sufficiently high (so that stable entanglements can form) and both the components are glassy, the block copolymer is likely to deform through the formation of localised zones, mainly in the form of crazes.

In a diblock copolymer, if one component is glassy and the other is rubbery at test temperature, cavitation of the rubbery phase is observed independent of the morphology type, leading to the formation of craze-like deformation zones [27, 28]. It should be noted that the deformation structures of the block copolymer are highly coupled to the chain architecture [14–16]. Figure 19.11 shows the electron micrographs of semi-thin sections (ca. 500 nm thick) of two different block copolymer architectures comprising glassy/rubbery components: One is a polystyrene-block-poly(n-butylmethacrylate) (PS-block-PBMA) diblock copolymer and other is an SBS triblock copolymer. The diblock copolymer deforms by forming craze-like deformation zones, as shown by a high-voltage electron microscope (HVTEM) image (Fig. 19.11, top). The deformation is characterised by cavitation of the rubbery poly(n-butylmethacrylate) phase accompanied by the plastic drawing of the surrounding glassy matrix. The formation of microvoids is indicated by areas that are more transparent to the electrons and appear brighter in the electron micrographs. On the other hand, in the SBS triblock copolymer with a cylindrical morphology, cavitation is not restricted to the rubbery phase but is instead mainly concentrated in the glassy PS phase. This situation arises from the fact that the chain ends are only located in the PS matrix [8, 29]. The craze-like deformation structures of the SBS triblock copolymer (Fig. 19.11, bottom) were studied with the aid of an environmental scanning electron microscope (ESEM). The semi-thin sections of the block copolymer were strained using a miniaturised stretching device prior to the microscopic examination. The ESEM technique also allows the inspection of the polymer deformation morphology without prior sample treatment.

The deformation structures formed in SBS block copolymers (where both the ends of the rubbery PB chains are crosslinked by glassy domains) depend largely on the morphology. If the glassy phase forms dispersed domains (such as PS cylinders), they undergo fragmentation at higher deformation, allowing the rubbery matrix to stretch [30]. On unloading the sample, the initial morphology is almost completely regenerated.

The processing-induced orientation of block copolymer nanostructures may have a large influence on the mechanical behaviour (especially anisotropic characteristics). The lamellar arrangements of SBS block copolymer nanostructures exhibit particularly interesting micromechanical properties. Equally important are their blends with polystyrene homopolymer. Figure 19.12 shows the deformation structures observed in a lamellar SBS triblock copolymer with a special molecular architecture (see sample SB2 in Fig. 19.6, and Fig. 19.10). The sample was oriented by applying steady shear and the resulting film was loaded parallel (see Fig. 19.12, top) and perpendicular (see Fig. 19.12, bottom) to the lamellae orientation direction. In contrast to the diblock

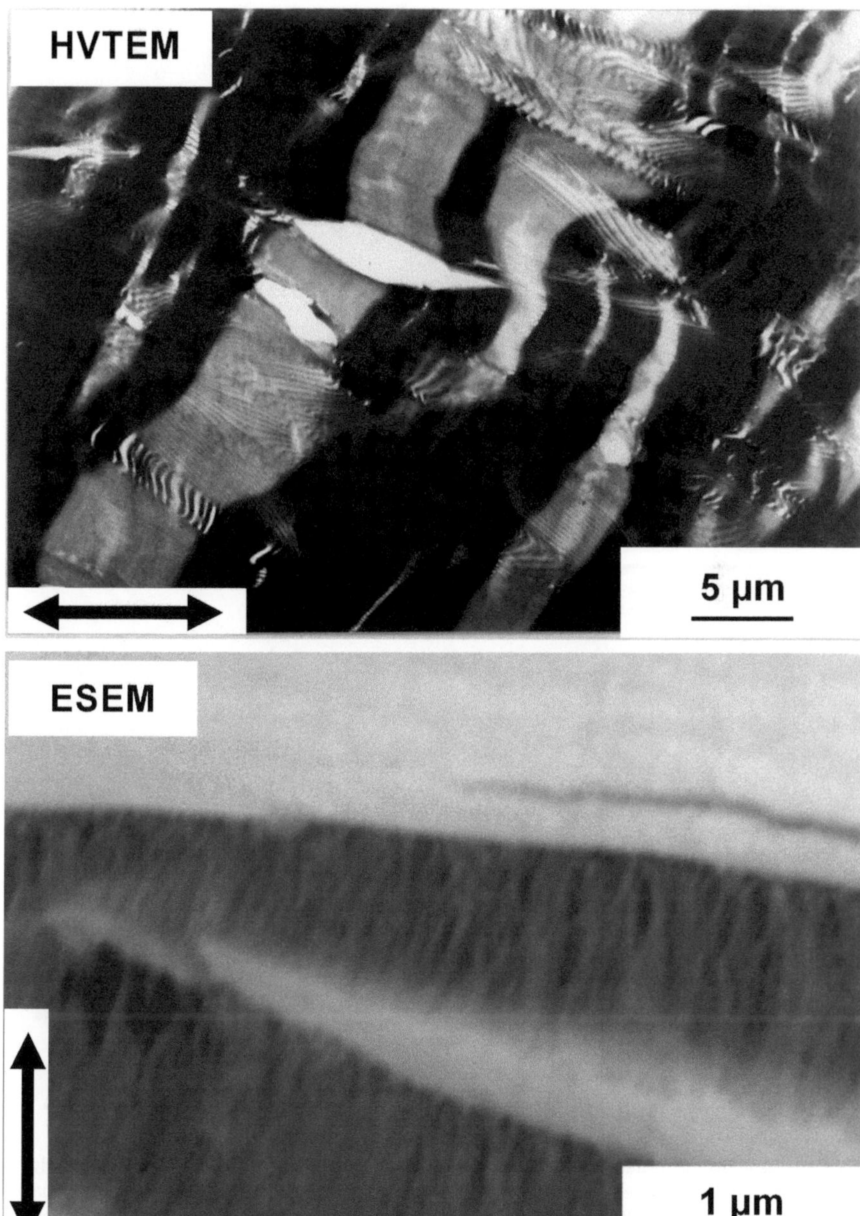

Fig. 19.11. Electron micrographs of semi-thin sections showing the deformation structures of block copolymers with cylindrical morphologies (PS content ~70%); *top*: PS-block-PBMA diblock copolymer; *bottom*: PS-block-PB-block-PS triblock copolymer; deformation direction shown by *arrow*

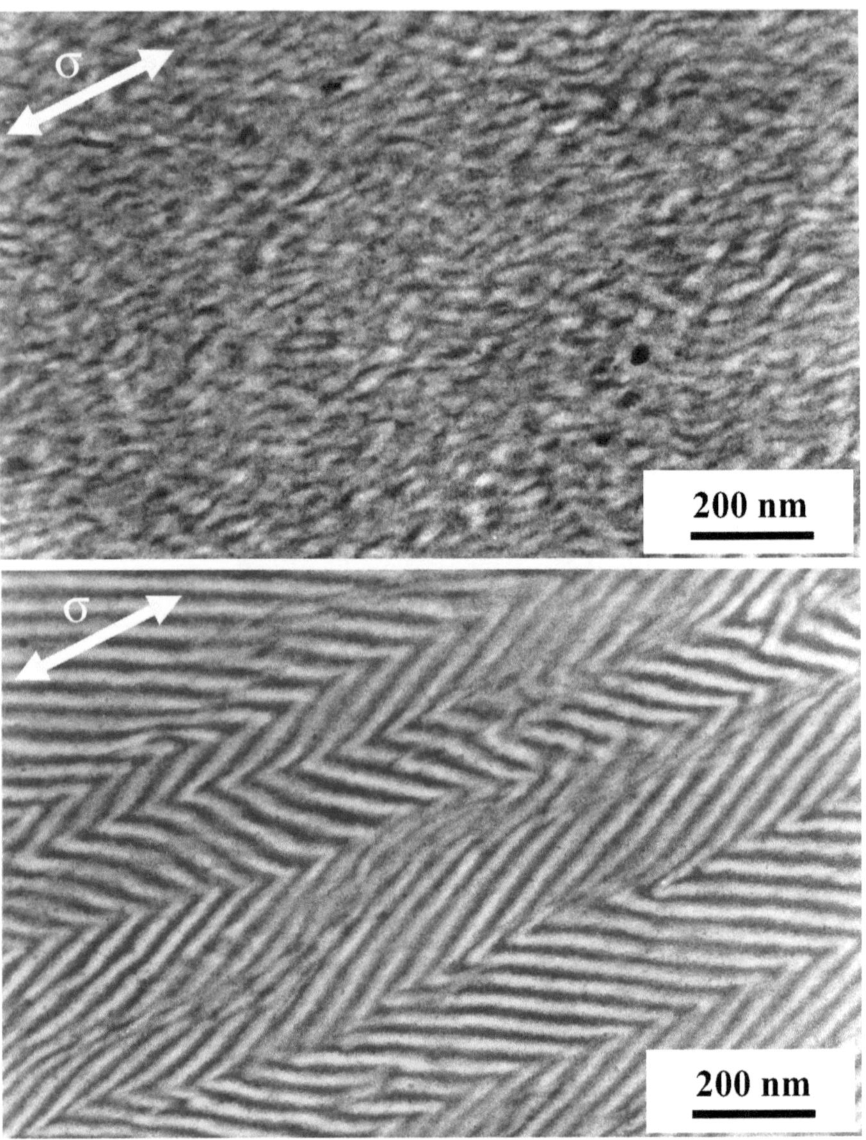

Fig. 19.12. TEM micrographs showing the deformation structures of a lamellar block copolymer produced upon loading the sample in a different direction from the lamellae orientation (deformation direction shown by *arrow*); *top*: parallel deformation leading to the thin-layer yielding mechanism; *bottom*: chevron morphology formation resulting from perpendicular deformation; the butadiene phase was stained by OsO_4

or SBS triblock copolymer with a PS matrix (see Fig. 19.11), where the deformation localisation in the craze-like zones is the principal deformation mechanism, no local deformation bands were observed. When subjected to uniaxial deformation parallel to the lamellae orientation direction (see Fig. 19.12, top), both the glassy and the rubbery lamellae were found to deform plastically at high deformation. The large plastic deformation of glassy PS lamellae was attested by a drastically reduced lamellae thickness. Based on the quantitative information obtained from TEM analysis of the deformation structures of the lamellar block copolymers, "thin layer yielding" has been proposed as an alternative toughness-enhancing mechanism for brittle polymers [22].

Figure 19.12, bottom, shows the deformation structures formed by an oriented lamellar block copolymer (the initial morphology of a neat sample is shown in Fig. 19.10) subjected to tensile deformation perpendicular to the lamellae alignment direction. The thin sections used for the TEM examination were prepared from a piece of sample cut close to the fracture surface of a deformed tensile specimen. It can be seen that the lamellae are folded into a fish-bone-like arrangement (the so-called "chevron morphology"). The reduction in thickness of the PB layers during the parallel deformation was found to be more pronounced. These observations indicate that the component that reacts to the applied stress earlier is the rubbery phase. The formation of a chevron morphology has been shown to occur in all lamellar-forming heterogeneous polymeric systems (see Chap. 16) [31].

19.4 Special Cases of Self-Assembly and Applications

Electron microscopy and scanning force microscopy have been used in the past to characterise the structures and properties of a large number of specific block copolymer systems. Recently, intensive studies on ABC triblock copolymers and even multicomponent multiblock systems have contributed a great deal to the understanding of self-assembly phenomena in complex soft matter [32,33]. For example, in simple two-component systems, the microphase separation behaviour is affected greatly by the physical confinement of the polymers within tiny pores [34, 35]. Other uses of block copolymers include their application as templates for the preparation of a variety of nanostructures, including quantum dots [36], fibres [37] and tubes [32]. Semicrystalline block copolymers [38] serve as model systems for the study of confined crystallisation. The preparation of organic/inorganic hybrid materials has opened up new approaches to controlling the mechanical properties of nanostructured polymer systems [39]. When used as surfactant molecules in solution, block copolymers form a wide variety of micelles that may be useful in the development of drug-delivery systems [40]. In all of these applications, EM plays a very important role in the structural characterisation of the block copolymer systems. In some cases, special preparation and staining techniques are needed. Two examples of current research topics are presented here: microphase separation in a confined environment (Fig. 19.13) and nanostructures of ABC block copolymers (Fig. 19.14).

PS-block-PMMA diblock copolymer (PS/PMMA volume fraction, 70/30; weight average molecular weight, ca. 74 000 g/mol; polydispersity index, 1.08) melt was introduced into porous alumina templates after annealing under vacuum at 180 °C for several hours [35]. Later the template was removed and the suspension of the

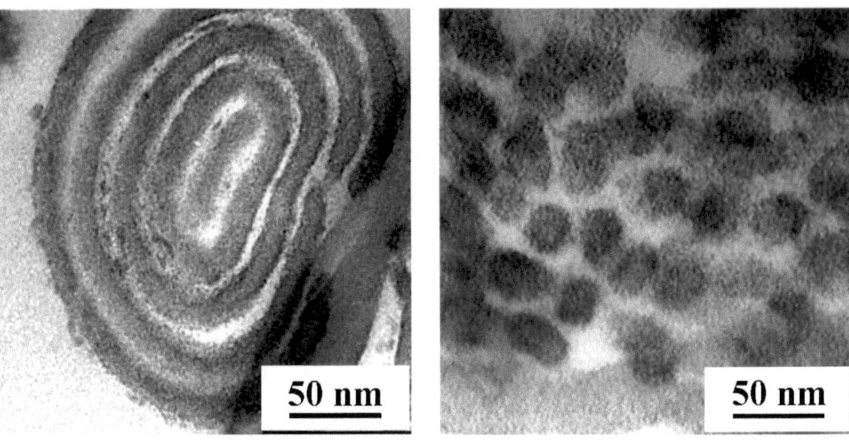

Fig. 19.13. TEM micrographs showing the morphology of a PS-block-PMMA diblock copolymer nanorod confined within a porous alumina template with different pore diameters; *left*: 400 nm; *right*: 25 nm (details in [35]); PS phase stained with RuO$_4$ appears *dark*

block copolymer nanostructures in ethanol was treated with RuO$_4$ vapour. Due to the large surface area of the nanorods, the staining process is completed within few seconds. Finally, the solvent was removed and the residue was embedded in an epoxy

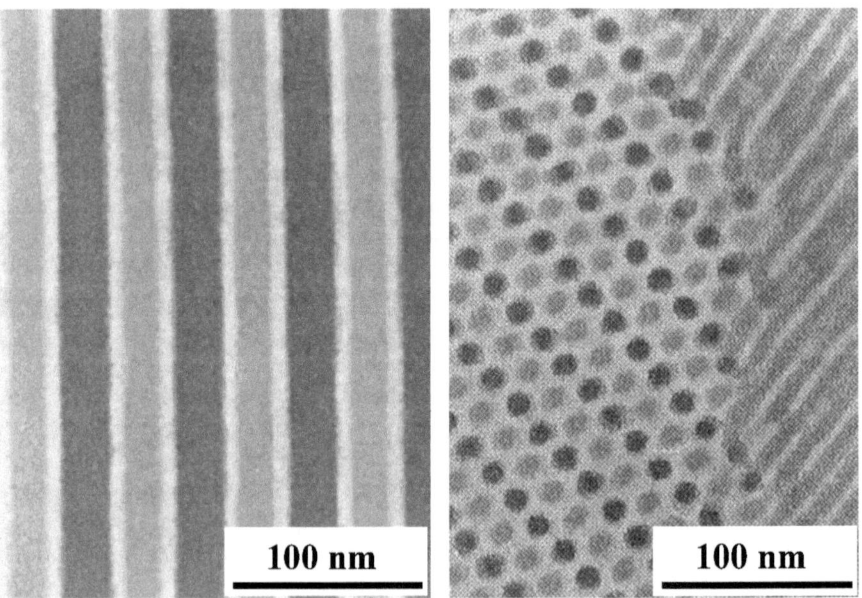

Fig. 19.14. TEM micrographs showing the morphologies of ABC triblock copolymers comprising polyisoprene (PI), polystyrene (PS) and poly-2-vinylpyridine (P2VP); staining with OsO$_4$ makes the PI phase appear *dark* while PS appears *white* and P2VP *grey* (reproduced from [33] with permission from the American Chemical Society)

resin. Thin sections for TEM were prepared by cryoultramicrotomy. The long axes of some nanorods were oriented perpendicularly, and those of some other nanorods were aligned parallel to the cutting direction. The PS phase, due to selective staining by RuO_4, appears darker than the PMMA. Figure 19.13, left, shows the morphology of a cross-section perpendicular to the long axes of the nanofibres released from an alumina template with a diameter of 400 nm, which depicts the concentric rings of alternating PS and PMMA lamellae (details in [35]). During the development of block copolymer structures confined to dimensions of less than the long period of the bulk sample, a drastic change in microphase separation behaviour was found to occur, i.e. a core–shell structure where the shell consists of PMMA and the core of PS was observed to form (see Fig. 19.13, right). In the absence of geometric confinement, this microphase separation would actually be expected for a diblock copolymer with PMMA as the major component.

New avenues to nanostructured materials with wider ranges of functionalities have been opened by ABC triblock copolymers, their star architectures and blends with other block copolymers and homopolymers. Due to the presence of three different interaction parameters (namely χ_{AB}, χ_{BC}, χ_{AC}), a richer spectrum of morphologies and properties may be expected in ABC systems [41]. Figure 19.14 shows the morphologies of two polyisoprene-block-polystyrene-block-poly(2-vinylpyridine) triblock copolymers (ISPs) with different compositions. The microphase separation of three components can be easily discerned from the TEM micrographs. Similar to the diblock copolymer, a lamellar morphology is formed if the volume fraction of each component is nearly equal (PI/PS/P2VP composition 35/34/33 in Fig. 19.14, left). If the middle block forms the matrix (for example in a PI/PS/P2VP composition of 12/76/12, see Fig. 19.14, right), superlattice structures consisting of cylindrical domains of PI and P2VP dispersed in the PS matrix evolve.

References

1. Quirk RP, Morton M (1998) In: Holden G, Legge NR, Quirk RP, Schroeder HE (eds) Thermoplastic elastomers, 2nd edn. Hanser, Munich
2. Davis KA, Matyjaszewski K (2002) Adv Polym Sci 159:107
3. Hamley IW (2004) Developments in block copolymer science and technology. Wiley InterScience, New York
4. Holden G (2000) Understanding thermoplastic elastomers. Hanser, Munich
5. Godehardt R, Rudolf S, Lebek W, Goerlitz S, Adhikari R, Allert E, Giesemann J, Michler GH (1999) J Macromol Sci Phys 38:817
6. Magonov SN (2000) In: Mayers RA (ed) Encyclopedia of analytical chemistry. Wiley, Chichester
7. Bar G, Brandsch R, Delineau L, Wangbo MH (2000) Langmuir 16:5702
8. Adhikari R, Michler GH (2004) Prog Polym Sci 29:949
9. Kloninger C, Knecht D, Rehahn M (2004) Polymer 45:8323
10. Simon P, Huhle R, Lehmann M, Lichte H, Mönter D, Bieber T, Reschetilowski W, Adhikari R, Michler GH (2002) Chem Mater 14:1505
11. Chou T-M, Libera M, Gauthier M (2003) Polymer 44:3037
12. Adhikari R, Godehardt R, Lebek W, Weidisch R, Michler GH, Knoll K (2001) J Macromol Sci Phys 40:833
13. Lee C, Gido SP, Poulos Y, Hadjichristidis N, Tan NB, Trevino SF, Mays JW (1997) J Chem Phys-107:6460

14. Adhikari R, Michler GH (2005) In : Michler GH, Baltá-Calleja FJ (eds) Mechanical properties of polymers based on nanostructure and morphology. Taylor & Francis, Boca Raton, FL, 3:81
15. Weidisch R, Gido SP, Uhrig D, Iatrou H, Mays J, Hadjichristidis N (2001) Macromolecules 34:6333
16. Mori Y, Lim L S, Bates FS (2003) Macromolecules 36:9879
17. Matsen MW (2000) J Chem Phys 113:5539
18. Milner ST (1994) Macromolecules 27:2333
19. Hasegawa H, Hashimoto T (1996) In: Aggarwal SL, Russo S (eds) Comprehensive polymer science: Suppl. 2. Elsevier, London, p 497
20. Koizumi S, Hasegawa H, Hashimoto T (1994) Macromolecules 27:4371
21. Sakurai S, Isobe D, Okamoto S, Nomura S (2002) J Macromol Sci Phys 4:387
22. Michler GH, Adhikari R, Lebek W, Goerlitz S, Weidisch R and Knoll K (2002) J Appl Polym Sci 85:683
23. Matyjaszewski K, Spunswick J (2004) In: Holden G, Kricheldorf HR, Quirk RP (eds) Thermoplastic elastomers, 3rd edn. Hanser, Munich, p 365
24. Bates FS, Koppi KA, Tirrell M, Almdal K, Mortensen K (1994) Macromolecules 27:5934
25. Cohen Y, Albalak RJ, Dair BJ, Capel MS, Thomas EL (2000) Macromolecules 33:6502
26. Bolker A, Knoll A, Elbs H, Abetz V, Mueller AHE, Krausch G (2002) Macromolecules 35:1319
27. Weidisch R, Michler GH (2000) In: Balta Calleja FJ, Roslaniec Z (eds) Block copolymers. Marcel Dekker, New York
28. Schwier CE, Argon AS, Cohen RE (1995) Polymer 26:1985
29. Koltisko B, Hiltner A, Baer E (1986) J Polym Sci Polym Phys 24:2167
30. Honeker CC, Thomas EL (2000) Macromolecules 39:9407
31. Krumova M, Henning S, Michler GH (2006) Phil Mag 86:1689
32. Abetz V, Goldacker T (2000) Macromol Rapid Comm 2:16
33. Mogi Y, Nomura M, Kotsuji H, Matsuhita Y, Noda I (1994) Macromolecules 27:6755
34. Xiang HQ, Shin K, Kim T, Moon SI, McCarthy TJ, Russell TP (2004) Macromolecules 37:5660
35. Sun Y, Steinhart M, Zschech D, Adhikari R, Michler GH, Gösele U (2005) Macromol Rapid Commun 29:269
36. Lopes WA, Jaeger HM (2001) Nature 414:735
37. Liu LG, Li Y, Yan X (2003) Polymer 44:7721
38. Loo Y-L, Register RA, Ryan AJ, Dee GT (2001) Macromolecules 34:8968
39. Simon PFW, Ulrich R, Spiess HW, Wiesner U (2001) Chem Mater 13:3464
40. Zhang L, Eisenberg A (1999) J Polym Sci Polym Phys 37:1469
41. Bates FS, Fredrickson GH (1999) AIP Phys Today 2:32

20 Rubbers and Elastomers

Elastomers are composed of a network of chemically or physically linked macromolecules and are characterised by low stiffness and large deformability. This chapter begins with a discussion of historic and recent applications of elastomers. TEM micrographs are used to show the arrangements of several rubber constituents and filler types using cryoultrathin sections. A special preparation technique, soft matrix fracture, is illustrated, which shows filler particles on fracture surfaces with improved contrast in the SEM. A description of how filler particles, which produce micro- and nanovoids in the rubber matrix, can be studied by in situ deformation tests in ESEM is then provided.

20.1 Overview

Elastomers form a group of materials that are structurally characterised by a cross-linked network of macromolecules. This linkage may be either physical or chemical in nature. Physical linking is related to the entanglements of polymer chains, absorption onto the surfaces of particulate fillers, the formation of crystallites, and the coalescence of ionic centres or of glassy blocks. Chemical linking involves randomly joining segments of pre-existing chains, random copolymerisation, or end-linking functionally terminated chain-ends. The most frequent methods of random crosslinking are sulfur curing, peroxide curing, and high-energy irradiation. From a mechanical point of view, elastomers show "rubber-like" behaviour, associated with low stiffness, large (up to tenfold) deformability, and recovery after unloading. Elastomers comprise about 15–20% of global polymer consumption.

The most common representative of the elastomer family is natural rubber obtained from the milk (latex) of rubber trees (*Hevea brasiliensis*) or rubber bushes like Guayule (*Parthenium argentatum*). Chemically, natural rubber is mainly composed of 1,4-*cis*-polyisoprene, and contains natural proteins and resins. Natural rubber was used by ancient Mesoamerican peoples before 1600 B.C. [1] to produce rubber bands, rubber human figures and other items. However, the main products made from rubber were hard rubber balls for ritual ballgames, which were made from heated latex. These varied between 25 and 30 cm in diameter and weighed 1.5–3 kg. The aim of the ballgames was to knock the ball without using the hands; only the hips, thighs and upper arms could be used to make the ball pass through a vertical stone ring at the side of the court. The game appeared in various myths and with associated with sacrifice and death (some references say that the prize for the winning team was to be deified by losing their heads). The English term "rubber" derives from one of the first applications of natural rubber in modern times (rubber = eraser).

Charles Goodyear's discovery of vulcanisation, which transforms soft natural rubber into a technically usable form, in 1839 was a milestone in the application of rubber. Vulcanisation, or sulfur curing, was named after the mythological Greek god of fire, Vulcan, since both heat and sulfur are involved in the curing process. Another milestone was the development of the synthetic rubber polybutadiene. This was achieved by polymerising butadiene using sodium (German: natrium) as a catalyst, a process that was given the tradename "Buna" (butadiene natrium). About 35% of all elastomer consumption is related to natural rubber, and about 65% to synthetic rubbers [2]. Besides natural rubber (NR), its synthetic version, polyisoprene rubber (IR), and butadiene rubber (BR), there are various copolymers of butadiene with styrene (styrene butadiene rubber, SBR). These rubber types are known as general-purpose rubbers. Nevertheless, general-purpose rubbers are unsuited to many applications due to their insufficient resistance to swelling, aging, or high temperatures. Speciality elastomers have been developed for special purposes, like polychloroprene (CR), acrylonitrile-butadiene rubber (NBR), butyl rubber (IIR), ethylene-propylenediene rubber (EPDM), silicon rubber (MQ), urethane rubber, and others.

A common feature of all of these rubbers is their enhanced mechanical properties resulting from the creation of a molecular network through vulcanisation or chemical crosslinking. Therefore, rubbers obtain their final shapes during manufacture. After vulcanisation, they cannot be melted again, and their shapes cannot be changed, in contrast to, say, thermoplastic polymers. The use of active fillers, especially carbon black and recently synthetic silica, together with accelerated sulfur vulcanisation has remained the fundamental technique used to achieve the incredible range of mechanical properties required for a great variety of modern rubber products [3]. The reinforcement mechanism is based on the Payne effect. The enhancement of mechanical properties in filled elastomers is the result of combining the properties of the elastomer network, the hydrodynamic effect, the elastomer–filler interaction and the filler–filler interaction. Active fillers form a three-dimensional network built on filler–filler interactions, which is responsible for the reinforcement mechanism and the hysteresis effects.

More recent elastomers include butadiene block copolymers (see Chap. 19) and the so-called thermoplastic elastomers. Their networks are formed through physical crosslinking, which can be broken down in the melt, thus making thermoplastic processing possible [4]. Typical examples include ethylene-propylene rubbers (EPR) and their blends with PP. In SB block copolymers, the flexible rubbery polybutadiene blocks ($T_g \sim -100\,°\mathrm{C}$) are anchored on both sides by glassy polystyrene blocks ($T_g \sim +100\,°\mathrm{C}$), so these materials behave like a crosslinked rubber under ambient conditions but allow thermoplastic processing at higher temperatures. Segmented polyurethanes with higher soft segment contents than hard segment contents as well as ethylene copolymers with high comonomer contents are also elastomers.

In most cases, rubber compounds are blends of different elastomer types that develop specific domain morphologies at the micro scale. Blends of general-purpose rubbers are mainly used for tyre production. Thus, a tyre is composed of natural rubber, synthetic butadiene rubber, styrene butadiene rubber, softener, vulcanising agents and different types of fillers, mainly carbon black and silica. A typical tyre blend contains about 27% natural rubber, 20% synthetic rubber, 21% chemicals and 30% filler particles. Today, tyres are made from high-tech compounds that exhibit efficiencies that were impossible to achieve years ago. However, the challenges in tyre development are still to minimise the discrepancy between dry and wet traction, decrease abrasion, increase the lifetime, and to reduce rolling resistance. These challenges demand that the length scales of the macromolecules (e.g. the distance between crosslinks), of the domain morphology, as well as of the fillers (e.g. the sizes of the primary particles, aggregates and agglomerates, and their distributions) are optimised. As filler concentration is increased, a flocculation of clusters and a continuous physically bonded filler network are attained. A deep understanding of the formation and structures of such filler networks can prove very useful for the design and preparation of high-performance elastomers applied not only in tyres but also in other dynamically loaded elastomer components [5–7]. Apart from the sizes of the filler particles and the agglomerates, the surface roughness of the particles and the rubber–filler interactions are also important. In the case of diene–rubber composites filled with carbon black, the polymer–filler interaction is generally rather strong due to the high affinity between the carbon surface and the rubber. In the case of silica, the elastomer chains are chemically bonded to the filler surface using special additives (silanes).

One practical procedure that can be used to characterise the polymer–filler interaction involves estimating the bound rubber, i.e. the amount of polymer tightly bound to the filler surface after mixing [7]. It is well known that this parameter increases with the molar mass of the polymer and the specific surface area of the filler particles [8]. While the incorporation of filler into the elastomer matrix (macrodispersion) can be determined by optical microscopy, the microdispersion of the filler can be characterised by TEM [6].

As well as the broad application of rubbers in the tyre industry (about 59% of all synthetic rubber is used for tyres [2]), rubbers have many other applications in the automotive industry, in technology, in households and, last but not least, as a modifier component in rubber-toughened polymers (see Chap. 18).

20.2 Morphology of Rubbers and Elastomers

The semicrystalline structure of natural rubber (1,4-*cis*-polyisoprene) has been studied using SEM after crystallisation from solution and after reacting it with OsO_4. It shows an arrangement of lamellae [9]. Thin films crystallised from the melt reveal spherulites and lamellae after the reaction with OsO_4 in the TEM [10].

Figure 20.1 shows the typical domain morphology of an elastomer blend composed of 40/60 NR/SBR, as observed in TEM. Visualisation of the domain morphol-

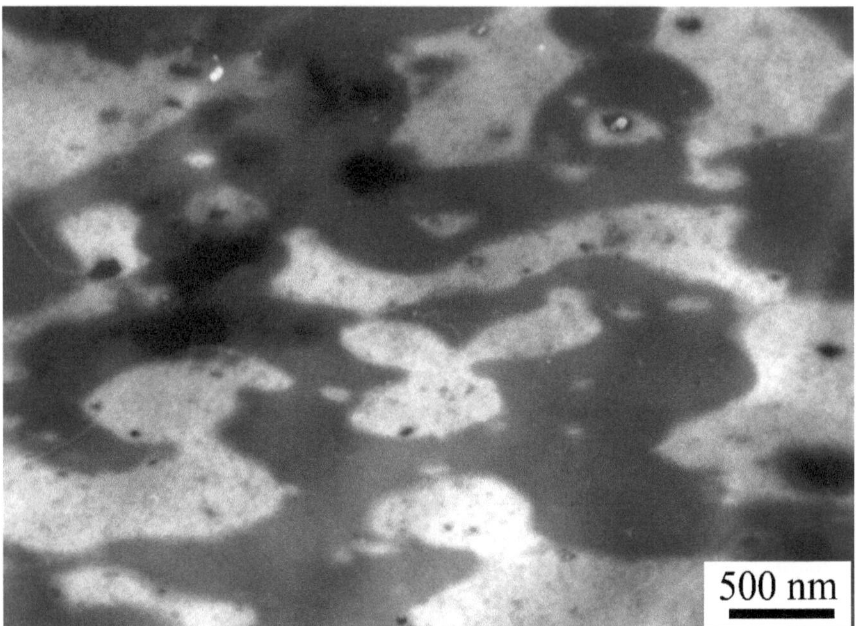

Fig. 20.1. TEM micrograph of a NR/SBR elastomer blend (40/60) showing a characteristic domain morphology after degradation of the NR phase for 20 min at 380 °C in vacuo. (J. Lacayo, with the permission of Continental AG)

ogy is possible after the degradation of the NR phase in the ultrathin section at 380 °C for 20 min in vacuo. Since SBR remain stable at temperatures of up to 420 °C, the selective etching of NR domains is a feasible way of inducing contrast. If the similar NR/SBR blend is filled with silica, the silica will create a three-dimensional network that interpenetrates the elastomer network. Characteristic arrangements of the most frequently used fillers in compounds with SBR (carbon black and silica) are shown in Fig. 20.2 in TEM micrographs of ultrathin sections. Due to the higher density of the filler particles, they appear with good contrast. The larger magnifications provided in Fig. 20.3 show individual carbon black particles about 30 nm in size and, particularly at a higher volume content (micrograph b), the agglomeration of the particles. Different grades of distribution of layered silicate are shown in Fig. 20.4 and, together with carbon black, in Fig. 20.5.

Larger agglomerates of particles are visible on brittle fracture surfaces of the compounds (see Chap. 9). However, the brittle fracture contours of the rubber matrix make it more difficult to see the particles. Here, the technique of high-temperature fracture (soft matrix fracture, see Sect. 9.4) yields a smoother matrix fracture with better visibility of carbon black, as demonstrated in the example of Fig. 20.6 [11].

The elastomers used as impact modifiers in rubber-toughened polymers are usually PB, SBR, NBR, EVAc, CPE, EPDM, EPR, etc. After mixing it with the polymer that needs a toughness enhancement, the structure and arrangement of the rubber

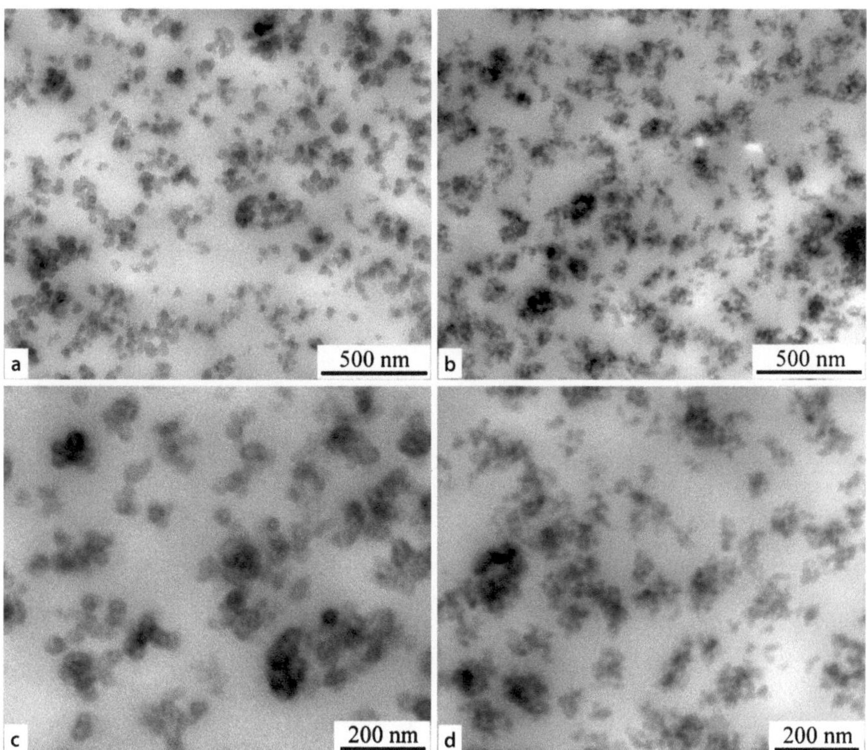

Fig. 20.2a–d. TEM micrographs of SB rubber filled with carbon black (**a,c**) and silica (**b,d**); cryoultrathin sections

phase can be easily studied in TEM using stained ultrathin sections (see Figs. 18.2 and 18.3), in SEM using the surfaces of OsO_4-stained blocks (see Fig. 18.4), or after surface etching (see Fig. 18.19).

Elastomers produced as block copolymers often show well-ordered nanostructures due to the intramolecular phase separation between the dissimilar chains linked together by covalent bonds. In SB block copolymers with low PS content, the PS comprises the disperse phase, in which it forms spheres or cylinders (see Figs. 19.2 and 19.3).

Segmented polyurethanes show favourable mechanical properties that are attributed to their domain structure, which can be directly observed by electron microscopy. Also, the presence of other morphological elements, mainly spherulites, globular structures and lamellae, has been revealed by light and electron microscopy [12–14]. Spherulites and their internal radial arrangements of lamellae can be made visible through the use of irradiation-induced contrast enhancement; see Fig. 13.12. Lamellae can be stained with a combined sample treatment of chlorosulfonic acid ($ClSO_3H$), osmium tetroxide (OsO_4) and formaldehyde solution (HCHO) [13,14]. Examples are shown in Fig. 20.7.

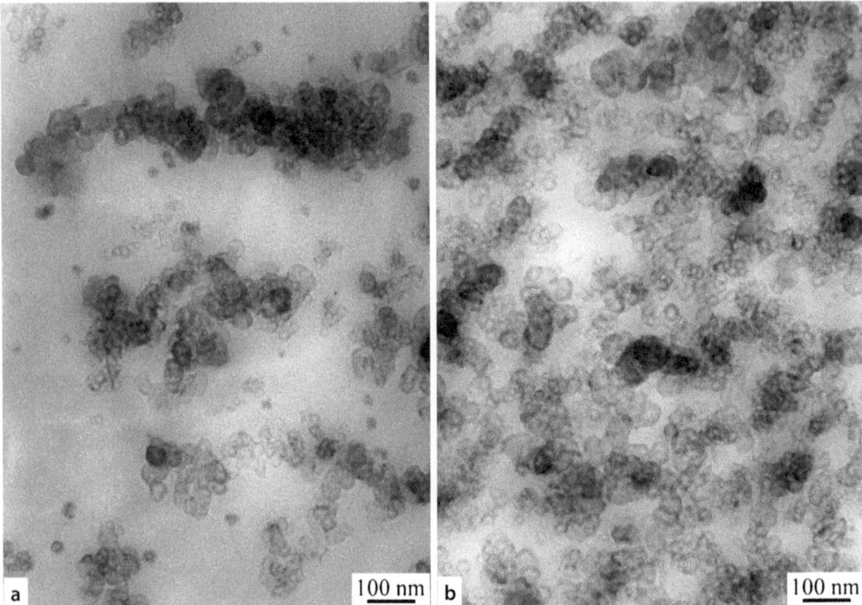

Fig. 20.3. TEM micrographs of SB rubber filled with 15% (**a**) and 50% (**b**) carbon black; cryoultrathin sections

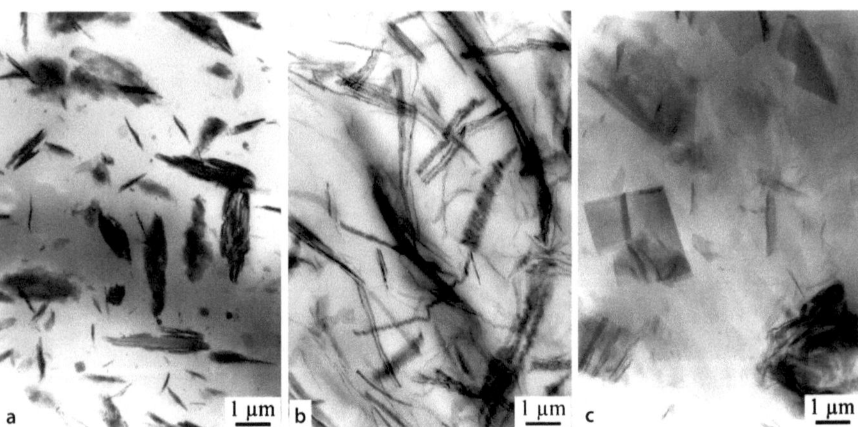

Fig. 20.4a–c. TEM micrographs of rubber filled with layered silicate (different grades of delamination); cryoultrathin sections

20.3 Micromechanical Deformation Behaviour

The characteristic mechanical behaviour of elastomers, known as rubber elasticity, is uniform elongation (by as much as 1000%) and subsequent complete recovery. Elastomers with morphological features, such as SB block copolymers and EPR/PP

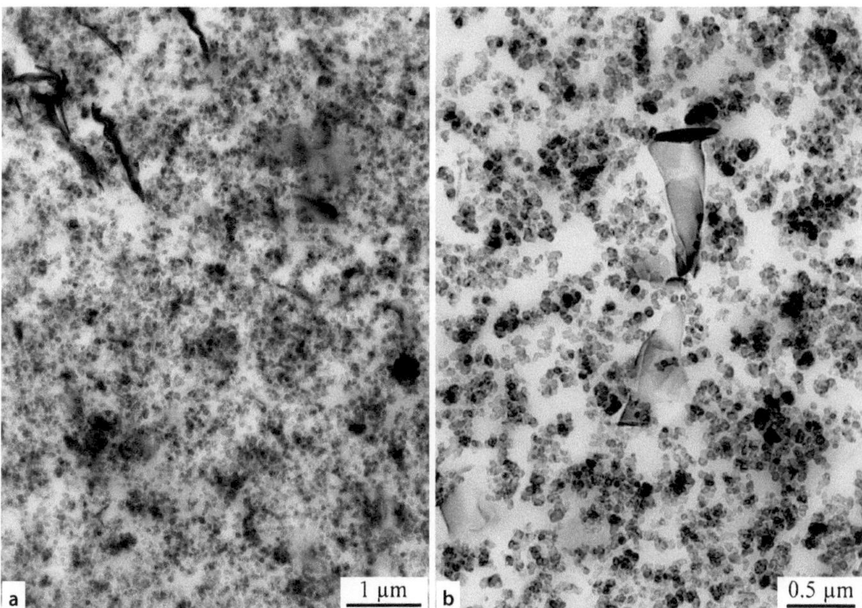

Fig. 20.5a,b. TEM micrographs of rubber filled with carbon black and layered silicate; cryoultrathin sections

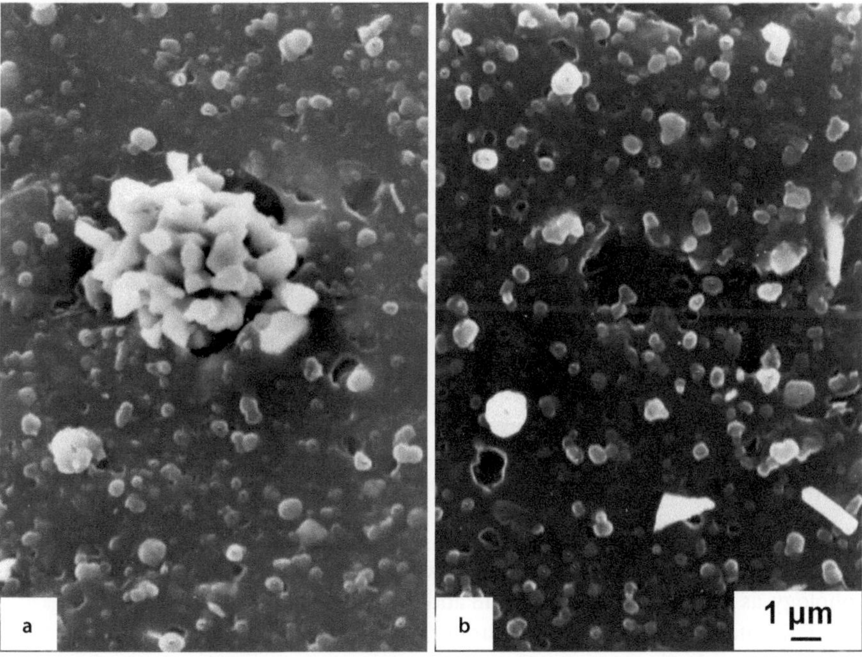

Fig. 20.6a,b. Rubber modified with carbon black, visible in SEM on a fracture surface, prepared at 150 °C (soft matrix fracture). (From [11], reproduced with permission from Chapman and Hall)

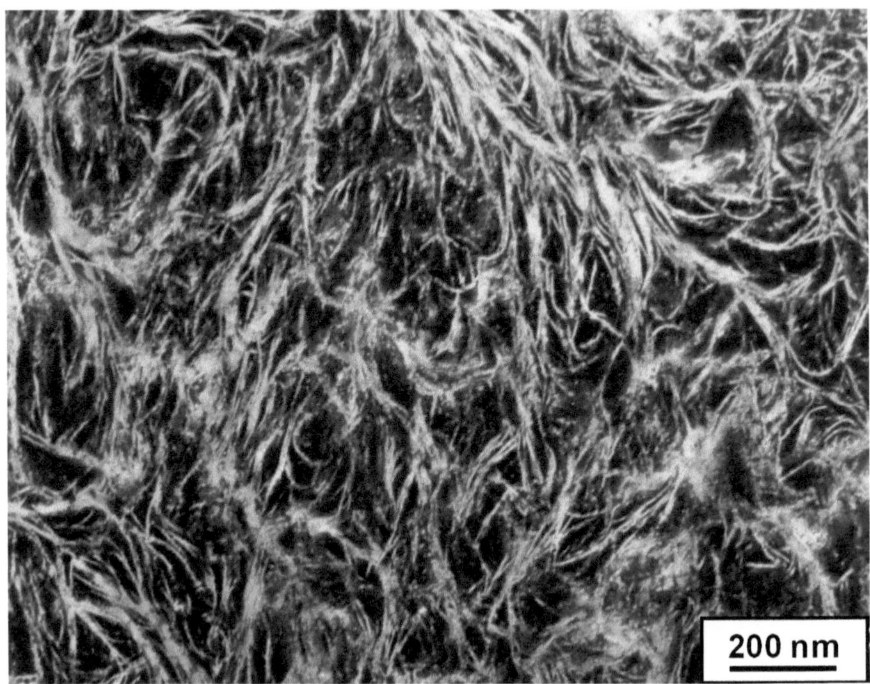

Fig. 20.7. Arrangement of lamellae in segmented PU with 57 wt% hard segments; staining with $ClSO_3H–OsO_4–HCHO/OsO_4$ treatment, ultrathin section, TEM

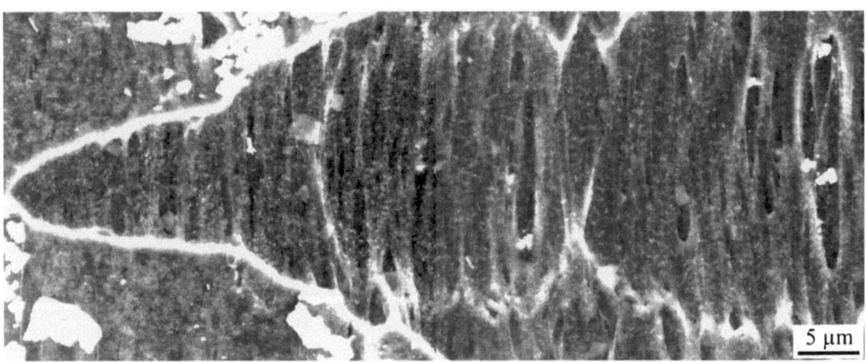

Fig. 20.8. ESEM micrograph of a surface and a propagating crack of an SB rubber filled with carbon black, layered silicate and zinc oxide that is stretched in situ

blends, show pronounced micromechanical effects, depending on the starting morphology, as presented in Chaps. 17, 18 and 19.

Polybutadiene rubbers deform uniformly. However, if they contain fillers, interesting localised effects appear. One micromechanical mechanism is the fracture of filler clusters and their re-aggregation during unloading [3]. The decrease in the elas-

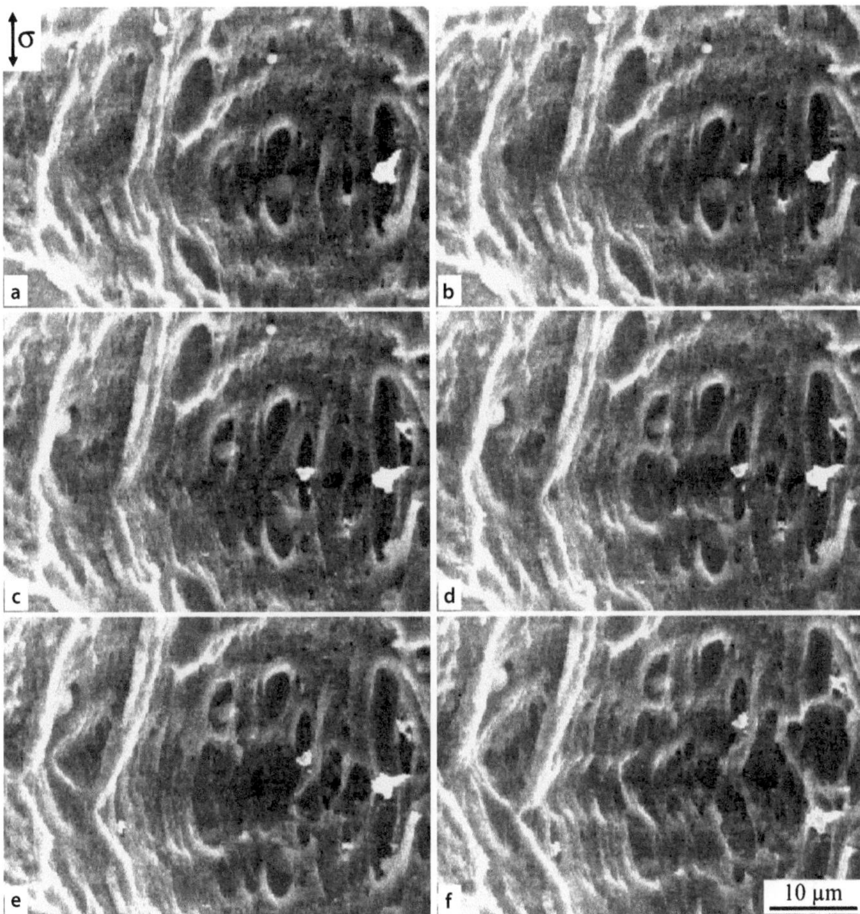

Fig. 20.9a–f. Sequence of micrographs of an in situ deformation test of SB rubber filled with carbon black, layered silicate and zinc oxide particles in ESEM. **a–f** show the same area of sample under increasing levels of strain

tic modulus after the beginning of strain is also connected to the successive breakdown of the filler network (Payne effect). Fundamental micromechanical concepts of nonlinear viscoelasticity have been developed that depend on different geometrical arrangements of the filler particles [6, 7]. The final rupture of the sample appears before the last filler–filler bond breaks down.

Larger particles or strongly bonded clusters seem to be the limiting factor for high elongations. This can be demonstrated by the in situ deformation of samples using a tensile or bending stage in ESEM. Figure 20.8 shows a crack tip (the crack runs from right to left) that is debonding the matrix from the inorganic particles, and the elongation of the microvoids. A PB rubber filled with carbon black, silicate and zinc oxide particles is shown as a sequence of in situ deformation tests in Fig. 20.9.

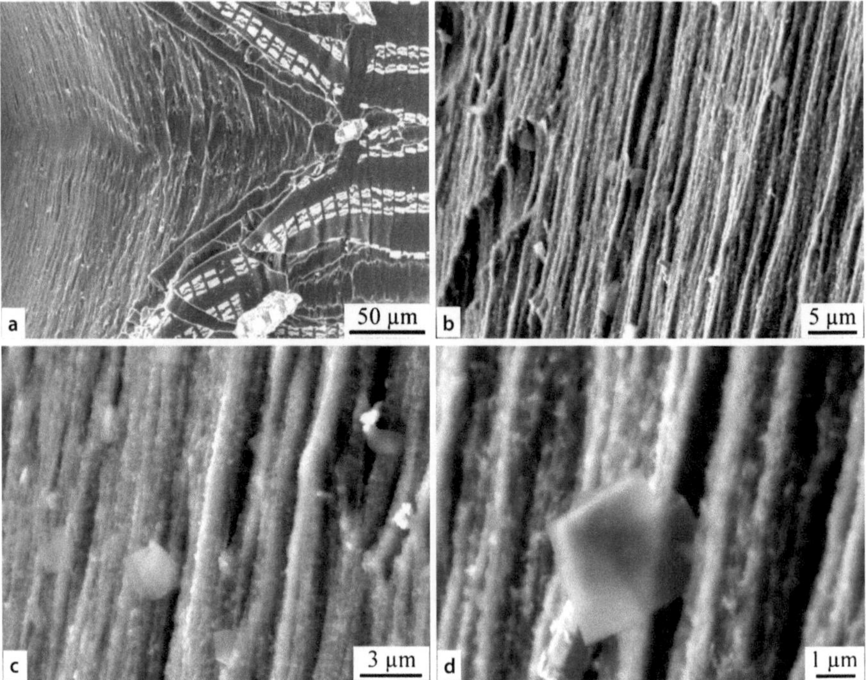

Fig. 20.10a–d. ESEM micrographs of the crack surface in a particle-filled SB rubber, revealing debonding around particles and high elongation of rubber fibrils and microvoids

With increasing strain, the connection between the rubbery matrix and the particles breaks, and microvoids appear around the particles.

The enlarged microvoids grow in size, the rubber is stretched until it breaks, and the crack propagates. At the tip of the crack, this mechanism repeats itself. This technique is a measure of the interfacial strength between the particles and the rubber matrix. It reveals the influence of particle size on crack initiation and propagation. Figure 20.10 shows the crack surface at higher magnifications in an ESEM micrograph. It demonstrates the role of larger particles (here zinc oxide) in producing larger voids in the rubber matrix and accelerating the crack propagation.

References

1. Hosler D, Burkett SL, Tarkanian MJ (1999) Science 284:1988
2. Ansarifar A, Wang L, Ellis RJ, Kirtley S (2006) GAK 59:769
3. Heinrich G, Klüppel M, Vilgis TA (2002) Curr Opin Solid State Mater Sci 6:195
4. Holden, G (2000) Understanding thermoplastic elastomers. Hanser Verlag, Munich
5. Klüppel M, Heinrich G (1995) Rubber Chem Technol 68:623
6. Klüppel M (2003) Adv Polym Sci 164:1
7. Heinrich G, Klüppel M (2002) Adv Polym Sci 160:1
8. Schuster RH (1989) Verstärkung von Elastomeren durch Ruß. WDK-Grünes Buch Nr 40
9. Xu J, Woodward AE (1986) Macromolecules 19:1114
10. Rench GJ, Phillips PJ, Vatansever N, Gonzalez A (1986) J Polym Sci Polym Phys 24:1943
11. Lednicky F, Michler GH (1990) J Mater Sci 25:4549

12. Foks J, Michler GH (1986) J Appl Polym Sci 31:1281
13. Foks J, Michler GH, Naumann I (1987) Polymer 28:2195
14. Foks J, Naumann I, Michler GH (1991) Angew Makromolekul Chem 189:63

21 Polymer Composites

Polymer composites are a well-established class of materials in which particles or fibres – used as reinforcing elements – are dispersed in the polymer matrix. In this chapter, the influence of morphology and deformation behaviour on the properties of these composites is demonstrated in more detail. The morphologies and deformation behaviours of particle-filled polymer composites (such as PP and PE with different filler contents of Al_2O_3 as well as PVC with $CaCO_3$) and fibre-reinforced polymer composites with long and short glass fibres have been investigated by SEM, TEM and HVTEM. Based on the experimental results obtained, it was confirmed that the overall properties of such morphologies depend strongly on the degree of interfacial strength (adhesion) and the interparticle distance. Large fillers and/or agglomerates readily induce large voids, which yield severe premature cracks and brittle fracture. It is shown that a uniform distribution of filler in the polymer matrix is highly desirable for improving the overall properties of the polymer composite.

21.1 Overview

Over the last few decades, composites have been rapidly developed to meet the demand for materials that provide higher standards of performance and in-service reliability. Composites are combinations of (at least two) materials: matrix and reinforcement [1–4]. The matrix material surrounds and supports the reinforcement materials by maintaining their relative positions. The reinforcements are embedded and arranged in specific internal configurations to obtain mechanical or other properties tailored to specific applications. The reinforcing materials (which consist of one or more chemically distinct constituents) can be metal, ceramic, or polymer. Due to the broad diversity of matrix and reinforcement materials available, the properties of the composite materials can combine the best features of each constituent to maximise a given set of properties (stiffness, strength-to-weight ratio, tensile strength, etc.) and minimise others (e.g. weight and cost). Typically, reinforcing materials are stiff and strong – long fibres, short fibres, whiskers, or particles (no long dimension) – while the polymer matrix is usually a ductile or tough material. If the composite is designed and fabricated correctly, it combines the strength of the reinforcement with the toughness of the polymer matrix to achieve a combination of desirable properties that cannot be achieved with any single constituent material. The advantages of composites can therefore be summarised as:

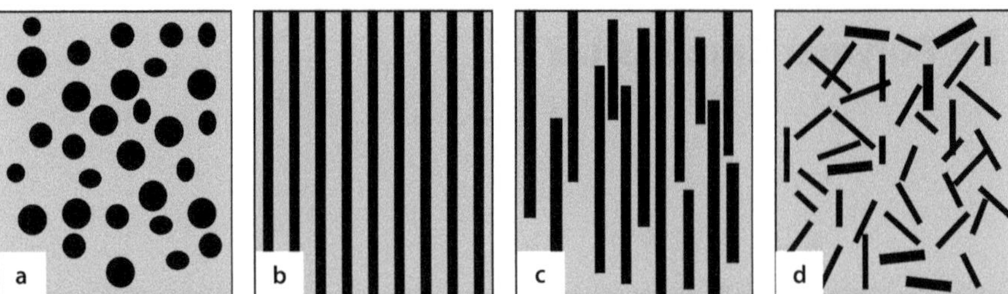

Fig. 21.1a–d. Schematic representations of polymer composites: **a** particle reinforcement **b** fibre reinforcement with continuous and aligned fibres; **c** with discontinuous and aligned fibres; **d** with discontinuous and randomly oriented fibres

- High strength-to-weight ratio (low density and high tensile strength)
- High creep resistance
- High tensile strength at elevated temperatures
- High toughness.

The strengthening mechanisms of the composites depend primarily on the amount, arrangement, geometry and type of reinforcement in the polymer matrix. Polymer composites can be divided into two main categories (Fig. 21.1), normally referred to as [2, 4]:

- Particle-reinforced polymer composites
- Fibre-reinforced polymer composites.

Because of their attractive specific stiffnesses and strengths, the mechanical behavior of polymer composites is of great interest to researchers and engineers in many science and engineering disciplines. Traditional continuum mechanics, based on continuity, isotropy and homogeneity of solids, is not directly applicable to heterogeneous composites, since microscopic fibres and particles are present within composites and have a significant effect on their overall properties. Thus, micromechanics is applied to analyse the relationship between material property performance and material structure on a finer scale (i.e. the microscale), which encompasses mechanics related to the microstructures of materials. The most valuable tools for studying the micromechanics are the (in situ) deformation test performed under a HVTEM (high-voltage transmission electron microscope; 1 MeV), and "fractography", where an SEM is used to study and analyse the events that occur during the failure of a material [5] (see Chap. 6).

21.2 Particle-Reinforced Polymer Composites

Particles used to reinforce polymers include ceramics and glasses such as small mineral particles (e.g. Al_2O_3, $CaCO_3$, talc etc.), quartz or glass powder, metal particles such as Ni, Cu, Ag, Al or Fe, and occasionally also organic particles of wood, rice hulls or starch [6–9]. Most mineral particles are typically used to increase the modulus of the matrix, to increase wear and abrasion resistance and surface hardness, to improve performance at elevated temperatures, to reduce friction and shrinkage, and to decrease the permeability of the matrix. Sometimes they are also just used to reduce the cost of the polymer. In contrast, metal particles are mainly used to improve the electrical conductivities of most insulating polymer matrices. Figure 21.2 shows an example of a particle-reinforced composite: a polypropylene composite with 50 wt% particles of chalk, talc and dolomite. The diameters of the particles vary from several µm up to <100 nm, and information on the shapes and types of particles present can be discerned. Such broad diameter distributions are advantageous for realising high filler contents, since the spaces between the large particles are filled with smaller particles. A general problem encountered during the preparation of highly filled polymers is incompatibility between fillers and polymers, which impedes the uniform dispersion of particles. This is illustrated in Fig. 21.3 with a HVEM micrograph of the phase morphology of a polypropylene composite with 40 wt% Al_2O_3 filler particles. In the sample, the average filler particle size is about 1 µm, and due to the high filler content a large particle agglomerate with a very low internal strength appears, yielding decohesion and microvoid formation.

To study the micromechanical deformation processes responsible for reinforcement, semi-thin sections about 1.5 µm thick were prepared at −80 °C using an ultramicrotome equipped with a diamond knife and then strained in situ in HVTEM.

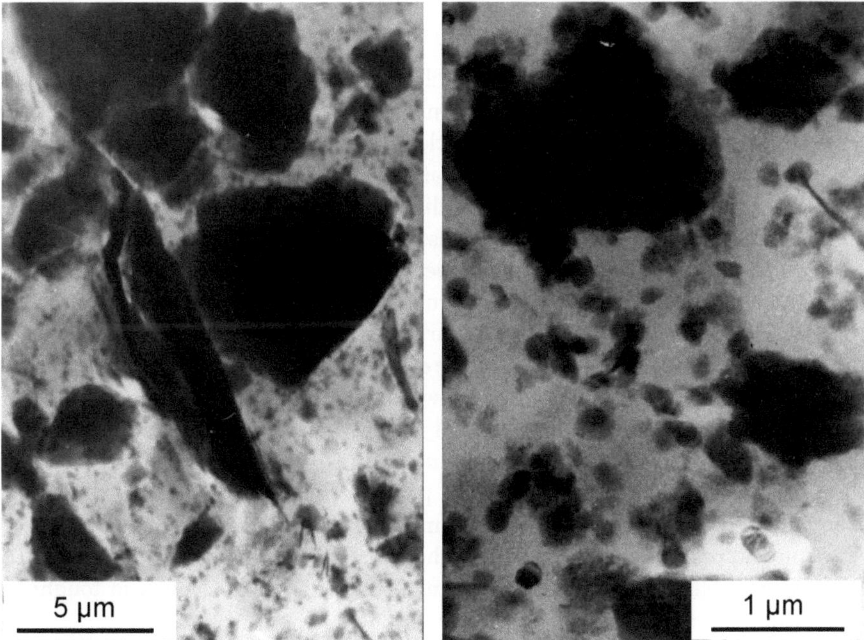

Fig. 21.2. Phase morphology of a PP composite filled with 50 wt% of different types of particles (chalk, talc, dolomite). (3 µm thick section, HVTEM; from [11], reproduced with the permission of Hanser)

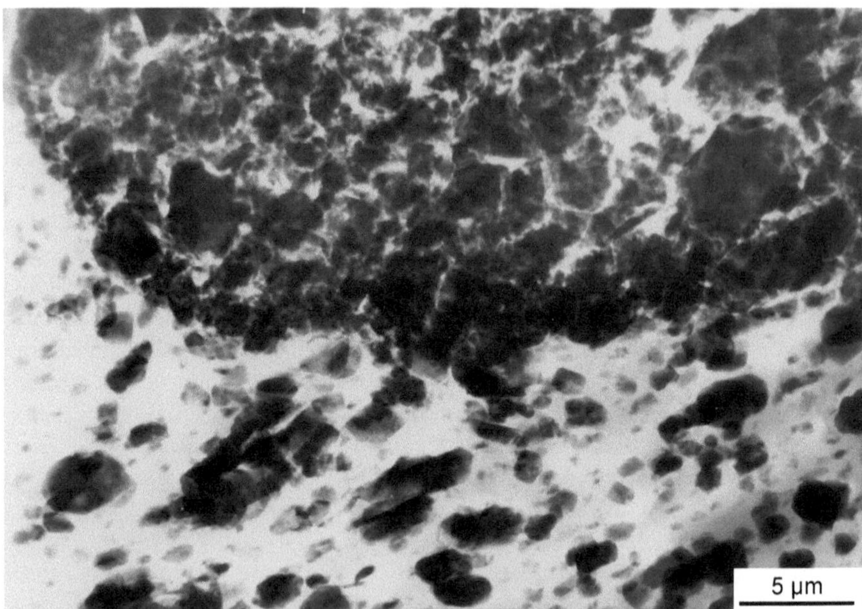

Fig. 21.3. Modified PP with 40 wt% Al_2O_3 filler particles, partly separated and partly agglomerated. (Semi-thin sections with 1.5 μm thickness, HVTEM; from [11], reproduced with the permission of Hanser)

Because of the rigidity of the inorganic filler particles, they cannot be deformed by external stress in the specimen but can only act as stress concentrators. Due to poor adhesion between the Al_2O_3 filler particles and the matrix, the debonding process takes place easily at both sides of the particles in the direction parallel to the applied stress (at the poles of the rigid modifier particles), where the maximum stress concentration is located (Fig. 21.4). In this polymer composite modified with 10 wt% Al_2O_3 filler particles, the matrix strands between the particles are of an optimum size such that the contraction of the specimen in the direction perpendicular to the applied stress occurs following necking. During these debonding processes, the matrix material between the voids deforms more easily to achieve intense shear yielding and improved toughness [10]. Figure 21.5 shows the deformation structures of a particle-reinforced PP composite with 60 wt% Al_2O_3 filler particles without large agglomerates. Here, the void formation appears in bands across the sample, with only very thin matrix strands present between the particles. Because the stress field interacts very intensely due to the high content of modifier particles, a craze-like deformation structure occurs. It is worth pointing out here that the overall deformation structures closely resemble those of rubber-toughened PP, and an enhancement in toughness was achieved for these PP composites.

Figure 21.6 demonstrates a negative effect on the fracture toughness caused by a broad size distribution of $CaCO_3$ particles dispersed in a PVC matrix. The larger particles readily debond under external stress, which leads to the formation of large

Fig. 21.4a,b. Deformation structures of modified PP with 10 wt% Al$_2$O$_3$ filler particles. (Semi-thin section with 1.5 μm thickness, HVTEM)

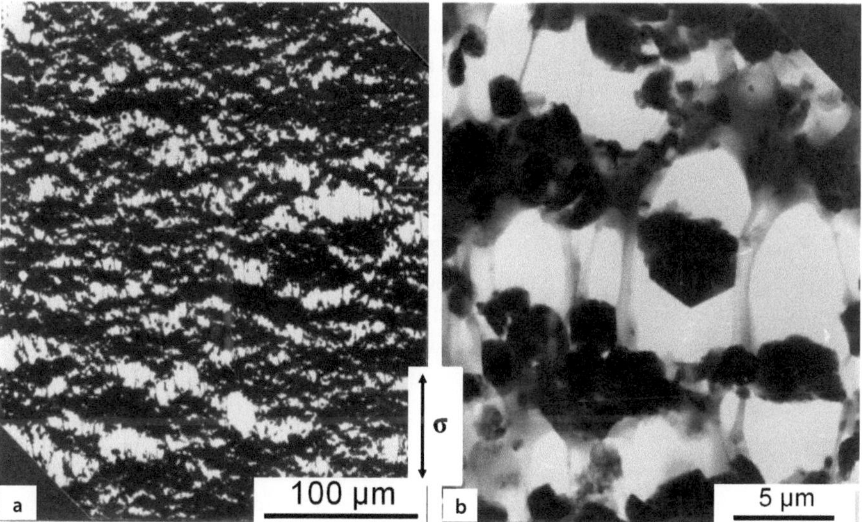

Fig. 21.5a,b. Deformation structures of modified PP with 60 wt% Al$_2$O$_3$ filler particles. (Semi-thin section with 1.5 μm thickness, HVTEM)

localised voids rather than uniform voids. As a consequence, this composite fails in a brittle manner compared with a composite modified with uniform size filler particles (see 21.4 and 21.5) [11]. A correlation between the size and the number of larger particles or agglomerates and the fracture toughness has been found by analysing brittle fracture surfaces in SEM [12].

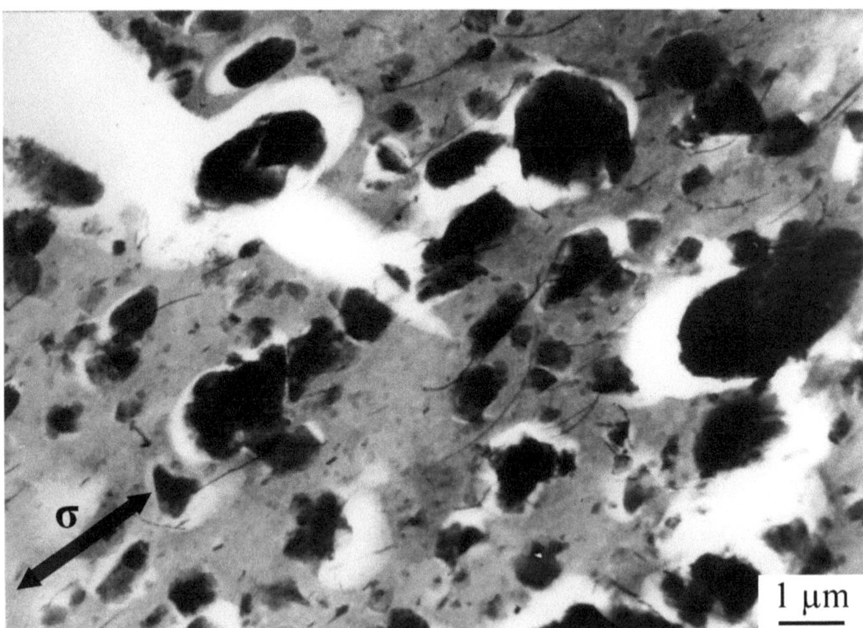

Fig. 21.6. Void formation at CaCO$_3$ filler particles in a PVC matrix and brittle crack propagation. (40 wt% of CaCO$_3$; semi-thin section with 3 μm thickness, for deformation direction see *arrow*, HVTEM; from [11], reproduced with the permission of Hanser)

Besides HVTEM, micromechanical deformation and fracture processes can be easily studied using SEM and ESEM. Figure 21.7 shows a sequence from an in situ deformation test of HDPE filled with 28 wt% of about 1-μm Al$_2$O$_3$ particles in SEM [13, 14]. The starting step is phase separation and void formation at the larger filler particles and agglomerates (a). Thin matrix strands between closely connected voids are plastically stretched and transformed into long fibrils (b). With increasing strain, craze-like deformation bands appear in the sample (c).

Toughening effects depend strongly on the interfacial strength, debonding, void formation, the size and volume content of the particles (both define the interparticle distance) and the deformability of the interparticle matrix strands [11]. To avoid agglomeration and to get a better dispersion of the particles in the matrix, surface modification of the particles is often employed. However, such a surface modifier can also change the properties of the matrix, as illustrated in Fig. 21.8. In an HDPE/chalk composite, the chalk was modified with stearinic acid to improve dispersion. However, during processing the stearinic acid also affected the HDPE matrix, yielding a higher deformability. Compared to the results obtained for a low content of stearinic acid (Fig. 21.8a), a higher content yields strongly plastically stretched matrix strands between particles and microvoids (Fig. 21.8b) [11].

Particularly in semicrystalline matrix materials, the filler particles can modify or alter the semicrystalline morphology. Small filler particles can be concentrated at

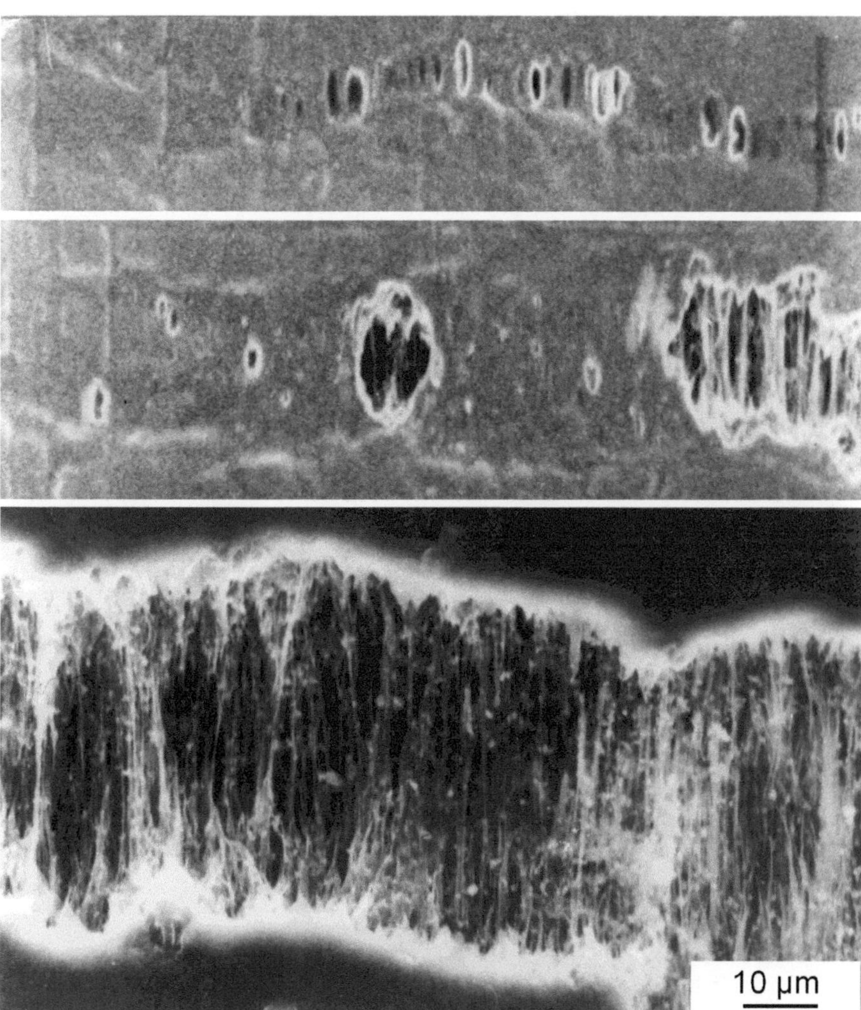

Fig. 21.7. Stepwise development of craze-like deformation bands in HDPE composite filled with 28 wt% Al$_2$O$_3$ particles. (In situ deformation in SEM, deformation direction vertical; from [11], reproduced with the permission of Hanser)

the boundaries of large spherulites in PP, initiating an interspherulitic brittle fracture [15]. Larger filler amounts suppress the formation of large spherulites. On the other hand, filler particles can act as crystallisation nuclei and modify the formation of lamellae. Epitaxial growth and orientation of lamellae in the matrix strands between particles has been discussed in connection with toughness enhancement in particle-filled HDPE and PP [16]. Enhancing the compatibility between the particles and the polymer matrix in order to improve the distribution of the particles can also be achieved through the additional mixing of a second polymer, usually

Fig. 21.8a,b. Comparison of fracture surfaces in SEM of $CaCO_3$ particle-filled HDPE composites with different contents of surface modifier (stearinic acid): **a** 0.9%; **b** 1.5%. (From [11], reproduced with the permission of Hanser)

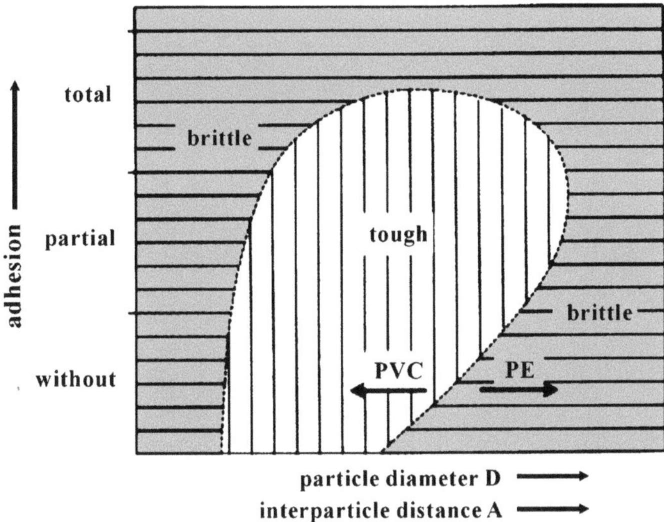

Fig. 21.9. Schematic representation of the influence of particle diameter (or interparticle distance) and the degree of adhesion (interfacial strength) on fracture behaviour

an elastomeric component, which wraps itself around the particles [17]. The effect on the distribution of the particles and the interfacial strength can be checked by performing electron microscopic investigations of deformed thin sections. The combined influence of the diameter and the volume content of the particles (and thus the interparticle distance) and the degree of interfacial strength (adhesion) on the mechanical behaviour (tough or brittle) is summarised in Fig. 21.9. In the case of missing or small interfacial strength, large particles create large voids, yielding severe premature cracks and brittle fracture. Small particles with a homogeneous distribution can initiate plastic deformation of the tiny matrix strands, yielding toughness. Partial adhesion between the particles and the matrix enhances the toughness of composites; in other words, it broadens the "tough window" to larger diameters. Total adhesion (high interfacial strength) avoids debonding, void formation and shear yielding of matrix strands, yielding stiff, strong, but brittle composites. The transition from ductile to brittle behaviour also depends on the deformability of the matrix polymer: ductile polymers with a pronounced strain hardening (such as PE) shift the window to larger particles, more brittle polymers (such as PVC) to smaller ones [11].

Some special polymer types are composites, too, such as flame-retardant polymers, electrically conductive polymers and bone cements (see Sect. 23.3.3). Here, it is doubly important to study the particle arrangement, size, and other parameters: first, to realise the specific function of the polymer; second, to maintain good mechanical properties.

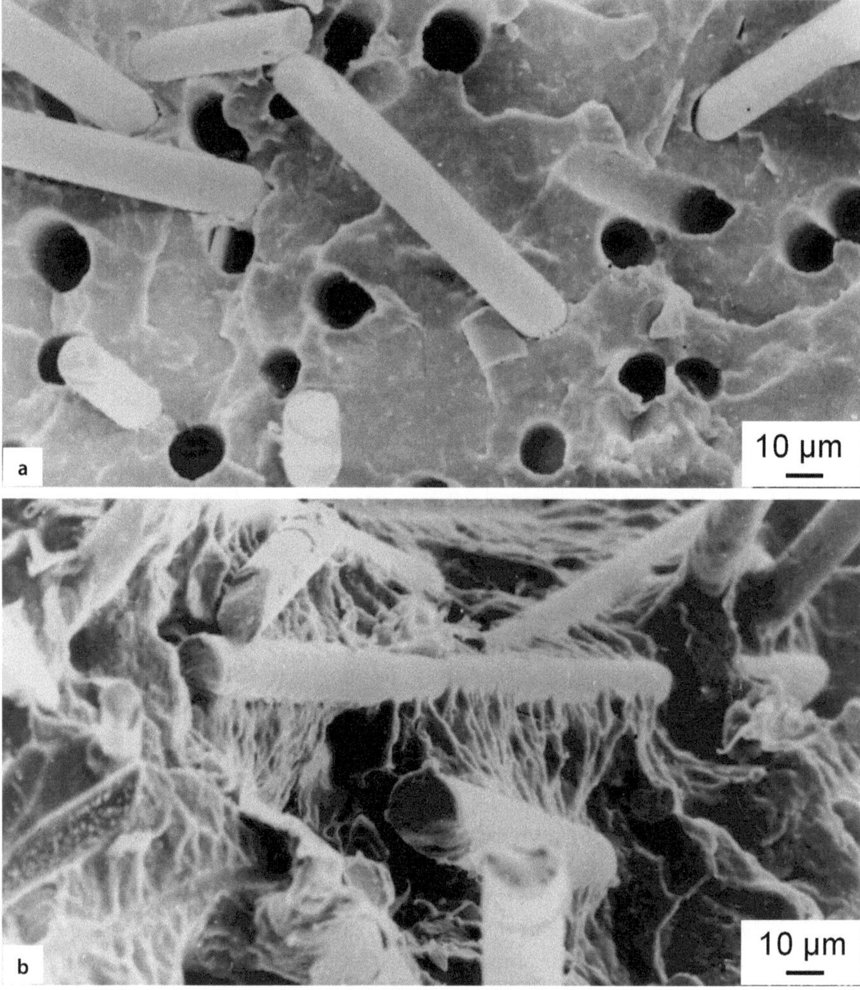

Fig. 21.10a,b. Fracture surfaces in SEM of glass fibre-reinforced PP composites: **a** with poor phase adhesion; **b** with good phase adhesion

21.3 Fibre-Reinforced Polymer Composites

Fibres, although strong and tough, are generally not very stiff because they are very small in diameter. Therefore, it would be impossible for a structure to be made only from small fibres. Adding a matrix material ties the fibres together to form a structure, so that stress can be transferred from one fibre to another, sharing the load. Adding reinforcing fibres increases the modulus of the matrix material. The arrangement or orientation of the fibres relative to one another, the fibre concentration and their distribution all have a significant influence on the strength and other properties of fibre-reinforced composites. Applications involving multidirectional applied

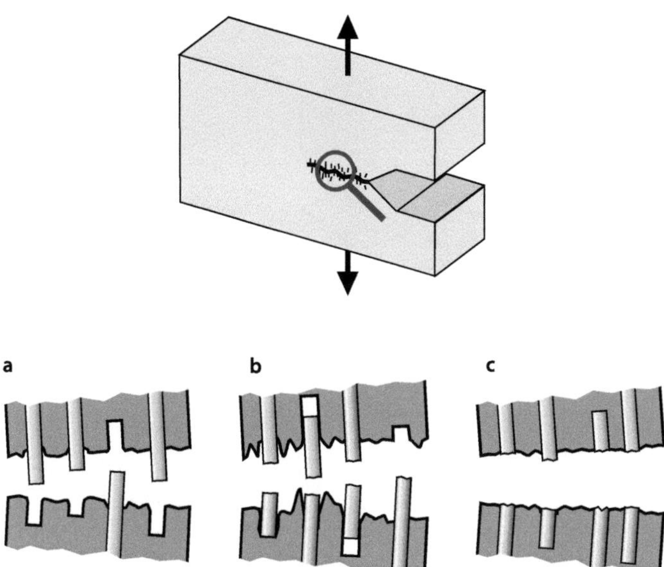

Fig. 21.11a–c. Schematic possible deformation processes in a fibre-reinforced composite where the fibres are arranged parallel to the direction of applied stress: **a**, debonding, gliding and pulling out of fibres; **b**, partial separation and pull-out of fibres, break-down of fibres; **c**, breakdown of fibres in the case of good phase adhesion

stresses normally use discontinuous fibres, which are randomly oriented in the matrix material [18].

The orientation and fibre length that should be selected for a composite depend on the level and nature of the stress to be applied as well as the fabrication cost. There are two main classes of fibrous composites: continuous (long fibre, Fig. 21.1b) and discontinuous (short fibre, Figs. 21.1c,d). Continuous fibre composites have much better stiffness and strength, but are more difficult to process within polymer matrices, which makes fibre-reinforced composites relatively expensive. In contrast, short fibres are favourable if we wish to form composites with intricate internal shapes (both aligned and randomly oriented). Therefore, short fibres are usually used for thermoplastic matrix materials and long fibres for epoxy or polyester resins.

The usual technique used to investigate the morphology and micromechanics of fibre composites is to study fracture surfaces in SEM. As an example, Fig. 21.10 shows fibre-reinforced PP composites with 30 wt% glass fibres that exhibit different levels of phase adhesion between the fibres and the matrix: Fig. 21.10a exhibits poor or missing phase adhesion between the fibres and the PP matrix, whereas in Fig. 21.10b the fibres oriented transverse to the fracture surface exhibit good phase adhesion. The well-defined interface between the fibres and the polymer matrix increases the mechanical properties of the polymer; tensile strength increases from 40 to 87 MPa and impact strength from 16 to 34 kJ/m^2 [19].

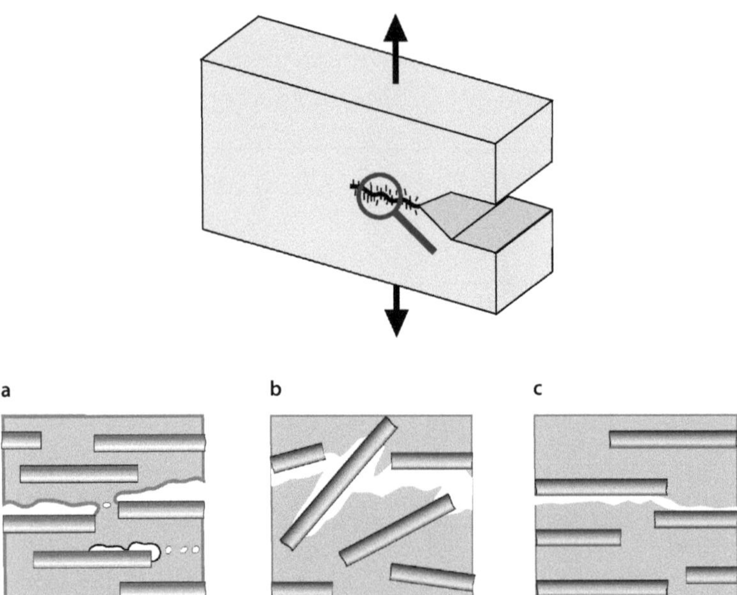

Fig. 21.12a–c. Schematic possible deformation processes in a fibre-reinforced composite where the fibres are arranged perpendicular to the direction of applied stress: **a**, poor phase adhesion, debonding fibres with void formation, interfibrillar fracture with coalescence of cracks at the ends of fibres; **b**, debonding of fibres or breakdown of the tilted fibres; **c**, matrix fracture along the fibre interface or within the matrix in the case of good phase adhesion

Micromechanical deformation processes that typically occur in fibre-reinforced polymer composites are schematically illustrated in Figs. 21.11 and 21.12; these were derived as result of fracture surface studies in SEM [19]. As mentioned above, the resulting mechanisms are strongly dependent on the degree of phase adhesion between fibre and matrix, as well as the orientation of the fibres with respect to the direction of applied stress. When the fibres are arranged parallel to the applied load direction, the micromechanical deformation process can be followed by debonding at the fibre/matrix interface, as caused by the pulling out of fibres from the matrix (Fig. 21.11a, if the interfacial adhesion is poor), fibre gliding (Fig. 21.11b, if partial interfacial adhesion is present), and fibre breakdown (Fig. 21.11c, if the interfacial adhesion is strong enough), which lead to the development of zigzag-shaped fracture surfaces during crack propagation. When the fibres are oriented perpendicular to or placed at a certain angle to the applied load direction, the fracture mechanisms can result in the following micromechanical deformation processes:

– If the interfacial adhesion is poor, microvoid formation can occur at the fibre interface or in the matrix between the fibres, which can have either positive or negative effects on the fracture toughness: facilitating the initiation of shear yielding or the crazing of the matrix material, respectively (Fig. 21.12a). Fibers tilted at a certain angle can separate from the matrix or break down with increasing load (Fig. 21.12b).

– If the interfacial adhesion is good, the crack propagates along the interface with the fibres or inside the matrix material; in this case a smooth fracture surface may be observed (Fig. 21.12c).

References

1. Klaus F, Stoyko F, Zhong Z (eds) (2005) Polymer composite, from nano- to macro-scale. Springer, Berlin
2. Rothon RN (ed) (2003) Particle-filled polymer composites, 2nd edn. Rapra Technology, Shrewsbury, UK
3. Hollaway L (ed) (1994) Handbook of polymer composites for engineers. Woodhead, Cambridge
4. Cripps A (2002) Fibre-reinforced polymer composites in construction. CIRIA, London
5. Zhang X, Knackstedt MA, Chan DYC, Paterson L (1996) Europhys Lett 34:121
6. Norman RH (1957) Conductive rubber. Maclaren, London
7. Delmonte J (1989) Metal-polymer composites. Kluwer, Norwell, MA
8. Razi PS, Raman A, Portier R (2000) J Compos Mater 34:980
9. Chaudhary DS, Jollands MC, Cser F (2004) Adv Polym Technol 23:147
10. Kim GM, Michler GH, Gahleitner M, Fiebig J (1996) J Appl Polym Sci 60:1391
11. Michler GH (1992) Kunststoff-Mikromechanik: Morphologie, Deformations- und Bruchmechanismen. Hanser, München, 10:320
12. Tovmasjan JM, Topolkarajev WA, Berlin AA, Schuravlev IL, Enikolopjan NS (1986) Vysokomolekul Soed SSSR A28:321
13. Michler GH, Tovmasjan JM (1988) Plaste Kautschuk 35:73
14. Michler GH, Tovmasjan JM, Topolkarajev WA, Dubnikova IL, Schmidt V (1988) Mech Kompos Mat Riga 2:221
15. Friedrich K (1978) Progr Colloid Polym Sci 64:103
16. Bartczak Z, Argon AS, Cohen RE, Weinberg M (1999) Polymer 40:2331
17. Pukanszky B, Tüdos F, Kolařik J, Lednicky F (1990) Polym Compos 11:98
18. De SK, White JR (eds) (1996) Short fibre-polymer composites. Woodhead, Cambridge
19. Michler GH (1992) Kunststoff-Mikromechanik: Morphologie, Deformations- und Bruchmechanismen. Hanser, München, 11:345

22 Polymer Nanocomposites

Polymer nanocomposites (PNCs) are a relatively new class of materials in which nanofillers – which have at least one dimension in the range of 1–100 nm – are dispersed in a polymer matrix. Due to the nanoscale of the fillers, TEM has largely been used for studies of morphological and micromechanical deformations in these materials. In this chapter we discuss representative polymer nanocomposites in relation to the dimensionality of the nanofiller (i.e. 0-D, 1-D, 2-D or 3-D nanofillers). The morphologies and deformation mechanisms of PNCs with POSS molecules, carbon nanotubes, layered silicates and SiO_2 nanoparticles are shown with EM. An example of a PNC of PMMA with ~25-nm SiO_2 nanoparticles showed excellent transparency and enhanced fracture toughness up to 20 wt% of filler. The dominant mechanism for improved fracture toughness was shown to be "nanoparticle modulated crazes".

22.1 Overview

Within the last decade, it has become apparent that nanoscale reinforcement is an attractive way of improving the properties and stability of polymers. This idea has led to a new and improved class of composites, the polymer nanocomposites (PNCs) [1–4]. The development of PNCs is now a rapidly expanding multidisciplinary global research activity. The nanoscale and high specific surface area of nanofillers (which exhibit at least one dimension in the range of 1–100 nm, such as nanoparticles or nanotubes) and the resulting predominance of interfaces in PNCs significantly affect the structure and morphology of PNCs at the molecular scale [5, 6], influencing their physical and material properties at scales that are inaccessible when traditional (e.g. micron-sized; see Sect. 21.2) filler materials are used. The resulting PNCs exhibit an excellent property profile that is applicable to a wide variety of industrial applications; for example, high stiffness, chemical and thermal resistance, dimensional stability, reduced water absorption, as well as improved electrical and optical properties, all of which are significantly different from those provided by conventional composites.

These synergisms originate from the nature of nanoparticles, and are associated with the following three concepts:

– Firstly, as shown in Figs. 22.1 and 22.2a, the number of filler particles at a given volume fraction (3% in the figures) rapidly increases as the size of particles decreases. As a consequence, compared to polymer composites with conventional

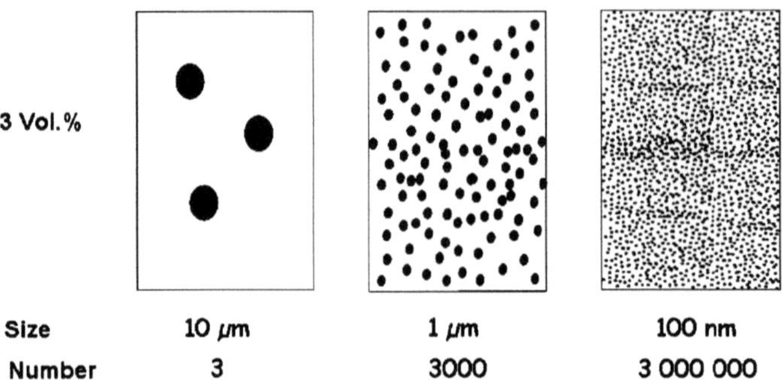

Fig. 22.1. Illustration of how the number of particles increases as the particle size decreases at a constant volume content

fillers (in the micron range), which nominally require loadings of 20 wt% or more, the distances between nanoparticles are drastically reduced to the nanometre range, even when relatively low concentrations of nanofillers are used.

- Secondly, the sizes of the nanoparticles are comparable to the radius of gyration of the macromolecules in the polymer matrix, so that the morphological development of polymer matrix can be substantially affected by the dispersed nanoparticles (*morphology under constraint condition*).
- Thirdly, the nanoparticles provide ultrahigh specific surfaces, and this huge interfacial area permits strong interactions with the polymer matrix. As a consequence, the amount of modified polymer interface relative to the total volume will be significantly increased, corresponding to the transition from a *polymer matrix material* to a *quasi-polymer interfacial material*.

Figure 22.2a illustrates that the specific surface (defined as the ratio of the surface area to the volume) and the number of filler particles increases with decreasing particle size. It is well established that the interparticle distance between dispersed particles has a crucial influence on the mechanical properties, in particular toughness, in heterogeneous polymer systems. This interparticle distance (*ID*) as function of particle size (*D*) and its volume fraction (ϕ) can be readily estimated by the Eq. 22.1 [7] (see also Fig. 22.2b). At a given particle size *D*, it is clear that the interparticle distance (*ID*) decreases as the filler volume fraction increases.

$$ID = D\left[\left(\frac{\pi}{6\phi}\right)^{1/3} - 1\right] \tag{22.1}$$

While PNCs are highly promising and some interesting properties have been already demonstrated for them, the resulting nanocomposites have yet to realise their full potential. For structural applications, the nanofiller may have two roles: to increase the stiffness of the polymer matrix and to enhance toughness by encouraging new energy dissipation mechanisms. Therefore, the development of successful PNCs with a good balance of properties demands a high degree of coupling between processing and morphological and micromechanical control. That is, the balance of

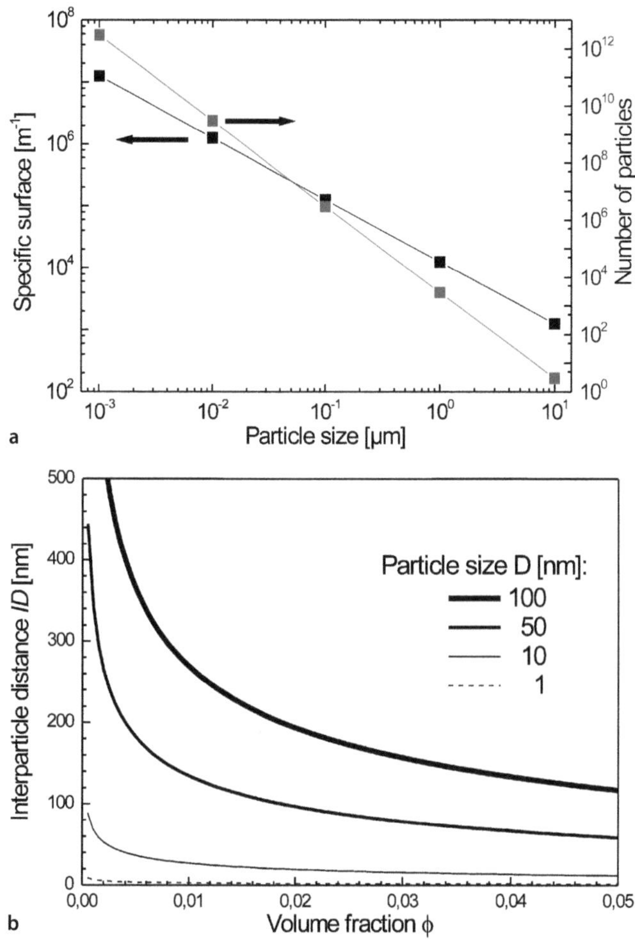

Fig. 22.2. **a** Variation in specific surface and number of particles with particle size (based on Fig. 22.1, i.e. reducing the size from 10 µm at constant content of 3 vol%). **b** Interparticle distance vs. volume fraction of particles for various particle sizes

properties must be optimised. To this end, at least two essential morphological requirements should be met in the rational design of new composites: the uniform dispersion of nanofillers to avoid large agglomerates and high stress concentrations, and good or optimum interfacial bonding between them and the polymer matrix to achieve effective load transfer across the filler–matrix interface [8–10]. Unfortunately, it has been extensively reported that nanofillers are often dispersed in the polymer matrix in the form of agglomerates. This strong tendency to agglomerate significantly reduces their ability to bond with the matrix since it reduces the contact area and decreases the effective aspect ratio of the reinforcement. Moreover, under external load the stress will be readily concentrated around such agglomerates, which in turn generally leads to the premature failure of the system.

All of these structural parameters that have an important influence on the properties of PNCs can be directly investigated by EM inspection. SEM is better at showing agglomerates and TEM is best for individual nanoparticles. Deformation tests can qualitatively reveal interfacial strength or debonding between the particles and the polymer matrix. To illustrate these possibilities, in this chapter we will discuss representative polymer nanocomposites containing different dimensionalities of common nanofillers, such as:

- Zero-dimensional filler particles: Polyhedral OligoSilSesquioxane (POSS)
- One-dimensional filler particles: multiwalled carbon nanotubes (MWCNTs)
- Two-dimensional filler particles: layered silicates (LSs)
- Three-dimensional filler particles: silica nanoparticles (SiO_2).

22.2 Examples of Different Classes of Nanocomposites

22.2.1 Polymer Nanocomposites Based on Zero-Dimensional Filler Particles (POSS)

POSS technology is based on revolutionary state-of-the-art synthesis using a so-called "bottom-up" process. Intensive studies on POSS-based systems have established that the introduction of such nanostructured POSS molecules within conventional polymers can lead to significant enhancements in matrix properties, such as increases in T_g, heat distortion temperature, modulus above T_g, and toughness [11–13]. This synergism in properties results from the ability of POSS molecules to disperse at a molecular level into the host polymer matrix due to their shape and size while still maintaining the processability and mechanical properties of the base resin. Due to the well-defined nanostructure of POSS and the relative ease of incorporation of POSS into conventional polymer matrices, there are already a variety of industrial applications of such polymers; i.e. in the aerospace industry due to the low density and long durability of POSS, radiation resistance for antenna, in solar sails and also in satellites.

As shown in Fig. 22.3, the POSS molecule presents a well-defined cage-like silicate structure and its chemical composition is a hybrid, intermediate ($RSiO_{1.5}$) between that of silica (SiO_2) and silicone (RSiO) in the range of $1 \sim 3$ nm. It is nearly equivalent in size to most polymer segments and coils, and thus, topographically speaking, can be considered to be a zero-dimensional nanoparticle in the sense of polymer composites. Each POSS molecule contains unreactive organic functionalities for solubility and compatibility with the matrix but also maintains one or more covalently bonded reactive functionalities that are suitable for polymerisation, grafting, and surface bonding.

As a representative example, Fig. 22.4 shows the phase morphology and deformation structure of a POSS-reinforced epoxy resin based on octaglycidyl epoxy polyhedral oligosilsesquioxane (OG-POSS) cured with a 100% stoichiometric ratio of 4,4'-diaminodiphenyl sulfone (DDS) (a high-performance hardener). In comparison with POSS-free epoxy, an obvious shift in the glass transition temperature to

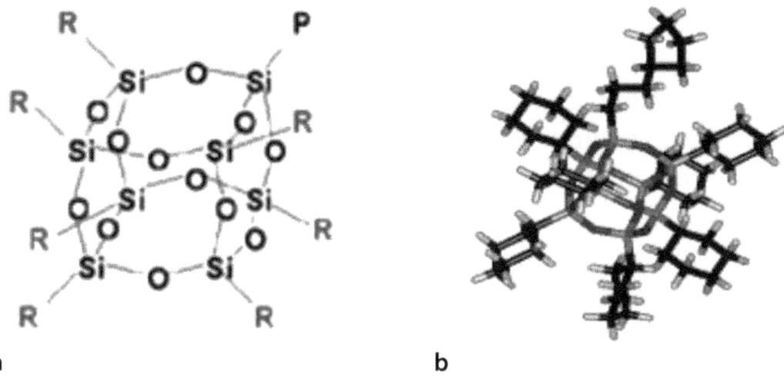

Fig. 22.3a,b. Cage-like structure of a polyhedral oligosilsesquioxane (POSS) molecule: **a** R are unreactive organic functionalities used for solubility and compatibility with the polymer matrix, and P are one or more reactive functional groups used for grafting or polymerisation. **b** 3-D model of the POSS structure, Si–Si distance 0.5 nm, R–R distance 1 ~ 3 nm

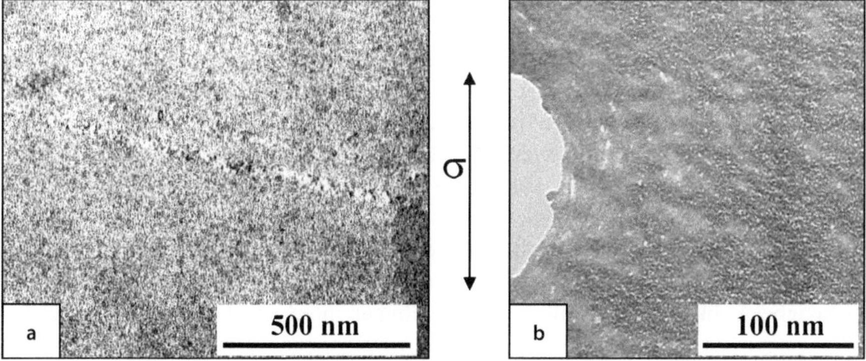

Fig. 22.4. **a** Morphology of OG-POSS-epoxy (100% stoichiometric ratio of DDS); **b** deformation structure at a crack tip. (Ultrathin sections with 50 nm thickness, TEM)

a higher value and an increase in the storage modulus at temperatures $T > T_g$ and $T < T_g$ were observed. Figure 22.4a indicates the phase structure of the resulting OG-POSS. After staining treatment with RuO_4, which is selective for POSS, the phase structures are clearly revealed in TEM micrographs. The OG-POSS molecules are well dispersed in the form of aggregates of size 15 nm or less. Direct observation of mechanical deformation using an in situ tensile test indicated that these POSS aggregates were clearly reflected in the deformation structure. To image the deformation structures, the bulk specimens were randomly ultramicrotomed; this process led indirectly to crack creation within the ultramicrosections, so that near-crack regions could be directly visualised. Figure 22.4b shows a TEM image exhibiting plastic deformation zones at a crack tip, where well-developed voids of size 8–12 nm are visible in the matrix surrounding the POSS moieties. Thus, "POSS aggregate-induced nanovoid formation" is a dominant deformation mechanism in this system [14]. The presence of such voids reduces the local build-up of hydrostatic stress at the crack tip and allows some shear yielding to occur around them in the matrix. Furthermore, nanovoids at the crack tip strongly interact with the yielding of the material at the front of the crack tip, blunting the cracks in the plastic deformation zone. The initiation of nanovoids and their subsequent growth should enhance matrix shear yielding, which is directly associated with energy dissipation during deformation, so that the toughness of the materials can be enhanced.

22.2.2 Polymer Nanocomposites Based on One-Dimensional Filler Particles (MWCNT)

As an example of this nanocomposite class [15–20], we prepared a nanocomposite based on polycarbonate (PC) with 4 wt% multiwalled carbon nanotubes (MWCNTs) by dilution of a master batch of 15 wt% MWCNT in PC (Hyperion Catalysis International, Inc., Cambridge, MA, USA) [21]. The MWCNTs are vapour-grown and typically consist of 8–15 graphitic layers wrapped around a hollow 5-nm core. As reported by Hyperion, the diameter of the MWCNTs used for the composites discussed here is in the range of 10–15 nm. They are produced as agglomerates and exist as curved intertwined entanglements. The ultrathin sections about 50 nm thick were prepared at −80 °C using a Leica Ultracut ultramicrotome equipped with a Diatome diamond knife (note that the carbon nanotubes often damage the diamond knife) and investigated the phase morphology of PC/MWCNT nanocomposite by TEM without any chemical staining. In Fig. 22.5a, one can clearly see that the MWCNTs are dispersed in the form of a highly entangled (interconnected) structure in the PC matrix. The MWCNTs used in this study exhibit distinct curved shapes that can be described as a "spaghetti"-like structure, ultimately forming an interlocked structure of MWCNT in the agglomerated state. Figure 22.5b shows a typical deformation structure produced during an in situ uniaxial tensile test in TEM, for which samples were microtomed at −80 °C with a thickness of about 1 μm. Although the pure PC deforms, in general, through shear yielding, the PC/MWCNT nanocomposite deforms by a fibril-

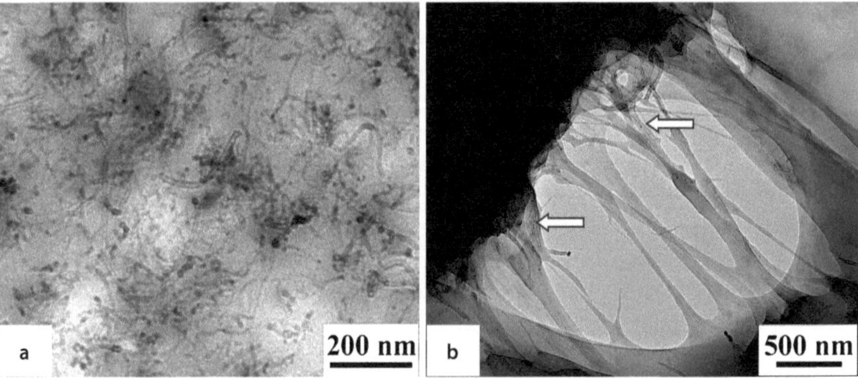

Fig. 22.5a,b. PC/MWCNT nanocomposite: **a** typical phase morphology (ultrathin section with 50 nm thickness, TEM); **b** deformation structure (*arrows* indicate the areas where the MWCNTs are embedded in fibrils, semi-thin section with 300 nm thickness, TEM)

lised crazing due to the highly entangled (interconnected) structure of the MWCNTs in the PC matrix. This indicates that the PC/MWCNT nanocomposite exhibits more brittle behaviour that the pure PC. It should be emphasised that no sign of displacement (debonding) of the embedded nanotubes from the fibre matrix was observed. This suggests that the individual MWCNTs aligned within the fibres interact with the surrounding polymer matrix not just through van der Waals interactions but possibly by bonding with the polymer. This indicates that the individual MWCNTs have an absorbed layer of PC around them, which results in enhanced polymer connectivity at the nanotube surface [22].

22.2.3 Polymer Nanocomposites Based on Two-Dimensional Filler Particles (MMT)

Polymer nanocomposites were synthesised by the melt blending of polyamide 6 (PA6) (of different molecular weights) with 3.2 wt% montmorillonite (MMT). The two resulting layered silicate polymer nanocomposites were designated HMW-NC and LMW-NC, based on whether the high or the low molecular weight PA6 grade was used, respectively [23–25].

Figure 22.6a and b show the typical morphologies of HMW- and LMW-NC, which are obtained from unstained ultrathin sections. Samples with a thickness of about 50 nm were prepared for TEM at −80 °C using a Leica Ultracut ultramicrotome equipped with a Diatome diamond knife. Due to the sufficient material contrast between MMT and PA6, it is not necessary to stain the ultrathin sections for TEM studies. In the TEM micrographs, the black lines with a thickness of about 1 nm represent the intersections of MMTs, while the grey part represents the nylon 6 matrix.

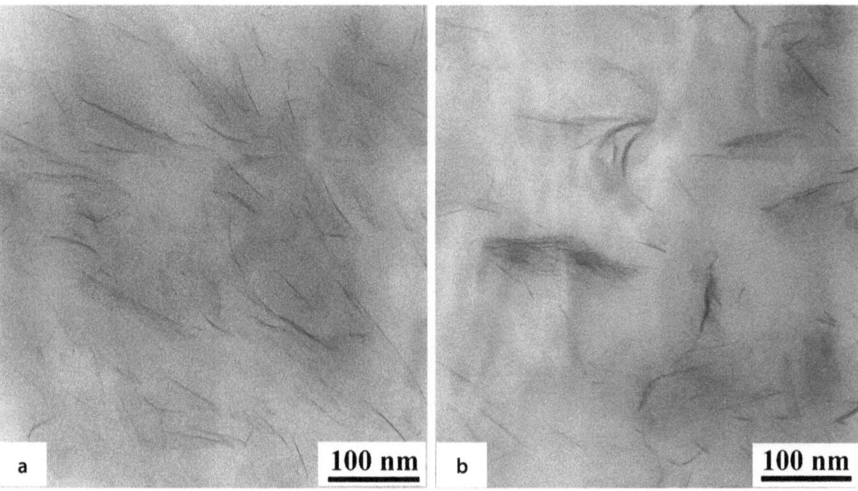

Fig. 22.6a,b. TEM micrographs of PA6 nanocomposites with 3.2 wt% MMTs without staining treatment: **a** HMW-NC; **b** LMW-NC. (Ultrathin sections with 50 nm thickness, TEM)

As seen in Fig. 22.6a for HMW-NC, the individual layered silicates (LSs), along with two- and three-layer stacks, are observed to be homogeneously well-dispersed (exfoliated) in the polymer matrix. In contrast, LMW-NC shows a mixed nanomorphology (Fig. 22.6b), with coexisting intercalated tactoids and a partially exfoliated morphology. It can be easily seen in the TEM micrographs that the LSs in HMW-NC are more likely to be oriented in the injection moulding direction, analogously to conventional glass fibre-reinforced polymer composites, whereas the LSs in LMW-NC are relatively heterogeneously dispersed in the matrix, i.e. they show no dominant direction of orientation. This forced orientation is a result of the high shear rates encountered during injection moulding. It is reasonable to expect that the higher molecular weight systems, due to their higher melt viscosities, transfer more stress or energy to achieve the separation of LSs. Therefore, it can be assumed that high levels of shear stress in HMW-NC aid help to break up the clay particles and ultimately improve the degree of exfoliation, yielding a fair number of single layers as well as a few stacks of LSs.

Figure 22.7a and b show HVTEM micrographs of the deformation structures observed at low magnification during in situ tensile tests of semithin sections (about 300 nm thick) of HMW- and LMW-NC, respectively. The deformation structures reveal well-developed microvoid formation in the intercalated tactoids as well as around the exfoliated layers in the plastically deformed specimens. Based on these observations, it can be concluded that the prominent deformation mechanism for PA6/LS nanocomposites is microvoid formation, which permits energy dissipation under external load during plastic deformation processes [5, 23].

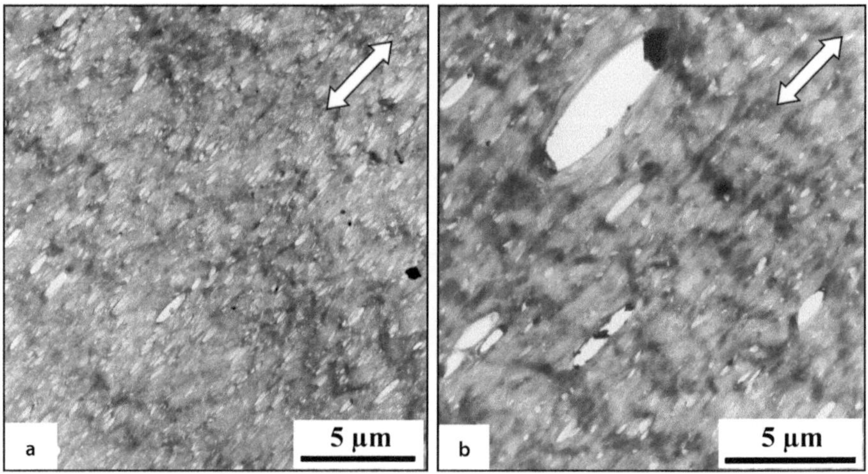

Fig. 22.7a,b. Characteristic deformation structures from in situ tensile experiments: **a** HMWNC; **b** LMW-NC. (Semi-thin section with 300 nm thickness, HVTEM)

22.2.4 Polymer Nanocomposites Based on Three-Dimensional Filler Particles (SiO$_2$)

The materials used here are PMMA composites modified with SiO$_2$ nanoparticles (provided by Röhm GmbH & Co. KG), which were prepared by solution blending, as described by Carotenuto et al. [26]. Figure 22.8a shows the phase morphology of such a PMMA/SiO$_2$ nanocomposite. The SiO$_2$ nanoparticles are uniformly dispersed in the PMMA matrix without any evidence of agglomerates. The size of the nanoparticles is quantitatively analysed from TEM micrographs with the aid of computerised image analysis. The average diameter is 26 nm, and the particle size exhibits a well-defined Gaussian distribution (Fig. 22.8b). As a consequence, this nanocomposite re-

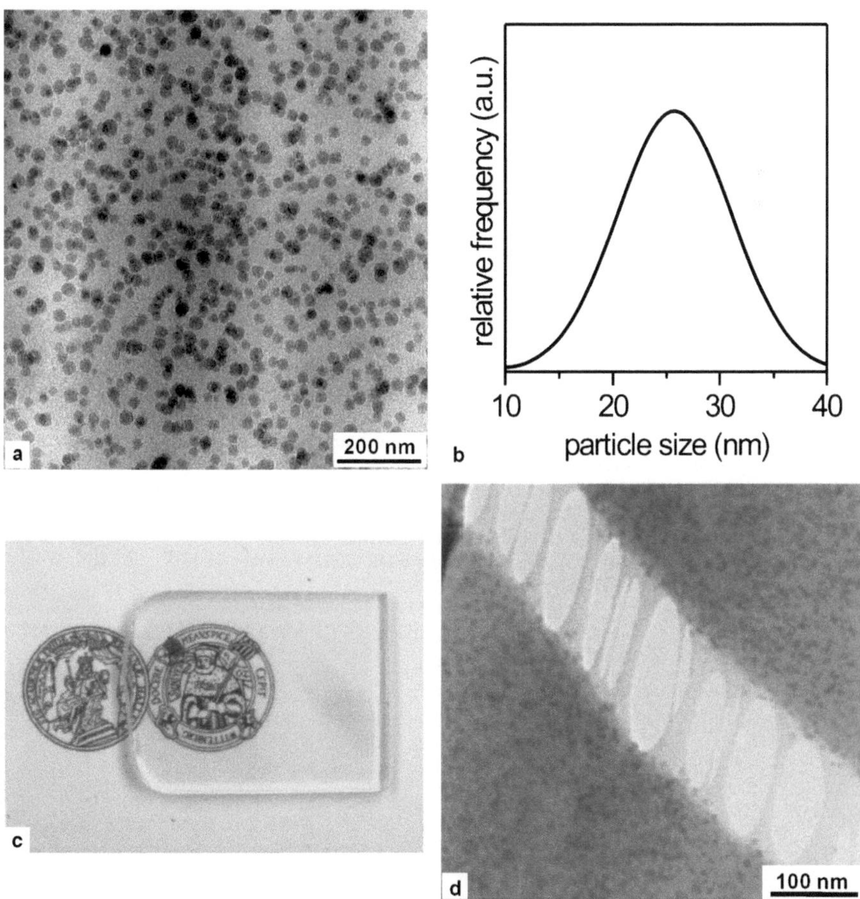

Fig. 22.8. **a** TEM micrograph of PMMA nanocomposite with 10 wt% SiO$_2$; **b** SiO$_2$ particle size distribution; **c** optical characteristics of PMMA nanocomposite with 20 wt% SiO$_2$; **d** deformation structure in PMMA with 10 wt% SiO$_2$ under uniaxial tensile load. (**a,d**: ultrathin sections with 50 nm thickness, TEM)

veals excellent optical properties. Light transmission is not affected by the introduction of SiO_2 nanoparticles, even up to 20 wt% of filler loading (Fig. 22.8c). Whereas this nanocomposite shows brittle behaviour, indicating no macroscopic yielding under tensile load, the nanocomposite with 10 wt% SiO_2 reveals well-defined crazes in the specimen deformed in situ (Fig. 22.8d). This type of "nanoparticle modulated craze" provides a source for the additional enhancement in fracture toughness [27].

References

1. Fukushima Y, Inagaki S (1987) J Inclusion Phenom 5:473
2. Giannelis EP (1996) Adv Mater 8:298
3. Zilg C, Reichert P, Dietsche F, Engelhardt T, Mülhaupt R (1998) Kunststoffe 88:10
4. Pinniavaia TJ, Beall G (2001) Polymer-clay nanocomposites. Wiley, New York
5. Kim GM, Lee DH, Hoffmann B, Kressler J, Stöppelmann G (2001) Polymer 42:1095
6. Creasy TS, Kang YS (2004) J Thermoplas Compos Mater 17:205
7. Wu S (1985) Polymer 26:1855
8. Thostenson ET, Zhifeng R, Chou-Tsu W (2001) Comps Sci Technol 61:1899
9. Andrews R, Jacques D, Minot M, Rantell T (2002) Macromol Mater Eng 287:395
10. Gojny FH, Nastalczyk J, Roslaniec Z, Schulte K (2003) Chem Phys Lett 370:820
11. Haddad TS, Lichtenhan JD (1995) Inorg J Organomet Polym 5:237
12. Lichtenhan JD, Noel CJ, Bolf AG, Ruth PN (1996) Mater Res Soc Symp Proc 435:3
13. Romo-Uribe A, Mather PT, Haddad TS, Lichtenhan JD (1998) J Polym Sci Polym Phys 36:1857
14. Kim GM, Qin H, Fang X, Sun FC, Mather PT (2003) J Polym Sci Polym Phys 41:3299
15. Sandler J, Shaffer MSP, Prasse T, Bauhofer W, Shulter K, Windle AH (1999) Polymer 40:5967
16. Pötschke P, Fornes TD, Paul DR (2002) Polymer 43:3247
17. Coleman JN, Curran S, Dalton AB, Davey AP, McCarthy B, Blau W, Barklie RC (1998) Phys Rev B 58:7492
18. Grimes CA, Mungle C, Kouzoudis D, Fang S, Eklund PC (2000) Chem Phys Lett 319:460
19. Kim HM, Kim CY, Lee CY, Joo J, Cho SJ, Yoon HS, Pejakovic DA, Yoo JW, Epstei AJ (2004) Appl Phys Lett 84:589
20. Curran SA, Ajayan PM, Blau WJ, Carroll DL, Coleman JN, Dalton AB, Davey AP, Drury A, McCarthy B, Maier S, Strevens A (1998) Adv Mater 10:1091
21. Ferguson DW, Bryant EWS, Fowler HC (1998) ANTEC'98, Atlanta, GA, 27 Apr–1 May 1998, p 1219
22. Kim GM, Michler GH, Pötschke P (2005) Polymer 46:7346
23. Fornes TD, Yoon PJ, Keskkula H, Paul DR (2001) Polymer 42:9929
24. Fornes TD, Yoon PJ, Hunter DL, Keskkula H, Paul DR (2002) Polymer 43:5915
25. Fornes TD, Paul DR (2003) Polymer 44:3945
26. Carotenuto G, Nicolais L, Kuang X, Zhu Z (1995) Appl Comp Mater 2:385
27. Lach R, Kim GM, Michler GH, Grellmann W, Albrecht K (2006) Macromol Mater Eng 291:263

23 Biomaterials

The specific challenge presented by polymeric biomaterials to electron microscopy is to monitor the interactions of the material with the biological environment. In this chapter, examples of electron microscopic methods that preserve the natural state of biomaterial and adhering host tissue are given. The ESEM technique, for instance, permits the imaging of samples that are not dehydrated, and AFM methods are used to image biomaterial samples in their original state. Various examples demonstrate the imaging of the interface between the biomaterial and the host tissue, the evaluation of the influence of the living environment on a polymeric material, and the analysis of polymer properties under physiological conditions. Typical problems, applications and results are discussed for natural (compact bone) as well as synthetic polymer-based biomaterials, such as UHMWPE, acrylic bone cements, composites and nanocomposites for bone replacement, dental composites, polymer fibres and nanofibres, scaffolds and meshes, stents and silicone-based voice prostheses.

23.1 Overview

Over the last few decades, significant advances have been made in the development of biocompatible and biodegradable materials (sometimes shortened to "biomaterials") for biomedical applications. Such materials are intended to replace parts of a living system or to function in intimate contact with living tissue. The term "biocompatibility" is generally used to refer to the ability of a biomaterial to provoke an acceptable cellular and biological response from the host environment, and in a clinical context this entails not producing a toxic or injurious reaction and not causing immunological rejection [1]. On the other hand, the term "degradable" relates to materials that disintegrate under environmental conditions within a reasonable and demonstrable period of time [2]. Several classes of materials, such as inorganic (metals, ceramics and glasses) and polymeric (natural and synthetic) materials, have been approved for medical use. Among them, polymeric materials are widely used in clinical applications because of their unique physical and chemical properties, their ability to be structurally and functionally modified, the easy control of their biodegradation and because they are relatively inexpensive to manufacture.

The biodegradation of polymers typically proceeds by hydrolysis and oxidation [3]. In general, a polymer based on the C–C backbone tends to be nonbiodegradable, whereas heteroatom-containing polymer backbones (e.g. hydrolysable and/or oxidisable linkages in the polymer main chain), suitable substituents, correct stereoconfiguration, balance of hydrophobicity and hydrophilicity, and conformational flexibility contribute to polymer biodegradability. Biodegradability can therefore be engineered into polymers through the judicious addition of chemical linkages such as anhydride, ester, or amide bonds, among others. The mechanism for degradation is by hydrolysis or enzymatic cleavage, resulting in a scission of the polymer backbone [4]. In order for it to be applicable for use in medical applications, the biodegradable polymer must be biocompatible, which includes considerations that go beyond nontoxicity to bioactivity (the ability of a material to interact with and, in time, be integrated into the biological environment), as well as other tailored properties (e.g., sterilization, mechanical stability and controlled degradation rate, etc.), depending on the specific in vivo application required.

Biodegradable and biocompatible polymers can be divided into two classes (see Fig. 23.1): natural polymers originating from plant or animal resources (e.g. starch, cellulose, albumin, collagen, chitosan, etc.) [5–7], and synthetic polymers produced by fermentation processes performed by microorganisms [e.g. polyhydroxy alkanoates (PHA)], as well as certain other polymers that possess biodegradable properties (e.g. polycaprolactone and polylactic acid) [8–10].

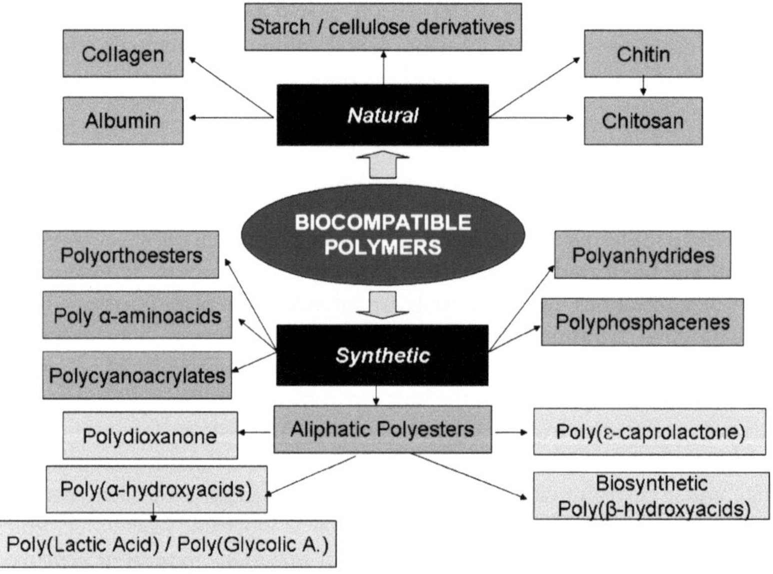

Fig. 23.1. Polymers applicable for use in biomedical applications

23.2 Electron Microscopy of Polymeric Biomaterials: Specific Problems and Solutions

Most polymers created for biomedical applications belong to one of the main groups of polymers that are discussed in the chapters of this book. Therefore, their morphological and micromechanical analyses are performed using the corresponding electron microscopic methods. The specific task in this case is to monitor the interactions of the polymeric biomaterial with the biological environment. In other words, one wishes to evaluate biocompatibility and biodegradation effects in the most undisturbed state. This challenge involves, for instance, performing the following actions:

- *Imaging the interface between the biomaterial and the host tissue.* In many cases, surface biocompatibility can be estimated roughly by inspecting the visible interactions at the contact area between the biomaterial and the living organism, such as: the response of bone to a biodegradable and/or bioactive implant (see Sect. 23.3.4); the mechanism of stent incrustation (see Sects. 23.3.7 and 23.3.8).
- *Evaluating the influence of the living environment on the material.* Any polymeric material will undergo changes as it is exposed to biological systems. There are polymers like PE and PMMA that are considered to be biologically inert. This means that any changes in their physical properties are marginal or so slow that they can be neglected. Other polymers will suffer severe changes in their bulk and/or surface properties, such as swelling, degradation, release of material constituents, biofilm formation, abrasion, and many more. Some examples are discussed in this chapter: the influence of friction and wear on the surface topography and the release of wear debris (see Sect. 23.3.2), and the degradation of polymeric materials in the living environment (see Sects. 23.3.4 and 23.3.8).
- *Analysing polymer properties under physiological conditions.* The properties of a polymeric material will strongly depend on the actual environment. Therefore, morphological and micromechanical characterisations of biomaterials should be performed under physiological conditions (i.e. $T = 37\,°C$, pH 7.4, aqueous environment). Since polymers will take up water, structures should be imaged in the swollen state. Drying will produce artefacts, especially at interfaces. Microcrack formation during the drying of composite materials is illustrated in Sect. 23.3.5.

The demands stated above can be fulfilled in two ways. First, it is desirable to apply electron microscopic methods that preserve the natural state of the biomaterial and adhering host tissue. The ESEM technique, for instance, allows the imaging of samples that are not dehydrated (see Chap. 4). AFM methods can be used to image biomaterial samples in the original state. One example of the investigation of an untreated bone sample is given in Sect. 23.3.1. Second, adequate preparation procedures must be applied to conserve the structures and to make the samples suitable for the high-vacuum conditions of conventional SEM and TEM. Typical preparation steps are fixation, dehydration, and defatting. Chemical fixation of the biological material that adheres to a sample is inevitable if it is under investigation for a long time. Section 23.3.4 provides an example.

It is not the intention of this chapter to describe any of the sophisticated fixation and staining procedures that are used in electron microscopic histology. The imaging of cells, cell components and biochemical interactions is a separate scientific field that is described in a number of textbooks.

23.3 Examples

23.3.1 Natural Biomaterials: Bone

It has been demonstrated for semicrystalline polymers (Chap. 16) and block copolymers (Chap. 19) that a layered arrangement of soft and hard materials can result in a good balance of stiffness and toughness. Though mineralised tissues like bone and teeth contain large amounts of inorganic material, their outstanding properties are largely due to the presence of a soft organic layer that separates the hard components. In this sense, bone can be treated as a natural nanocomposite material consisting of hard hydroxy apatite (HA) platelets that are separated by a polymeric, fibrous, and elastic matrix (collagen). Figure 23.2a shows the periodic arrangement of HA nanoparticles in an ultrathin section prepared from human cortical bone (femur). The preparation included dehydration in ethanol, embedding in PMMA, and ultramicrotomy using a diamond knife. Since the sample is not chemically stained, the collagen matrix is relatively electron-transparent. The HA platelets, due to their higher density, give mass thickness contrast and are visible as dark objects. There is a periodicity in the HA arrangement because biomineralisation (i.e. the precipitation of HA nanocrystals in the collagen matrix) occurs preferably at certain positions of the protein that have a period of approximately 64 nm. A similar view is derived from

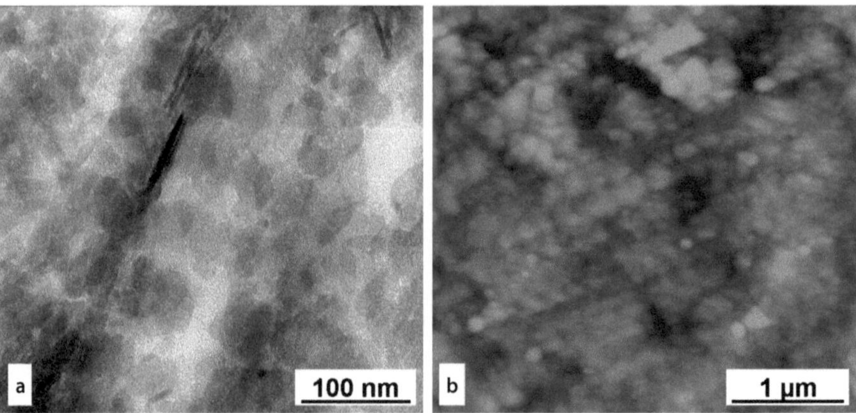

Fig. 23.2a,b. Nanostructures of bone: **a** periodic arrangement of hydroxy apatite particles in the collagen matrix (EFTEM, ultrathin section); **b** TMAFM image of an untreated bone sample

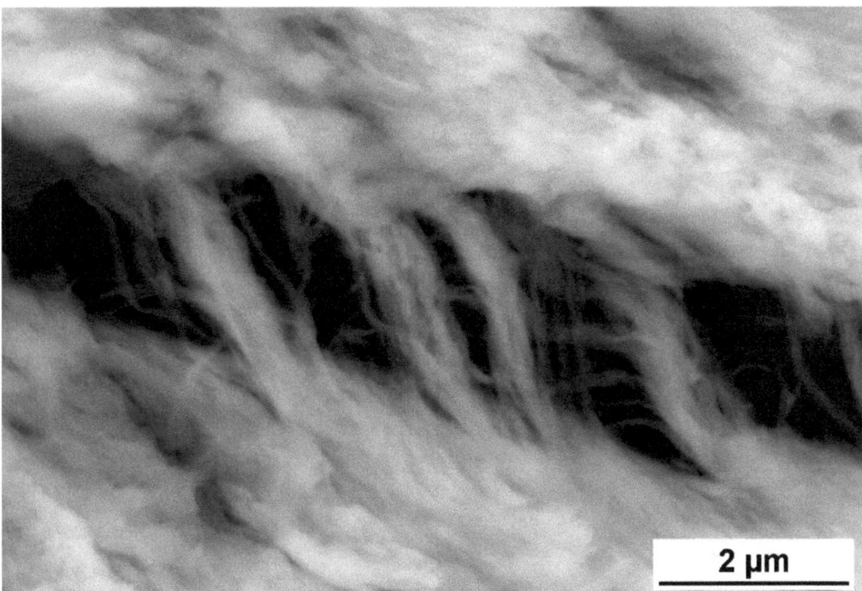

Fig. 23.3. Craze-like deformation in a microcrack tip of cortical bone. (Deformed semi-thin section, ESEM-BSE image)

TMAFM investigations (Fig. 23.2b). Here, the contrast is generated by differences in the stiffness of the components. The sample is prepared from the compacta of a rabbit femur by performing sawing, grinding and polishing steps without prior dehydration or chemical fixation.

By following the preparation procedures for in situ tests that are specified in Chap. 12, it is possible to elucidate the micromechanical mechanisms of bone. Figure 23.3 gives an example of the craze-like nature of microcracks that are observed in cortical bone. The tensile experiment was performed using a semi-thin section of cortical bone attached to a ductile copper mesh.

23.3.2 UHMWPE in Orthopaedics

UHMWPE is the material of choice for low-friction and long-lasting total hip and knee replacements. It is used for the bearing of the metallic balls of the artificial femoral head or for the so-called inlays that separate the upper and lower metallic parts of the artificial knee. With these metallic partners it exhibits a very low friction coefficient, exceptional wear resistance, and long-term dimensional and chemical stability [11]. Although UHMWPE is considered to be bioinert, showing no undesirable side effects on the host tissue, severe damage is sometimes observed, leading to failure of the implant and revision surgery. The reason for this is either the mechanical disintegration of the implant itself or an inflammatory response to wear debris encapsulated in the tissue near the implant.

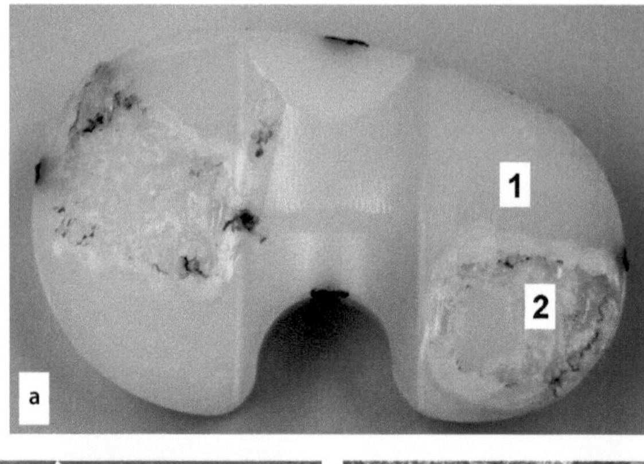

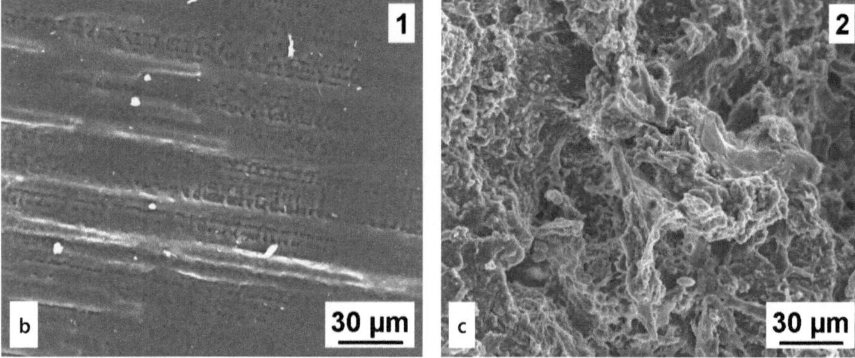

Fig. 23.4a–c. Wear surface of an UHMWPE inlay from a total knee prosthesis: **a** photographic image of the explanted inlay; **b** detail from region 1 showing scratches, **c** detail from region 2 that is strongly eroded. (SEM images)

One of the main tasks is to clarify the correlations between processing, crosslinking, semicrystalline morphology, biomechanical loading conditions, and the wear of these UHMWPE implants. Whereas the morphological analysis is performed by following the procedures that are described, for instance, in Chaps. 13 and 16, the investigation of UHMWPE wear surfaces, fragments, and debris requires additional preparative steps. The sample that is presented in Fig. 23.4a first had to be disinfected to ensure safe handling. Second, the sample was dehydrated in ethanol. After drying and carbon coating, the whole knee inlay was inspected for wear marks and surface defects using a conventional SEM. There are different types of wear patterns, such as scratches (region 1, Fig. 23.4b), pits (region 2, Fig. 23.4c) and delamination zones. The latter can be correlated to wear debris that is found in the joint.

23.3.3 Acrylic Bone Cements

Bone cements are self-curing materials used for the fixation of implants into bone or for the repair of bone defects. Nearly all of the commercial materials that are used

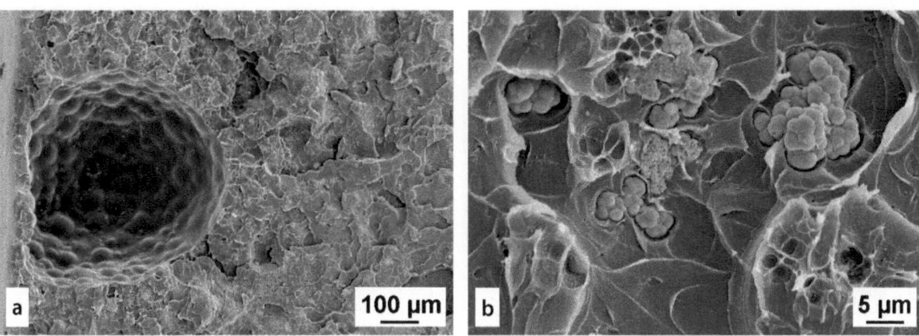

Fig. 23.5a,b. Morphological defects in hand-mixed acrylic bone cements: **a** large pores produced during the mixing of the components; **b** large agglomerates of the X-ray opacifier. (SEM images of fracture surfaces)

in orthopaedic surgery are based on PMMA, methyl methacrylate monomer and suitable activator/initiator systems. After mixing a powder of PMMA beads or flakes with liquid MMA monomer, the dough is brought to a prepared cavity in the bone, and the setting reaction starts in vivo. The system usually contains particulate fillers, such as antibiotics and X-ray opacifiers (i.e. high-density compounds for enhancing the visibility of the cement in X-ray diagnostics). The specific task is to investigate the influence of morphological defects like pores, filler agglomerates, and microcracks on the long-term stability of the cement. Figure 23.5 shows typical morphological features of acrylic bone cements that act as defects. Large pores are introduced during the mixing of the components (Fig. 23.5a), and X-ray opacifier particles form large agglomerates that have no adhesion to the polymer matrix (Fig. 23.5b). After thousands of loading cycles, such defects may result in the disintegration of the cement mantle and implant loosening [12–15].

Another example is presented in Fig. 9.5 (see Chap. 9). The secondary electron image in Fig. 9.5a allows us to identify the shapes of the primary PMMA pearls that are embedded in a matrix of PMMA that is polymerised after mixing the components. Figure 9.5b is recorded using backscattered electron material contrast. Here, the X-ray opacifier particles appear as bright dots. It is apparent that these particles were supplied with the liquid portion, so they are all located in the secondary polymerised matrix.

23.3.4 Bioactive Composites for Bone Replacement

In contrast to conventional acrylic bone cements, which are considered to be bioinert, bioactive systems should have a desirable influence on the host tissue, i.e. they should initiate the formation of new bone. Moreover, a partial degradation of the matrix polymer should guarantee a good interlocking between bone and implant. When evaluating such bioactive systems it is crucial to conserve the bone–implant interface. In this example, the composite consists of a self-curing, partially degradable matrix polymer (see Sect. 16.4, last paragraph) that is filled with bioactive hydroxy apatite (HA) particles. After 24 weeks of implantation in rabbit femora, the whole fe-

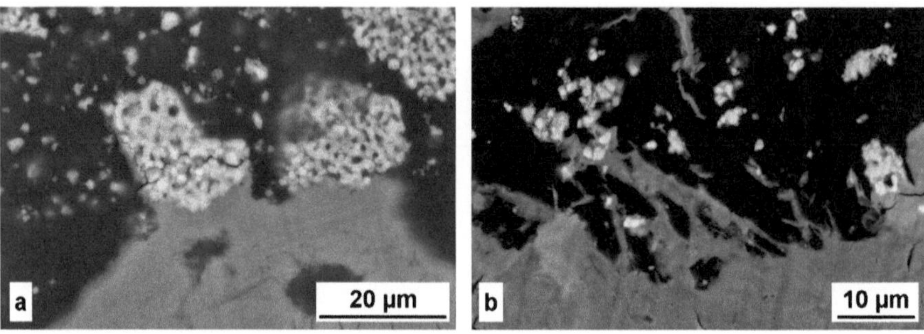

Fig. 23.6a,b. Imaging of the bone–implant interface: **a** the formation of new bone is directed toward the bioactive component (HA); **b** partial degradation of the matrix polymer is followed by the ingrowth of new bone. (ESEM-BSE composition images, see text)

mur carrying the implant was dehydrated in ethanol, embedded in PMMA, roughly prepared by sawing, ground, and finally polished using 0.25-µm diamond paste. The best results are achieved using the BSE material contrast in the ESEM (Fig. 23.6). Due to the different densities, the components of the complex system can be clearly distinguished. The polymer matrix solely consisting of light elements appears black, hydroxy apatite consisting of the elements P and Ca appears white, and bone is represented by intermediate shades of grey.

It can be demonstrated that the interface between bone and implant is free of connective tissue. There is close contact, and newly formed bone is able to interlock with the implant, yielding good mechanical stability of the interface. The ESEM images allow a rough classification: the material is biocompatible and has osteoinductive properties.

23.3.5 Dental Composites

A number of modern dental composites that are used in restorative dentistry consist of a light curable polymeric matrix (e.g. methacrylates) and glass or glass–ceramic particles. In most cases, several types of particulate filler are used to achieve a multimodal particle size distribution and desired properties like wear resistance, processability, colour, gloss, and others. The typical morphology of a dental composite is shown in Fig. 23.7. The SEM–BSE material contrast image shows inorganic particles that are embedded in a polymeric matrix. The particles, mainly consisting of the elements Si and Al, appear in brighter shades of grey.

Under physiological conditions, dental composites take up a certain amount of water. Repeated drying and wetting cycles result in microcrack formation and accumulation at the interfaces between the polymer matrix and the inorganic filler. This may lead to misinterpretation of electron microscopic images. Figure 4.25 (see Chap. 4) demonstrates that the drying process generates artefacts. Microcracks that one would observe under high-vacuum conditions are probably produced during drying and will not exist in the original material.

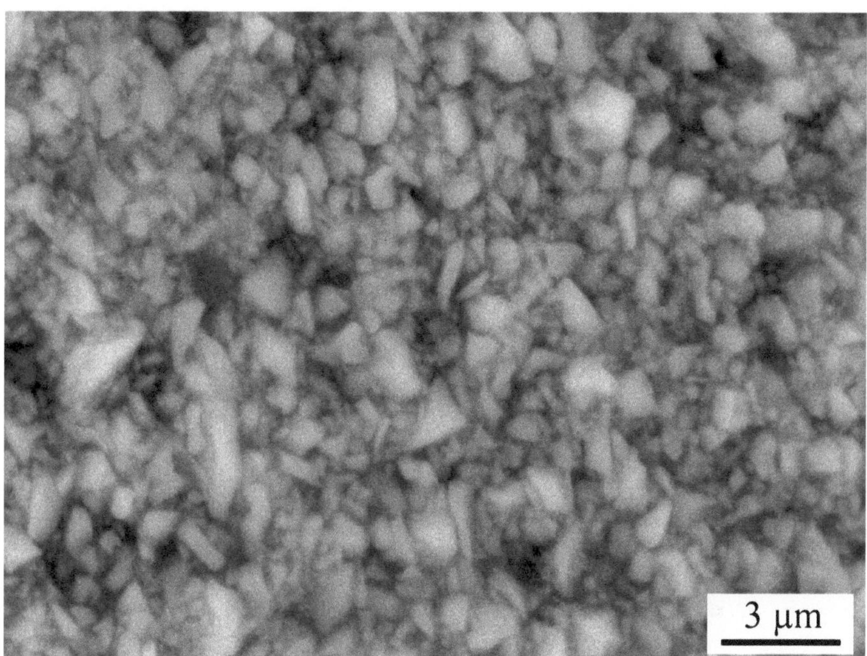

Fig. 23.7. Morphology of a dental composite; SEM-BSE image of a fracture surface

23.3.6 Sutures, Scaffolds and Meshes

The imaging of surfaces of sutures, scaffolds and meshes provides information on fibre diameters, mesh sizes, degradation, and biofilm formation. Figure 23.8a shows a typical woven suture before use. On the same material a biofilm has formed after

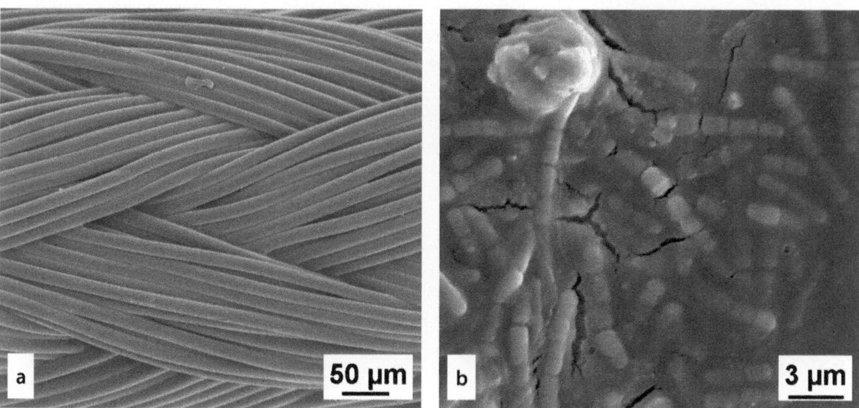

Fig. 23.8a,b. Woven surgical suture: **a** in the original state; **b** after few days of service in the oral cavity. (SEM images)

several days in the oral cavity. The suture was allowed to dry in air and was imaged in a conventional SEM after gold coating. Although no special preparation procedure was applied, it is possible to discern microorganisms on the surface (Fig. 23.8b).

Figure 23.9 provides a comparison of two kinds of mesh-like structures. Polypropylene structures like those shown in Fig. 23.9a are used for mesh prostheses that support overstretched and weakened tissue (e.g. the treatment of abdominal hernias). In this case the material must be bioinert. A completely different system consists of much finer micro- or nanofibres (Fig. 23.9b). Such scaffolds are used as substrates for tissue engineering. They are, for instance, produced by electrospinning (cp. Chap. 24) and consist of bioresorbable polymers like PHB and PLLA (see Sect. 16.4). Due to their small diameters they have a relatively large surface area. The bioactivity can be improved by functionalisation or by adding drugs or other bioactive components (Fig. 23.9c).

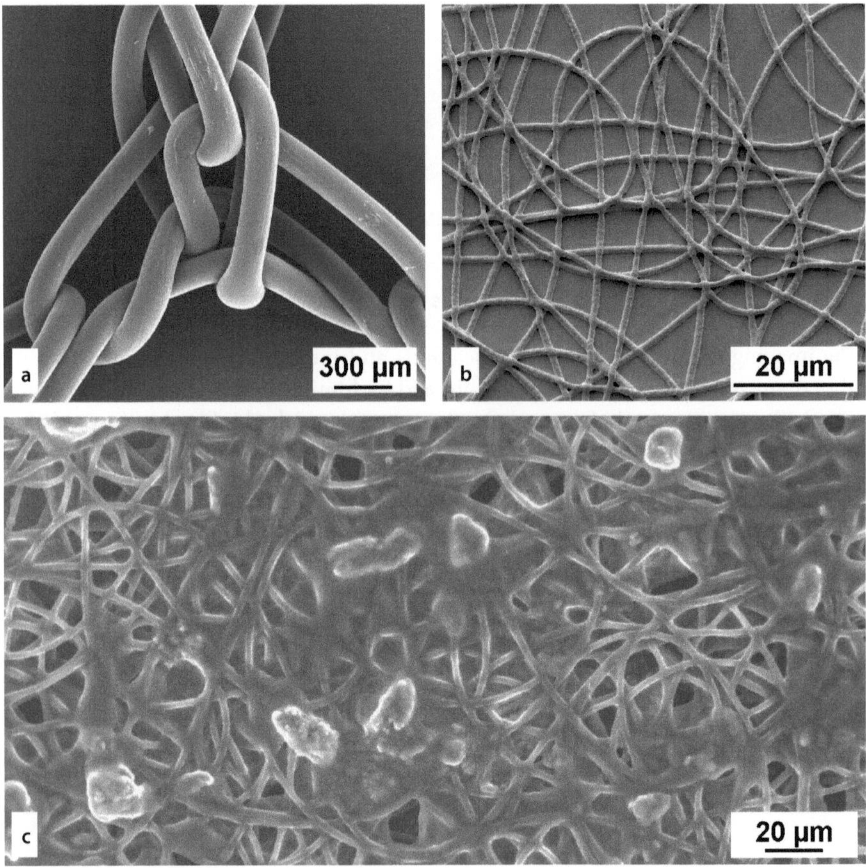

Fig. 23.9a–c. Comparison of different mesh-like structures: **a** surgical PP mesh; **b** nonwoven PHB scaffold produced by electrospinning; **c** electrospun scaffold of PHB nanofibres showing the residual water film and chondrocytes that are attached to the porous material. ((E)SEM images)

23.3.7 Ureter Stents

Ureter stents are placed in the ureter between the kidney and bladder to treat or to prevent obstructions [16]. Sometimes severe incrustations occur after relatively short service times, making replacement necessary (Fig. 23.10a). The crust that is shown in Fig. 23.10b mainly consists of the elements Ca, P, Mg, Na, K and Cl that are present in the urine. Fig. 23.10c proves that the crust also contains microorganisms. Again, no special preparation procedure was applied. The explanted stent was allowed to dry, was coated with carbon and was then observed in a conventional SEM.

The cross-section of such a stent that was produced using a microtome shows that the polymeric base material is highly filled with inorganic particles (Fig. 23.11). Similar to the inorganic fillers in some bone cements, these particles are incorporated for improved visibility of the stent during X-ray diagnostics. In this case, the sharp-edged particles penetrate the surface of the material. Such a rough surface may cause irritations of the host tissue and may support biofilm and crust formation.

23.3.8 Silicone-Based Tracheal Stents and Voice Prostheses

Silicone rubber [e.g. poly(dimethyl siloxane), PDMS] is the base material for tracheal stents (e.g. the Montgomery type) and voice prostheses (e.g. PROVOX) that are used for rehabilitation after throat surgery. Again, the material itself is considered to be bioinert, which means it should have no undesirable effects on the host tissue and it should not be influenced by the specific environment. Nevertheless, material degradation, biofilm formation, and incrustation are observed after a relatively short time of implantation [17–19]. For instance, after a few weeks tracheal stents start to discolour. SEM inspection shows that the material is degraded. Figure 23.12a shows a deep circular pit or crater on the surface of the stent. Other regions of the same implant carry a crust that is firmly attached to the silicone. The SEM overview image of the cross-section of such an area indicates that the plaque forms a dense layer up to 30 μm thick (Fig. 23.12b). The EDX mappings for the elements Si, Ca and P that are derived from the same area at higher magnifications (Fig. 23.13a) reveal that there is indeed intensive interlocking between the silicone base material (Si, Fig. 23.13b) and the crust. The detection of the elements Ca and P and their localisation in the plaque (Fig. 23.13c,d) lead to a hypothesis for the incrustation process. First, a mixed biofilm consisting of yeasts and bacteria is formed. The microorganisms start to degrade the silicone rubber, and they are able to intrude into the material close to the surface. Second, the biofilm is calcified by the precipitation of Ca and P [20].

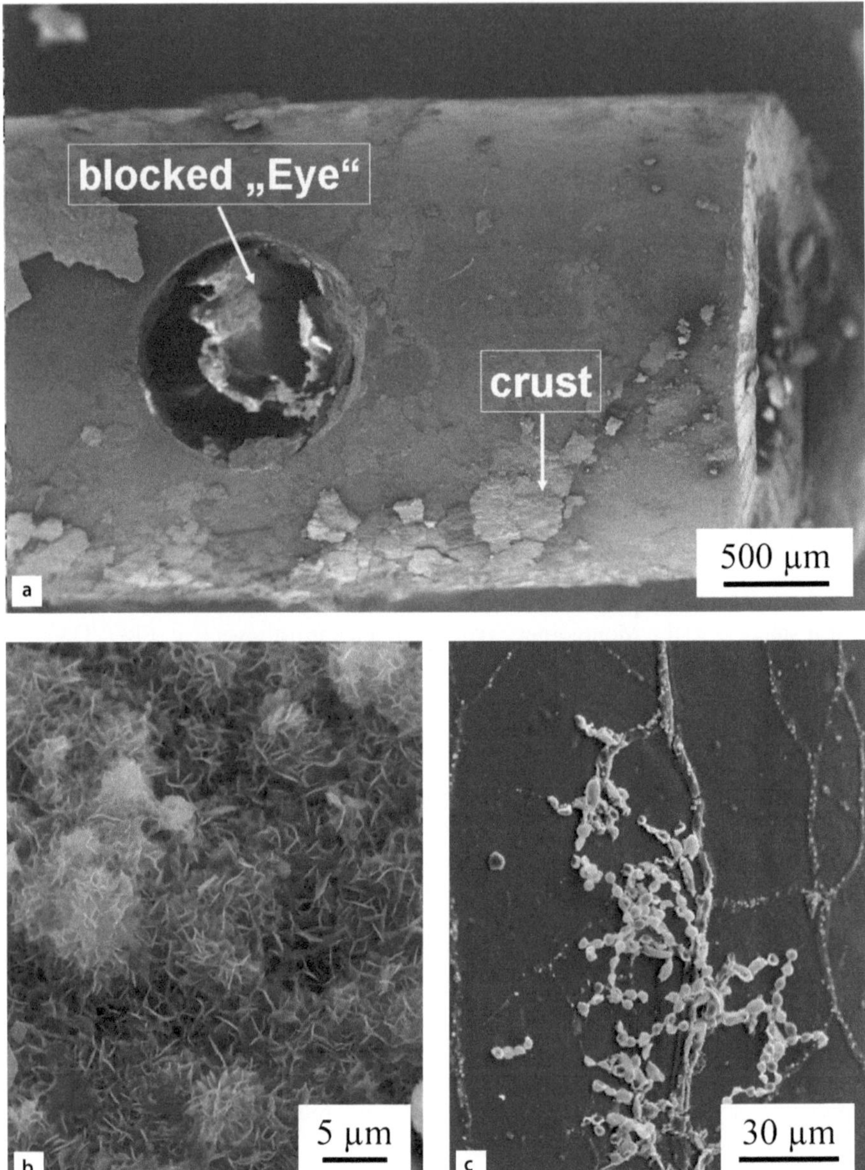

Fig. 23.10a–c. Incrustation of ureter stents: **a** overview image displaying the blocking of a so-called eye and the formation of a compact crust; **b** detail from (**a**) showing the surface of the crust; **c** microorganisms on the stent surface. (SEM images)

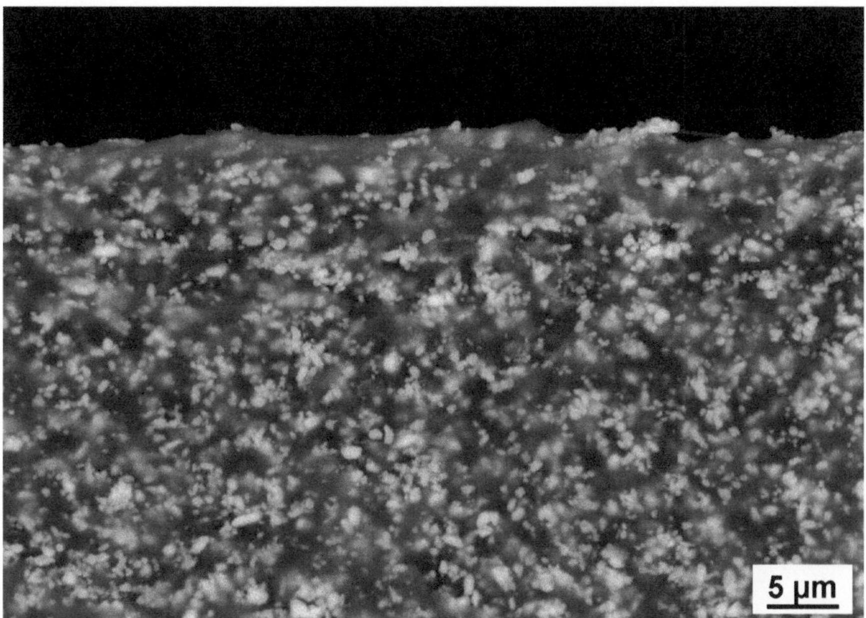

Fig. 23.11. Cross-section of an ureter stent showing X-ray opacifier particles that are reaching the surface of the stent

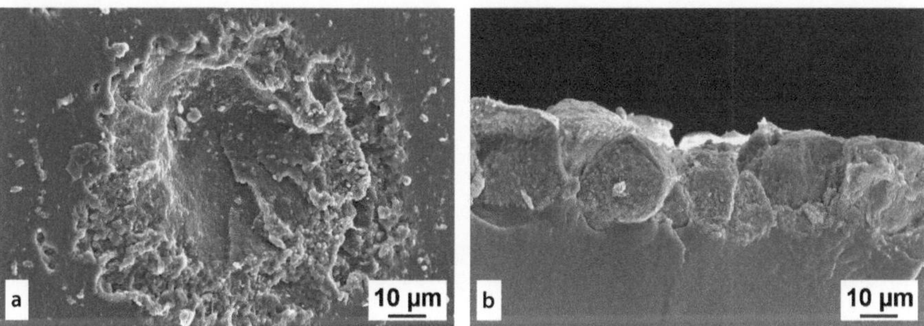

Fig. 23.12a,b. Degradation zones in a tracheal stent of the Montgomery type made of silicone: **a** surface showing a circular pit; **b** cross-section of the base material and the adhering plaque. (SEM images)

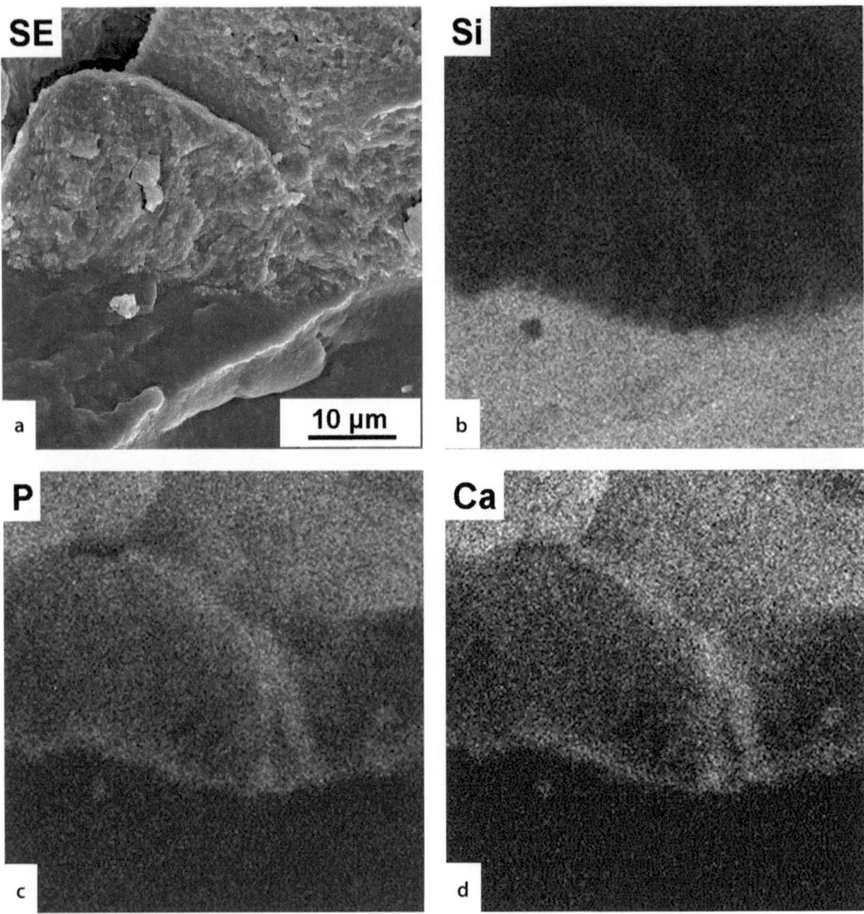

Fig. 23.13a–d. SEM-SE image and element mappings of the interface between a tracheal stent and incrustation (see Fig. 23.12b): **a** SE image showing the intense interlocking; **b–d** distributions of the elements Si (silicone rubber), Ca and P (crust)

References

1. Williams DF (ed) (1987) Blood compatibility, vol 1, CRC Press, Boca Raton, FL
2. Narayan R (1990) Introduction. In: Barenberg SA, Brash JL, Narayan R, Redpath AE (eds) Degradable materials: perspectives; issues and opportunities. CRC, Boston, MA, p 1
3. Swift G (1990) In: Glass JE, Swift G (eds) Agricultural and synthetic polymers: biodegradability and utilization. American Chemical Society Press, Washington, DC, 1:2
4. Huang SJ, Edelman PG (1995) In: Scott G, Gilead D (eds) Degradable polymers: principles and applications. Chapman & Hall, London, 2:18
5. Sealey JE, Samaranayake G, Todd JG, Glasser WG (1996) J Polym Sci Polym Phys 34:1613
6. Matthews JA, Wnek GE, Simpson DG, Bowlin GL (2002) Biomacromolecules 3:232
7. Ma L, Gao C, Mao Z, Zhou J, Shen J, Hu X, Han C (2003) Biomaterials 24:4833
8. Gilmore DF, Fuller RC, Lenz R (1990) In: Barengerg SA, Brash JL, Narayan R, Redpath AE (eds) Degradable materials: perspectives, issues and oppurtunities. CRC, Boston, MA, p 481
9. Marletta G, Ciapetti G, Satriano C, Perut F, Salerno M, Baldini N (2007) Biomaterials 28:1132
10. Wim H, De J, Bergsma JE, Robinson JE, Bos RRM (2005) Biomaterials 26:1781
11. Kurtz SM (2004) The UHMWPE handbook: ultra-high molecular weight polyethylene in total joint replacement. Elsevier, San Diego, CA
12. Kühn KD (2000) Bone cements. Up-to-date comparison of physical and chemical properties of commercial materials. Springer, Berlin
13. Draenert K, Draenert Y, Garde U, Ulrich C (1999) Manual of cementing technique. Springer, Berlin
14. Kühn KD, Ege W, Gopp U (2005) Orthop Clin N Am 36:29
15. Murphy BP, Prendergast PJ (2000) In: Prendergast PJ et al. (eds) Proceedings of the 12th Conference of the European Society of Biomechanics. Royal Academy of Medicine in Ireland, Dublin, p 235
16. Mattei A, Danuser H (2003) Therapeutische Umschau 60(4):233
17. Neu TR, Dijk F, Verkerke GJ, Van der Mei HC, Busscher HJ (1992) Cells Mater 2(3):261
18. Neu TR, Van der Mei HC, Busscher HJ, Dijk F, Verkerke GJ (1993) Biomaterials 14(6):459
19. Eerenstein SEJ, Grolman W, Schouwenburg PF (1999) Clin Otolaryngol 24:398
20. Henning S, Haberland EJ, Kluge A, Burkert S, Michler GH, Bloching M (2005) Biomaterialien 6(S1):74

24 Special Processing Forms

This chapter provides an overview of new developments and trends in the field of special micro- and nanostructured polymers. Multilayered films are produced by coextrusion and consist of up to several thousand layers containing two or more polymers with layer thicknesses in the micrometre and nanometre range. Self-reinforced polymers are made of highly oriented polymer fibres or films of the same polymers, which are linked together by applying heat and pressure, in a process called hot compaction. Nanofibres with diameters ranging from micrometres to nanometres – i.e. several orders of magnitude smaller than those made by conventional fibre spinning – can be produced via electrospinning. Processing small amounts of polymers via microextrusion or microinjection gives microformed samples with structures and properties that are often different from those of bulk pieces; such samples are used for applications in electronics or medicine. Typical examples of the morphologies and micromechanical effects of these four special processing forms are illustrated via TEM and SEM.

24.1 Overview

In the preceding chapters, the application of electron microscopic techniques to the characterisation of morphological and deformation phenomena in various classes of conventional polymers, blends and composites was discussed. There are, however, new developments and trends in the field of polymer development; some specific micro- and nanostructured polymers can be produced by special processing routes, and electron microscopy has contributed a great deal to the structural characterisation of these novel polymers. Micro- and nanostructured biomaterials such as bones, mollusc shells, fibres produced by different insects, etc. have also been attracting increasing levels of interests from materials scientists, who are aiming to develop new advanced functional materials that are based on the unique architectures of these natural composites [1,2]. In this chapter, we will deal with four special processing forms: multilayered films, high-strength self-reinforced materials produced by the hot compaction technique, electrospun nanofibres and microsamples realised through microprocessing techniques.

24.2 Multilayered Films

24.2.1 Introduction

As shown in Chap. 19, materials comprising layered morphologies can be obtained by the copolymerisation of polymers into block copolymers. The length scales of these structures are determined by the sizes of the individual molecules. So far, the nanostructures that have been created have been quite uniform but it has proven very difficult to align the molecules macroscopically in a defined direction. The technique of multilayer coextrusion can be used to produce two-dimensional continuous layers of different polymers via the process of "forced assembly" [1]. Microlayer coextrusion combines two or more polymeric materials into layered configurations which can be used to produce new architectures with improved mechanical, electrical, barrier and adhesive properties. Layered composites comprising up to 8000 continuous layers can be successfully produced, with the thicknesses of individual layers ranging from several microns down to a few nanometres [1, 3]].

This coextrusion technique consists of two or more (depending on the number of polymers to be coextruded) single-screw extruders that produce a melt with a specified thickness. The melt goes through a series of multiplying elements. Each multiplying element first slices the melt vertically and subsequently spreads it horizontally. The flowing streams recombine, doubling the number of layers. Each multiplying die doubles the number of layers. The technique is not limited to two polymers, and can be extended to different polymers (as well as three or more polymers), when their viscosities are adjusted [1, 3]].

The motivation for polymer microlayer coextrusion originated in metal/metal systems where improvements in the mechanical properties of thin-layered structures were first discovered. For example, antique laminated or layered steels, known as Damascus sabre steels, show extraordinary mechanical properties with exceptionally sharp cutting edges. Many layers of hard but brittle steel (with a high carbon content) and softer, tough steel (with a low carbon content) were forged together alternately using a sophisticated thermomechanical forging and annealing treatment [4, 5]. On the other hand, layered composites that occur in different forms and at various length scales in nature (e.g. in abalone shells) are also of special interest. All of these have one thing in common: they possess excellent mechanical properties. Inspired by these composite structures that occur in nature, polymer scientists are now trying to create biomimetic morphologies [1, 2].

24.2.2 Morphology

Using the multilayer coextrusion technique, it is possible to prepare laminates with different polymers and with various thicknesses of each polymer. As an example, Fig. 24.1 provides micrographs showing multilayered tapes of polycarbonate (PC) and polyethylene terephthalate (PET) with three different PC/PET volume fractions. The extruded films consisted of 4096 layers [6,7]. To make the individual layers visible in the transmission electron microscope (TEM), the specimen was treated with RuO_4

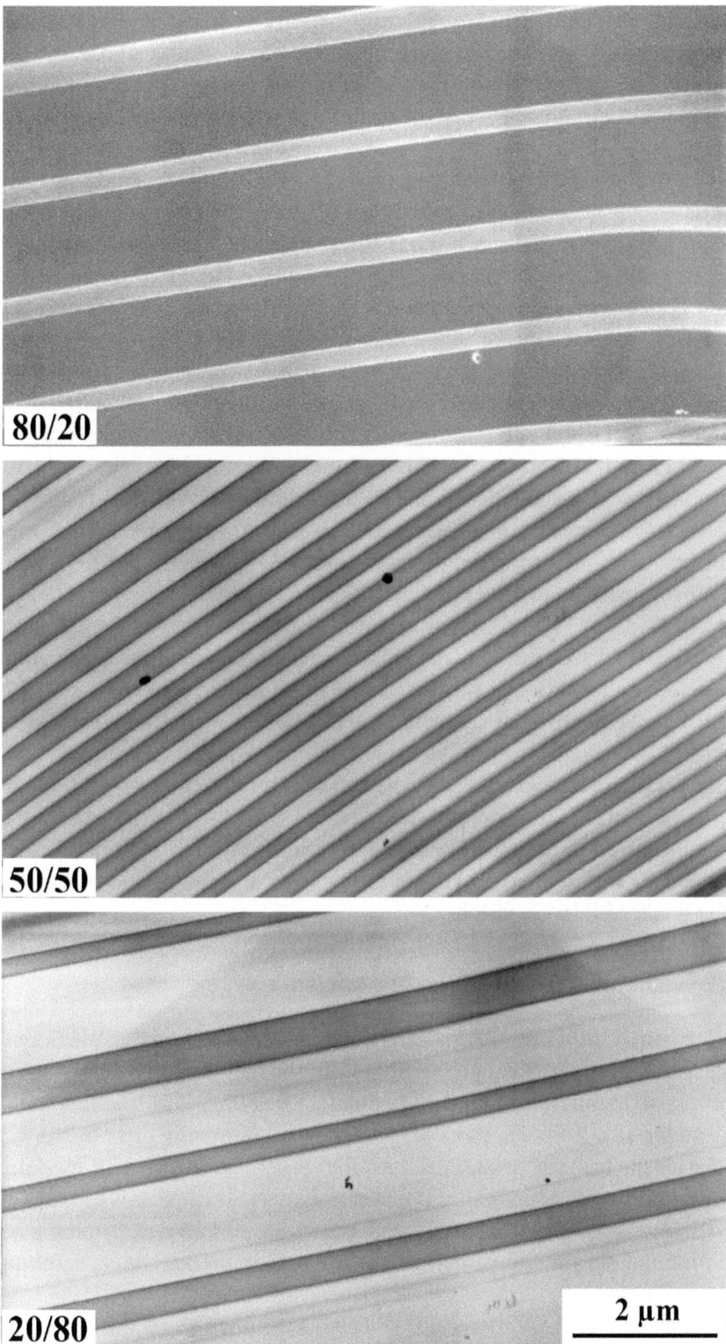

Fig. 24.1. TEM micrographs showing the morphologies of 4096-layered PC/PET tapes having with different compositions; the PC layers appear *dark* due to staining by RuO_4

vapour for several hours. Due to the lower density of polycarbonate compared to that of PET, the staining agent preferentially penetrated into the PC phase. As a result, the PC phase appears grey in the electron micrographs. One can see that the multilayered tapes consist of continuous layers which extend practically up the length of the extruded film.

Also, most of the layers are uniform. Depending on the extrusion conditions and especially on how thin the layers are made, layers of one of the polymers may break into droplets during the coextrusion process. These droplets then form a dispersed phase embedded in the matrix of the other polymer [7] (see Fig. 24.2). In particular, polymer laminates comprising very thin layers of a crystallisable polymer are very useful for studying various physical phenomena, such as the influence of narrow interfaces and structural confinement on the nanostructure evolution of semicrystalline polymers, diffusion processes occurring at interfacial regions, etc. [3].

Because they have a similar chemical nature (polyesters), the polymer pairs presented in Figs. 24.1 and 24.2 are compatible to some extent in this type of system. Hence, the problem of adhesion between the layers is eliminated. Under certain processing conditions, the polyesters may even react with each other to form a copolymer that acts as an internal binder between the polymers [1, 3]. However, in the coextrusion of highly incompatible polymer pairs such as polystyrene (PS) and polymethylmethacrylate (PMMA) or PS and polypropylene (PP), despite the very small layer thickness achieved, the laminates may delaminate easily, worsening the mechanical stability overall. A weak interfacial contact can often be made visible by a stronger staining effect: Fig. 24.3 shows an example of a PP/PS film with a composition of 90/10 and with 1024 layers. After RuO_4 treatment, the tiny PS layers are stained grey, with particularly dark interfaces apparent between the layers [Scholtyssek St, Michler GH, Baer E, Hiltner A, 2007, unpublished results]. This local strong staining is due to a high concentration of defects, a large free volume or a low material density and therefore reveals soft or weak interfaces. The chemical staining also makes the lamellae in the thicker PP layers visible.

24.2.3 Micromechanical Deformation Mechanisms

The architecture consisting of alternating layers of different polymers provides a synergistic combination of mechanical properties [3]. In the following, we demonstrate the positive effect of microlayering on the toughness, using PS/PMMA laminate systems as an example [8]. Both PS and PMMA are well known for their brittle behaviour. Under tensile loading conditions, localised deformation zones in the form of single, isolated crazes are observed in the neat polymers. These crazes are usually long and thin and grow quickly to macroscopic cracks, leading to catastrophic fracture before additional deformation zones can be initiated in the specimen, resulting in the brittle behaviour.

A high-voltage electron microscopy (HVTEM) examination of the deformed PS/PMMA microlayered samples has provided the physical explanation for the mechanical property improvements mentioned above. Semi-thin sections (ca. 500 nm thick) were stretched with the aid of a special miniaturised tensile devise before

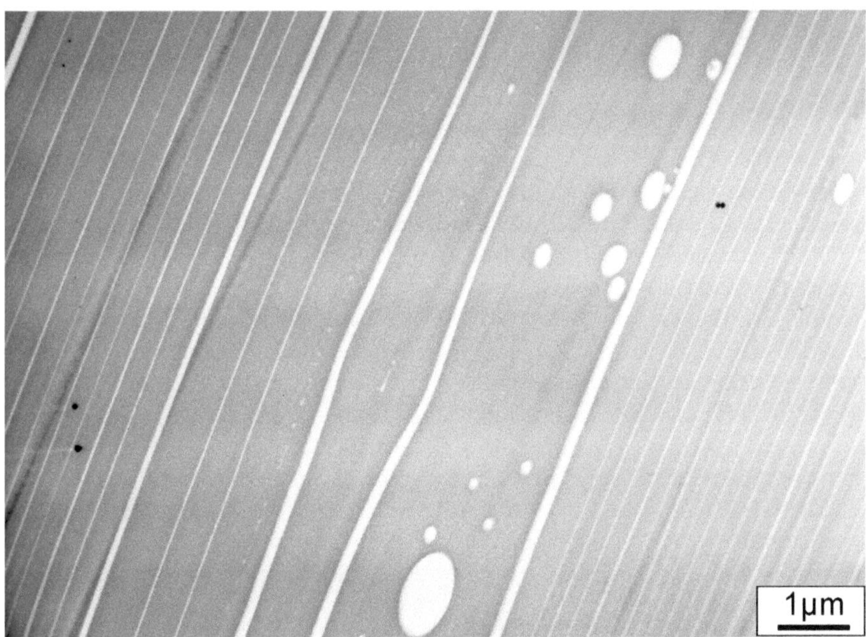

Fig. 24.2. A TEM micrograph showing droplets of PET phase embedded in continuous layers of PC in a 4096-layered PC/PET (90/10) tape; the PC layers appear *dark* due to staining by RuO_4

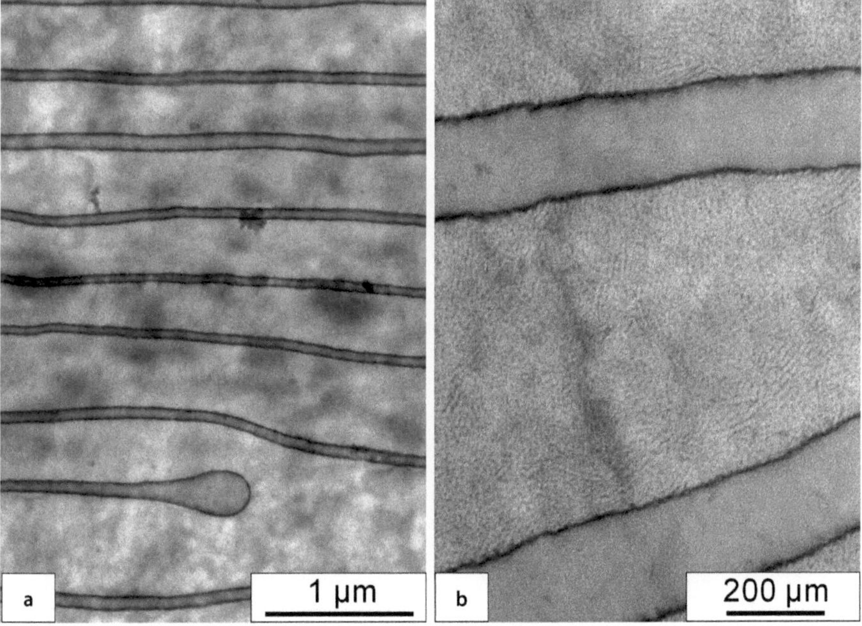

Fig. 24.3a,b. TEM micrographs of 1024-layered PP/PS (90/10) tape with a particularly strong staining effect at the interfaces due to RuO_4 between the layers; PS layers appear *grey*

the electron microscopic study. After the deformation, the PS phase was selectively stained with RuO_4. Figure 24.4 shows the strain-induced deformation structures observed in PS/PMMA tapes consisting of 64 (Fig. 24.4a,b) and 4096 (Fig. 24.4c,d) layers. The decrease in layer thickness as the total number of layers in the tape is increased is obvious in the micrographs. In the 64-layered sample with layer thicknesses in the range of μm, the crazes run in the PS layers and their tips are terminated in the adjacent PMMA layers. Due to the lower yield stress of polystyrene (ca. 55 MPa) compared to that of PMMA (ca. 75 MPa), the crazes are initiated in the former and extend into the neighbouring PMMA layers at the locations of stress concentrations. The feasibility of stress transfer indicated by the movement of craze tips from PS layers to PMMA layers and the absence of delamination are the signs of sufficient adhesion between the layers. When the internal details of the crazes were investigated (Fig. 24.4b), it was found that the layers retain the properties of the individual neat polymers – the formation of fibrillated crazes in PS and in PMMA layers. However, due to termination of the crazes at the PMMA layers, their lengths are reduced, the transformation into long, dangerous crazes is prevented, and numerous crazes can be created as the thickness of the individual layers is decreased: the micromechanism changes from single crazing to multiple crazing, which enables more energy to be absorbed.

Figure 24.4c,d examines the situation when the layer thickness is below 100 nm (nanolayers). In this case, the crazes are not localised in a single layer but are in-

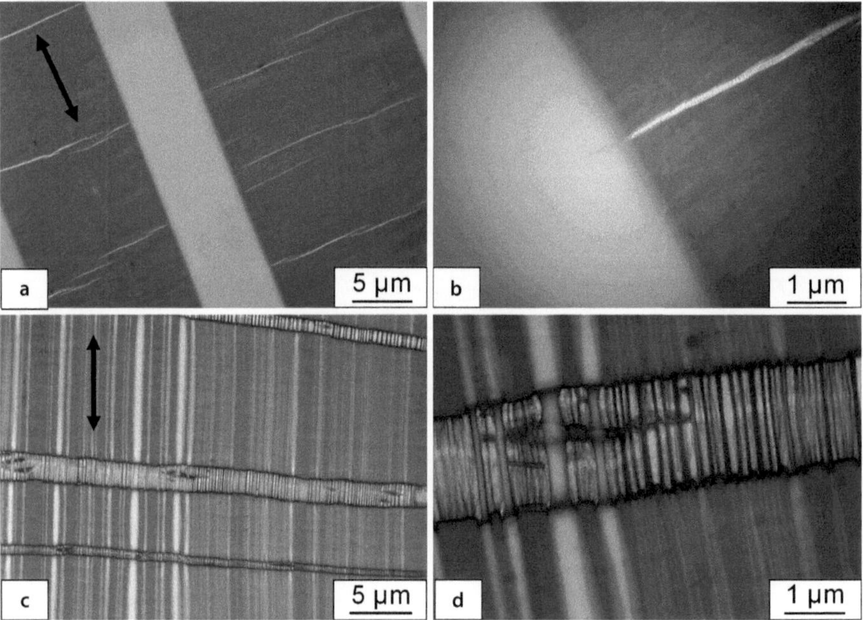

Fig. 24.4a–d. HVEM micrographs showing deformation structures in multilayered PS/PMMA tapes: **a,b** 64-layer sample; **c,d** 4096-layer sample. **a,c** Lower magnification; **b,d** higher magnification. (Deformation direction is shown by *arrows*; PS phase appears *dark* due to staining by RuO_4)

stead spread throughout many layers. For a 4096-layered sample, both the lengths of the crazes and the total volume of crazes are increased, leading to a larger volume of yielded material. As a result, microlayered samples with thousands of individual layers have a higher toughness than their bulk counterparts [3, 8].

To conclude, nanolayer processing via multilayer coextrusion facilitates the creation of new hierarchical systems and is a highly flexible tool for the fabrication of unique architectures of various classes of polymers [1]. Besides TEM, other microscopic techniques, such as optical microscopy, SEM and SFM, are often used to elucidate the structures and properties of multilayered composites, and the results are often supplemented by scattering studies [9].

24.3 Hot-Compacted Self-Reinforced Polymers

24.3.1 Overview

As mentioned frequently in this book, polymeric materials are characterised by long-chain molecules that can be organised into a number of architectures, leading to a variety of morphologies and properties. Blending with other polymers or inorganic materials and optimising the processing conditions can help to control the structures and properties of polymers. A conventional way of improving the mechanical performance of the polymers is to turn them into composites, e.g. to reinforce the plastic matrix with stiff fibres (normally made of glass or carbon), thus enhancing the properties of the material such as its strength and toughness (see Chap. 21). Such composites are becoming increasingly popular for instance in the automotive industry, as they exhibit a good combination of physical properties and offer advantages over metal components, such as lower weight, corrosion resistance and good impact performance. A disadvantage of such heterogeneous composites is that they are not very compatible with common processing conditions (thermoforming) and are mostly unrecyclable. In order to produce reinforced polymeric systems while retaining their recyclability and ease of thermoforming, new kinds of self-reinforced polymer compositions (popularly called "green composites", "single-polymer composites" or "monocomposites") are being developed [10–12]. In this section, we will discuss how electron microscopy can be used to characterise the structures of such composites and to correlate them with their properties.

The idea behind the self-reinforcement strategy is to make high-strength compositions starting from highly oriented anisotropic fibres or tapes of a single polymer. Thus the polymer is first turned into highly oriented fibres or tapes (i.e. they have very high values of stiffness and strength in the orientation direction but low values of them in the perpendicular direction). Then the fibres or tapes are connected to woven mats and the mats are placed on top of each other. The decisive step is to connect the single fibres or tapes through the application of heat and pressure without melting the whole fibres or tapes. The technologies used to produce single-polymer compositions are called "hot compaction". Extensive work has been carried out to develop the technology of hot compaction for polyolefin fibres and tapes by researchers at Leeds

University and the DaimlerChrysler Company [10–12]. However, the technique has also been extended to other polymeric systems [13,14]. Indeed, self-reinforced PP is commercially available, and it is used in the construction of automotive undershields, hoods for passenger cars, mid-range loudspeaker cones, sports goods, etc.

A literature survey reveals that monocomposite production can be realised via two different routes:

- Direct compaction of a pile of the woven mats consisting of oriented fibres or tapes where only the outer skins of the resulting mesh are made to melt [10,11,13, 15].
- Incorporation of an intermediate ("tie") polymer layer between the woven mats prior to hot compaction (film stacking) [12].

To follow the first route, the oriented tapes or fibres are initially woven into a mat. Several mats are then compacted (or thermoformed) under pressure and heating. It is possible to find suitable conditions of temperature and pressure so that only a thin layer on the surface of each fibre or tape melts to form a cohesive structure with a strong bond between them. As only thin skins of the fibres or tapes are melted, a substantial fraction of the original oriented phase is retained, so that the composition can achieve sufficiently high stiffness and strength [11,13].

Alternatively, one can pile up the mats made from the high-melting polymer, interspersing them with thin tie layers of a separate film made up of the same polymer but with a lower melting point. The hot compaction process is then applied to the woven tape mats at a temperature that is well below the melting point of the "tape core" but is high enough to cause the tie layer to melt, leading to strong bonding between the tape cores [12].

24.3.2 Morphology

Electron microscopy has provided not only morphological information on the mono-composites but also the structural basis for the mechanisms of self-reinforcement. The special features of their morphology are illustrated using the following two examples.

Figure 24.5 shows the morphology of a hot-compacted two-dimensional woven mat of high-modulus polyethylene (PE) fibres investigated by means of a scanning electron microscope (SEM) following the permanganate etching procedures developed by Olley and Bassett [17]. The sample used for the SEM studies was prepared by cryomicrotoming the hot-compacted sample, which was achieved via the first procedure mentioned above, i.e. by directly melting a pile of mats consisting of high-modulus PE fibres [16]. The subsequent melting and recrystallisation of thin skins of fibres gives rise to the matrix of the composite. As the low-density part of the composite is preferentially removed during the etching process, the matrix phase degrades much faster than the denser cores, leading to contrast in the electron micrograph. The micrograph presented in Fig. 24.5 shows that the high-modulus PE fibres are embedded in the matrix of the melted skin layer.

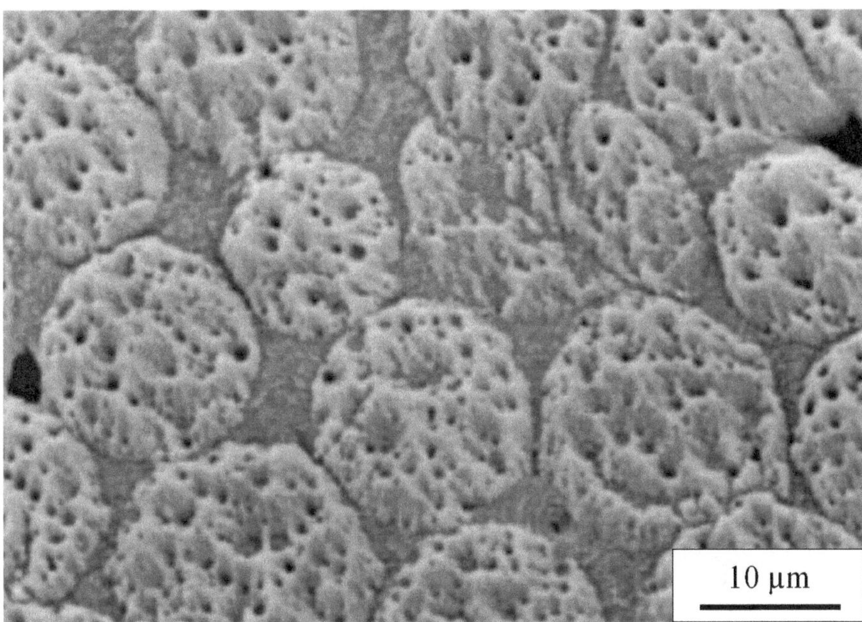

Fig. 24.5. SEM micrograph showing the morphology of hot-compacted two-dimensional woven high-modulus polyethylene fibres (from [15], reproduced with the permission of Chapman and Hall)

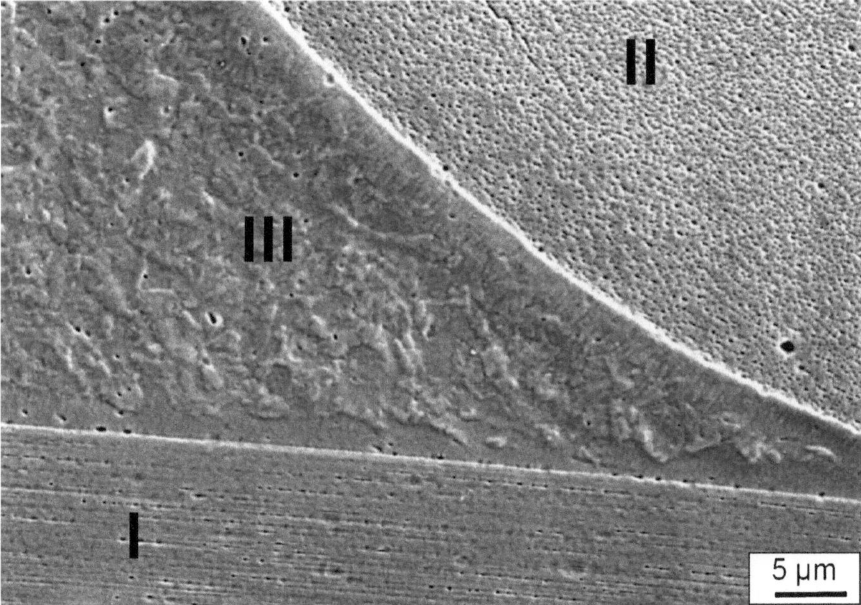

Fig. 24.6. SEM micrograph of hot-compacted PP consisting of woven PP tapes (*I*, parallel; *II*, perpendicular to the image plane) with intermediate PP film (*III*)

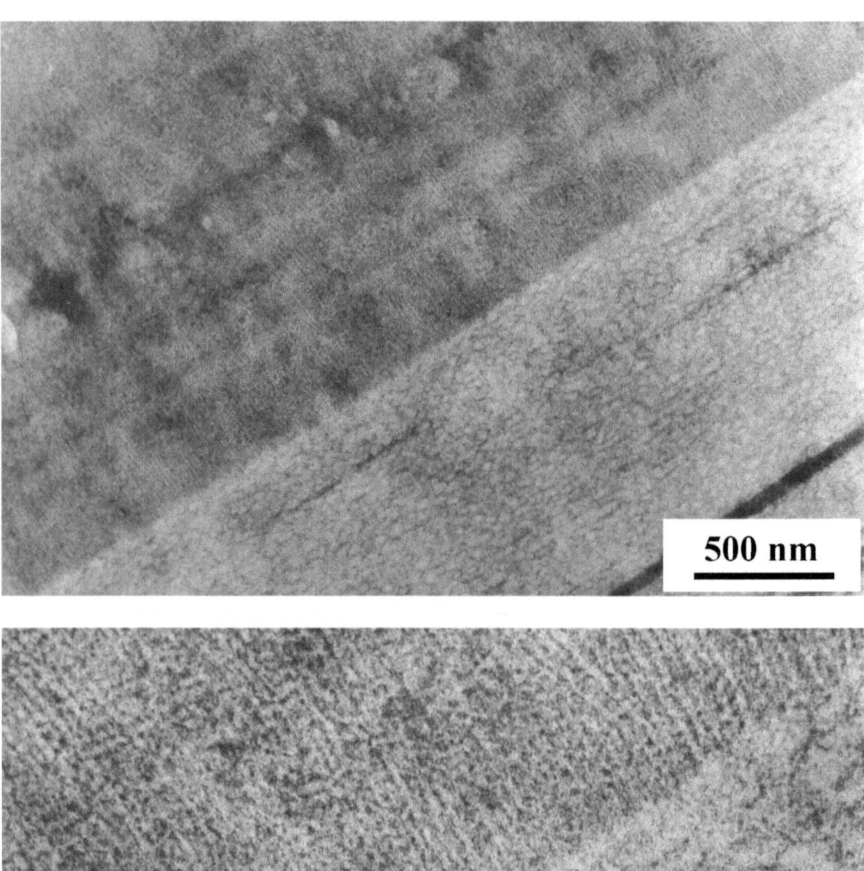

Fig. 24.7. Lower (*top*) and higher (*bottom*) magnifications of TEM micrographs showing the morphology at the tape/tie layer interface of hot-compacted polypropylene tapes; iPP lamellae epitaxially grown onto the surface of an oriented PP tape are visible

Systematic studies by Ward and coworkers have established the structural basis for the self-reinforcement mechanisms [10,11,13]. It was demonstrated that by choosing the processing conditions (i.e. temperature, pressure) appropriately, a thin skin on the surface of each fibre can be selectively melted, which recrystallises on cooling and binds the fibres together. It was found that good fibre-to fibre bonding can be achieved by melting between 10 and 20% of the original fibre, with the result that the compacted sheets retain most of the properties of the original fibre.

TEM studies confirmed that the melted material had recrystallised on cooling epitaxially onto the original fibre or tape backbone, forming a coherent network that binds the oriented fibres or tapes into a continuous structure [10, 11, 13]. The resulting hot-compacted material is therefore composed of a single, uniform, polymeric material, and as such has often been termed "self-reinforced".

Figures 24.6 and 24.7 show micrographs of hot-compacted polypropylene composites [12, 18]. The SEM micrograph in Fig. 24.6 shows a woven mat of oriented PP tapes and an intermediate PP film. The specimens for TEM were prepared by ultramicrotomy of the compacted sample, which had previously been treated with ruthenium tetroxide (RuO_4) in order to stain the amorphous parts of the sample. In Fig. 24.7, the lower right parts of the TEM micrographs show the oriented polypropylene, which has a domain-like appearance. The surface of the oriented PP tape shows the recrystallised lamellae of the PP interlayer, which are characterised by a typical crosshatched structure of the α-form of isotactic PP.

24.4 Nanofibres by Electrospinning

24.4.1 Overview

Recently, electrostatic fibre spinning, or "electrospinning", has attracted more and more interest in materials science alongside the rapid growth of nanotechnology, because polymer fibres prepared by this straightforward technique achieve diameters ranging from nanometres to micrometres, which is several orders of magnitude smaller than those made by conventional fibre spinning [19–21]. The reduction in the fibre diameter from the micrometre to the nanometre range gives rise to numerous unique characteristics, including an enormous surface area per unit volume, variations in wetting behaviour, modifications of the release rate, and a strong decrease in the number of structural defects per unit volume [22], enhancing the strength of the fibres [23, 24]. Electrospun polymer fibres are thus of interest for a wide variety of applications, including filters [25, 26], composite applications [27, 28], drug-delivery systems, as well as scaffolding for tissue engineering [29, 30]. A typical electrospinning setup is demonstrated in Fig. 24.8. To form such fibres using electrospinning, a polymer solution or melt is first forced through a capillary, producing a drop of polymer solution at the tip of the capillary. Then a high voltage is applied between the tip and a grounded collection target. When the electric field strength overcomes the surface tension of the droplet, a polymer solution jet is initiated and accelerated towards the collection target. As the jet travels through the air, the solvent evaporates and a nonwoven polymeric fabric is formed on the target surface.

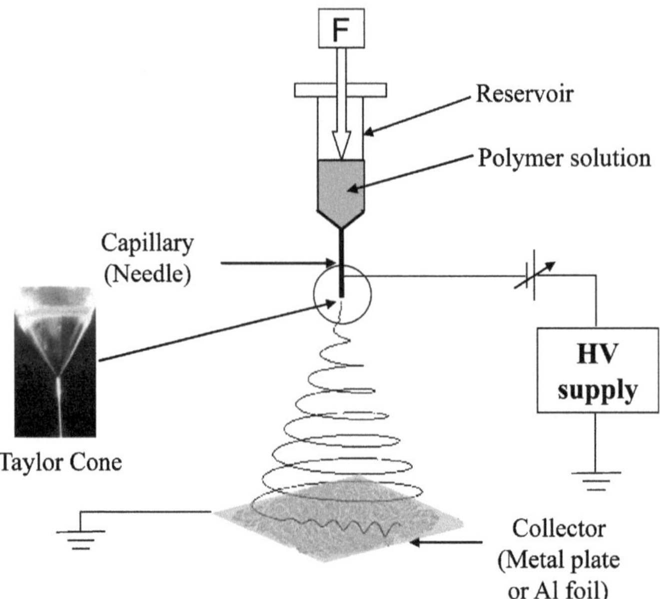

Fig. 24.8. Typical electrospinning setup

24.4.2 Morphology

Electrospinning is a very effective tool for producing polymer nanocomposite fibres with improved separation of nanoparticles. As an example, a PMMA nanocomposite containing 5 wt% sodium montmorillonite (Na-MMT) was synthesised by a dispersion polymerisation and used as the base material. To ensure a homogeneous solution for electrospinning, the PMMA/Na-MMT nanocomposite was dissolved in chloroform at room temperature with gentle stirring for ten hours. Electrospinning was carried out at ambient temperature in a vertical spinning configuration using a 1 mm inner diameter flat-end needle with a spinning distance of 5 cm. The applied voltage was 3–20 kV, as driven by a high-voltage power supply. To investigate the morphologies of polymer nanocomposite fibres, the fibres were directly electrospun onto either Cu-grids for TEM or onto slide glasses for field-emission gun environmental scanning electron microscopy (FEG-ESEM, Philips ESEM XL 30 FEG). Figure 24.9a shows a typical electrospun nanocomposite fibre from a PMMA nanocomposite with 5 wt% Na-MMT. The electrospun fibres observed had relative uniform diameters of less than 500 nm and exhibited nanoporous morphology on the fibre surface. The pores have an extremely narrow size distribution and the average pore size is on the order of 100 nm in width and 200 nm in length, and are arranged along the fibre axis. The porous structure seen on the fibre surface during the electrospinning process arises from a rapid phase separation during the electrospinning process [31–34]. Figure 24.9b shows a cross-section of an

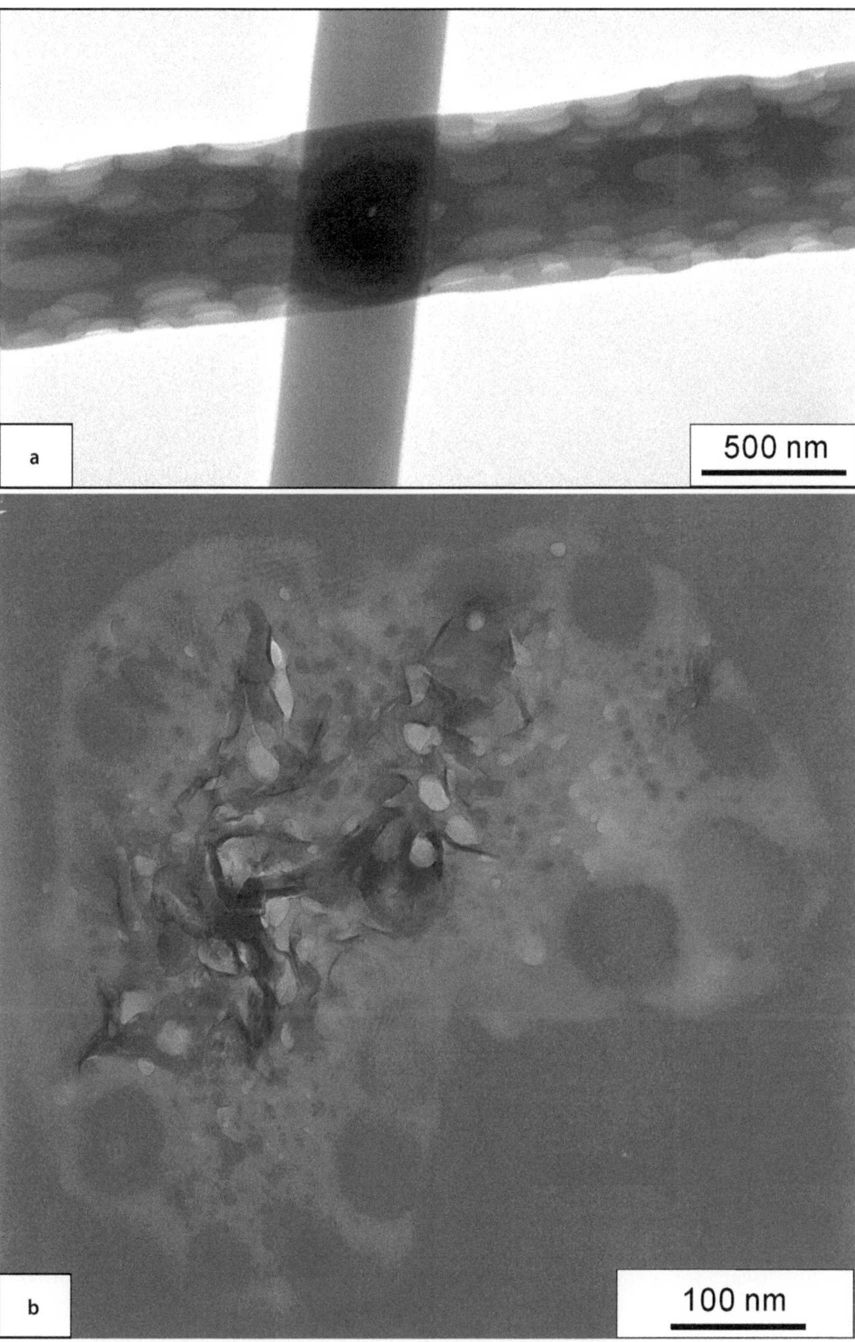

Fig. 24.9a,b. TEM micrographs: **a** nanoporous structure of the electrospun polymer PMMA/Na-MMT nanocomposite fibre (directly electrospun on a Cu-grid); **b** a cross-sectional view through it. (The fibres were embedded in epoxy and then cryomicrotomed to prepare ultrathin sections of thickness 50 nm)

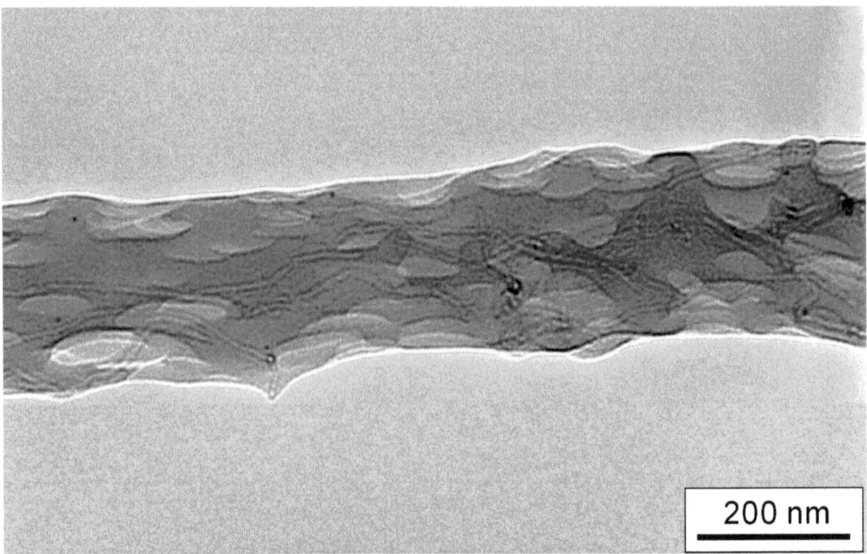

Fig. 24.10. TEM micrograph of nanoporous structure on the surface of an electrospun nanocomposite fibre of PC with 4 wt% MWCNT, showing highly aligned MWCNTs within a fibre. (Prepared using a 5 cm working distance at 8 kV)

electrospun fibre. The pores appear at the border of the fibre, and layered silicates are clearly present in the middle of the fibre. Another example of an electrospun fiber is demonstrated in Fig. 24.10. In this case, the fibres were electrospun from a nanocomposite of polycarbonate (PC) with 4 wt% multiwalled carbon nanotubes (MWCNTs). TEM measurements indicate that the MWCNTs are well-dispersed in the polymer matrix without significant aggregation. They are found to orient parallel to the fibre direction. This alignment is obviously associated with the extremely high longitudinal strain rate of the jet during the electrospinning process, which may arise from disentanglement or the pulling out of curved nanotubes under high shear force. The electrospun fibres were circularly shaped and submicron in diameter (about 350 nm on average). They exhibited well-developed nanoporous morphology on the fibre surface. The occurrence of this porous structure on the fibre surface during the electrospinning process may be associated with a rapid phase separation.

24.4.3 Micromechanical Deformation Mechanism

To investigate the deformation mechanism of the electrospun polymer nanocomposite nanofibres, a single fibre is stretched in situ under TEM (Fig. 24.11). Early on in the deformation process the electrospun fibre will be slightly elongated along the fibre axis due to stress concentrations caused by the nanopores (Fig. 24.11a,b), which leads to a reduction in the fibre diameter. When the stress reaches a certain critical

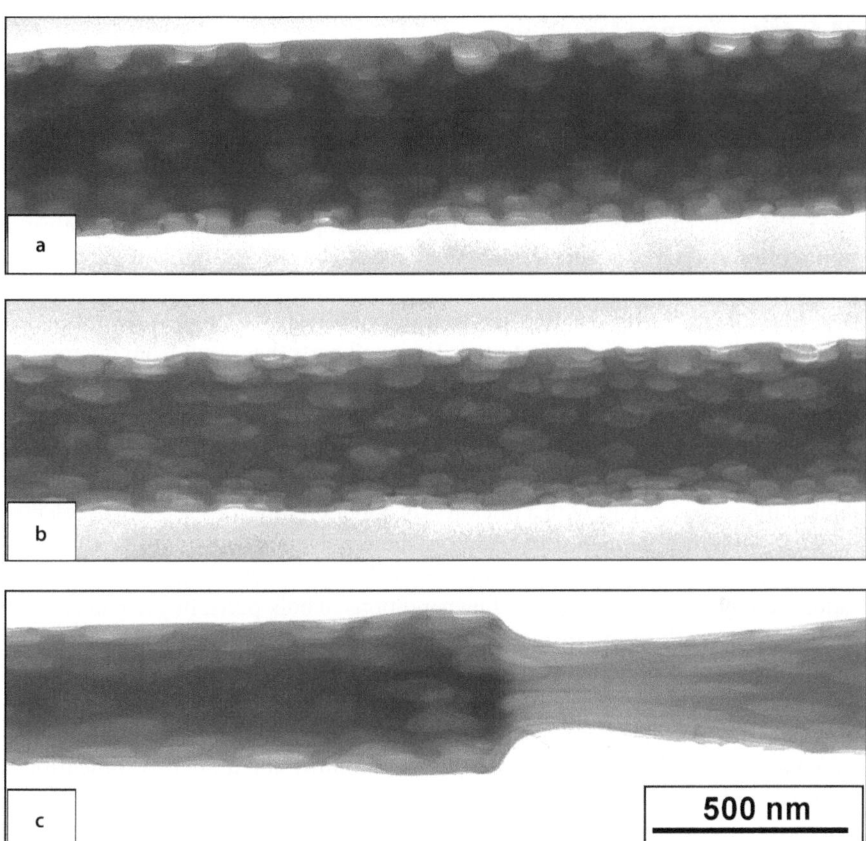

Fig. 24.11a–c. Sequence of TEM micrographs showing a mechanical deformation process: **a** before deformation of a nanoporous electrospun PMMA/Na-MMT nanocomposite; **b,c** deformed states below and above a critical strain under uniaxial tensile load, respectively

value, the nanoporous fibre deforms due to shear yielding in the form of necking (Fig. 24.11c). With a further increase in strain, the necking progresses along the fibre until the fibre completely fails. While the bulk nanocomposite deforms in a brittle manner (i.e. by crazing, see Chap. 22), the electrospun nanoporous fibre deforms in a ductile manner (i.e. by necking), with a considerably enlarged strain at break. This homogeneous stretching is similar to the effect of thin-layer yielding in layered block copolymers (Chap. 19) and this indicates that the electrospinning process provides an alternative way to manipulate the inherently brittle common nanocomposite into a ductile form without significantly sacrificing other attractive properties. This unique synergistic energy-absorbing mechanism in nanocomposite nanofibres presents a highly promising route to reinforcing polymer composites with potentially a great deal of mechanical applications [27, 28, 34, 35].

24.5 Microformed Materials

24.5.1 Overview

There are several reasons for processing small amounts of polymeric materials via microextrusion, microinjection or some other microforming technique. One reason is that only small amounts of new polymer samples are synthesised in research laboratories, and these small amounts then need to be processed into different sample shapes for additional testing methods. Another reason is that polymers are also used in mini- or microparts (e.g. in electronics or medicine) and so the properties of such small samples need to be tested. The small parts that result from microforming processes show morphologies and mechanical properties that are often different from those of the corresponding bulk pieces. This is understandable, since extruded or injection-moulded parts show an oriented surface layer that is different from the unoriented ones inside. If the oriented surface layers are, for example, about 50 μm thick, a microformed part with a thickness of about 0.1 mm consists only of oriented material. Structure formation in semicrystalline polymers depends strongly on the cooling rate from the melt. Small and tiny-walled parts with large surface/volume ratios are subjected to different cooling conditions to bulk parts. In particular, larger cooling rates result in the formation of smaller spherulites and smaller details inside [36]. Spatial confinement on a microscale influences the development and final form of spherulites [37], while confinement on a nanoscale affects the lamellar structure [38]. Therefore, there is a need to achieve a fundamental understanding of scale-dependent properties as bulk polymers become thinner and more two-dimensional. Some examples are discussed in the following.

24.5.2 Several Examples

PP Microtensile Bars

A series of homopolymer PPs covering a wide molecular weight $\overline{M_w}$ range from 101 000 to 1 600 000 g mol^{-1} were manufactured into micro dumbbell specimens via a microinjection moulding process [38]. The length of each micro specimen was 25 mm and the cross-section of the small, parallel zone was 1.25×0.5 mm^2. The samples showed an oriented skin layer and a spherulitic structure in the core with increasing thickness of the skin layer from 30 μm for the lowest $\overline{M_w}$ up to 170 μm for the sample with the highest $\overline{M_w}$. In this sample, the skin layers make up almost 70% of the overall cross-section (total thickness = 500 μm). TEM investigations of the skin layers revealed an oriented superstructure in the form of shish-kebab structures; Fig. 24.12. This highly oriented structure causes the micro dumbbell specimens to display exceptionally good mechanical properties (increases in modulus from 1270 MPa to 2330 MPa and in strength from 33 MPa up to 100 MPa) [39].

This shows that micro parts can exhibit outstanding properties and much higher mechanical performances than the corresponding bulk materials.

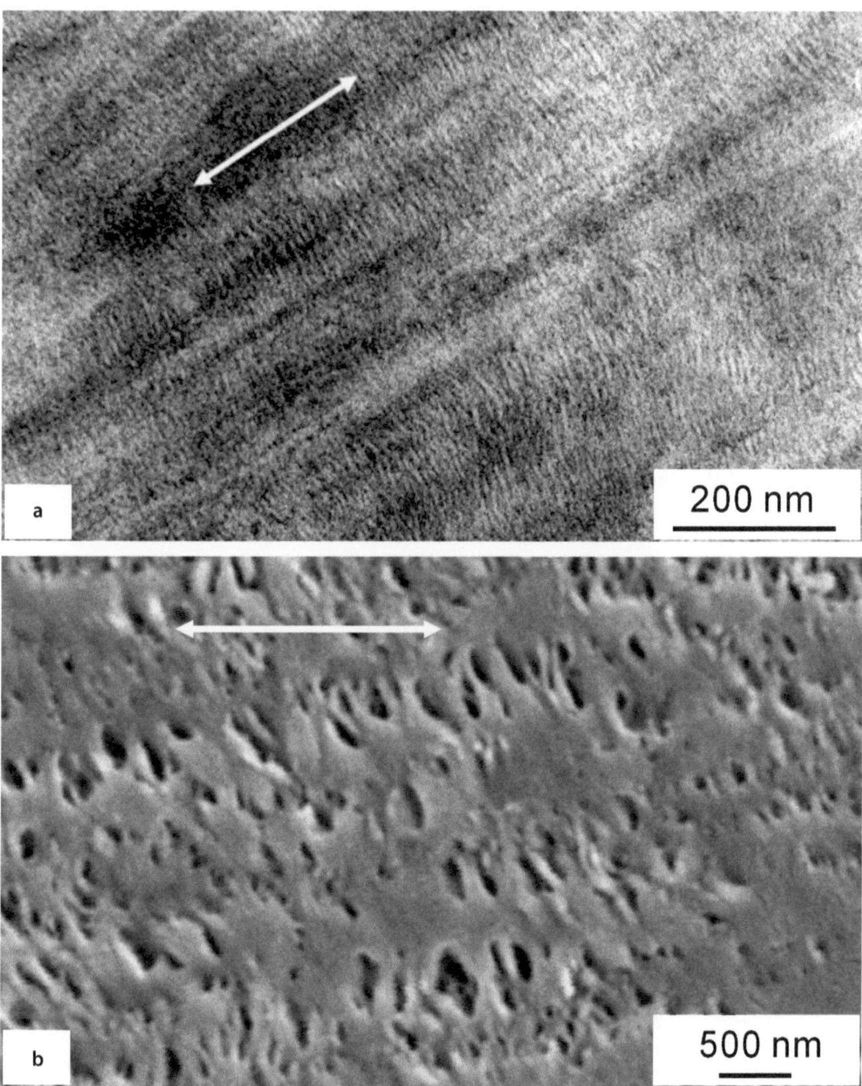

Fig. 24.12a,b. PP micro injection bar ($\overline{M_w} = 1.2 \times 10^6$, 500 μm thick). **a** Morphology, showing shish kebab structure in the skin layer that makes up most of the sample cross-section; the *arrow* indicates the machine direction. (TEM image of an ultrathin section after ruthenium tetroxide staining.) **b** Multiple microvoid formation at initial stages of tensile deformation due to the separation of crystalline lamellae oriented perpendicular to the strain direction (see *arrow*). (SEM image of a microtomed tensile bar after permanganic etching)

Lamellar Organisation in Multilayers

Forced assembly by layer-multiplying coextrusion, as discussed in Sect. 24.2, makes it possible to create polymer layers with thicknesses of just a few nanometers, and so such samples are very useful for studying the effects of spatial confinement [Scholtyssek St, Michler GH, Baer E, Hiltner A, 2007, unpublished results]. These layers are thinner than the usual interface between the coextruded polymers, which means that the layers consist only of interphase material [40]. It is known that glass transition temperature, crystallinity, macromolecular conformation and other properties can change significantly as the layer thickness decreases down to the nanometre scale.

There are several multilayered systems that involve the confined crystallisation of at least one semicrystalline polymer. In HDPE/PS multilayered systems [41, 42], lamellae are organised into flattened spherulites or "discoids". If the layer thickness drops below 100 nm, the morphology changes from discoids to long bundles of lamellae. The crystallinity of HDPE decreases from 60% to almost 30% for layers 10 nm thick. In PP/PS multilayers [38], the lamellar texture of PP varies, the crystallinity decreases from 58% to 46% and the lamellae thickness is reduced. In PET/PC systems, the PET layers were found to undergo cold crystallisation and to form lamellae when the samples were annealed [7]. TEM and AFM were used to reveal the morphology and also the micromechanical deformation behaviour of such systems [7].

References

1. Baer E, Kerns J, Hiltner A (2000) In: Cunha AM, Fakirov S (eds) Structure development during polymer processing. Kluwer, London, p 327
2. Adhikari R, Henning S, Michler GH (2006) Macromol Symp 233:26
3. Bernal-Lara T, Ranade A, Hiltner A, Baer E (2005) In: Michler GH, Baltá Calleja FJ (eds) Mechanical properties of polymer based on nanostructure and morphology. Taylor & Francis, Boca Raton, FL, 15:629
4. Verhoeven JD (2002) Steel Res 73:356
5. Levin AA (et al.) (2005) Crystal Res Technol 40:905
6. Ivankova EM, Michler GH, Hiltner A, Baer E (2004) Macromolec Mater Eng 289:787
7. Adhikari R, Lebek W, Godehardt R, Henning S, Michler GH, Baer E, Hiltner A (2005) Polym Adv Technol 16:95
8. Ivankova E, Krumova M, Michler GH, Koets PP (2004) Colloid Polym Sci 282:203
9. Baltá-Calleja FJ, Ania F, Puente Orench I, Baer E, Hiltner A, Bernal T, Funari SS (2005) Progr Colloid Polymer Sci 130:140
10. Ward IM (1998) Mater World 6:608
11. Ward IM, Hine PJ (2004) Polymer 45:1413
12. Bjekovic RE (2003) Monocomposite Schichtwerkstoffe auf Basis von Polypropylen. VDI Verlag GmbH, Düsseldorf, Reihe 5 Nr. 683
13. Hine PJ, Ward IM (2005) In: Michler GH, Baltá-Calleja FJ (eds) Mechanical properties of polymers based on nanostructure and morphology. Taylor & Francis, Boca Raton, FL, 16:683
14. Rassburn J, Hine PJ, Ward IM, Olley RH, Bassett DC, Kabeel MA (1995) J Mater Sci 30:615
15. Rein DM, Vaykhansky L, Khalfin RL, Cohen Y (2002) Polym Adv Technol 13:1046
16. Hine PJ, Ward IM, Abo El Maaty MI, Olley RH, Bassett DC (2000) J Mater Sci 35:5091
17. Olley RH, Bassett DC (1982) Polymer 23:1707
18. Bjekovic RM, Bledzki AK, Michler GH (2002) In: 60th ANTEC, San Francisco, CA, 5–9 May 2002
19. Reneker DH, Yarin AL, Fong H, Koombhonges S (2000) J Appl Phy 87:4531
20. Bognitzki M, Hou H, Ishaque M, Frese T, Hellwig M, Schwarte C, Schaper A, Wendorff JH, Greiner A (2000) Adv Mater 12:637
21. Shin YM, Hohman MM, Brenner MP, Rutledge GC (2001) Polymer 42:9955
22. Gordon JE (1984) The new science of strong materials. Princeton Univ Press, Princeton, NJ
23. Dresslhaus MS, Dresslhaus G, Avouris Ph (eds) (2000) Carbon nanotubes. Springer, Berlin
24. Koski A, Yim K, Shivkumar S (2004) Mater Lett 58:493
25. Doshi J, Reneker DH (1995) J Electrostat 35:151
26. Gibson PW, Shreuder-Gibson HL, Rivin D (1999) AIChE J 45:190

27. Bergshoef MM, Vancso GJ (1999) Adv Mater 11:124
28. Kim JS, Reneker DH (1999) Polym Compos 20:124
29. Buchko CJ, Chun LC, Shen Y, Martin DC (1999) Polymer 40:7397
30. Boland ED, Wnek GE, Simpson DG, Pawlowski KJ, Bowlin GL (2001) J Macromol Sci Pure Appl Chem A 38:1231
31. Megelski S, Stephens JS, Chase DB, Rabolt JF (2002) Macromolecules 35:8456
32. Bognitzki M, Czado W, Frese T, Schaper A, Hellwig M, Steiner M, Greiner A, Wendorff JH (2001) Adv Mater 13:70
33. Kim GM, Lach R, Michler GH, Chang YW (2005) Macromol Rapid Commun 26:728
34. Kim GM, Michler GH, Pötschke P (2005) Polymer 46:7346
35. Kim GM, Lach R, Michler GH, Pötschke P, Albrecht K (2006) Nanotechnology 17:963
36. Michler GH, Tiersch B, Purz HJ (1979) Acta Polyme 30:529
37. Piorkowska E, Billon N, Haudin JM, Gadzinowska K (2005) J Appl Polym Sci 97:2319
38. Jin Y, Rogunova M, Hiltner A, Baer E, Nowacki R, Galeski A, Piorkowska E (2004) J Polym Sci Polym Phys 42:3380
39. Stern Cl, Frick AR, Weickert G, Michler GH, Henning S (2005) Macromol Mater Eng 290:621
40. Liu RYF, Ranade AP, Wang HP, Bernal-Lara TE, Hiltner A, Baer E (2005) Macromolecules 38:10721
41. Bernal-Lara TE, Masirek R, Hiltner A, Baer E, Piorlowska E, Galeski A (2006) J Appl Polym Sci 99:597
42. Bernal-Lara TE, Liu RYF, Hiltner A, Baer E (2005) Polymer 46:3043

Subject Index